D0712954

Atomic Masses and Fundamental Constants 4

The four previously published conferences on Atomic Masses are:

Nuclear Masses and Their Determinations, H. Huitenberger (ed.), Pergamon Press, London (1957).
Proceedings of the International Conference on Nuclidic Masses, H. E. Duckworth (ed.), University of Toronto Press, Toronto (1960).
Nuclidic Masses, Proceedings of the Second International Conference on Nuclidic Masses, W. H. Johnson, Jr. (ed.), Springer-Verlag, Wien-New York (1964).
Proceedings of the Third International Conference on Atomic Masses, R. C. Barber (ed.), University of Manitoba Press, Winnipeg (1968).

Atomic Masses and Fundamental Constants 4

Proceedings of the Fourth International Conference on Atomic Masses and Fundamental Constants held at Teddington England September 1971

Edited by

J. H. Sanders

University of Oxford
Department of Physics
Clarendon Laboratory
Oxford

and

A. H. Wapstra

Instituut voor Kernphysisch Onderzoek
Ooster Ringdijk 18
Amsterdam 1006

PLENUM PRESS · London - New York · 1972

Plenum Publishing Company Ltd
Davis House
8 Scrubs Lane
London NW10 6SE
Telephone 01-969 4727

U.S. Edition published by
Plenum Publishing Corporation
227 West 17th Street
New York, New York 10011

ISBN 0-306-35084-X
LCC 72-80246

Printed in Great Britain by
Page Bros. Ltd., Norwich

Preface

The Conference on Nuclear Masses and their Determination which was held at the Max Planck Institute in Mainz from 10 to 12 July 1956 resulted in the formation by the International Union of Pure and Applied Physics of a Commission on Atomic Masses and Related Constants. Under the auspices of this Commission conferences covering these subjects were held in Hamilton, Ontario (12-16 September, 1960), in Vienna (15-19 July 1963) and in Winnipeg (28 August-1 September 1967).

After the last of these conferences the Commission felt that the position regarding nuclear masses was reasonably good and that as a consequence the fundamental constants should get more emphasis in future conferences. For this reason they were very pleased to accept an offer from the National Physical Laboratory to accept the main burden of organizing the Fourth International Conference on Atomic Masses and Fundamental Constants. An Organizing Committee was appointed with the members:

J. H. Sanders, *Chairman*
B. W. Petley, *Secretary*
A. Horsfield, *Treasurer*
P. Dean
A. H. Wapstra.

They were assisted by a Program Committee consisting of:

A. H. Wapstra, *Chairman*
E. R. Cohen
A. Horsfield
W. H. Johnson, Jr.
J. H. Sanders
J. Terrien.

The Committees were greatly assisted at the National Physical Laboratory by Dr B. P. Kibble, the secretaries Miss P. Fowler and Mrs R. C. Gordon-Smith, and by Messrs. G. J. Hunt, D. J. McCauley and K. Morris who assisted in many ways, not the least being the arrangement of the ladies' program. Thanks to their efforts the Committees look back on a successful and smoothly organized Conference.

The extent to which the Commission has carried out its intention regarding the emphasis of the Conference can be judged from the fact that the number of papers on fundamental constants rose to about thirty, compared with about ten in each of the two preceding conferences and very few in the first. The number of contributions on atomic masses stayed at about thirty and the program was full, but not uncomfortably so.

The Organizing Committee was fully aware of present difficulties in providing travel funds for the Conference participants, and they are all the more grateful to the following generous sponsors:

Bureau International des Poids et Mesures
British Petroleum
CODATA
The European Physical Society
EURATOM
The International Atomic Energy Agency
Imperial Chemical Industries Limited
The International Union of Pure & Applied Chemistry
The International Union of Pure & Applied Physics
The National Bureau of Standards
The National Physical Laboratory
N. V. Philips' Gloeilampenfabrieken
Physikalish-Technische Bundesanstalt
The Royal Society
Shell International Petroleum Co. Ltd.

A summary of the conference is given in the last paper; two points can, however, be emphasised here. The first is that in the field of atomic mass determinations the precision of mass spectroscopic and nuclear reaction measurements has now reached the point where corrections for atomic electron binding energies are necessary both for very light and for very heavy nuclides. As far as fundamental constants are concerned recent determinations involving quantum electrodynamics have proved to be very consistent and show that the theory on which they are based is valid to a high precision.

The Editors of these Proceedings have kept in mind the value of rapid publication, and in securing this they would like to thank the authors for their cooperation, and the Plenum Publishing Company for their invaluable collaboration and for maintaining a high standard of production.

A. H. WAPSTRA
J. H. SANDERS

Welcome to the 4th International Conference on Atomic Masses and Fundamental Constants

John V. Dunworth, C.B., C.B.E.

Director, National Physical Laboratory, Teddington, England

I have never forgotten the moment when the late Dr. Bhabba who, in 1957 in his opening address to the first International Conference on Nuclear Power held at Geneva commented that most people who stood up had nothing to say. In consequence, he observed, there was no obvious criterion by which they could decide when to stop speaking. However, this morning I have a clear task to perform, which I do with much pleasure. Firstly I welcome those of you who have come from overseas to the United Kingdom. Secondly I welcome all of you to the National Physical Laboratory.

The NPL is the designated authority in Great Britain for measurement science, and we serve government, industry, the universities, and in fact all concerned with science, engineering and education. Although our duty here is primarily to serve those who live and work in the United Kingdom, we are almost unique in that many of our activities in measurement science are subject to international discussion, and ultimately to agreement on an international basis insofar as their technical content is concerned. The International Standards Organisation, the Bureau of Weights and Measures at Sèvres, the International Scientific Unions and Codata are amongst those organisations who are concerned in such discussions. Despite overlap in that individual decision may affect two or more of these bodies, there is rarely any real difficulty in reaching good agreement. The system works well, and it is by no means certain that an administratively tidier arrangement would be more effective.

The National Physical Laboratory was established on its present site in 1900, though certain component parts had existed elsewhere in the United Kingdom prior to this date. The maturity of the site is a source of pleasure to the staff and to our many visitors. There are, unfortunately, very few laboratories which have been established for a sufficiently long time and with sufficient imagination in their site planning for one to be able to make an equivalent comment. The site has the attractiveness of an older town, where there is a somewhat random mixture of old and new buildings, and where the road system has developed from a mixture of pragmatism and planning. One particular feature of the Teddington site is the very beautiful garden adjoining Bushy House. I invite you all to take a stroll round it during the breaks in your sessions. I am sure you will find it relaxing and a pleasant place in which to hold some of those informal discussions which are such an important part of any conference.

Finally I have been asked by Professor N. Kurti, the Chairman of the British National Committee of the Royal Society for Physics of which I am a member, to extend a welcome to you all on behalf of the Committee, and this I do with much pleasure. It only remains for me to wish you all a very successful conference.

(In reply to this opening address, Professor Dr. A. H. Wapstra, president of the Commission on Atomic Masses and Related Constants of the International Union of Pure and Applied Physics, thanked Dr. Dunworth for his kind words and expressed the pleasure of all participants to find themselves at the National Physical Laboratory in Teddington).

List of Participants

ALLISY, A.
AMBLER, E.
BAIRD, J. C.
BARBER, R. C.
BARGER, R. L.
BAY, Z.
BERGKVIST, K. E.
BONSE, U.
BOS, K.
BOWER, V. E.
BOYNE, H. S.
BREITENBERGER, E.
BROWNE, C. P.
BUECHNER, W. W.
CERNY, J.
COHEN, E. R.
CONNOR, R. D.
CSILLAG, L.
CZOK, U.
DESLATTES, R. D.
DRATH, P.
DUCKWORTH, H. E.
ELIEZER, I.
EVENSON, K. M.
EWALD, H.
FARLEY, F. J. M.
FINNEGAN, T. F.
FRANK, L. E.
FREEMAN, J. M.
GERSTENKORN, S.
GIACOMO, P.
GOVE, N. B.
GRENNBERG, B.
HARA, K.
HART, M.
HELMER, R. G.
HILF, E. R. H.
JAFFE, A. A.
JANECKE, J.
JENNINGS, J. C. E.
JOHNSON JR, W. H.
KAYSER, D. C.
KESSLER JR, E. G.
KLAPISCH, R.
KNOWLES, J. W.
KOSE, V. E.
KPONOU, A.
LANGENBERG, D. N.
LEA, K. R.
LOWRY, R. A.
MACFARLANE, J. C.
MASUI, T.
McDONALD, D. G.
MILNE, A. D.
MONTAGUE, J. H.
MULLER, J. W.
MYERS, WM.
NIELSEN, O. B.
OGATA, K.
OLSEN, P. T.
PARKINSON, W. C.
RICH, A.
ROESLER, F. L.
RYDER, J. S.
RYTZ, A.
SAKUMA, A.
SANDERS, J. H.
SAUDER, W. C.
SCHULT, O. W. B.
SCOTT, D. K.
SEEGER, P. A.
SEN, S. K.
SERIES, G. W.
SMITH, L. G.
SMITH, S. J.
SQUIER, G. T. A.
STAUB, H. H.
STILLE, U.
TAYLOR, B. N.
TELEGDI, V. L.
TERRIEN, J.
THOMAS, B. W.
TINKER, M. H.
VAN ASSCHE, P.
VAN DER LEUN, C.
VON GROOTE, H.
WAPSTRA, A. H.
WAY, K.
WILLIAMS, P.
ZELDES, N.

Contents

Part 3 Mass Spectroscopy

Part 4 Coulomb Energies

Part 5 Mass Formulae and Mass Calculations

Part 6 Velocity of Light

Part 7 Wavelength Comparisons

Part 8 Fine Structure Constant

Part 9 2e/h

Part 10 Rydberg Constant

Part 11 Magnetic Moments

Part 12 Miscellaneous Constants

Part 13 Evaluation

Part 14 Conference Summary

Part 1 Particle Energies

New Alpha Energy Standards

A. Rytz, B. Grennberg,* and D. J. Gorman

Bureau International des Poids et Mesures, F-92 Sèvres, France

I. Introduction

The measurements of absolute energy of α-particles carried out in the past are only few in number. This is due to the fact that in α-ray spectroscopy only a moderately accurate energy scale is actually needed. Further, an absolute spectrometer has necessarily such a low transmission that it is of little use in more advanced α-ray work. Nevertheless, there are several good reasons why reliable α-energy standards are of considerable importance.

- They are calibration fixpoints for all α-spectra observed with highly resolving spectrometers. Identification of weak lines from daughter substances or impurities is uncertain or impossible without a reliable energy scale.

- Similarly, precise Q value determinations of nuclear reactions are often based on α-energy standards.

- Atomic mass differences may be calculated very accurately from α-energies.

- The mere fact that a physical quantity can be measured to six or seven significant figures presents a metrological interest and calls for an investigation of how far the accuracy may be increased.

The number of compilations of α-energies found in the literature exceeds considerably the number of times absolute energy measurements have been carried out. It is interesting to follow the evolution with time of the energy values adopted for some widely used standards. Fig. 1 shows the values given by some compilers along with the oldest (1) and the most recent (6) direct measurements. The following facts may be learned from this figure:

1) The excellent measurements by Rosenblum and Dupouy (1) are all low by once or twice the error quoted (about 1 part in 2 000). The fact that they used only a linear extrapolation of the line shapes observed might account, at least partially, for this systematic error.

2) Part of Briggs' careful analysis (2) of the then available data was based on but one precise measurement of ^{214}Po which later on

turned out to be low by as much as seven times the error quoted. This accounts for the rather low accuracy of many of the values stated and carried over by Strominger, Hollander, and Seaborg (3).

3) It was not until new measurements were available that a more accurate set of recalibrated values could be presented by Wapstra (4).

4) Since ^{218}Po, ^{222}Rn, ^{224}Ra, and ^{226}Ra have all frequently been used as standards, many measurements relative to these energies are known. However, with the exception of Rosenblum's ^{218}Po result, no absolute measurement with these α-emitters can be found before 1970. It should be noted how far off the earlier adopted values were.

It has repeatedly been suggested that α-energies should be measured on a relative scale only, using weak sources in a high-transmission spectrometer. Only one single α-emitter would have to be involved in a particularly good absolute measurement. All the other values would then be referred to this single standard value. However, we do not agree with this suggestion but feel that a rather large number of nearly equally reliable absolute results give a greater chance to obtain a consistent set of accurate standards. The examples just given confirm our view that relative energy measurements are even more subject to systematic errors. This is so because in each measurement two different sources and line shapes come into play. The absence, in our experiments, of a correlation between measured value and source strength does not suggest that weak sources would give much better results.

In 1964, a number of α-ray spectroscopists attending an international conference in Paris (7) expressed their wish that a committee consisting of F. Asaro, S.A. Baranov, R.J. Walen, and A. Rytz should choose the best standard values for some easily accessible α-emitters. However, since mere compiling and adjusting old results did not seem to be very promising in this case, the Bureau International des Poids et Mesures decided to take up new absolute measurements. Thus a 180° uniform field magnetic spectrograph was constructed with emphasis on absolute energy measurements of highest accuracy. Since 1969 thirty-one new results from twenty α-emitters have been obtained all of which have been published recently (6).

II. Description of the measurements

Since the experimental set-up has already been described at length, only a short survey of the method will be given here.

A. Magnetic field

The homogeneous field of up to 1 tesla in the 70 mm wide gap of a large electro-magnet has been corrected by shimming and stabilized by proton resonance to about 1.5×10^{-6}. Before and after each run the field was measured at 31 points along the particle paths to be determined and which had radii between 36.0 and 47.5 cm. All field measurements could be carried out inside the spacious vacuum chamber.

The mean effective field was calculated according to the well-known Hartree formula. However, for the high accuracy sought in our work, more elaborate field measurements and calculations were needed. Weak transverse fields in known directions applied to a small proton resonance probe permitted the direction of the main field to be determined. Finally the following corrections were applied to each mean effective field value:

1) A higher order "Hartree correction" of up to 50 eV, depending on radial field gradients (always positive),
2) A differential "Hartree correction" of up to 70 eV, accounting for deviations of field direction from a perpendicular to the plane of motion (always negative),
3) A focussing correction (mostly below 50 eV, but up to 150 eV, for large radii), due to field gradients (always positive).

Preliminary results obtained by running the same source at different field-settings are not in disagreement with these corrections. Some more such experiments will be made in a near future. However, the uncertainty of these corrections is difficult to evaluate and is accounted for by an additional systematic error of at least 50 eV.

B. Length measurements

The α-particles were detected with a narrow strip of nuclear track plate. The plate holder and the defining entry slit were rigidly joined to form a length standard. Twenty-two radioactive markers gave a series of fine reference marks on each plate. The distances between the slit and each marker were measured periodically with the aid of a classical comparator and two standard metre bars, one of which had been calibrated by interferometry. The temperature of the length standard was controlled by a thermostated water circuit. The radii of curvature of the particle paths were determined with a relative accuracy similar to that of the magnetic field. It must be emphasized that accurate energy measurements are only possible if the spectrograph is designed in such a way as to form a suitable length standard. Therefore, a wide enough gap and a spacious vacuum chamber are essential.

C. Sources and line shape

Three types of sources were used, depending on the radionuclides concerned:

a) Active deposit (daughter activity from Rn isotopes),

b) Volatilization in vacuum,

c) Recoil collection.

The source backings were made of platinum or stainless steel. The sources covered a surface of 5 x 2.5 mm. They may be thick but must not be covered by an inert layer.

The analytical form of the high-energy edge of a line can be calculated as follows (see Fig. 2). The centre of a particle orbit reaching the detector between E and Q lies inside the shaded area the sharp ends of which are separated by a distance $f = (4\rho\lambda - \lambda^2)^{1/2}$, ρ being the radius of the particle orbits and λ the distance between E and Q.

The number of particles per second reaching the detector between λ and $\lambda + \Delta\lambda$ is proportional to an annular segment of chord f and width $\Delta\lambda$. Since $\rho \gg \lambda$, we may write for the intensity obtained with a mono-layer source

$$\Delta I(\lambda) \sim \lambda^{1/2} \Delta\lambda \ .$$

However, in practical cases, the sources are always much thicker than just a single layer of active atoms. Thus, the total intensity may be calculated by integrating over the source thickness and we get

$$I(\lambda) \sim \lambda^{3/2} \text{ , near the endpoint.}$$

The photographic plates were scanned with a special microscope mounted on a Zeiss comparator. The tracks were counted in bands of 0.01 mm width. The energy defining endpoint was obtained by fitting a straight line to $[I(\lambda)]^{2/3}$ and extrapolating to the background level. Poor source quality or bad field stabilization showed up by a marked trend of the endpoint position with the number of points used in the extrapolation. Although we have no direct proof of the absence of inactive layers on the sources, it seems rather improbable that results from different sources would cluster so closely around a single value. Moreover, the observed smallness of trend in extrapolation is an argument of considerable strength.

III. Selection of α-emitters; transmission of the spectrograph

Table I shows a list of the measurements already finished and of the ones in progress. We have also listed other possible α-emitters, part of which we hope to be able to measure later.

Table I

List of α-emitters for absolute measurement with the BIPM spectrograph

A. Already measured

Th family : ^{228}Th, ^{224}Ra, ^{220}Rn, ^{216}Po, ^{212}Bi, ^{212}Po,

Ra family : ^{226}Ra, ^{222}Rn, ^{218}Po, ^{214}Po,

Ac family : ^{227}Th, ^{223}Ra, ^{219}Rn, ^{215}Po, ^{211}Bi,

others : ^{253}Es, ^{244}Cm, ^{242}Cm, ^{241}Am, ^{238}Pu.

B. In progress or planned for measurement in a near future

^{232}U, ^{240}Pu, ^{148}Gd, ^{212}Po (long range).

C. Some other possible α-emitters for later work

^{246}Cm, ^{243}Cm, ^{241}Cm, ^{240}Cm, ^{243}Am, ^{241}Pu,

^{236}Pu, ^{230}U, ^{229}Th, ^{226}Th,

^{225}Ac, ^{225}Ra,

^{210}Po, ^{209}Po, ^{208}Po.

Clearly, many α-emitters have too short a half-life or are not available in sufficient quantity or purity for such measurements. For long-lived substances an upper limit of the half-life can be given beyond which no line would be observed. This limit depends on the transmission of the spectrometer (mean effective solid angle divided by 4π). By the way, it should be noted that the term "resolution" has no useful meaning in the geometry used here. We know from experience that only the utmost 0.1 mm of a line recorded on the photographic plate is used in the extrapolation. Therefore, the effective solid angle comprises the particle paths reaching the plate between

$\lambda = 0$ and $\lambda = 0.1$ mm. Fig. 2 explains the geometry of the solid angle calculation. A particle leaving the source at P and passing as close as possible to S reaches the plate at E. Since the intensity at E is infinitely small, this point has to be determined by an extrapolation to zero of the intensity distribution between E and Q.

Elementary geometrical considerations give the values of x and y as functions of ρ, h, and λ (see Fig. 2) and of the source coordinate q. Further we determine the angle φ and average over q. For $\rho = 400$, $h = 30$, $\lambda = 0.1$ (all in mm), and assuming a detector width, in the direction of the magnetic field, of 5 mm, we get a transmission of $T = 0.74 \times 10^{-6}$.

Assuming further a background of G particles per interval of 0.1 mm and time unit, and a number N_s of active atoms in the useful part of the source, we get an upper limit of the half-life still granting detectable lines,

$$T_{1/2}^{max} = \frac{N_s \, T \ln 2}{G \, n \, p} \, .$$

Here n is the signal-to-background ratio, p is a factor between unity and about 10, depending on self-absorption. With reasonable assumptions we get $T_{1/2}^{max} \leq 4 \times 10^4$ a. The longest half-life which we have used so far is 1.6×10^3 a. Yet we hope to get closer to the limit with ^{240}Pu, the measurement of which will start soon.

IV. Results

From the directly measured kinetic energy of the α-particles E_α we obtain the (atomic) Q value of the decay by adding the recoil energy,

$$Q = E_\alpha \, (1 + 4.001\,506/M_R) + 0.15 \text{ keV}.$$

Here M_R is the atomic mass of the daughter atom, in u. The last term takes into account the binding energy of the electrons in ^4He (0.08 keV) and the adiabatic energy loss to excitation and ionization (8). In the cases of ^{223}Ra and ^{241}Am, the excitation energy (5) of the final state has to be added. Table II presents all the measured E_α values for transitions to the ground state (with the two exceptions just mentioned) and the corresponding Q values calculated as explained above. The last column gives the differences between these Q values and the output values of the 1967 mass table (9). The general trend of these differences may of course already be guessed from Fig. 1.

Table II

Summary of the results

Parent atom	E_α (keV)	standard error (keV)	Q (keV)	$Q - Q_{67}$ (keV)
^{211}Bi	6 623.1	0.6	6 751.3	+ 1.5
^{212}Bi	6 090.06	0.08	6 207.38	+ 1.3
^{212}Po	8 784.37*	0.07*	8 953.46	- 0.4
^{214}Po	7 687.09	0.06	7 833.73	- 1.2
^{215}Po	7 386.4	0.8	7 526.6	+ 3.1
^{216}Po	6 778.5	0.5	6 906.6	+ 1.5
^{218}Po	6 002.55	0.09	6 114.94	+ 2.9
^{219}Rn	6 819.29	0.27	6 946.36	+ 2.1
^{220}Rn	6 288.29	0.10	6 404.93	+ 0.3
^{222}Rn	5 489.66	0.30	5 590.57	+ 3.7
^{223}Ra (α_{159})	5 716.42	0.29	5 979.6	+ 3.0
^{224}Ra	5 685.56	0.15	5 789.12	+ 1.8
^{226}Ra	4 784.50	0.25	4 871.67	+ 4.4
^{227}Th	6 038.21	0.15	6 146.70	+ 2.0
^{228}Th	5 423.33	0.22	5 520.35	- 0.6
^{238}Pu	5 499.21	0.20	5 593.38	+ 1.3
^{241}Am (α_{60})	5 485.74	0.12	5 638.03	+ 0.1
^{242}Cm	6 112.92	0.08	6 215.83	- 1.2
^{244}Cm	5 804.96	0.05	5 901.87	- 0.1
^{253}Es	6 632.73	0.05	6 739.44	- 6.6

* This is a new result, almost identical with that of ref. (6) but more precise.

References

* Present address: Institute of Physics, University of Uppsala, Sweden.

(1) S. Rosenblum and G. Dupouy, J. Phys. Radium 4, 262 (1933).

(2) G.H. Briggs, Rev. Mod. Phys. 26, 1 (1954).

(3) D. Strominger, J.M. Hollander, and G.T. Seaborg, Rev. Mod. Phys. 30, 585 (1958).

(4) A.H. Wapstra, Nucl. Phys. 57, 48 (1964).

(5) C.M. Lederer, J.M. Hollander, and I. Perlman, Table of Isotopes (John Wiley and Sons, Inc., New York, 6th ed., 1967).

(6) B. Grennberg and A. Rytz, Metrologia 7, 65 (1971).

(7) Comptes Rendus du Congrès International de Physique Nucléaire, 30e anniversaire de la découverte de la radioactivité artificielle; éditions C.N.R.S. Paris (1964) Vol. I, p. 447.

(8) J.O. Rasmussen, in Alpha-, Beta- and Gamma-Ray Spectroscopy edited by K. Siegbahn (North-Holland Publishing Company, Amsterdam, The Netherlands, 1965) p. 703.

(9) A.H. Wapstra, in Proceedings of the Third International Conference on Atomic Masses, Winnipeg, 1967, edited by R.C. Barber (Univ. of Manitoba Press, Winnipeg, Canada, 1967) p. 153.

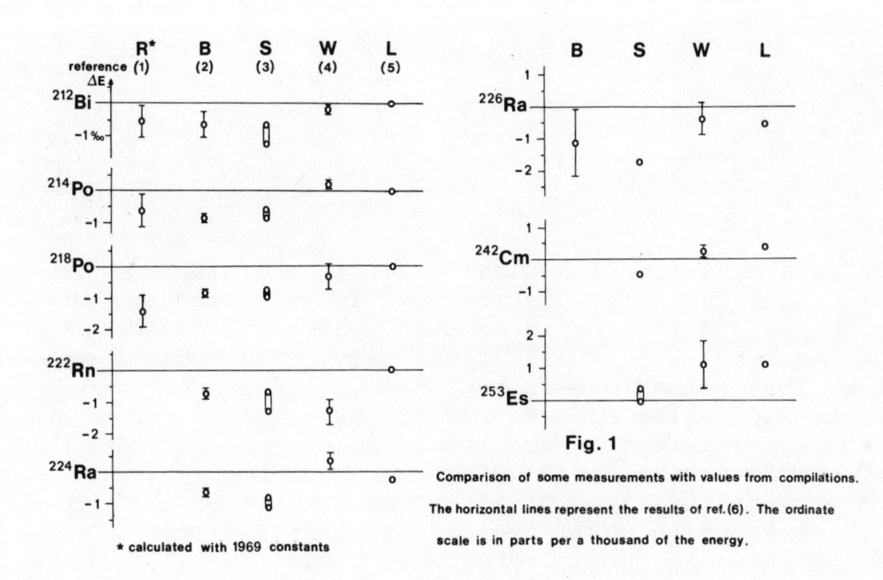

Fig. 1

Comparison of some measurements with values from compilations. The horizontal lines represent the results of ref. (6). The ordinate scale is in parts per a thousand of the energy.

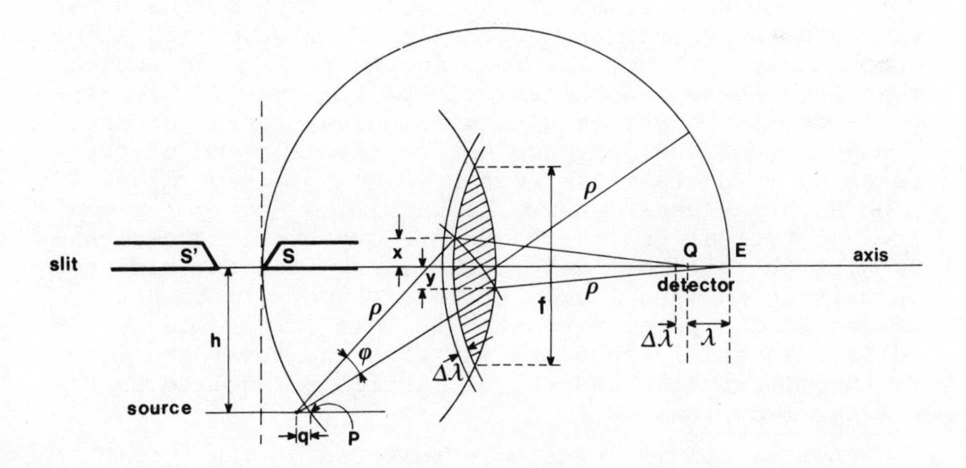

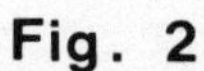

Fig. 2

Schematic view of the geometry of the spectrograph

Influence of Atomic Effects on Nuclear Reaction Energies

Hans H. Staub

University of Zürich, Switzerland

The determination of nuclear reaction energies constitutes, besides mass spectroscopy, an important and powerful method for obtaining highly accurate values of the mass and hence the binding energy of a nuclear species. This method offers not only an independent check of the mass value of a stable or unstable nucleus with rather long life time but it is the only method by which one can determine the binding energy of highly unstable nuclei with lifetimes of a few nuclear periods. Since such experiments until now were never performed with all the reaction partners beeing bare nuclei and since present day accuracy is in many instances well below 1 keV attention has to be given to the unavoidable interaction of the electronic neighbourhood of the interacting nuclei.

Broadly speaking, we can distinguish two effects. First the interactions of the incoming and outgoing particles with the electronic shells of neighbouring atoms whose nuclei are not involved in the particular nuclear reaction. These are the effects of the unavoidable finite target thickness resulting in energy losses of the projectile at the instance of the reaction and of the reaction products whose energy we want to determine. In high accuracy work, where the particles undergo only a few non nuclear collisions before or after the nuclear process the discreetness of the energy losses leads to peculiar phenomena (Lewis effect) which will be discussed in the paper presented to this conference by J.W. Müller. We shall therefore restrict our considerations to the electrons bound to the particles involved in the nuclear reaction.

Let us consider a simple reaction of the type:

$$1 + A \rightarrow 2 + B$$

which might proceed via direct or compound interaction. The Q value of the reaction is defined as the difference of the initial masses $M_1 + M_A$ of the neutral <u>atoms</u> 1 and A and the final masses $M_2 + M_B$ of the neutral atoms 2 and B in their atomic groundstate and their nuclear state under consideration. It can be determined from the kinetic energies of particles 1, A and say 2 since momentum has to be conserved. But we also have to know the states of the electronic configuration of all four particles. This is usually quite simple and accurate for 1 and A, mostly also for particle 2 since it might be a light bare nucleus. But about the state of B to which we might count nucleus B and all the electrons having been present very little can be directly observed and detailed calculations of this complicated many body problem are difficult. In one respect however there is a simplification of the atomic collision problem in as far as collisions leading to a nuclear reaction have zero impact parameter. On the other hand one usually is not faced with a clear limiting situation where the velocity of the projectile or the reaction products is very large or very small compared with the velocities of the electrons. In any case we would of course expect for a given state and fixed kinetic energy T_1 and T_A of particles 1 and A to observe at a given reaction angle a fine structure in the energy of particle 2 which should be observable with present day resolution of the detectors. However this fine structure will be partially smeared out due to the finite energy spread of particle 1 (projectile) and often even more by the random thermal motion of the target atoms A. But in any case even the completely washed out structure of the spectral distribution of particle 2 will be shifted by a non negligible amount. These average energy shifts have been calculated by Christy (1) under certain simplifying assumptions as for instance that all the electrons are associated to particles A and B. Christy considers separately three processes responsible for the energy shift:

1) The nuclear recoil of particle B imparts some of its momentum to the electrons thereby exciting higher bound or unbound states of B.

2) Excitations of higher atomic states of B through change of the nuclear charge from Z_A to Z_B (monopole charge effect).

3) Excitation of electronic states by the passage of charged particles 1 or 2 through the electron configuration around A (dipole charge effect).

The calculations of Christy showed that the average energy shift in all practical cases is of the order of 0.1 keV and was therefore at the time of that publication of no significance. They also showed that the fine structure in the energy of particle 2 is too small to be resolved even with present day detecting instruments.

However, the average shifts can be taken from Christy's work and should be applied to Q value determinations if one aims at an accuracy of better than 1 keV. To the author's knowledge no fine structure in the energy of 2 has so far been observed. There might however exist some very special cases where it could be observed.

A somewhat different situation is encountered when one wants to determine the mass of a relatively long lived unstable compound nucleus. Obviously effects due to the atomic electrons are of importance only if the nuclear lifetime τ of the compound state is sufficiently large to make the width $\Gamma = \hbar/\tau$ of the state comparable to atomic excitation energies. A large number of such long lived compound states have been observed even at high excitation energies mainly in the region of medium heavy nuclei by elastic scattering of protons and α particles where they manifest themselves through large and characteristic anomalies caused by the interference of the coherent Coulomb and nuclear resonance scattering in the neighbourhood of such a state. This energy dependence of the scattering cross section should again exhibit a fine structure caused by the different electronic states of the compound atom C in the reaction

$$1 + A \rightarrow C \rightarrow 1 + A$$

Atoms 1 and A might change their electronic configuration in the scattering process. Such processes however would represent an inelastic scattering and would therefore contribute only little to the scattered intensity since they are to be added incoherently.

Up to the present day such fine structure or Coulomb hyperfine structure has only been observed in one case reported by the author at the last A.M.F.C. conference in 1967 (2). It occurs in the elastic scattering of α particles in helium, $^4He(\alpha,\alpha)^4He$ leading to the groundstate of ^8Be which is unstable against α decay by 92 keV and has a width of about 7 eV. If the reaction is carried out with bare α particles 1 and neutral helium as target particle A the compound system C is a doubly ionized $^8Be^{2(+)}$. Its first excited atomic state lies about 120 eV above the ground state. The energy separation of the first atomic states is therefore large compared to the width of the single nuclear state and the fine structure is therefore well resolved for these states. The energy spectrum of the scattering cross section had been calculated by Meyer et al. (3) under the assumption that the passage of the α particle through the helium atom was sufficiently fast to make process 3 negligible compared to the effects of process 1 (recoil of the Be ion) and 2 (monopole charge effect). Likewise the excitation of states in the continuum was neglected. The calculation of the distribution over the first few levels carried out to second order gave 36 % to the ground state $(1s)^2$, 39 % to the combined group 1s2s, 1s2p and 24 % to $(2s)^2$, 2s2p. The experiment showed very clearly resonances corresponding to the ground state and the first excited state and some indication of the group $(2s)^2$, 2s2p with the correct energy separation well within the observational errors. The fit of the observed pattern required a distribution of 29 % to the ground state and 50 % and 21 % resp. to the first excited groups of states. In view of the approximations made this result is quite satisfactory particularly in view of the fact that in the experiment particle 1 was not a bare α particle but a He^+ ion carrying a third electron into the compound system. Since the third electron has to go into one of the higher states or the continuum it can be expected that its contribution will simply broaden the observed resonances by an amount of the order of some 10 eV.

During the past years we have made a cursory examination of elastic proton scattering measured with high resolution, approx. 100 eV, on several medium heavy nuclei where quite often groups of narrow resonance scattering anomalies occur with separations of the order of some 100 eV. We have however found no case where such a fine structure could be caused by excitation of the electronic shell in conjunction with a single nuclear level. If this were the case the structure of groups separated by an amount of energy which is small compared to the energy of the incident proton would have to be the same. Likewise if we observe elastic proton scattering on two isotopes at about the same energy of the incident proton the spectrum of the nuclear states will be very different. In contrast a fine structure caused by the electrons would have to be practically identical since the mass difference of the nuclei produces only insignificant changes in the electronic configurations. With this aim we have recently examined the elastic scattering of protons on ^{54}Fe and ^{56}Fe with an energy around 2 to 4 MeV. In the case of ^{56}Fe a large number of partly unresolved scattering anomalies are found but the structures of "groups" are very different. In the case of ^{54}Fe one observes well resolved individual narrow resonances with no characteristic structure. We conclude therefore that in these processes the splitting of the levels due to the shell electrons is very small and smeared out.

The influence of the electrons will therefore simply consist in a broadening and shift of the nuclear scattering anomaly. This result is not surprising. First of all the recoil velocity of the compound atom C is rather small due to the heavy target nucleus. Secondly the innermost electrons of the K shell whose excitation energy would be large, of the order of some keV, see only a relative change of the nuclear charge of $1/Z_A$. In contrast the outermost electrons which encounter nearly doubling of the effective charge can go to any one of the numerous unoccupied levels at energies of a few eV above the ground state. This consideration shows that the case of α-He elastic scattering is indeed a very special one due to the fact that only two K electrons are involved and the energy of the first excited state of a two electron system is high above the ground state and quite close to the first ionization energy.

We intend however to investigate proton scattering on other nuclei where conditions for the observations of resolved fine structures caused by the electronic shell might be more favorable.

1. CHRISTY, R.F., Nucl. Phys., 22, 301 (1961).

2. BENN, J., DALLY, E.B., MUELLER, H.H., PIXLEY, R.E., STAUB, H.H., and WINKLER, H., Nucl. Phys. A106, 296 (1968).

3. MEYER, V., MUELLER, H.H., and STAUB, H.H., Helv. Phys. Acta 37, 611 (1964).

Gamma Ray Energy vs. Particle Energy Measurements*

C. P. Browne

Department of Physics,
University of Notre Dame, Notre Dame, Indiana

1. Introduction

This work is concerned with the consistency of energy scales used in nuclear reaction energy measurements. It is part of our contiuing program on the comparison of energy standards. In earlier work we compared the energy of the alpha particles from ^{210}Po with the threshold energy of the ^{7}Li(p,n)^{7}Be threshold and with other energies. You may recall that an apparent discrepancy of mass values obtained from mass spectroscopy measurements with those from Q-value measurements was resolved by bringing the Po-α energy into agreement with the Li(p,n) threshold. This was accomplished both through our direct comparison and through improved measurements made in other laboratories on the absolute values of these two energies. Figure 1, taken from my report [1] at the Second International Mass Conference reviews the situation. The lower portion of the figure shows values from measurements of the Po-α energy, the upper portion shows comparisons with the energy of α-particles from RaC', and the central portion shows comparisons with the ^{7}Li(p,n)^{7}Be threshold. The absolute value of the energy of the α-particles from ^{210}Po is known to 0.01% and this is consistent with the ^{7}Li(p,n) threshold to the same accuracy.

At the time this figure was presented, I stated that the point labeled D(Mg^{24}, γ) indicated an apparent discrepancy between the gamma-ray energy scale and the now consistent particle energy scale. There was however a disagreement [2] in the γ-ray energy measurements of the decay of the first state in ^{24}Mg. We wished to conclude our comparisons of energy standards by making a direct comparison of the γ-ray energy with the Po-α energy.

But aren't gamma-ray energies very well known? Ge(Li) detectors have very high resolutions. Energy values measured with such detectors are quoted ± 0.01 keV. A report [3] on these devices was made at the Third Conference on Atomic Masses by R. L. Heath and a further report is to be made at this conference. Many "standard" energies are listed in the tables [4] by Marion, with uncertainties of a small fraction of a keV. It is clear that γ-ray energies may be compared with very high accuracy.

For an absolute energy measurement, the energy will be related to fundamental constants either through the relation $E = hc/\lambda$, from measurements of λ in meters or through the relation $E = m_0c^2$ by comparing electron

*Work supported by the National Science Foundation. The author thanks the Foundation for a Travel Grant which allowed attendance at the Conference.

TABLE I

^{198}Hg Gamma transition energy

Groups	Instrument	Value (1961 constants)
Lind and Hedgran[a]	β-ray spectrometer	411.838 ± 0.041
Muller et al.[b]	crystal spectrometer	411.806 ± 0.036
P. Bergvall[c]	crystal spectrometer	411.773 ± 0.033
Murray et al.[d]	β-ray spectrometer	411.799 ± 0.007
	Murray et al. (1963 constants)	411.795 ± 0.009
	Murray et al. (1965 constants)	411.795 ± 0.007

a D. Lind and A. Hedgran, Ark. Fys. 5, 29 (1952).
b D. E. Muller, H. C. Hoyt, D. J. Klein, and J. W. M. DuMond, Phys. Rev. 88, 775 (1952).
c P. Bergvall, Ark. Fys. 17, 125 (1960).
d See text Ref. [5].

momenta from external conversion with electron momenta from conversion of annihilation radiation. It is customary to use the 412 keV γ-ray from ^{198}Hg as a primary standard. Measurements of this energy by both methods agree to better than 0.01%. This is illustrated in Table I which is taken from the paper [5] by Murray, Graham and Geiger. Two of these measurements are essentially a measurement of wave length and two are essentially a comparison with rest mass energy. For the work reported here, we do not need to be concerned with the small effects of changes in fundamental constants.

2. Comparisons of Excitation Energy Measurements

Now let us consider the consistency of γ-ray energies with particle energies. With the Po-α energy known to an absolute accuracy of 0.01% and the γ-ray energies known to an accuracy better than this it would seem that the two energy scales must agree to high accuracy. Agreement is usually assumed and there is much indirect evidence of agreement. I shall now present an example of such indirect evidence by comparing two excitation energies which have been measured both with charged particle reactions and with γ-rays.

The results are shown in Table II. The ^{16}O measurement was made some time ago by Browne and Michael [6] at Notre Dame. The 3.4 keV spread in the γ-ray results should be noted. The energy standard for the particle measurements is Po-α and the standards used for the gamma measurements are indicated in the table. The values for this 6.1-MeV excitation energy measured against the two scales are seen to agree well, within the rather large uncertainties.

The more accurate measurement of the excitation energy of the ^{11}B state has just been done at Notre Dame by Browne and Stocker [7]. The

TABLE II

Excitation energy from
particle measurement vs. gamma ray measurement. Two examples

Nucleas	Particle Energy Measurement		Gamma Ray Measurement	
	Reaction	Excitation Energy	Excitation Energy	Standard
			6131.22 ± 0.46[c]	Co and Na sources
			6129.1 ± 1.2[d]	Co, Cs, ThC"
^{16}O	$^{14}N(^{3}He, p)^{16}O$	6131 ± 4[a]	6130.2 ± 0.4[e]	m_0c^2, H capture,
			6129.0 ± 1.0[f]	^{24}Na
^{11}B	$^{11}B(p, p')^{11}B$	7978.1 ± 1.7[b]	7978.1 ± 1.9	of level separations[h]

[a] See text Ref. [6]. [b] See text Ref. [7]. [c] See text Ref. [14].
[d] R. F. Berg and F. Kashy, Nucl. Instr. Methods, 39, 169 (1966).
[e] R. C. Greenwood, Phys. Letters, 23, 482 (1966).
[f] R. E. Cote, quoted in text Ref. [14]. [g] See text Ref. [8].
[h] Separations up to 6.8 MeV excitation from $^{10}B(d, p)^{11}B$, $^{9}Be(^{3}He, p)^{11}B$ and $^{13}C(d, \alpha)^{11}B$.

γ-ray measurements have been carried out recently by Alburger and Wilkinson [8]. Their energy standard was actually separations of levels up to 6.8 MeV measured [9] at Notre Dame with the reactions shown in the table. The uncertainty is that of the particle energy measurements, not the γ-ray comparisons. The extrapolation of the smaller separation energies to the excitation energy of nearly 8 MeV gives a comparison of the two methods of measurement. The resulting values are seen to agree and the uncertainties are of the order of 0.03%.

With the small uncertainties of the absolute energy scales and the good evidence of agreement in indirect comparisons, the reason for our direct comparison may perhaps be considered more of a check on our methods and techniques, and the removal of the earlier apparent discrepancy. Our direct comparison of the gamma-ray energy scale with the Po-α energy, using the excitation energies of the first excited states of ^{24}Mg and ^{12}C will now be described.

3. Excitation Energy of $^{24}Mg(1.37)$

For the charged particle measurement an excitation near the 5.3 MeV energy of Po-α would be desirable, whereas a low excitation energy is best for measuring the γ-ray wave length or comparing with annihilation energy. The first excited state of ^{24}Mg has been accurately compared to m_0c^2 or hc/λ by more than one other laboratory. Another reason for choosing this state is the earlier apparent discrepancy mentioned above. The level energy is somewhat low for the particle measurement but is one of the highest energies that has been well measured with γ-rays on an absolute basis.

Two very accurate measurements of the energy of the γ-decay of ^{24}Mg(1.37) have been reported. An iron free $\pi/\sqrt{2}\beta$-spectrometer was used by Murray, Graham and Geiger [10] to measure the momentum of electrons from the external conversion of the γ-ray. The energies of two γ-rays in ^{208}Pb and two γ-rays in ^{60}Ni were used as secondary standards in addition to making a direct comparison with the primary standard, ^{198}Hg. Values for the secondary standards were obtained in terms of ^{198}Hg in the same work. The result from these five comparisons differed by at most 0.052 keV and the γ-ray energy is given as 1368.526 ± 0.044 keV.

J. J. Reidy at Michigan [11] used a curved crystal spectrometer to compare this ^{24}Mg γ-ray with the one from ^{198}Hg. Quartz and Ge crystals are used and diffraction angles compared with a precision lead screw mechanism. Thus a comparison of wave length is made. The γ-ray energy is found to be 1368.73 ± 0.11 keV. Thus the results from these two extremely careful measurements using quite different methods, agree within 0.2 keV. Although this is larger than the combined uncertainties, it is smaller than the uncertainty in our present particle measurement. We choose to take the unweighted average of the two measurements and adopt 1368.67 keV as the γ-ray value for the excitation energy of the level.

Next I shall describe our recently completed measurement of the excitation energy of this level using charged particles. This work was done by Stocker, Rollefson, Hrejsa and Browne and the report [12] is in press. The energies of deuterons and alpha particles were measured with the same 50 cm broad range spectrograph used for the energy standard work, reported on at the earlier mass conferences. An essential feature of the measurement is the use of a fixed magnetic field. Bombarding energy, reaction angle and field were chosen so that deuterons elastically scattered from ^{24}Mg fell at the same point near the top of the focal surface as the alpha particles from ^{210}Po. The inelastically scattered deuterons leaving ^{24}Mg in the first excited state then fell about 1/6th of the way up the focal surface. Deuterons were accelerated to 3.34 MeV with the 4 MV accelerator and the observation angles were 116° or 115°.

The crux of the measurement is conversion of the observed distance along the focal surface between the elastic and inelastic groups, into an energy difference. Use of the fixed magnetic field and a calibrating Po-α group virtually overcomes the problem of differential hysteresis. The usual question remains however, whether the positions of the two groups is correctly converted to a difference in trajectory radius when a calibration curve obtained by varying the magnetic field is used for the fixed field measurement. Of course the magnetic field must be varied when using Po-α particles to find the trajectory radius as a function of position. We attempted to verify to more than the usual precision that this curve is valid for a fixed field.

Targets with mass numbers ranging from 6-197 were used to scatter 2.5 MeV deuterons at 90°. The group from ^{197}Au then fell near the plate position used for the elastically scattered particles in the data runs, whereas the group scattered from ^{6}Li fell at the position of the inelastic group. Figure 2 shows the results demonstrating that the calibration obtained with varying fields holds for constant field. The points at plate distance 11 cm and 70 cm represent deuterons scattered from ^{6}Li and ^{197}Au respectively. The Po calibration was used to calculate a bombarding energy and scattering angle from the position of these groups. Then the expected trajectory radii for groups scattered from ^{7}Li, ^{12}C, ^{16}O, ^{19}F and ^{24}Mg at this energy and angle were calculated. The differences between these calculated radii and those given by the calibration curve are plotted in the figure. The average deviation is 0.002 cm. This is 0.01% of the radius and is much smaller than the usual variation between calibration runs.

Table III lists uncertainties for a given run. With the fixed field method and simultaneous calibration with the Po-αs, uncertainties in the input energy and the reaction angle are seen to have a very small effect. Although the position of the beam spot was observed under the microscope after each run, and the calibrating Po source mounted at the same position, we nevertheless assigned a one-quarter millimeter uncertainty to the object position. The difference between the magnetic fields measured with the flux meter and those calculated from the position of the Po-α groups led us to assign an uncertainty of 1.5G to the field. In spite of the verification discussed above the largest known uncertainty comes in the determination of the difference in trajectory radii for the elastic and inelastic groups. The third column of the table shows the resulting uncertainties in the excitation energy. Adding these in quadrature gives 1.5 keV as the internal error for a given run.

TABLE III

Uncertainties in ^{24}Mg(d, d')^{24}Mg measurement. From text Ref. [12]

Quantity	Uncertainty	ΔE_X (keV)
Trajectory radius	0.15 mm	1.1
Magnetic field	1.5 G	0.7
Object position	0.25 mm	0.6
Calibration	0.5 keV	0.4
Reaction angle	5'	0.2
Input energy	4 keV	0.03

Net ΔE_i for a given run 1.5 keV

TABLE IV

Excitation energy of ^{24}Mg 1368-keV-state

Method	Energy Standard Source	Value	Excitation Energy (keV)	σ_m
Present d, d' 10 runs	^{210}Po	331.767 kG cm	1368.2 ± 0.5	0.2
Present α, α' 1 run	^{210}Po	331.767 kG cm	1369.1 ± 1.5	
γ-ray	^{198}Hg	411.795 keV	1368.67 ± 0.08	

One run was made with 7.0 MeV $^4He^{++}$ particles accelerated with the FN Tandem. Table IV shows our results as well as the value adopted for the gamma-ray measurements. The standard deviation of the mean for the ten runs using deuterons is 0.2 keV; the single run using α-particles has an uncertainty of 1.5 keV but agrees very well with the deuteron runs. The internal error for the particle measurement is 0.5 keV. The final value is seen to be almost 0.5 keV lower than the number from the γ-ray measurements but agrees within the errors. I point out that if the γ-ray values were weighted in accordance with the stated uncertainties of the measurements, the value would be approximately 0.1 keV less than the value chosen and would agree even better with our result. The Po-α energy scale is thus found to be consistent with that used for γ-ray measurements to a precision better than 0.04%.

4. Excitation Energy of the First Excited State in ^{12}C

A second comparison of the energy scales is obtained from the recent measurement [13] by Stocker, Rollefson and Browne, of the energy of the first two excited states of ^{12}C. The Notre Dame FN Tandem accelerator provided proton and ^3He beams of up to 20 MeV energy. The work provides a good example of the problems encountered in using a tandem accelerator for precision energy measurements. It was inspired by the importance of the energy of the second excited state in stellar processes but the energy of the first excited state (4.4 MeV) allows a comparison with precise γ-ray measurements. In this case the γ-ray measurements are not absolute but are based on secondary standards, namely γ-rays from ^{56}Co. Figure 3 shows the layout of the tandem accelerator, the beam line leading to the 50 cm spectrograph that has been used for all the measurements, and some of the features of beam transport that are necessary in the use of tandem accelerators. The necessity of producing negative ions for injection into the accelerator and of then stripping the electrons from these ions in the high voltage terminal, results in a beam that is quite large and/or divergent in angle. Beam intensities are considerably lower than those from single ended machines. This combination of low quality and weak beam requires rather sophisticated beam transport equipment to produce sufficient intensity on a small enough target spot for precision measurements. We have three quadrupole lenses which focus the beam, in addition to the beam analyzing magnet and the switching magnet. This does not include the focussing in the ion source and the electrostatic lens ahead of the accelerator. There are also beam deflecting devices ahead of the accelerator, following the accelerator, and following the switching magnet. We have tried to illustrate in the figure the varying dimensions of the beam as it passes through these many elements. To transmit all the beam, the diameter just ahead of the last quadrupole must be as much as 3.5 cm. Focussing this down to the 0.5 mm target spot results in a large convergence angle. In these measurements, the beam was confined by slits ahead of this last quadrupole so that the convergence angle was about 0.5°. Unless steps are taken to center the beam within these wide slits, the actual reaction angle may vary by a fair fraction of the convergence angle.

In addition to changing the reaction angle, motion of the beam changes the position of the spectrograph object, even though the beam is defined by slits only 2.5 cm in front of the target. The problem has been dealt with in other laboratories by initial tuning, using additional defining slits placed in the target position or behind the target position. We wish to continuously monitor the beam throughout the run and use beam profile monitors at the points indicated in the figure. With these monitors, we can determine the midpoint of a two cm wide beam to $\pm$ 0.1 cm. Before taking data the accelerated beam is tuned to center the trace on the beam profile monitor ahead of the last quadrupole, in addition to balancing the currents on the slit jaws immediately ahead of the target. No adjustments in steering or focussing were made during the data runs. While targets were inserted, the beam was interrupted by a shutter beyond the control slits of the beam analyzer. The reaction angle as well as the bombarding energy was measured for each run by recording the particles elastically scattered from ^{197}Au and from ^{12}C targets with the same spectrograph field. Data runs were usually long enough so that counting in only one field of view (0.5 mm by 0.5 mm) at the center of the exposed zone gave sufficient statistical accuracy. When this was not possible, a correction was made for the kinematic broadening of the group. Counting was done in 1.5 mm wide strips and a shift in the distance scale computed for each strip to compensate for the change in outgoing momentum with the change in angle. Counts were added from as many of these shifted strips as needed to give statistical accuracy. It is an advantage of the single focussing broad range spectrograph that kinematic broadening may be greatly reduced by this procedure.

The procedure in the ^{12}C measurement is essentially the same as that used for ^{24}Mg and the sources of errors are similar. In this work, however, we did not take a Po-α run with each data run so we have added an uncertainty for the shift of the whole calibration curve. The most significant items are again the conversion of the observed distance between the

TABLE V

Excitation energy of ^{12}C 4442-keV-state

Method	Energy Standard Source	Energy Standard Value	Excitation Energy (keV)	σ_m
Present p, p'	^{210}Po	331.767 kG cm	4441.7 ± 2.1	0.6
Present ^{3}He, ^{3}He'	^{210}Po	331.767 kG cm	4442.7 ± 2.1	0.3
Average			4442.2 ± 1.5	
Chasman et al. γ-rays	^{56}Co	3253.7 keV 3452.1 keV	4439.8 ± 0.3	
Kolata et al. γ-rays	^{56}Co	3254.5 keV 3451.1 keV	4440.0 ± 0.5	

groups to a trajectory radius and in this case with the tandem beam, the uncertainty in the object position.

Our results are summarized in Table V which also shows results of γ-ray measurements. The average of six runs using protons agrees with the average of six runs using ^{3}He ions within 1.0 keV. The overall uncertainty for the twelve runs is 1.5 keV. Our result is seen to be 2.3 keV higher than the average of the results of Chasman et al. [14] and Kolata et al. [15]. This is somewhat outside the stated uncertainties. It is most interesting to note, however, that although both groups doing the gamma-ray measurements used ^{56}Co as the energy standard, they used values disagreeing by 1.0 keV for the 3.4 MeV γ-ray. This value has the dominant effect on the final γ-ray energy value. If the same standard value is used the two γ-ray measurements disagree by 1.2 keV which is outside the stated errors. It is then not clear that the γ-ray measurements and the particle measurements disagree. A third γ-ray measurement is needed. It may also be noted that in the ^{12}C comparison, the particle energy measurement gave the higher excitation energy whereas for ^{24}Mg, the particle energy measurements gave the lower excitation energy, compared to the γ-ray measurements.

To summarize, I have compared four excitation energies measured with charged particle reactions, to excitation energies deduced from γ-decay energies. In two cases the Po-α energy scale is compared to the ^{198}H energy scale by measuring the excitation energy of a nuclear state with elastic and inelastic scattering of charged particles and with γ decay of that state to the ground state. In a third case spacings of energy levels deduced from particle measurements are used as calibrating energies for γ decay from a higher lying state and the result is compared to the energy difference measured with charged particles. The results based on the ^{210}Po-α energies and the ^{198}Hg γ energy agree within 0.03% and 0.05% respectively. We have demonstrated the consistency of energy scales commonly used in nuclear reaction measurements and excitation energies deduced from particle and γ-ray measurements to a precision of approximately 0.03% or 1.5 keV in the Po-α energy.

References

1. BROWNE, C. P., in Proceedings of the Second International Conference on Nuclidic Masses, edited by W. H. Johnson (Springer Verlag, Wien, 1964), p. 234

2. MARION, J. B., Revs. Mod. Phys. 33, 139 (1961)

3. HEATH, R. L., in Proceedings of the Third International Conference on Atomic Masses, edited by R. C. Barber (Univ. of Manitoba Press, Manitoba, 1967), pp. 251-277

4. MARION, J. B., Gamma-Ray Calibration Energies, Univ. of Maryland Dept. of Physics and Astronomy, College Park (1968), Tech. Report 656

5. MURRAY, G., GRAHAM, R. L., and GEIGER, J. S., Nucl. Phys. 45, 177 (1963)

6. BROWNE, C. P. and MICHAEL, I., Phys. Rev. 134, B133 (1964)

7. BROWNE, C. P. and STOCKER, H., Phys. Rev. C4, 1481 (1971)

8. ALBURGER, D. E. and WILKINSON, D. H., Phys. Rev. C3, 1492 (1971)

9. BROWNE, C. P., MAILLE, G., TARARA, R. and DURAY, J. R., Nucl. Phys. A153, 289 (1970)

10. MURRAY, G., GRAHAM, R. L. and GEIGER, J. S., Nucl. Phys. 63, 353 (1965)

11. REIDY, J. J. (to be published)

12. STOCKER, H., ROLLEFSON, A. A., HREJSA, A. F. and BROWNE, C. P., Phys. Rev. 4, 930 (1971)

13. STOCKER, H., ROLLEFSON, A. A. and BROWNE, C. P., Phys. Rev. C4, 1028 (1971)

14. CHASMAN, C., JONES, K. W., RISTINEN, R. A. and ALBURGER, D. E. Phys. Rev. 159, 830 (1967)

15. KOLATA, J. J., AUBLE, R. and GALONSKY, A., Phys. Rev. 162, 957 (1967)

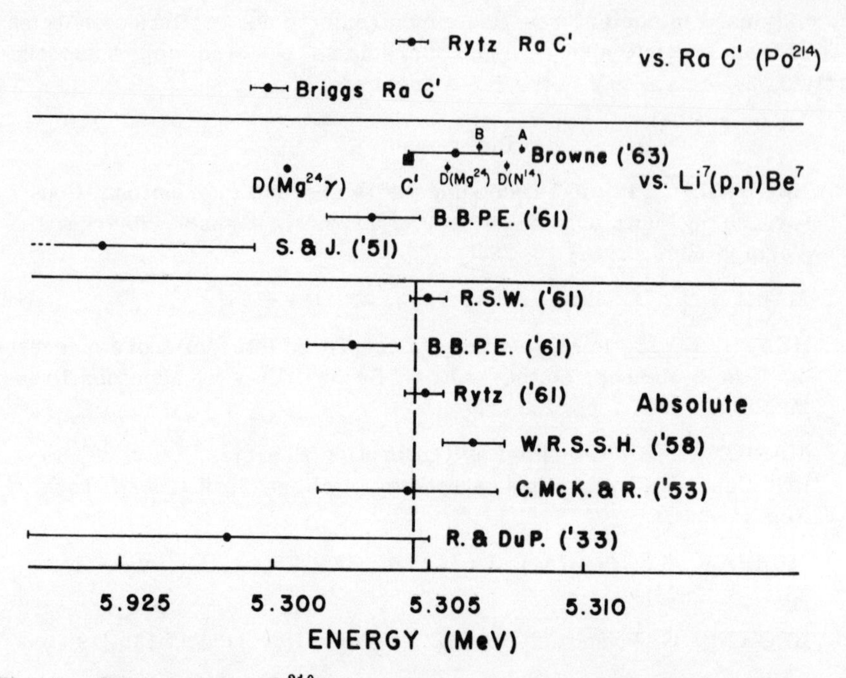

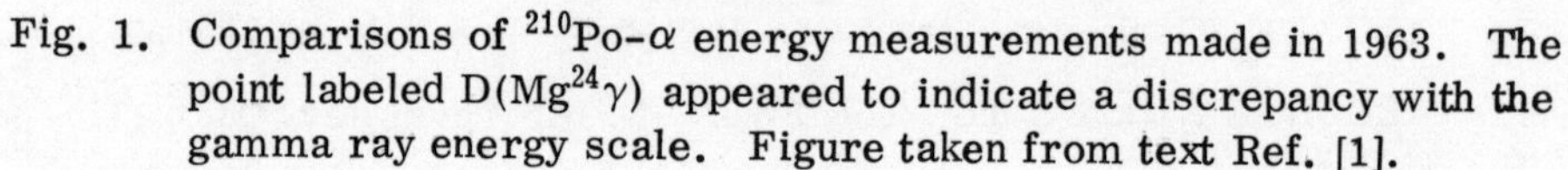
Fig. 1. Comparisons of ^{210}Po-α energy measurements made in 1963. The point labeled D($Mg^{24}\gamma$) appeared to indicate a discrepancy with the gamma ray energy scale. Figure taken from text Ref. [1].

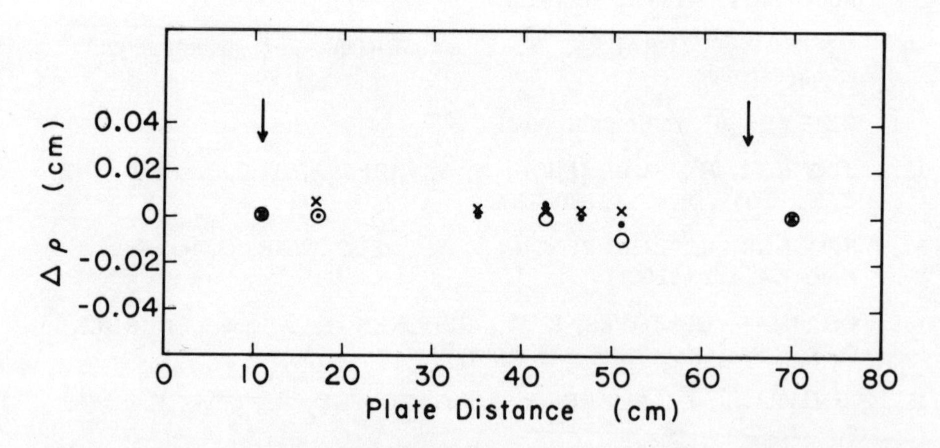

Fig. 2. Difference in trajectory radii obtained from the calibration curve and from elastic scattering of 2.5 MeV deuterons. Input energy and angle were adjusted to place the extreme points on the axis. Arrows show approximate plate position for groups from data runs. From text Ref. [12].

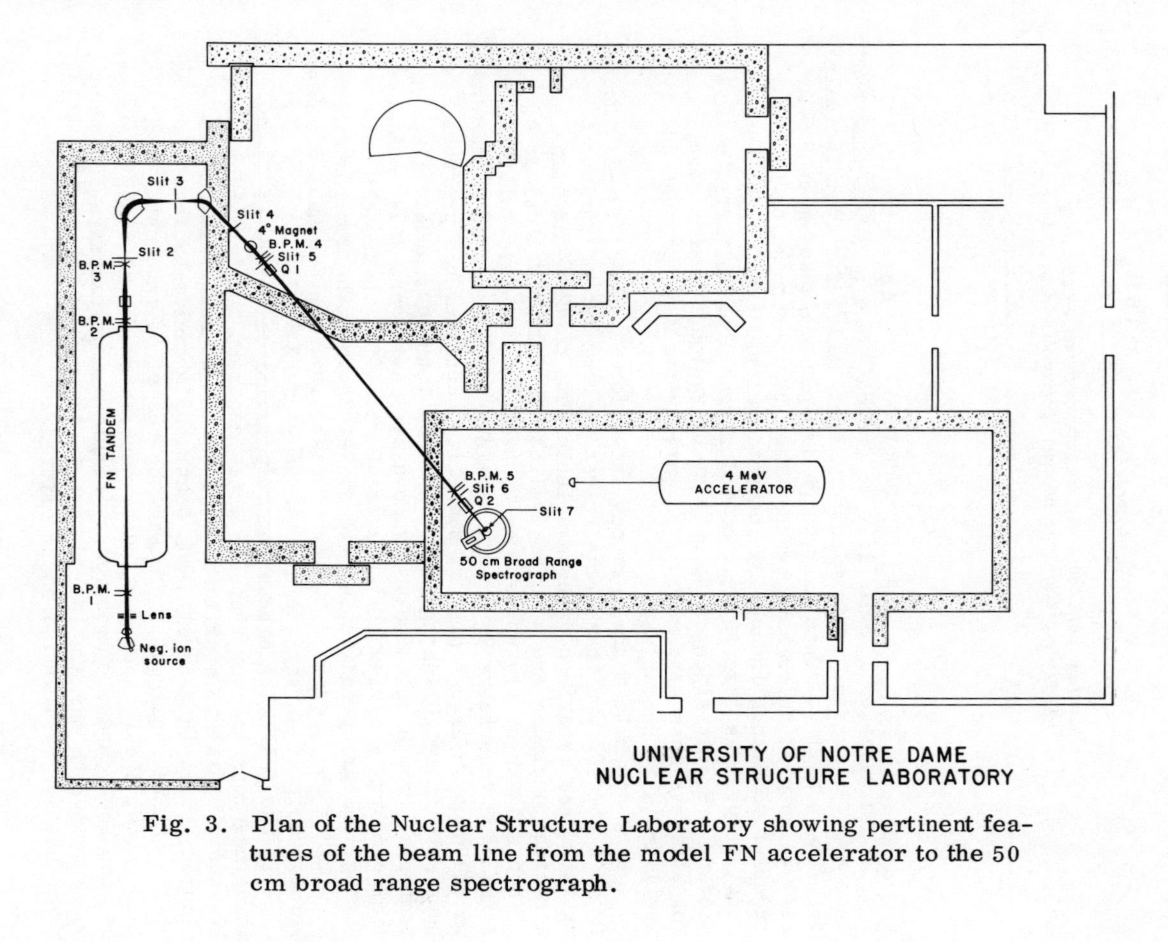

Fig. 3. Plan of the Nuclear Structure Laboratory showing pertinent features of the beam line from the model FN accelerator to the 50 cm broad range spectrograph.

Masses of Light Nuclei Far from Stability*

Joseph Cerny

*Department of Chemistry and Lawrence Berkeley Laboratory,
University of California, Berkeley, California 94720, USA*

1. Introduction

In this review I would like to indicate the progress that has been made in mass determinations of light nuclei far from the valley of beta-stability. Generally, high-isospin nuclei of mass $5 \leqslant A \leqslant 40$ will be covered which have been studied by in-beam multi-neutron transfer reactions induced by conventional projectiles and heavy ions, as well as through fragmentation reactions initiated by GeV protons. Recent attempts to utilize heavy-ion transfer reactions for accurate mass measurements of various nuclides through the $f_{7/2}$ shell are also summarized.

Many accurate masses are now known for $T_z = (N-Z)/2 = -3/2$ nuclei from ^{7}B to ^{37}Ca and these results can be compared to various theoretical mass relations. The available data on these neutron-deficient nuclides are used to evaluate the mass predictions of Kelson and Garvey (1) as well as to test the isobaric multiplet mass equation (since these $T_z = -3/2$ nuclides complete isospin quartets). On the neutron-excess side of stability, considerable progress has been made in establishing additional neutron-rich, nucleon-stable isotopes, though relatively few accurate masses are known. Figure 1 presents an overview of the current situation in the elements through sulfur by indicating all nuclei known to be nucleon stable as well as the predicted limits of stability from the nuclidic mass relationship of Garvey, Kelson and co-workers (1,2).

In the following discussion only nuclei which are at least three neutrons lighter than (or three neutrons heavier than) the lightest (heaviest) stable isotope of each element will be considered. This in general means that complex multi-nucleon transfer reactions are required to produce the nuclei of interest. With this basis, then, those particular neutron-excess nuclides whose masses can be accurately measured by simple charge-exchange reactions on neutron-rich targets [e.g., ^{18}N via $^{18}O(t,{}^{3}He)$ (3) or ^{26}Na via $^{26}Mg({}^{7}Li,{}^{7}Be)$ (4)] will also be omitted.

2. Neutron-Deficient Light Nuclei

Table I (5-10) presents a summary of the known neutron-deficient nuclei which meet the restrictions noted above. As yet no nuclei with $T_z = -2$ or greater have been characterized. The results of all the various mass measurements on both nucleon-stable and -unstable nuclei are included in the table; a weighted average is also shown.

A general approach employed for accurate mass-measurements of neutron-deficient nuclei has been to utilize the $({}^{3}He,{}^{6}He)$ and

TABLE I

Masses of known $T_z = -3/2$ nuclei far from stability

Nuclide	Nucleon-stable	Reaction	Lab	Mass-excess (MeV ± keV)	Reference[a]	Adopted mass-excess (MeV ± keV)
^{7}B	No	$^{10}B(^{3}He,^{6}He)$	LBL-67	27.94 ± 100	5	27.94 ± 100
^{9}C	Yes	$^{12}C(^{3}He,^{6}He)$	LBL-64	28.99 ± 70	5	28.910 ± 3[b]
		$^{7}Be(^{3}He,n)$	CIT-67	28.916 ± 5	5	
		"	CIT-69	28.907 ± 4	6	
		$^{12}C(^{3}He,^{6}He)$	MSU-70	28.911 ± 9	7	
^{13}O	Yes	$^{16}O(^{3}He,^{6}He)$	LBL-66	23.11 ± 70	5	23.105 ± 10
		"	LBL-70	23.107 ± 15	8	
		"	MSU-70	23.103 ± 14	7	
^{17}Ne	Yes	$^{16}O(^{3}He,2n)$	BNL-67	16.47 ± 250	5	16.479 ± 49
		$^{20}Ne(^{3}He,^{6}He)$	LBL-70	16.479 ± 50	8	
^{19}Na	No	$^{24}Mg(p,^{6}He)$	LBL-69	12.974 ± 70	9	12.974 ± 70
^{21}Mg	Yes	$^{24}Mg(^{3}He,^{6}He)$	LBL-68	10.889 ± 40	8	10.908 ± 16
		"	MSU-70	10.912 ± 18	7	
^{23}Al	Yes	$^{28}Si(p,^{6}He)$	LBL-69	6.766 ± 80	9	6.766 ± 80
^{25}Si[c]	Yes	$^{28}Si(^{3}He,^{6}He)$	LBL-70	3.817 ± 50	8	3.831 ± 12
		"	MSU-71	3.832 ± 12	10	
^{37}Ca	Yes	$^{40}Ca(^{3}He,^{6}He)$	LBL-68	-13.23 ± 50	5	-13.23 ± 50

[a] Whenever possible, reference is made to review articles.

[b] See ref. 6.

[c] These measurements assume that only the ^{25}Si ground state is populated.

$(p,{}^6He)$ reactions. These reactions possess high negative Q-values ($\sim$ -24 to -38 MeV) and low cross-sections ($\sim$ 0.06 to 3 μb). Until recently, most of these investigations were performed at Berkeley using counter-telescope techniques; however, this past year a group at Michigan State University has begun similar measurements using a magnetic spectrograph with a position-sensitive detector in the focal plane (7).

Figure 2 presents the experimental approach used in the Berkeley measurements. Due to the low cross-sections for these reactions, two four-counter semiconductor-telescope, particle-identifier systems are employed. Basically, in each telescope two particle-identifications of all events of interest are performed and compared using signals from the two successive differential-energy-loss detectors--denoted $\Delta E2$ and $\Delta E1$, respectively--and the third E detector. (The fourth detector rejects any events traversing the first three.) Further, to eliminate background due to pile-up within a single beam burst, time-of-flight measurements over the 51 cm flight path between the target and the $\Delta E2$ counter as well as subnanosecond pile-up detection using the signal from this counter are utilized. Those events in each system which are of interest are sent via an analog-to-digital converter, multiplexer system to an on-line computer. Six parameters are recorded for each event: $\Delta E2$, $\Delta E1$, E(total), particle identification, time-of-flight and pile-up detection. The details of our set-up and calibration procedures are discussed elsewhere (see 8, 9 and references therein). It appears that such an overall system is capable of studying highly endothermic nuclear reactions with cross-sections as low as 10 nb/sr.

Figures 3 and 4 present ${}^6He$ energy spectra from the ${}^{20}Ne({}^3He,{}^6He){}^{17}Ne$ and ${}^{28}Si({}^3He,{}^6He){}^{25}Si$ reactions (11) induced by 62.6 MeV ${}^3He$ ions from the Berkeley 88-inch cyclotron. In both cases the $({}^3He,{}^6He)$ reactions on ${}^{12}C$ and ${}^{16}O$ are also shown; since the masses of ${}^9C$ and ${}^{13}O$ are accurately known (particularly the former--see Table I), these reactions provide a valuable reference standard. Both the reactions on ${}^{20}Ne$ and ${}^{28}Si$ have cross-sections $\sim$ 0.5 to 1 μb/sr. Transitions to several excited states of ${}^{17}Ne$ can be seen in Fig. 3.

Although the $({}^3He,{}^6He)$ reaction can be used to study all the members of the $T_Z = -3/2$, A = 4n+1 mass series through the calcium isotopes (and in fact only the ${}^{29}S$ and ${}^{33}Ar$ masses are presently unknown), a different reaction is required to investigate the comparable $T_Z = -3/2$, A = 4n+3 mass series (since only ${}^7B$ and ${}^{11}N$ can be reached via the above reaction). For this latter mass series the $(p,{}^6He)$ reaction can be utilized, and Fig. 5 presents ${}^6He$ energy spectra from this reaction induced by 54.7 MeV protons on several targets (9). The $(p,{}^6He)$ reaction on ${}^{24}Mg$ and ${}^{28}Si$ was used to determine the masses of ${}^{19}Na$ and ${}^{23}Al$ given in Table I, while the ${}^{14}N(p,{}^6He){}^9C$ reaction was used as a calibration. Cross sections at forward angles for production of ${}^9C$, ${}^{19}Na$, and ${}^{23}Al$ are 160, 120, and 60 nb/sr, respectively. These measurements established ${}^{23}Al$ as the first nucleon-stable member of this mass series. No further measurements of members of this $T_Z = -3/2$, A = 4n+3 mass series have been reported, though ${}^{15}F$, ${}^{27}P$, ${}^{31}Cl$, and ${}^{35}K$ remain measurable by this technique.

Table II (12-14) compares the measured mass-excesses of all these $T_Z = -3/2$ nuclei to various theoretical predictions. First, since these nuclides complete isospin quartets in which the other three members are accurately known, one can investigate two aspects of the isobaric multiplet mass equation (IMME), $\Delta M = \underline{a} + \underline{b}\, T_Z + \underline{c}\, T_Z^2$. (This relation among the masses of an isobaric multiplet is derived by treating any two-body charge-dependent forces as a first-order perturbation to a charge-independent nuclear Hamiltonian. See, e.g., refs. 5, 12). One of these aspects is its predictive ability: in Table II are shown the predicted mass-excesses for these nuclides obtained from using the known masses of the other three members of the various quartets in the IMME. In general excellent first-order agreement can be noted over the entire mass range, regardless of whether the $T_Z = -3/2$ nuclide is nucleon-bound or -unbound. Such results suggest that the IMME can be reliably used to predict the masses of other undiscovered light, neutron-deficient isotopes.

Since the isobaric multiplet mass equation does so well, considerable recent interest has been attached to determining at what level and in which isobars significant deviations from this quadratic form appear. This deviation is normally parameterized by an additional term $\underline{d}\, T_Z^3$ in the mass equation (it arises as the next term in a second order perturbation treatment (see, e.g. ref. 15)). Values for $\underline{d}$ for these isobars are also given in Table II. Apart from the accurate non-zero value for the A=9 isobar, all the values of $\underline{d}$ are essentially consistent with zero, including the equally accurate result for A=13 (the uncertainty noted in footnote $\underline{f}$ of the table precludes taking the A=25 data to be of comparable importance). These results are in agreement with two recent theoretical treatments (16,17), both of which find generally small values for $\underline{d}$ ($\lesssim 1$ keV). At present, the A=9 result appears to be anomalous, and experimental remeasurement as well as more theoretical work seem necessary to clarify the situation.

In addition, Table II also compares the experimental masses to the predictions of Kelson and Garvey (1) which arise from a relation based fundamentally on the charge symmetry of nuclear forces. One in general finds excellent overall agreement between their predictions and experiment (by far the largest discrepancy occurs for ^{13}O). Finally, the relevant masses are compared with those predicted from a systematic study of Coulomb displacement energies in the $1d_{5/2}$ shell (14); in this case extremely good agreement can be seen between the predicted and experimental masses.

3. Neutron-Excess Light Nuclei

Table III (18-24) presents a summary of the known nucleon-stable, neutron-excess nuclei--which meet the restrictions noted in the Introduction--along with their mode of production. All available mass-excesses are tabulated; however, when only the nucleon-stability of a series of isotopes has been established, then solely the heaviest known neutron-stable isotope is listed (see also Fig. 1).

Although few accurate mass-excesses for very neutron-rich nuclei are available (the ^{29}Mg measurement will be discussed in the next section), the nucleon-stability of many such isotopes has been determined. The various approaches for investigating these latter nuclei are indicated in Table III and its references, but a particular

TABLE II

Experimental and predicted mass-excesses of neutron-deficient nuclides

$T_z = -3/2$ Nuclide	Experimental mass-excess (MeV ± keV)	IMME[a] prediction (MeV ± keV)	d (keV ± keV)	Kelson-Garvey[b] (MeV)	$d_{5/2}$ - shell Coul. calc.[c] (MeV ± keV)
^{7}B	27.94 ± 100	27.87 ± 150	-11 ± 30		
^{9}C	28.910 ± 3	28.956 ± 22[d]	8.0 ± 3.7[d]	28.88	
^{13}O	23.105 ± 10	23.102 ± 14	-0.5 ± 2.9	23.52	
^{17}Ne	16.479 ± 49	16.514 ± 21	5.8 ± 8.9	16.63	
^{19}Na	12.974 ± 70	e	e	12.87	12.965 ± 25
^{21}Mg	10.908 ± 16	10.940 ± 24	5.3 ± 4.9	10.79	10.916 ± 7
^{23}Al	6.766 ± 80	6.699 ± 77	-11.2 ± 18.5	6.71	6.743 ± 25
^{25}Si[f]	3.831 ± 12	3.796 ± 17	-5.8 ± 3.5	3.77	3.828 ± 8
^{37}Ca	-13.23 ± 50	-13.198 ± 91	5.3 ± 17.3	-13.17	

[a] Required data were taken from refs. 5, 8, and 12 insofar as possible. Also see ref. 13.

[b] See ref. 1.

[c] See ref. 14.

[d] See ref. 6.

[e] The lowest T = 3/2 state in ^{19}Ne has not been established.

[f] These results assume that only the ^{25}Si ground state was populated in the $^{28}Si(^{3}He,^{6}He)^{25}Si$ reaction.

TABLE III

Nucleon-stable, neutron-excess nuclides far from stability

Nuclide[a]	T_z	Reaction	Lab	Mass-excess (MeV ± keV)	Ref.[b]	Garvey et al.[c] (MeV)
^{8}He	2	$^{26}Mg(^{4}He,^{8}He)^{22}Mg$	LBL-66	31.60 ± 115	5	29.7
		$^{16}O,^{12}C(\pi^-,^{8}He)$	JINR-66		5	
^{11}Li	5/2	^{238}U + 5.3 GeV p	LBL-66	$\leqslant$ 41.1[d]	5	42.0
^{12}Be	2	$^{15}N(p,4p)^{12}Be$	BNL-65	$\leqslant$ 28.3[d]	5	25.0
^{15}B	5/2	^{238}U + 5.3 GeV p	LBL-66		5	stable
^{16}C	2	$^{14}C(^{3}H,p)^{16}C$	AWRE-61	13.693 ± 16	5	13.67
^{19}C	7/2	^{197}Au + 3 GeV p	PPA-70		18	33.7[e]
^{21}N	7/2	^{232}Th + 174 MeV ^{22}Ne	JINR-70		19	stable
^{24}O	4	" "	"		19	"
^{25}F	7/2	" "	"		19	"
^{25}Ne	5/2	^{181}Ta + 180 MeV ^{22}Ne	JINR-71	(-2.3 ± 300)[f]	20	-1.9
^{26}Ne	3	^{232}Th + 174 MeV ^{22}Ne	JINR-70	$\leqslant$ 5.8[d]	21	-0.9
^{27}Na	5/2	^{238}U + 24 GeV p	CERN-68	-7.0 ± 500	22	-6.6
^{31}Na	9/2	" "	CERN-69		22	stable
^{29}Mg	5/2	$^{26}Mg(^{11}B,^{8}B)^{29}Mg$	AERE-71	(-12.33 ± 160)[f]	23	-12.56

[a]The heaviest known neutron-stable isotope is tabulated. Lighter isotopes are listed only if their mass-excess has been determined.
[b]Whenever possible, reference is made to review articles.
[c]See ref. 2.
[d]A limit is given only when the mass of the (1 or 2)-neutron decay channel is known.
[e]See ref. 24.
[f]Preliminary values are enclosed in parentheses.

new technique deserves mention. Volkov and collaborators at Dubna (19,21) have recently identified eleven new neutron-rich isotopes through multi-nucleon transfer reactions of heavy ions on thorium targets; the reaction products were identified using a combined magnetic analysis-counter-telescope detection system.

Also listed in Table III are the mass predictions of Garvey, Kelson and collaborators (2) based on an independent-particle model; where appropriate, either the expected neutron-stability or the predicted mass-excess of the various nuclides is given. In general, excellent agreement between their prediction of nucleon-stability and experiment is observed (see also Table II of ref. 5). Hopefully, it will soon be possible by the techniques of Volkov et al. (19,21) and those of Klapisch et al. (22) to establish experimentally the limits of neutron stability, at least in the elements through sodium, and thereby provide a severe test of this mass relationship. Additional accurate mass data for neutron-excess nuclei are required to test their detailed predictions. Although generally good agreement can be noted in Table III, the predicted mass of ^{8}He is considerably in error, and both ^{11}Li and ^{19}C, though observed to be nucleon stable, were in fact predicted to be unbound (2).

4. Future Prospects

In this section I would like to indicate some useful extensions of present techniques for accurate mass measurements as well as to mention a few of the newer approaches being attempted, particularly those using heavy ions as projectiles.

On the neutron-deficient side of stability, two lines of development are of interest: the determination of masses of nuclei with $T_Z = -2$ and greater below mass 40, and the measurement of masses of $Z > N$ nuclei in the $f_{7/2}$ shell, since almost nothing is known concerning these latter nuclides above the titanium isotopes. Of immediate value would be the determination of the masses of all the $T_Z = -1/2$ nuclides in the $f_{7/2}$ shell. This would permit accurate mass predictions for many highly neutron-deficient nuclei in that shell following the procedure used by Kelson and Garvey (1) in the lighter nuclei.

Table IV (25-28) presents a few representative reactions capable of reaching these neutron-deficient nuclei. Such multi-neutron transfer reactions as (^{4}He,^{8}He) and (^{3}He,^{8}He) can produce very neutron-deficient nuclides, though at present only a limit can be set for their yield. Hopefully, the increasing use of large solid-angle magnetic spectrometers will soon make these investigations feasible. The ^{40}Ca(^{12}C,t)^{49}Mn ($T_Z = -1/2$) reaction has been successfully observed (26), so that either similar heavy-ion transfer reaction measurements, or extension of the (^{3}He,^{6}He) and (p,^{6}He) studies discussed in Section 2, could determine the masses of all the $T_Z = -1/2$ nuclei in the $f_{7/2}$ shell. In addition, a recent attempt to study the $T_Z = -1$ nuclide ^{50}Fe is listed in the table.

Many masses of neutron-rich isotopes are yet unknown. Table IV presents two quite recent, potentially general studies capable of producing such nuclides. A four-neutron transfer ^{18}O(^{18}O,^{14}O)^{22}O reaction is under investigation at Brookhaven, and measurements of the three-neutron transfer (^{11}B,^{8}B) reaction have been successfully

TABLE IV

Recent in-beam transfer reactions employed in attempts to determine accurate masses of light high-isospin nuclei far from stability

Nuclide	Reaction	Beam Energy (MeV)	Q-value (MeV)	$d\sigma/d\Omega$ (μb)	Ref.
$(^{20}Mg)^a$	$^{24}Mg(^{4}He,^{8}He)^{20}Mg$	96	-61	$\lesssim 0.025$	25
^{49}Mn	$^{40}Ca(^{12}C,t)^{49}Mn$	27.5	-12	~ 1.0	26
$(^{50}Fe)^a$	$^{40}Ca(^{16}O,^{6}He)^{50}Fe$	65	-23	< 0.3	27
$(^{22}O)^{a,b}$	$^{18}O(^{18}O,^{14}O)^{22}O$	60	-23	< 0.5	28
^{29}Mg	$^{26}Mg(^{11}B,^{8}B)^{29}Mg$	114.5	-18	~ 3.6	23

[a]Reactions leading to nuclides enclosed in parentheses can at present only set upper limits for their production via this mechanism.
[b]^{22}O is known to be nucleon stable.

reported at Harwell. Figure 6 presents an energy spectrum from the $^{26}Mg(^{11}B,^{8}B)^{29}Mg$ reaction due to Scott and collaborators (23). Although these are only preliminary results, exploitation of this reaction will give valuable mass information over much of the nuclidic chart; such data will be extremely useful in verifying the quantitative predictions of various mass relations such as those of Garvey et al. (2).

References

1. KELSON, I. and GARVEY, G. T., Phys. Letters, 23, 689 (1966).
2. GARVEY, G. T. and KELSON, I., Phys. Rev. Letters, 16, 197 (1966); GARVEY, G. T., et al., Rev. Mod. Phys., 41, S1 (1969).
3. STOKES, R. H. and YOUNG, P. G., Phys. Rev., 178, 1789 (1969).
4. BALL, G. C., DAVIES, W. G., FORSTER, J. S., and HARDY, J. C., Bull. Am. Phys. Soc., 16, 536 (1971).
5. CERNY, J., Ann. Rev. Nucl. Sci., 18, 27 (1968).
6. MOSHER, J. M., KAVANAGH, R. W., and TOMBRELLO, T. A., Phys. Rev., C, 3, 438 (1971).
7. TRENTELMAN, G. F., PREEDOM, B. M., and KASHY, E., Phys. Rev. Letters, 25, 530 (1970); Phys. Rev., C, 3, 2205 (1971).
8. MENDELSON, R., et al., Phys. Rev. Letters, 25, 533 (1970).
9. CERNY, J., et al., Phys. Rev. Letters, 22, 612 (1969).
10. TRENTELMAN, G. F. and PROCTOR, I. D., Phys. Letters, 35B, 570 (1971).
11. WOZNIAK, G. J., et al., to be published.

12. GARVEY, G. T., in Nuclear Isospin, edited by Anderson, J. D., Bloom, S. D., Cerny, J., and True, W. W. (Academic Press, Inc., New York, 1969), 703.
13. More recent data used in the IMME predictions of Table II:
A=17. McDONALD, A. B., ALEXANDER, T. K., and HAUSSER, O., Bull. Am. Phys. Soc., 16, 489 (1971).
A=23. SHIKAZONO, N. and KAWARASAKI, Y., Phys. Letters, 32B, 473 (1970).
A=25. BOHNE, W., et al., Nucl. Phys., A131, 273 (1969).
BERMAN, B. L., BAGLAN, R. J., and BOWMAN, C. D., Phys. Rev. Letters, 24, 319 (1970).
DETRAZ, C. and RICHTER, R., Nucl. Phys., A158, 393 (1970).
14. HARDY, J. C., BRUNNADER, H., CERNY, J., and JANECKE, J., Phys. Rev., 183, 854 (1969).
15. JÄNECKE, J., in Isospin in Nuclear Physics, edited by Wilkinson, D. H. (North-Holland, Amsterdam, 1969), 297.
16. HENLEY, E. M. and LACY, C. E., Phys. Rev., 184, 1228 (1969).
17. BERTSCH, G. and KAHANA, S., Phys. Letters, 33B, 193 (1970).
18. RAISBECK, G. M., et al., in High Energy Physics and Nuclear Structure, edited by Devons, S. (Plenum Press, New York, 1970), 341.
19. ARTUKH, A. G., et al., Phys. Letters, 32B, 43 (1970).
20. TARANTIN, N. I., et al., Proc. of the Inter. Conf. on Heavy Ion Physics, Dubna, Feb. 1971, (in preparation).
21. ARTUKH, A. G., et al., Phys. Letters, 31B, 129 (1970).
22. KLAPISCH, R., et al., Phys. Rev. Letters, 23, 652 (1969).
23. SCOTT, D. K. and CARDINAL, C. U., private communication.
24. THIBAULT-PHILIPPE, C., Thesis, Orsay (1971).
25. CERNY, J., et al., unpublished data.
26. CERNY, J., et al., Phys. Rev. Letters, 25, 676 (1970).
27. HARNEY, H. L., et al., unpublished data.
28. LEVINE, M. J., private communication.

* Work performed under the auspices of the U. S. Atomic Energy Commission.

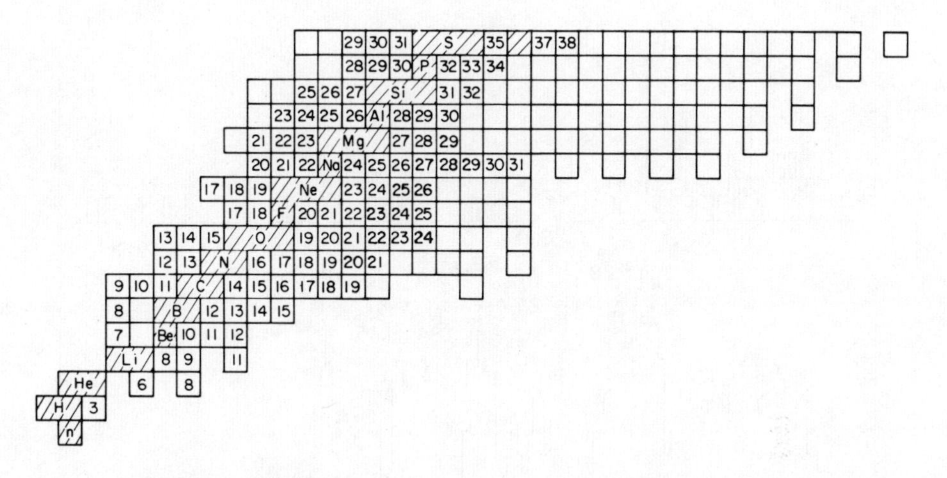

Fig. 1. Nucleon-stable nuclei through the sulfur isotopes. Unfilled squares represent the predictions of Garvey, et al.

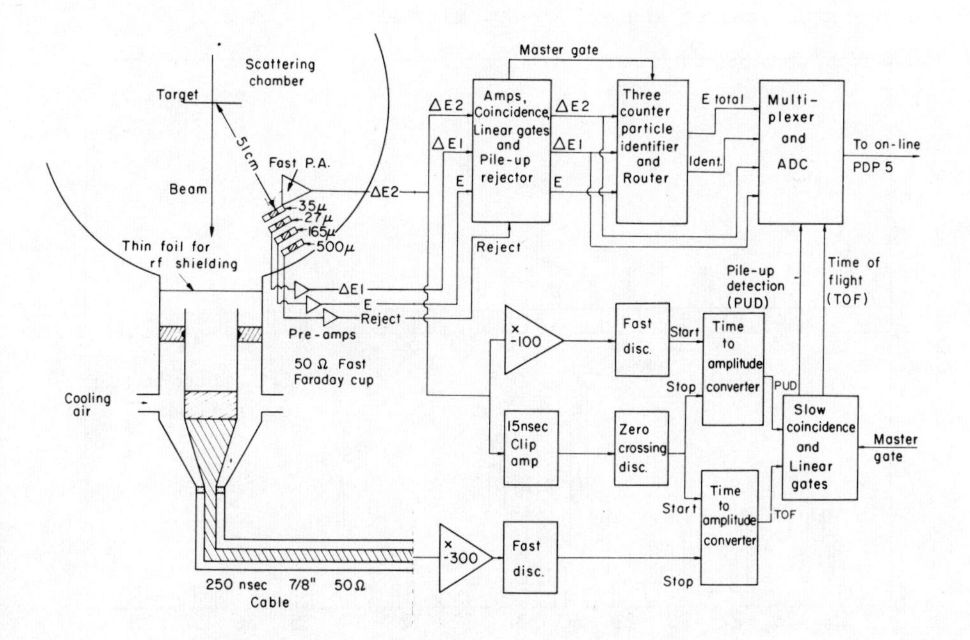

Fig. 2. An abbreviated diagram of the experimental layout for one of the two similar detection systems employed in the Berkeley measurements.

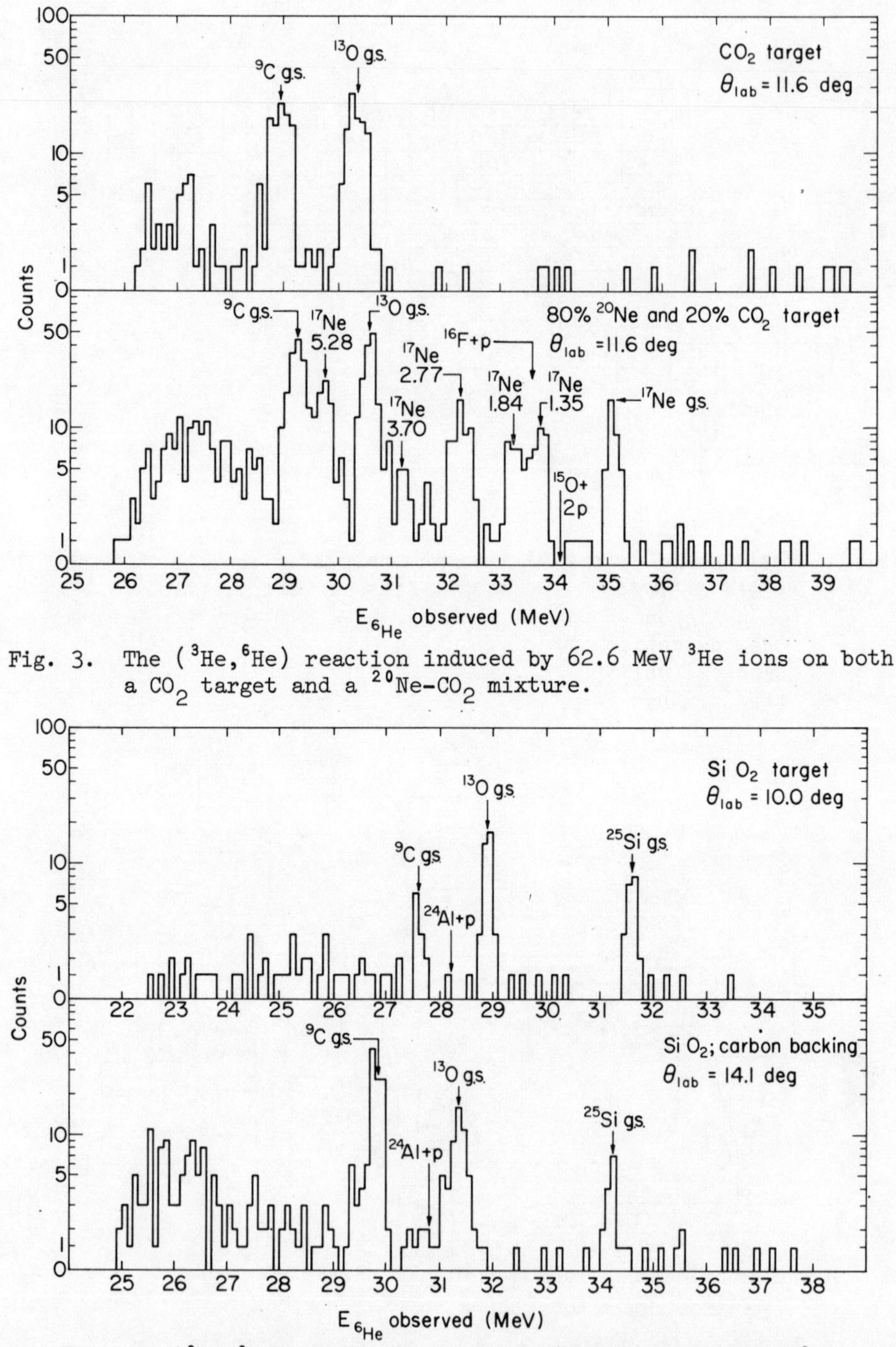

Fig. 3. The (^{3}He,^{6}He) reaction induced by 62.6 MeV ^{3}He ions on both a CO_2 target and a ^{20}Ne-CO_2 mixture.

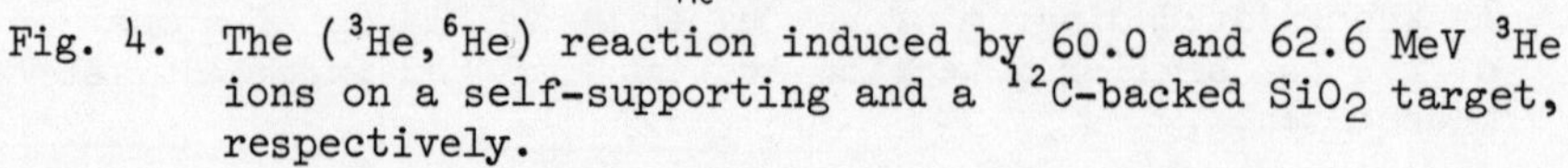

Fig. 4. The (^{3}He,^{6}He) reaction induced by 60.0 and 62.6 MeV ^{3}He ions on a self-supporting and a ^{12}C-backed SiO_2 target, respectively.

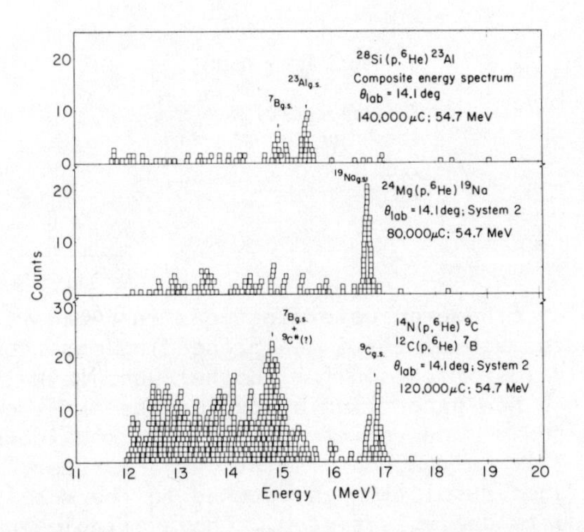

Fig. 5. The energy spectra from the (p, ^{6}He) reaction on natural silicon (top), ^{24}Mg (middle), and adenine (bottom). Each block is one count and the block width is 80 keV. Data from detection system 2 only are shown for the last two targets, while data from both systems are combined to produce the $^{28}Si(p,^{6}He)^{23}Al$ spectrum.

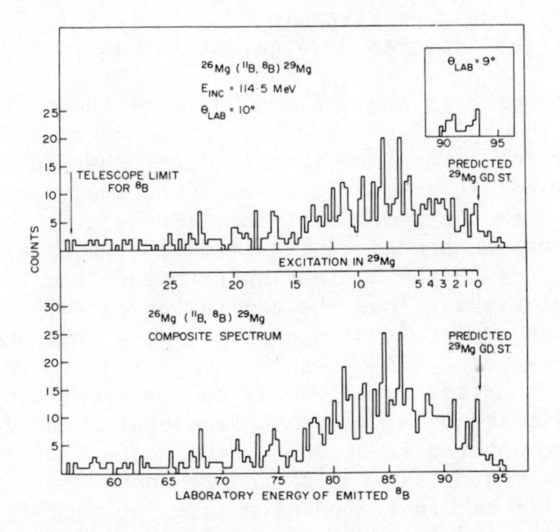

Fig. 6. Energy spectra from the $^{26}Mg(^{11}B,^{8}B)^{29}Mg$ reaction. The top spectrum shows the data at an angle of 10°, while the inset is the spectrum at 9° with tighter gates around the ^{8}B region of the identifier spectrum. The bottom spectrum shows the data from both angles kinematically compensated to 10°.

Energy Determinations in Medium and Heavy Nuclei

W. C. Parkinson

The University of Michigan,
Ann Arbor, Michigan, USA

The problem of energy determinations in nuclear reactions has been discussed in each of the three preceeding mass conferences. Perhaps the only justification for another such talk is found in the fact that the new generation of cyclotrons and their associated equipment provided a wide variety of beams at considerably higher energies and really quite good resolutions. Further, in some laboratories this good resolution is matched to the resolving power of a reaction products analysis system. The available energy permits the study of reactions having large negative Q-values and large Coulomb barriers, such as (α,t) and $(\alpha,{}^6He)$, and the acceleration of heavy ions with a high energy per nucleon opens up the possibility of mass determinations far from the region of stability. New techniques of instrument calibration are required since the old standard $^{210}Po(\alpha)$ and threshold measurements are no longer suitable. Further, the available accuracy suggests that the effects of electron binding energies, thermal motions of the target nuclei, and target thickness should be re-explored. In fact it is time to ask once again what the measured Q-value determines.

I don't propose to present a summary of the more recent mass determinations in medium and heavy nuclei since many here keep much better track of the literature than I do. Rather let me assess what I think can now be achieved in the way of accuracy in atomic mass determinations with the new cyclotrons. Perhaps at the outset I should admit that the primary interest of the group at The University of Michigan centers on the nuclear spectroscopy required for nuclear structure calculations. Thus the excitation energies and character of nuclear states are of more interest to us at the moment than ground-state Q-values. But since this does require instrumentation of good resolution with a good energy calibration, accurate measurements of ground-state Q-values can be a useful by-product. Many of the problems encountered in accurate Q-value determinations (1) are common to all laboratories, but of course each laboratory has problems unique to its own instrumentation, so let me talk in terms of the instrumentation at The University of Michigan.

I. The Instrumentation and Its Calibration

A description of the 83-inch cyclotron facility has been given previously (2), and some details on the performance for high-resolution spectroscopy, including part of the calibration technique, have been described (3). The floor plan of the magnet system including the cyclotron is shown in Fig. 1. To indicate the scale, two photographs of the system are shown in Figs. 2 and 3. Figure 2 shows the cyclotron and first beam preparation magnet while Fig. 3 shows the scattering chamber and the three analyzer magnets. The function of the cyclotron and beam preparation system is to provide at the target position a beam of particles well defined in energy and angular convergence. The quality of this beam depends not only on the characteristics of the beam preparation system, but also on the characteristics of its effective source, the beam from the cyclotron. It is perhaps worth describing this part of the system in some detail since the accuracy of the Q-value measurements is determined in part by the characteristics of the beam striking the target. The relevant parameters are defined in Fig. 4. The cyclotron provides an effective source (S) for the two dispersive elements P_1 and P_2 (the beam preparation magnets), and slits w_1 and w_2 are placed at the image surfaces of P_1 and P_2 respectively. It is the beam through w_2 which impinges on the target at the center of the scattering chamber. The effective source S is characterized by an emittance, and therefore has some radial width S and angular divergence. In considering the effect of S on the resolving power of the system and the line shape at the target, two cases are of interest: 1) S is incoherent and particles of all energies are emitted from each element of S with the number N of particles per unit time per unit ΔE constant; and 2) the source is coherent so that the energy E of particles passing through each infinitesimal element of S is mono-energetic and the energy changes monotonically across S. Resolving power is here defined as the fractional change in energy required to move an image out of a slit of width w so that it will not be confused with an adjacent image, thus $R = E/\Delta E$. The expression for R depends on the relative values of S, w_1 and w_2, and is derived in ref. 2, where it is shown that for $S \leq w_1$ the resolving power for an incoherent source is

$$R_i = E/\Delta E = \frac{\left[M_2 r_1 \frac{(1+M_1)}{(1-n_1)} + \frac{(1+M_2)}{(1-n_2)} r_2\right]}{2(w_2 + S\, M_1 M_2)}$$

while for a coherent source

$$R_c = \frac{M_1 M_2 r_c}{w_2} + M_2 \left[1 + \frac{M_1}{(1-n_1)}\right] \frac{r_1}{2w_2} + \left[\frac{(1+M_2)}{(1-n_2)}\right] \frac{r_2}{2w_2}$$

where the subscripts 1 and 2 refer to the magnets P_1 and P_2 and r, M and n are the radii of curvature, the magnification, and the field index $[n = -(r/B)\, dB/dr]$ respectively, and r_c is the radius of curvature of the ion at the time of extraction from the cyclotron. The magnets are designed to have $n = 1/2$ and $M = 1$ and $r = 2$ meters. Thus

$$R_i = \frac{4(r_1+r_2)}{2(w_2+S)} = \frac{8000}{(w_2+S)}$$

or for a line source, and $w_2 = 0.8$ mm, $R_i = 10^4$, or at $E = 20$ MeV, $\Delta E = 2$ keV. But if S has a radial width of say 0.8 mm, then $R_i = 5\times10^3$ and $\Delta E = 4$ keV. The resolving power with a coherent source is independent of S and is

$$R_c = \frac{1}{w_2}\left[2(r_1+r_2) + r_c\right] \simeq \frac{8093}{w_2}\,.$$

The relative improvement in resolving power with a coherent source is

$$\frac{R_c}{R_i} = (1+\frac{S}{w_2})\ [1 + \frac{r_c}{2(r_1+r_2)}].$$

Since S is measured to be approximately 0.5 mm, and $r_c \simeq 93$ cm, a coherent source improves the resolving power by more than a factor of 2 for the same value of w_2, and more as w_2 is slitted down. My point here is simply that for any spectrometer in general the resolving power depends on S, but that with a properly designed and adjusted cyclotron, S can be made coherent, so that the resolving power becomes independent of the source.

For certain measurements, such as excitation functions, the instrument is operated with a slit w_2 as indicated above. But for other types of measurements such as Q-value measurements there is an advantage to removing w_2 on two counts. For a system with at least two dispersive elements and an intermediate focus, it is possible to eliminate nearly all slit scattering by removing w_2. The thickness of the slit jaws of w_1 are made just thick enough to degrade the energy of the unwanted portion of the beam sufficiently so that the second dispersive element sweeps it off the target. This results in a clean beam. The second advantage results from the fact that the sum of the dispersions of the three reaction products analyzer magnets A_1, A_2 and A_3, was made equal but opposite to the sum of the dispersions of the two beam preparation magnets, P_1 and P_2. It is easily shown that for a coherent beam transported through the five magnets the width of the final image is

$$Y = S\ \{1 - [\frac{(1-n_P)}{(1-n_A)}]\ [\frac{1+|M_A|}{1+|M_P|}]\ \frac{\Sigma r_A}{\Sigma r_P}\ \}$$

and is independent of the width of the slits w_1 and w_2. Because $n_A = n_P = 1/2$, $M_A = M_P = -1$ and $\Sigma r_A = \Sigma r_P = 400$ cm, then $Y = 0$. If S is not coherent, the width of the final image is not zero but is the width of the effective source from the cyclotron. If S is coherent, the width of the image is equal to the width of the slit in the ion tower at the center of the cyclotron, reduced by the phase space compaction due to the acceleration of the beam.

The line shape as well as the resolving power is an important factor in determining the accuracy with which Q-values can be measured. The calculated and measured line shapes at the center of the scattering chamber (the image surface of P_2) for various values of w_1 (w_2 removed) is shown in Fig. 5. The value of S used in calculating the line shape was the measured value S = 0.46 mm. One measurement of the line shapes along the focal surface of the third analyzer magnet for various values of w_1 (again with w_2 removed) is shown in Fig. 6. Note that the intensity is essentially proportional to w_1. The peak for $w_1 = 6$ mm is 60 times more intense than for $w_1 = 0.1$ mm. Thus there is great value in dispersion matching. The lower limit on width, 1.9 mm, (corresponding to 1.15 mm normal to the optic axis) is due mainly to the source. This was confirmed by a subsequent direct measurement at the machine exit which, for the tuning condition of this particular run, yielded S = 1.4 mm, in good agreement with the above value. This again emphasizes the importance of the contribution of the source to resolution. As will be evident later such a poor line shape would make it difficult to correct for the effect of finite target thickness. Instrumentation has since been developed to allow continuous monitoring of the source size and position during experimental runs.

For precision nuclear spectroscopy it is essential that the magnetic field rather than the current of each of the magnets be regulated and that hysteresis and saturation effects, hopefully small, are known. The magnetic field strengths of the cyclotron magnet and each of the five spectrometer magnets are regulated using NMR fluxmeters; the same fluxmeters are also used for determining the magnitudes of the fields. Because the fields fall off with radius (n=1/2), the NMR probes are located in radius near the inside edge of the gap at the turn-over point where the field is a maximum and has no radial component, and in azimuth near the entrance where, because of the divergence of the beam, it does not reduce the solid angle.

The important considerations for spectroscopy are 1) the absolute calibration of the field averaged over the particle path at all fields, 2) the reproducibility of the fields from day-to-day, and 3) differential hysteresis along the magnetic path. The second beam preparation magnet P_2 has been calibrated in terms of the T=3/2 isobaric analogue resonance in ^{13}N which occurs at a laboratory energy of 14.233 ± .008 MeV in the $^{12}C(p,p)$ excitation function (4). The resonance has been measured several times, first in May, 1966 and most recently in June, 1969. A typical result is shown in Fig. 7. The width of the resonance, peak to valley, is nearly 3 keV and the peak-to-valley ratio is approximately 1.8. These data were taken with a 40 μg/cm² (1.4 keV) carbon target and an energy spread of the incoming beam on the target of 1.8 keV. The points plotted with circles were taken by varying the magnetic field of the second beam preparation magnet, and those plotted with X were taken keeping the magnetic field constant and varying the dc potential of the target. The position of the resonance is reproducible within the accuracy of the measurement (≈ 200V) independently of the past history of the magnetic field. No shift in position as large as 200V was detected when the field was brought from zero to the resonance value, or from zero to full excitation then down to the resonance value, or with small changes across the resonance. Thus, it is concluded that the average field is reproducible to better than 1 part in 20,000 and any uncertainties due to differential hysteresis are considerably less than 1 keV. We believe the reproducibility and lack of significant differential hysteresis is due to the conservative design of the magnets, which operate well below saturation.

The three reaction products analysis magnets were calibrated using this well defined beam. Thus, the entire system is calibrated at one energy in terms of the T=3/2 resonance in ^{13}N (but which is of itself uncertain to 0.056%).

To calibrate the magnet system and determine its linearity over a wide range of excitation, well defined beams at higher energies are needed. While the H_2^+ ion consisting of two protons of essentially equal energy plus an electron would appear to offer a calibration at twice the energy of H_1^+ at the T=3/2 resonance (corrected of course for the electron mass), it is well known that the 2.65 electron volt dissociation energy of the H_2^+ molecule ion causes a spread in kinetic energy of the two protons on break-up of the H_2^+. This amount to approximately 42 keV at this energy, which is more than enough to mask the resonance (5). However, while the resonance cannot be detected with H_2^+, a calibration can be made by adjusting the two beam preparation magnets to transport H_2^+ to the scattering chamber through small slit openings. Then a thin stripping foil is inserted at the intermediate focus of the two-magnet system to dissociate the H_2^+ into $H_1^+ + H_1^+$. The field of the second magnet must then be reduced from the corresponding 28.474 MeV to 14.233 MeV to bring the H_1^+ into the scattering chamber. The relation between the two energies is well defined, and this serves to calibrate the magnet system at essentially twice the energy of 14.233 MeV. The currents to the Faraday cup as a function of the NMR frequency of the second magnet (which is proportional to the magnetic field and hence to momentum) is shown in Fig. 8 for H_2^+ and H_1^+. The line shape for H_2^+ is that expected from the geometry (3). The full width of the H_1^+ curve is 18.7 kHz which, when corrected for target thickness and the energy spread in the target, leads to the value of ΔE = 42 keV as the spread in energy resulting from dissociation. (It is interesting to note that this corresponds to a potential energy of the repulsive state of the H_2^+ of 7.8 e.v. Further, the ratio of the intensities measured using first the H_2^+ beam then the $H_1^+ + H_1^+$ beam suggest that the break-up channel is $H_2^+ \rightarrow H_1^+ + H_1^+$, with essentially none of the $H_2^+ \rightarrow H_1^0 + H_1^+$ component. It is also interesting to note that from a careful measurement of the line shape, the probability distribution for the spatial configuration of the two protons in H_2^+ could be unfolded.)

If the magnetic field scaled accurately with the NMR frequency, the ratio of the frequencies would be

$$\frac{f(H_2^+)}{f(H_1^+)} = \frac{B\rho(H_2^+)}{B\rho(H_1^+)} = \frac{M(H_2^+)}{M(H_1^+)} = \frac{2m_p+m_e}{m_p} = (2 + \frac{m_e}{m_p}) = 2+5.45\times10^{-4}.$$

Experimentally, when the (H_1^+) frequency is corrected for the energy loss in the carbon stipping foil, the difference of the calculated and measured values is 11.2 kHz at 24.06 MHz or the deviation from perfect scaling of the magnetic field corresponding to 14.233 MeV protons and 28.474 MeV protons is $\Delta f/f$ = 1/2140 or less than 0.05%. An accurate scaling factor for other fields can be measured using $^3He^+$ and $^3He^{++}$ where the energy spread due to molecular dissociation does not enter, but this has not yet been done. With these energies well defined in terms of the ^{13}N, T=3/2 resonance, additional calibration points can be obtained in terms ofother known reaction Q-values.

As a supplement to and a check on this method a wider and fuller range of calibration points can be obtained using the method of momentum matching. This method was first used some years ago (6) to determine the beam energy of and to calibrate the spectrometer associated with our former 42-inch cyclotron. The method has been applied by Bardin and Rickey (7) and more recently by the group at Michigan State University (8,9) who have used the technique to make an accurate determination of their beam energy and a calibration of their magnetic spectrograph in the region of 720-1120 kG-cm. The principle (6) is that for a particular target nucleus the Q of the charged-particle reaction, say a (p,d) reaction, insures that at some angle of observation the deuteron momentum is equal to the momentum of the proton elastically scattered by the same nuclide. Evidently the angle at which the particles have equal momenta depends on the beam energy and therefore can be used to measure it. In practice the momenta need only be approximately equal provided both types of particles are brought to a focus at the image surface on the same detector, (a nuclear emulsion or position-sensitive detector, for example) and provided the dispersion function for the magnet is well known. Some possible reaction pairs for calibration are given in Table I. Again, calibration points at intermediate as well as higher and lower values of magnetic rigidity can be obtained by observing other reactions whose Q values are well established using the beam energy precisely determined by the momentum-matching method. The uncertainty in the beam energy due to uncertainties in the angle of observation, the Q-values, and the determination of the energy of the particle groups at the image surface can be computed from the relativistic kinematics. To avoid displaying the long algebraic formulas, consider the two reactions $^{12}C(d,d)^{12}C$ and $^{12}C(d,t)^{11}C$ (Q = -12.462 MeV). The reaction products have equal magnetic rigidity at $\psi = 15.0°$ for $E_d = 38.502$ MeV. The uncertainty in the incoming deuteron energy is very nearly equal to the uncertainty in the reaction particle energy varying from essentially unity at 0° to approximately 1.5 at 90°. At 15° $\partial E_1/\partial E_3 = 1.014$ for deuterons and 1.015 for the tritons. The uncertainty due to Q is $\partial E_1/\partial Q \cong -1.44$ for the (d,t) reaction at 15° and increases to -1.22 at 90°. It is the uncertainty in the scattering angle that poses the problem. For the elastically scattered deuterons $\partial E_1/\partial\psi = 58$ keV/degree at $\psi = 15°$, and increases to 229 keV/degree at $\psi = 90°$, while for the tritons $\partial E_1/\partial\psi = 69.2$ keV/degree at $\psi = 15°$ and 227 keV/degree at $\psi = 90°$. Thus there is good reason to choose as small a scattering angle as practical; even so the scattering angle must be determined to a small fraction of a degree if an accurate calibration is to be obtained. By slitting down the exit aperture of P_2 the angle of convergence of the beam onto the target can be made small (less than 0.05 degrees). The zero degree beam direction is then determined by measuring the elastic scattering intensity at several angles on both sides of zero degrees. The kinematic relations of energy versus angle can also be used as a check on the mechanical measurement of the scattering angle.

Because $\partial E_3/\partial\psi = 0$ at $\Psi = 0°$, the requirement of the precise measurement of the scattering angle can be relaxed if reactions can be measured at zero degrees. This is practical when two or more analyzer magnets are used in series although it does preclude elastic scattering as one of the reactions. For example, the reaction pair $^{12}C(d,d')^{12}C^*$ ($E_x = 4.4398 \pm 0.0003$ MeV) and $^{12}C(d,t)^{11}C$ provides a momentum match at 0° with a deuteron beam energy of 28.785 MeV, and $\partial E_1/\partial E_3, \partial E_1/\partial Q$ are essentially unity and $\partial E_1/\partial\psi \equiv 0$. Thus, at a

TABLE I*

Possible reaction pairs for momentum match energy calibrations. Calculations are for $\theta_{lab} = 15.0°$.

Reaction 1	Excitation energy 1 (MeV ± keV)	Reaction 2	Excitation energy 2 (MeV ± keV)	Beam energy[a] (MeV ± keV)	Beam particle	Magnetic rigidity of outgoing particles (kG·in)
$^{12}C(p,p)^{12}C$	0.0	$^{12}C(p,d)^{11}C$	0.0	33.691 ± 2.2	p	332.256
$^{12}C(p,p)^{12}C^*$	4.4398 ± 0.3	$^{12}C(p,d)^{11}C$	0.0	29.009 ± 2.2	p	282.884
$^{16}O(p,p)^{16}O$	0.0	$^{16}O(p,d)^{15}O$	0.0	27.336 ± 2.5	p	299.009
$^{16}O(p,p)^{16}O$	0.0	$^{16}O(p,d)^{15}O^*$	6.180 ± 4.0	40.106 ± 8.0	p	363.381
$^{16}O(p,p)^{16}O^*$	6.1305 ± 0.4	$^{16}O(p,d)^{15}O$	0.0	20.981 ± 2.5	p	219.312
$^{16}O(p,p)^{16}O^*$	6.1305 ± 0.4	$^{16}O(p,d)^{15}O^*$	6.180 ± 4.0	33.607 ± 8.0	p	299.600
$^{12}C(d,d)^{12}C$	0.0	$^{12}C(d,t)^{11}C$	0.0	38.502 ± 3.0	d	498.783
$^{12}C(d,d)^{12}C^*$	4.4398 ± 0.3	$^{12}C(d,t)^{11}C$	0.0	29.203 ± 3.0	d	398.802
$^{16}O(d,d)^{16}O$	0.0	$^{16}O(d,t)^{15}O$	0.0	28.867 ± 4.0	d	431.974
$^{16}O(d,d)^{16}O^*$	6.1305 ± 0.4	$^{16}O(d,t)^{15}O$	0.0	16.283 ± 4.0	d	254.009
$^{16}O(d,d)^{16}O$	0.0	$^{16}O(d,t)^{15}O^*$	6.180 ± 4.0	48.096 ± 12.0	d	558.980
$^{16}O(d,d)^{16}O^*$	6.1305 ± 0.4	$^{16}O(d,t)^{15}O^*$	6.180 ± 4.0	35.232 ± 12.0	d	433.230
$^{15}N(p,p)^{15}N$	0.0	$^{15}N(p,d)^{14}N$	0.0	17.446 ± 1.6	p	238.220
$^{15}N(p,p)^{15}N$	0.0	$^{15}N(p,d)^{14}N^*$	2.3128	22.174 ± 1.6	p	268.898

[a] Error reflect uncertainty in the masses.

* From G. F. Trentleman and E. Kashy, Nucl. Inst. & Meth. 82, 304 (1970).

scattering angle of 0° the requirement of an accurate measure of the angle is considerably relaxed.

Uncertainties due to the effects of target thickness and non-uniformity can be essentially eliminated from both the calibration and the measurement of Q-values. For a given target thickness the most probable energy loss and the statistical fluctuations in the energy loss can be calculated (10). It appears possible for the thin targets of interest here to calculate the distribution near the point of zero energy loss, and thus determine from the measured shape that point on the image surface that corresponds to zero target thickness. This problem is being investigated. The line shape at the image surface is determined by the natural ion-optical line shape taken in quadrature with the shape due to the fluctuations, and it is for this reason that a knowledge of the ion-optical line shape is of some importance.

It is important when using solid targets, such as evaporated films, that the target material have electrical conductivity, for otherwise it may assume some indefinite but large positive potential which will alter the kinetic energy of the incoming and outgoing projectiles unequally if they differ in Z. In a (h,d) reaction, for example, depending on the nature of the target, this could well amount to several kilovolts.

Let me now make a remark about Q-values, and what is actually measured. The nuclear Q-value Q_n is by definition $Q_n=[m_1+m_2-m_3-m_4]c^2$ where m_1, m_2, m_3 and m_4 are the nuclear masses, i.e., the rest masses of the bare nuclei for the projectile, the target, the outgoing light particle, and the residual nucleus, respectively. The atomic Q-value Q_A is defined as $Q_A=[M_1+M_2-M_3-M_4]c^2$ where M_i are the atomic masses. Since $M_i=m_i+Zm_e-B/c^2$ then $Q_n=Q_A+(B_1+B_2-B_3-B_4)$ where the B_i are the total electronic binding energies associated with each M_i. As is well known, the electron rest masses drop out. One estimate of the total electronic binding energy, based on the Fermi-Thomas model of the atom is (11) $B=20.83Z^{7/3}$ eV and may be good to 20% for the heavier nuclei. For light nuclei the value of the coefficient has to be reduced to 15.73 to agree with the available experimental data. While we are not able to calculate B_i with the precision we would like, we can make reasonable estimates of their effects. For the heavier nuclei, B_i is large; for example for $^{208}_{82}$Pb it is about 608 keV. The differences in the binding energies $(B_1+B_2-B_3-B_4)$ for the $^{208}_{82}$Pb(h,d)$^{209}_{83}$Bi reaction is approximately 17.4 keV which is hardly negligible when we talk of determining atomic masses to the order of one kilovolt. Even for the reaction ^{12}C(d,α)^{10}B, the difference is about 0.48 keV. [Of course, if there is no change in Z, as in a (d,p) reaction, the difference in the binding energy is zero.] But the point I wish to make here is that what we measure in a nuclear reaction is neither Q_n nor Q_A but something intermediate and the problem is to relate the measured Q_m to Q_A. The complexity of the problem can be illustrated by considering a (h,d) direct reaction with a target nucleus of charge Z in its atomic ground state. Assume for the moment that neither the helion nor the deuteron interacts with and causes excitations of the atomic electrons in passing through the electron cloud. One of the protons in the helion is left in the nucleus, changing the charge of the nucleus to (Z+1), as the deuteron leaves. The nuclear interaction occurs in a time of the order of 10^{-21} seconds, and during this time the residual nucleus acquires linear momentum. But in approaching the nucleus the helion is in the Coulomb field of the atomic electrons and similarly for the deuteron on the way out, and as a result their energies will be altered. Consider the two extreme cases, namely: 1) the velocities of the helion and deuteron are large compared to the velocities of the electrons in their orbits (the impulse limit); and 2) the velocities are small compared to the electron velocities (the adiabatic limit). In the impulse approximation the Z atomic electrons are left in the same spatial configuration but with a binding energy B_i corresponding to (Z+1) charges in the nucleus. This results because the time required for the rearrangement of the electrons is assumed long compared to the time of interaction. After a relaxation time of something longer than $\tau = \sqrt{\frac{m}{2}}\frac{\pi e^2}{W_i^{3/2}}$ seconds, where W_i is the binding energy of the i-th electron, and varies from 10^{-16} for the lightest nuclei to 5×10^{-20} seconds for the K-electrons in lead, the electrons will rearrange into the ground state of the atom of (Z+1) with the emission of energy in the form of X-rays and light quanta of magnitude $(B_i - B_4) = \Delta E_2$ where B_i and B_4 are the total binding energies of the intermediate state and final states, respectively. This energy ΔE_2 is in no way involved in the measured Q_m. Rather it is the difference $(B_2 - B_i) = \Delta E_1$ between the original and intermediate states that appears in Q_m.

A rough estimate of ΔE_1 indicates* that it should be approximately $\Delta E_1 \simeq \frac{z}{Z} B(Z)$ where z is the change in charge of the target nucleus and B(Z) is the total electronic binding energy of the target nucleus. This estimate is based on the simplified argument that $(\partial E/\partial Z) = \frac{\partial}{\partial Z}\left[20.83\, Z \cdot Z_1^{4/3}\right]_{(Z_1=\text{const})}$ where $Z_1^{4/3}$ is the contribution of the electrons to the Coulomb energy (Fermi-Thomas) and in our approximation does not change, and Z is the contribution due to the nuclear charge. Thus $\partial E/\partial Z \simeq 20.83\, Z_1^{4/3}$ or $\Delta E_1 \simeq \frac{z}{Z}$ (B(Z). For $_{82}$Pb, B(Z) ≃ 608 keV and for the (h,d) reaction z = +1, thus $\Delta E_1 \simeq \frac{1}{82} \times 608 \simeq 7.4$ keV and this appears in the measured Q_m. The atomic Q-value, Q_A, the quantity of interest would be, in terms of the measured Q_m

$$Q_A \approx Q_m + \{B_3 - B_1 + B_2 \left[\frac{4}{3}\left(\frac{z}{Z_2}\right) + \frac{14}{9}\left(\frac{z}{Z_2}\right)^2 + \text{---}\right]\}. \quad (1)$$

At the other extreme, the adiabatic limit, the helion and deuteron velocities are small so that the electrons readjust during the interaction and the measured Q_m differs from Q_A only by the electronic binding energies of the helion and deuteron. Clearly the truth lies somewhere between these limits. Projectiles in the 20 - 50 MeV energy range have velocities smaller than but of the same

* I am indebted to A. C. T. Wu and J. P. Draayer for many helpful discussions on these questions and for suggesting the form of ΔE_1.

order as the K-shell electron velocities. Thus one might reasonably expect the effect to be appreciable.

The Q of a reaction can also be expressed as $Q=T_3+T_4-T_1-T_2$, where the T_i are the kinetic energies. Neglecting thermal motions of the target nuclei for the moment $T_2=0$, and T_1 and T_3 are the quantities normally measured, presumably as accurately as we wish. The value of T_4 is usually computed from T_1 and T_3 on the basis of the conservation laws. But this requires a knowledge of the masses m_1, m_3 and m_4. We know the masses of the observed ions m_1 and m_3, but we may not know m_4.

The question arises as to what are the correct masses, nuclear or atomic, to use in the determination of Q. This must depend on the nature of the reaction. Again consider the two extremes: 1) The nuclear interaction is so short that no interaction between the struck nucleus and the atomic electrons or the Coulomb field can occur. Clearly here the correct mass is the nuclear mass. 2) The nuclear interaction is slow, so slow that the atomic electrons are in good communication with the nucleus, and move happily with it. Clearly in this case the correct mass is the atomic mass. For intermediate interaction times the effective mass m^* must be $m < m^* < M$. For the light ions, m_1 and m_3, stripped of their electrons, only the nuclear mass, of course, is involved. Neglecting thermal motions the mass of the target does not enter. The expression for the effective mass m_4^* of the residual nucleus can be estimated. Let $m_4^*=m_4+\Delta m_4$, then

$$\Delta m_4 \approx \sum_{n\ell} 2(2\ell+1)(m_e-B_{n\ell}/c^2) \int_0^{c\Delta t} R_{n\ell}^2(r)dr / [\int_0^{\infty} R_{n\ell}^2(r)dr]$$

where the sums are taken over the occupied levels. Here c is the velocity of light and Δt is the time of the interaction. The integral $\int_0^{\infty} R_{n\ell}^2(r)dr$ is unity if the wavefunction is normalized and $\sum_{n\ell} 2(2\ell+1)$ is the total number of electrons Z. For a direct reaction where $\Delta t \simeq 10^{-21}$ seconds, only the K-shell electrons make any significant contribution and even this is small, of the order of 0.1% of the masses of the K-shell electrons. Thus for practical purposes in computing T_4 for a direct reaction, the nuclear mass m_4 can be used without introducing significant error.

There is in addition the possibility that the helion on the way in will induce electronic excitations and the deuteron on the way out will do likewise. These are just the excitations Professor Staub (12) reported at the last conference. As he indicated, these are of relatively low energy and would contribute an additional but small correction to the measured Q-value.

The effect of the thermal energies of the target nuclei remains small at normal target temperatures, but does add a contribution to the line shape. As examples, for the $^{12}C(p,d)^{11}C$ reaction with $E_d \simeq 30$ MeV, $\delta E_d \simeq 0.6$ keV, while for $^{208}Pb(h,d)^{209}Bi$ with $E_d \simeq 30$ MeV, $\delta E_d \simeq 0.16$ keV. Since δE is to be added in quadrature to the energy distribution at the image surface of the reaction products analyzer, it remains a relatively small effect.

The question is, can these effects be measured? In principle if T_4^* could be determined accurately and simultaneously with T_1 and T_3, then m_4^* could be determined. But this would have to be measured in a time short compared to the relaxation time of T_4^* to T_4. If, however, the total energy associated with T_4 were measured and if the de-excitation energy could be measured separately, then the value of Q_A could be determined directly. Because of the times involved and the difficulty of measuring the total de-excitation energy accurately, this does not seem feasible at present. However, with the accuracy now being obtained in atomic mass measurements, particulary by Smith (13), a comparison with accurately measured Q-values would indicate the magnitude of these effects and how they vary with Z and the time of interaction. It would be particularly useful if the mass-spectroscopy measurements could be extended to the region of ^{208}Pb with comparable accuracy. But it does appear that the obtainable accuracy in Q-value measurements would make such comparisons meaningful.

REFERENCES

1. The problems of accurate Q-value determinations were discussed in detail in the paper of BROWN, A. G., SNYDER, C. W., FOWLER, W. A., and LAURITSEN, C.C., Phys. Rev. 82, 159 (1951).

2. PARKINSON, W. C., and TICKLE, R. S., Nucl. Inst. & Meth. 18, 19, 93 (1962).

3. PARKINSON, W. C., and BARDWICK, J., Nucl. Inst. & Meth. 78, 245 (1970).

4. TEMMER, G., Nuclear Structure (International Atomic Energy Agency, Vienna, 1968) p. 258; MARION, J. B., Rev. Mod. Phys. 38, 660(1966).

5. HERRING, D. F., et al., Phys. Rev. 100, 1239A (1955); ANDERSEN, S. L., et al., Nucl. Phys. 7, 384 (1958); DAHL, P. F., et al., Nucl. Phys. 21, 106 (1960); WALTERS, W. L., et al., Phys. Rev. 125, 2012 (1962); PURSER, K. H., et al., Phys. Letters 6, 176 (1963). The first two references attribute the effect to internal motion of the protons rather than to the Coulomb energy available at the instant the electron is stripped.

6. BACH, D. R., CHILDS, W. J., HOCKNEY, R. W., HOUGH, P. V. C., and PARKINSON, W. C., Rev. Sci. Inst. 27, 516 (1956).

7. BARDIN, B. M., and RICKEY, M. E., Rev. Sci. Inst. 35, 902 (1964).

8. TRENTELMAN, G. F., and KASHY, E., Nucl. Inst. & Meth. 82, 304 (1970).

9. TRENTELMAN, G. F., PREEDOM, B. M., and KASHY, E., Phys. Rev. 3, 2205 (1971).

10. There is an extensive amount of literature on the subject, and a summary is given by MACCABEE, H. D., "Fluctuations of Energy Loss by Heavy Charged Particles in Matter", UCRL-16931 (1966).

11. FOLDY, L. L., Phys. Rev. 83, 397 (1951).

12. Proc. Third Intnl. Conf. on Atomic Masses, Winnipeg (1967), p. 499.

13. SMITH, L. G., Phys. Rev. 4C, 22 (1971).

NOTE ADDED AFTER THE CONFERENCE: Professor Staub and Professor A. A. Jaffe called my attention to papers by Christy [Nuc. Phys. 22, 301 (1961)] and Serber and Snyder [Phys. Rev. 87, 152 (1952)], respectively, in which each discusses the effects of atomic binding on Q-value measurements. Their results differ from the equation for Q_A above (eqn. 1) in that in their derivation the first order term in (z/Z) vanishes. The reason is to be found in the value they use for the electrostatic potential $\phi(Z)$ at the position of the nucleus due to the electron cloud. They obtain $\phi(Z)$ by differentiation of the Fermi-Thomas expression for the total electronic binding energy. But since the binding energy includes the kinetic energy of the electron cloud (which does not change in the impulse approximation), such a differentiation does not lead to the correct potential. In effect they are calculating the rate of change in binding energy with Z for the adiabatic case and their value of ΔE must in the limit approach zero. While the expression for ΔE_1 used to obtain eqn. 1 is at best a rough estimate, it should be closer to the correct value in the impulse limit.

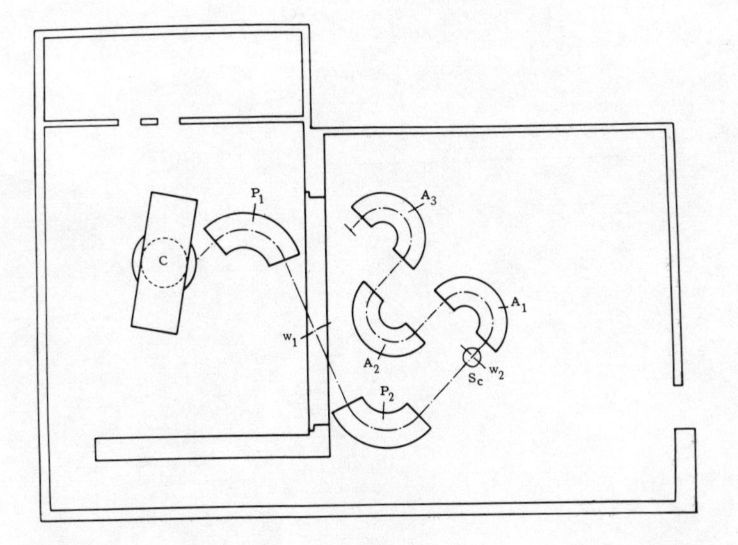

Fig. 1. Floor plan of the magnet system for the 83-inch cyclotron installation.

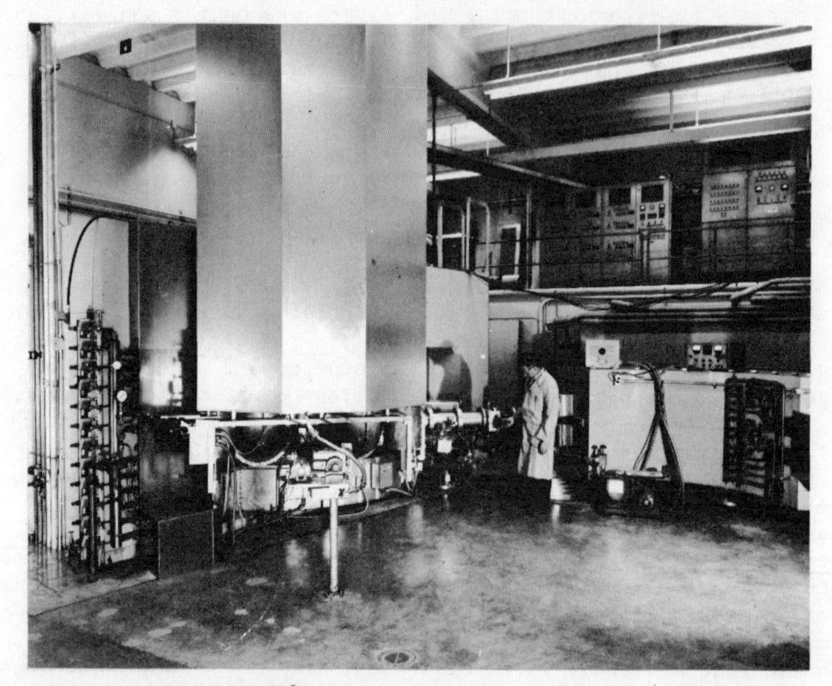

Fig. 2. A view of the 83-inch cyclotron and first beam preparation magnet.

Fig. 3. A view of the scattering chamber and three analyzer magnets.

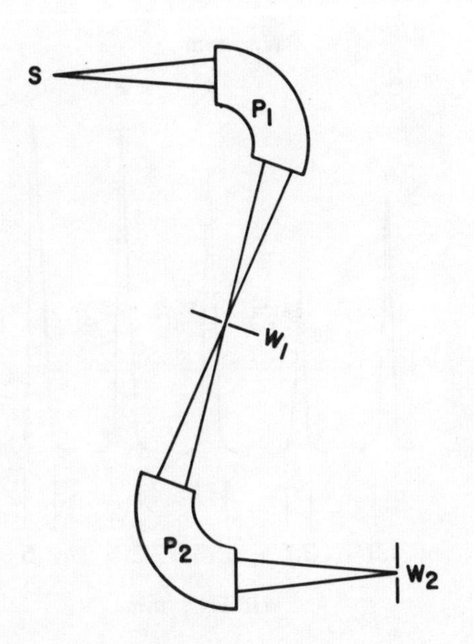

Fig. 4. Schematic of the beam preparation system at The University of Michigan 83-inch cyclotron. Note that the beam extracted from the cyclotron enters the first dispersing element without passing through a defining slit.

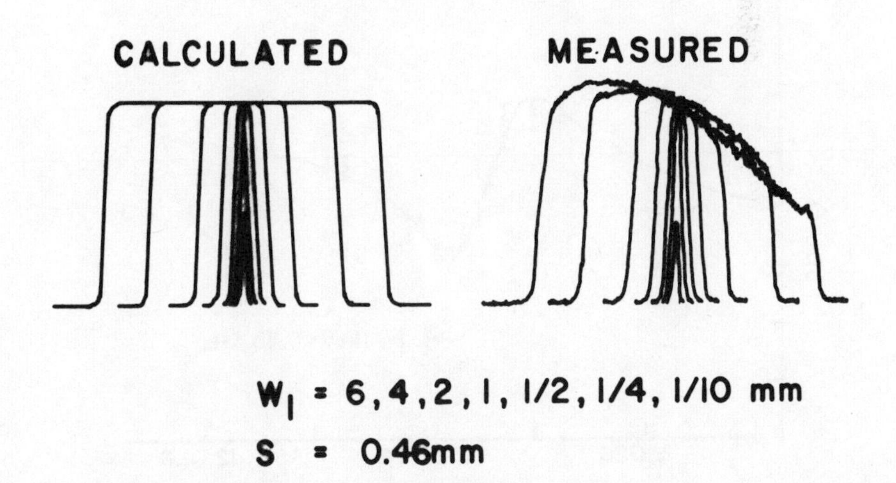

Fig. 5. Calculated and measured line shapes at the focal surface of P_2, the center of the scattering chamber, for several values of w_1.

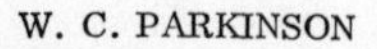

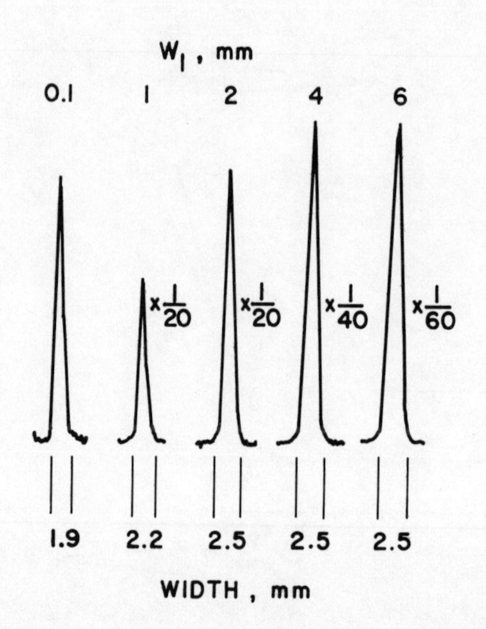

Fig. 6. Line-shapes at the focal surface of the third analyzer. Note that the intensities are essentially proportional to w_1. The peak for w_1 = 6 mm is very nearly 60 times more intense than for w_1 = 0.1 mm.

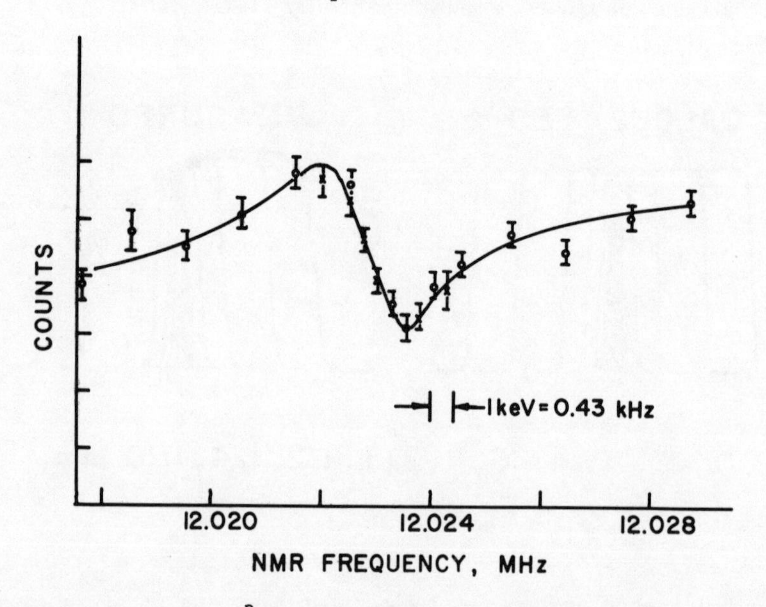

Fig. 7. The ^{13}N T = $\frac{3}{2}$ resonance at E_p = 14.233 MeV in the excitation function for ^{12}C(p,p).

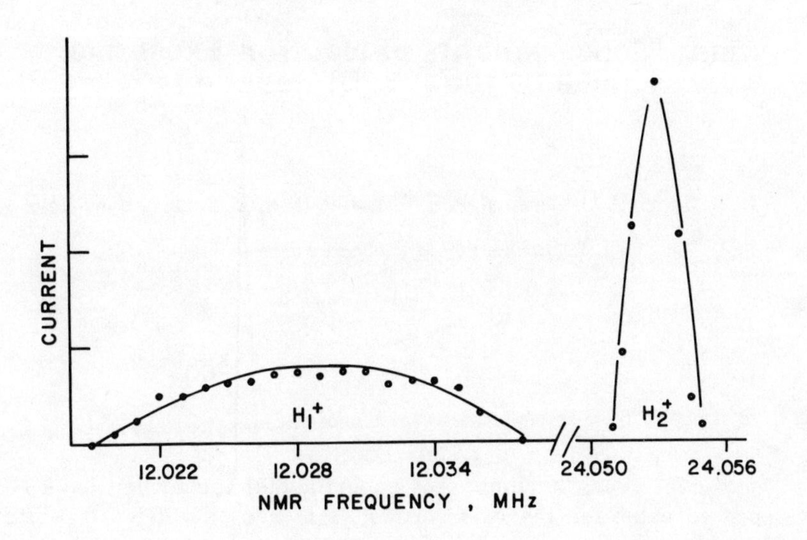

Fig. 8. Current vs the NMR frequency of the second beam preparation magnet, for H_1^+ and H_2^+.

Mass Measurements of Neutron Excessive Nuclei by the (^{11}B,^{8}B) Reaction

D. K. Scott, C. U. Cardinal, P. S. Fisher, P. Hudson and N. Anyas-Weiss

Nuclear Physics Laboratory, Oxford

1. Introduction

In recent years a number of experimental techniques have been developed to establish the existence of light nuclei which are particle stable on the neutron excessive side of the valley of beta-stability (1). Figure 1 shows the present status of observed particle stable nuclei below silicon, together with the boundaries predicted by two theoretical calculations (2,3). Most of the new nuclei have been produced in the bombardment of heavy targets with high energy protons or heavy ions. In the work of Poskanzer's group (4,5,6) a uranium target was bombarded with protons of several GeV, and the new isotopes of ^{11}Li, ^{14,15}B and ^{17}C were detected in a telescope of solid state counters. A similar technique has been used by Thomas et al. (7) to demonstrate the stability of ^{19}N, ^{21}O and recently (1) they may have observed ^{19}C. A further eleven new isotopes, ^{18}C, ^{20,21}N, ^{22,23,24}O, ^{23,24,25}F and ^{25,26}Ne were reported by Artukh et al. (8,9,10), formed in the bombardment of Th by ^{18}O. The stability of ^{11}Li and of five neutron rich isotopes of Na has been demonstrated by Klapisch et al. (11) using an on-line mass spectrometric technique. As illustrated in figure 1, none of these methods is able, at the present time, to yield accurate information on nuclear masses, and such information is essential for the development of theoretical predictions on the limits of nuclear stability. For example, the nuclei ^{11}Li and ^{19}C were initially predicted to be unbound by the Garvey-Kelson mass relations (2,12) although these nuclei are now known to be particle stable. Figure 1 shows further that Garvey and Kelson predict the last stable isotope of oxygen to have mass 28, whereas Vinogradov and Nemirovsky predict (3) the limit at mass 24. Similar discrepancies exist for other isotopes. It seems likely that further progress will only be made by making accurate mass measurements of neutron rich nuclei in order to make a systematic comparison with theoretical predictions. In this paper we report on the feasibility of using the three neutron transfer reaction (^{11}B, ^{8}B) as a technique for direct mass measurements, and apply the method to a determination of the mass-excess of the T=5/2 nucleus ^{29}Mg.

2. Experimental Method

The three neutron pick-up reaction (^{3}He, ^{6}He) on $T_z=0$ targets has been used with great success (13,14) in producing proton-rich nuclei with cross sections in the region of a few $\mu b/sr$. Neutron-rich nuclei produced in the equivalent stripping reaction, e.g. (^{12}C, ^{9}C), are inherently much more difficult to identify, because the ^{9}C has to be detected in a particle identifier on the low mass side of the much more abundant elastically and inelastically scattered ^{12}C, or ^{11}C and ^{10}C produced in the one and two neutron transfer reactions. Among the possible three neutron transfer reactions, e.g. (^{16}O, ^{13}O), (^{12}C, ^{9}C), the reaction (^{11}B, ^{8}B) has the advantage that ^{9}B, formed in the two neutron transfer reaction, is particle unstable. The ^{8}B particles can therefore be more reliably identified, since they would not be obscured in a particle identifier spectrum by the "tail" of the inherently more abundant ^{9}B particles. However, even given this advantage, the systematics of cross sections for highly endothermic multinucleon transfer reactions indicate that the cross section for $^{26}Mg(^{11}B, ^{8}B)^{29}Mg$, which has a predicted ground state Q-value of ~ -18 MeV, will be so small that refined identification procedures must be used (13-17).

A $^{11}B^{4+}$ beam with intensity 20nA and energy 114.5 MeV from the A.E.R.E. Harwell Variable Energy Cyclotron was used to irradiate $100\mu g/cm^2$ targets of ^{26}Mg, enriched to 99.5%, evaporated on $20\mu g/cm^2$ carbon backings. The reaction products were identified in a telescope of solid-state, surface barrier detectors comprising two ΔE-detectors, denoted ΔE_1 and ΔE_2 of thickness 75μ and 50μ, an E-detector of thickness 2 mm, and a rejection detector for long range reaction products traversing the first three counters. Two independent identifications of each particle stopping in the system, and satisfying a fast coincidence requirement of 100 ns resolving time, were made using the method of Fisher and Scott (18). The method is based on a simultaneous measurement of the energy loss ΔE in a thin counter and the residual energy E, from which a characteristic of particle type can be computed as $(E + \Delta E)^n - E^n$, where n is a constant. This exponent was chosen to be 1.67, which is appropriate (19) for heavy ions in the energy range of 80 to 115 MeV encountered in the experiment. The first identification PI1 was based on ΔE_1 as the ΔE-signal and the summed pulse $(\Delta E_2 + E)$ as the signal, while the second identification, PI2, was based on ΔE_2 and E. An example of the spectrum of signals from particle identifier 1 (PI1) is shown in the upper half of figure 2. The position of ^{8}B, indicated by the arrow, was predicted using that fact (20) that in a power law identifier, the PI signal is proportional to $M^{n-1}Z^2$, where n is the power index, in this case 1.67. The positions of particle groups were also located by using pulse generators to simulate the appropriate energy losses in the three counters. The calculations on energy loss were based on the tabulation of Northcliffe and Schilling (19). The figure shows the ^{8}B

group superimposed on the tails of the more abundant ^{10}B and ^{11}B groups. The tails are due to anomalous energy loss in the ΔE counter, and since the probability of such anomalous energy loss being repeated by the same particle in a second ΔE counter is small, the requirement that a particle should be identified twice, greatly improves the reliability of separation. This effect is illustrated in the bottom half of figure 2, which shows the effect of gating PI2 by a single channel analyser window, set in the region of ^{8}B in PI1. Those ^{10}B and ^{11}B particles, which resembled ^{8}B in PI1, are now clearly separated from the true ^{8}B events. This separation was achieved with only a few percent rejection of genuine data. The effect of gating in the region of ^{10}B in PI1 is also illustrated. A second set of single channel analyser windows were set on the groups in the PI2 spectrum. Coincident events in the two windows set on the PI spectra were used to generate pulses which routed the corresponding energy spectra into one of eight 1024 channel subgroups of an 8K Laben pulse height analyser.

3. Results

The detectors were calibrated provisionally with the aid of a 5.48 MeV, Americium 241 alpha source and an Ortec precision pulse generator, which also enabled the channel corresponding to zero energy to be accurately established. At the same time as the ^{8}B data was recorded, the other subgroups of the analyser were used to record spectra from the following reactions: $^{12}C(^{11}B, ^{11}B)^{12}C^{*}$, $^{26}Mg(^{11}B, ^{11}B)^{26}Mg^{*}$, $^{12}C(^{11}B, ^{10}B)^{13}C$, $^{26}Mg(^{11}B, ^{10}B)^{27}Mg$, $^{12}C(^{11}B, ^{10}Be)^{13}N$, $^{26}Mg(^{11}B, ^{10}Be)^{27}Al$, $^{12}C(^{11}B, ^{9}Be)^{14}N$ and $^{26}Mg(^{11}B, ^{9}Be)^{28}Al$. These reactions yielded twelve clearly identifiable groups, in the energy range 95 to 115 MeV, leading to known final states. The data was fitted by least squares fitting to yield an accurate calibration relating channel to energy in a linear fashion. In the fitting procedure the channel corresponding to zero energy was taken as that of the pulser calibration, but the beam energy was left as a free parameter since this was not known to great precision. The accuracy of the calibration was checked in an independent experiment involving a ^{12}C beam on a ^{12}C target leading to a variety of final products. The calibration was then used to determine the energy of the ^{8}B groups, and hence the mass-excess for the nucleus ^{29}Mg.

An ^{8}B spectrum at 10^{o}(lab) is shown in figure 3. The most energetic peak is close to the predicted location of ^{29}Mg (g.s.), based on the Garvey-Kelson predicted mass excess (21). The background events of higher energy are almost certainly due to a chance coincidence of lighter particles which can simulate ^{8}B particles in both identifications. This effect is well known for lighter ions (15), but it seems clear that such simulations are also possible with heavier ions, since the group above ^{10}Be in figure 2 is unlikely to correspond to ^{11}Be and ^{12}Be.

There was no evidence for background at the lower end of the spectrum, which cuts off at a point corresponding to the range of ^{8}B ions in the combined thickness of the two ΔE-counters, and the threshold set on the E-counter. The inset in figure 3 shows part of a spectrum at 9° in which tighter single channel analyser windows were set in the regions of ^{8}B in both identifier spectra, and in this case the higher energy background events do not appear. The lower figure is a composite spectrum, kinematically corrected to 10°.

The energy resolution in this initial work was 300 keV FWHH, determined primarily by the incident beam and kinematic effects. However in spite of the relatively poor resolution it is likely that the observed state corresponds exclusively to the ground state of ^{29}Mg, since the energy levels of this nucleus should resemble those of ^{31}Si and ^{33}S formed by addition of successive proton pairs. These nuclei (22) have their first excited states at 0.75 and 0.84 MeV ($1/2^{+}$) and second excited states at 1.69 and 1.97 MeV ($5/2^{+}$). The ground states have spins of $3/2^{+}$. We therefore anticipate a level spacing of about 1 MeV, and indeed there is evidence for excited states of ^{29}Mg at this spacing, the relative strengths of which are not inconsistent with the statistical weights associated with the predicted spins. There was no evidence for the ground state of ^{15}C which could have been produced in the ^{12}C(^{11}B, ^{8}B)^{15}C reaction on the target backing. If it is present the low yield could be due to the smaller cross section anticipated for this reaction, which has a higher negative Q-value ($\approx$ - 24 MeV) and a low ground state spin ($1/2^{+}$). It would be located in the 10° spectrum at an equivalent excitation of 10.5 MeV. However the adjacent lower energy group does correspond to the predicted location (23) of the 0.75 MeV level in ^{15}C which has a spin of 5/2. The broadening in the composite spectrum is consistent with a reaction on ^{12}C, since the kinematic shift applied in adding was appropriate only to a target of mass 26. Measurements on other targets are in progress.

Centroid analysis of the ^{29}Mg peak, after subtracting the estimated background, yields a mass excess of - 12.41 MeV from the 9° spectrum and - 12.25 MeV from the 10° spectrum. The best value was calculated as - 12.33 MeV with an error of $\pm$ 0.16 MeV, computed from the least squares fit to the calibration data. This represents a deviation of 230 keV from the Garvey-Kelson prediction (21) of - 12.56 MeV.

4. Conclusion

Our results demonstrate the feasibility of direct mass measurements of neutron excessive nuclei by the (^{11}B, ^{8}B) reaction. The accuracy obtainable with the equipment can be improved to better than 100 keV. The measurement of nuclear masses through charged particle transfer reactions of the type (^{11}B, ^{8}B) has the advantage that information on excited states is obtained simultaneously. Figure 3 shows that, with improved energy resolution, detailed information on highly excited states in nuclei far removed from stability could be obtained. If the c.m. cross section of 3.6 μb/sr observed for the ^{26}Mg(^{11}B, ^{8}B)^{29}Mg reaction is typical of three neutron transfer reactions, other nuclei which could be readily studied are ^{12}Be, ^{14}B, ^{17}C, ^{18}N and ^{21}O. It may also be possible to observe a four neutron transfer reaction, e.g. (^{13}C, ^{9}C), although the more negative Q-value for these reactions (typically -40 MeV) is likely to decrease the cross sections by at least a factor of ten (24). More hopeful perhaps is the study of two proton and three proton transfer reactions, which have the advantage that the detected particle is not an isotope of the nuclei in the incident beam, and therefore the problem of detecting low yield reaction products in the presence of elastic scattering is less severe. Thus the mass of ^{12}Be could be determined by two proton transfer on ^{14}C, and of ^{11}Li by three proton transfer. These are the last stable isotopes of Li and Be observed so far and there is strong evidence that ^{13}Be and ^{14}Be are unstable (25). Similarly the last stable isotope of Boron could be observed through three proton transfer on ^{18}O. The nuclei accessible by 3n, 4n, 3p and 4p transfer are indicated in figure 1. Work on these reactions is in progress, and should permit mass measurements to be made at the very limit of nuclear stability for a systematic comparison with existing theories.

The authors wish to acknowledge the interest and support of Professor K.W. Allen, and the generous assistance of Mr. R.W. McIlroy and Mr. E.J. Jones at the A.E.R.E. Cyclotron Laboratory.

References

1. KLAPISCH, R., International Conference on the Properties of Nuclei far removed from the region of β-stability (Leysin 1970) p.21

2. GARVEY, G.T. and KELSON, I., Phys. Rev. Lett., 16, 197 (1966); ibid. 23, 689 (1966).

3. VINOGRADOV, B.N. and NEMIROVSKY, P.E., Sov. J. of Nuc. Phys., 10, 290 (1970).

4. POSKANZER, A.M., COSPER, S.W., HYDE, E.K. and CERNY, J., Phys. Rev. Lett., 97, 1271 (1966).

5. POSKANZER, A.M., BUTLER, G.W., HYDE, E.K., CERNY, J., LANDIS, D.A. and GOULDING, F.S., Phys. Lett., 27B, 414 (1968).

6. POSKANZER, A.M., BUTLER, G.W. and HYDE, E.K., Phys. Rev., C3, 882 (1971).

7. THOMAS, T. D., RAISBECK, G. M., BOERSTLING, P., GARVEY, G. T. and LYNCH, R. P., Phys. Lett., 27B, 504 (1968).

8. ARTUKH, A. G., GRIDNEV, G. F., MIKHEEV, V. L. and VOLKOV, V. V., Nuc. Phys., A137, 348 (1969).

9. ARTUKH, A. G., AVDEICHIKOV, V. V., GRIDNEV, G. F., MIKHEEV, V. L., VOLKOV, V. V. and WILCZYNSKI, J., Phys. Lett., 31B, 129 (1970).

10. ARTUKH, A. G., AVDEICHIKOV, V. V., CHELNOKOV, L. P., GRIDNEV, G. F., MIKHEEV, V. L., VAKATOV, V. I., VOLKOV, V. V. and WILCZYNSKI, J., Phys. Lett., 32B, 43 (1970).

11. KLAPISCH, R., THIBAULT-PHILLIPPE, C., DETRAZ, C., CHAUMONT, J., BERNAS, R. and BECK, E., Phys. Rev. Lett., 23, 652 (1969).

12. GARVEY, G. T., Ann. Rev. of Nuc. Science, 19, 433 (1969).

13. BUTLER, G. W., CERNY, J., COSPER, S. W. and McGRATH, R. L., Phys. Rev., 166, 1096 (1968).

14. TRENTELMAN, G. F., PREEDOM, B. M. and KASHY, E., Phys. Rev., C3, 2205 (1971).

15. CERNY, J., COSPER, S. W., BUTLER, G. W., PEHL, R. A., GOULDING, F. S., LANDIS, D. A. and DETRAZ, C., Phys. Rev. Lett., 16, 469 (1969).

16. CERNY, J., MENDELSON, R. A., WOZNIAK, C. J., ESTERL, J. F. and HARDY, J. C., Phys. Rev. Lett., 22, 612 (1969).

17. CERNY, J., CARDINAL, C. U., JACKSON, K. P., SCOTT, D. K. and SHOTTER, A. C., Phys. Rev. Lett., 25, 676 (1970).

18. FISHER, P. S. and SCOTT, D. K., Nuc. Inst. and Meth., 49, 301 (1967).

19. NORTHCLIFFE, L. C. and SCHILLING, R. F., Nuclear Data, A7, 233 (1970).

20. ARMSTRONG, D. D., BEERY, J. G., FLYNN, E. R., HALL, W. S., KEATON, P. W. and KELLOG, M. P., Nuc. Inst. and Meth., 70, 69 (1969).

21. GARVEY, G. T., GERACE, W. J., JAFFE, R. L., TALMI, I. and KELSON, I., Rev. Mod. Phys., 41, No. 4, Part II (1969).

22. ENDT, P.M. and VAN DER LEUN, C., Nuc. Phys., A105, 1 (1967)

23. AJZENBERG-SELOVE, F., Nuc. Phys. A152, 1 (1970)

24. ARTUKH, A.G., AVEDICHIKOV, V.V., ERÖ, J., GRIDNEV, G.F., MIKHEEV, V.L., VOLKOV, V.V. and WILCZYNSKI, J., Nuc. Phys. A160, 511 (1971).

25. ARTUKH, A.G., AVDEICHIKOV, V.V., ERÖ, J., GRIDNEV, G.F., MIKHEEV, V.L., VOLKOV, V.V. and WILCZYNSKI, J., Phys. Lett., 33B, 407 (1970).

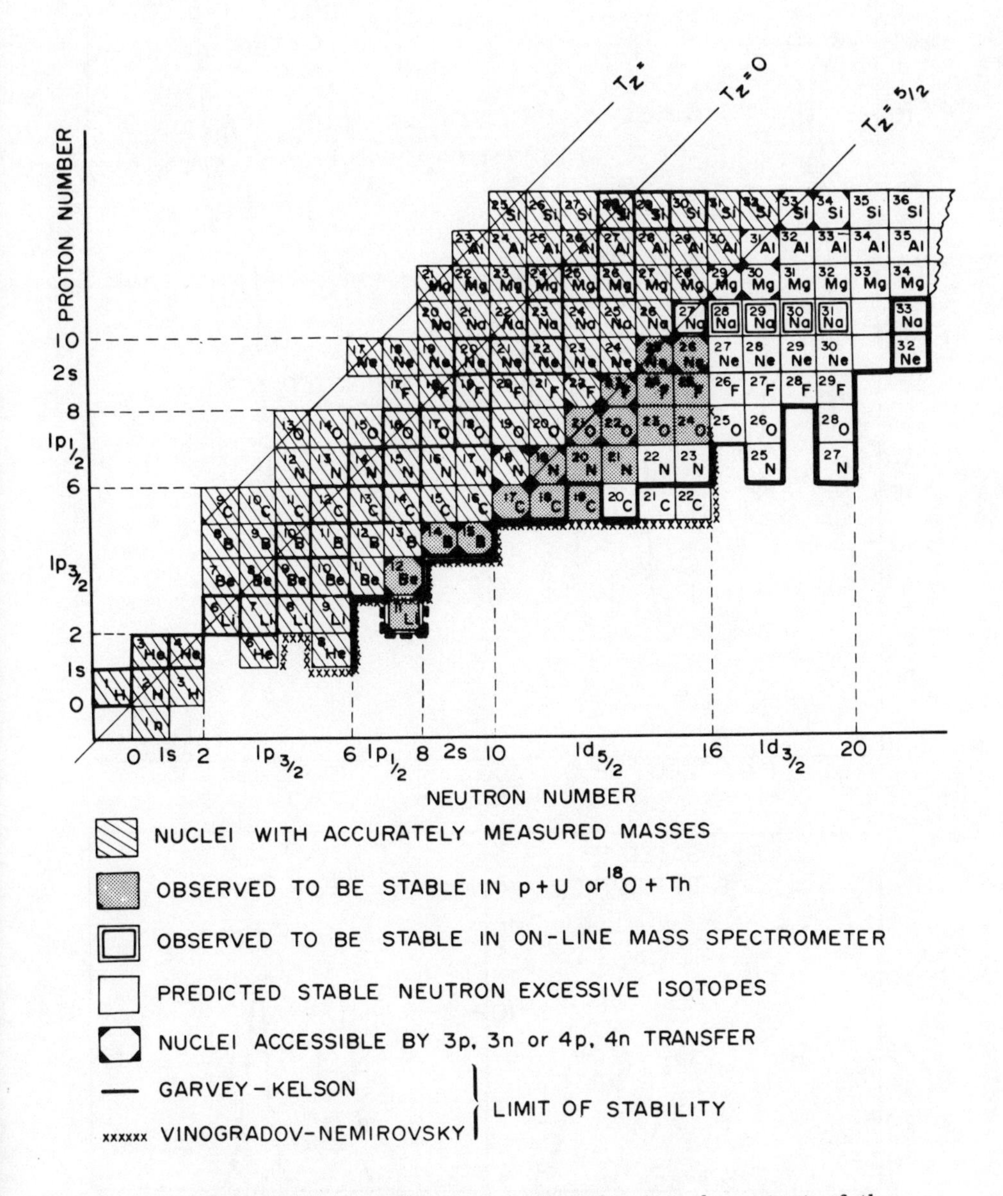

Fig. 1. Chart of the nuclides up to Si, showing the extent of the observed particle stable nuclei, and the predicted theoretical limits.

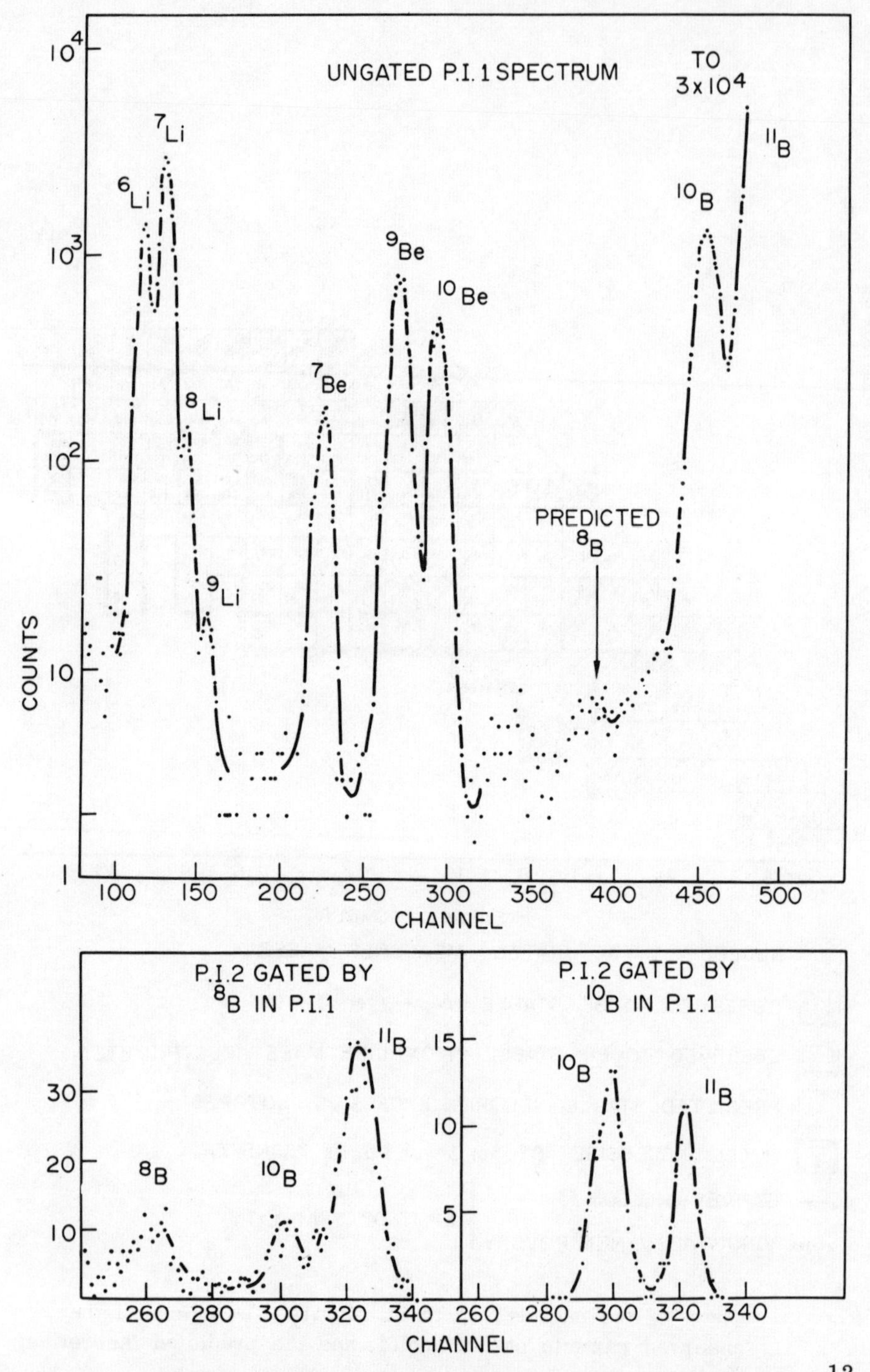

Fig. 2. Particle identifier spectra taken in the bombardment of ^{12}C and ^{26}Mg with 114.5 MeV ^{11}B ions.

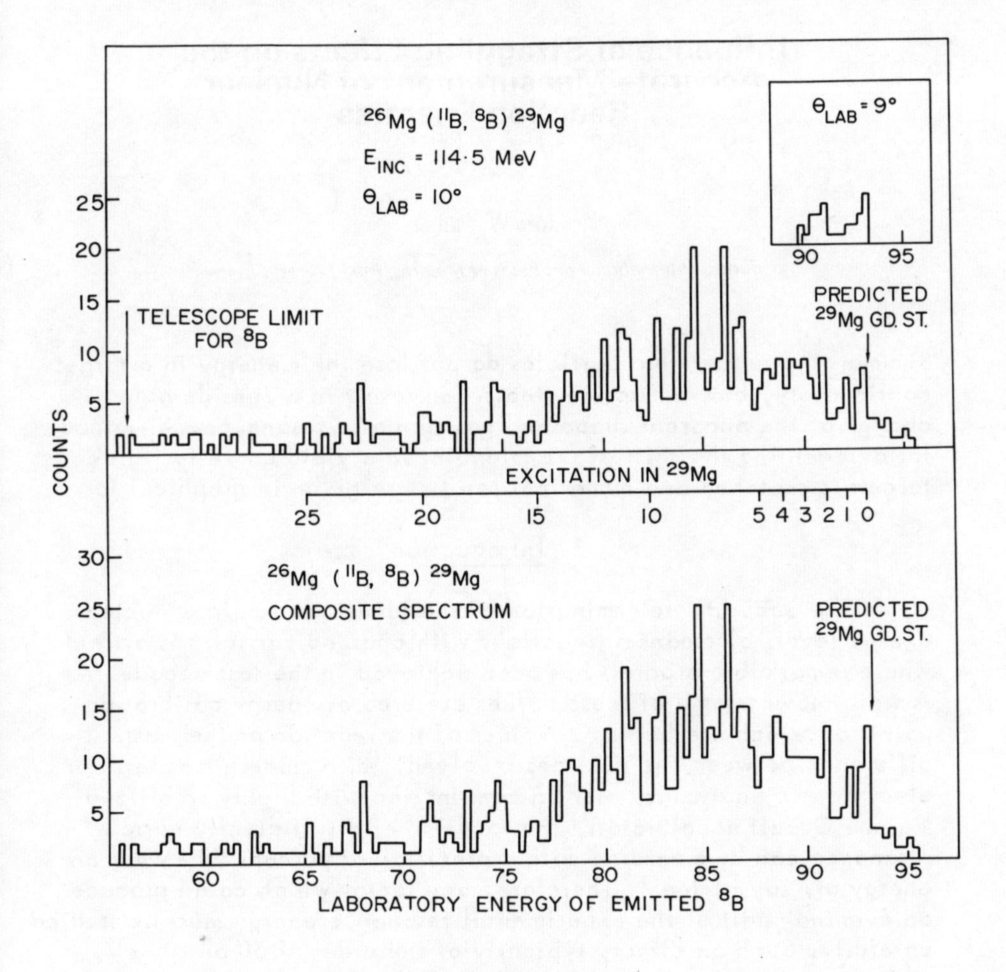

Fig. 3. Energy spectra of ^{8}B particles observed in the reaction ^{26}Mg(^{11}B, ^{8}B)^{29}Mg.

Influence of Straggling Effects on the Accurate Measurement of Nuclear Reaction Energies

Jörg W. Müller

Bureau International des Poids et Mesures, F-92 Sèvres, France

Summary: The fact that particles do not lose their energy in a target continuously, but in discrete steps, can result in a considerable change of the apparent shape and position of a resonance. A method for determining the shift of the midpoint in a yield curve for thick targets is sketched and numerical results are given in graphical form.

1. Introduction

The accurate determination of the location of narrow nuclear energy levels by means of reactions with charged particles is a field where remarkable progress has been achieved in the last decade. As is well known, some of these values are used as energy calibration points on which are based Q values of the reaction or the mass difference between the nuclides involved. With modern magnetic or electrostatic analyzers, used in conjunction with highly stabilized Van de Graaff accelerators, the location of a sufficiently narrow resonance can be measured with a precision of perhaps 50 eV for an energy of, say, 1 MeV. Therefore, any factor which could produce an eventual shift of the experimental resonance energy must be studied carefully. Such an effect, typically of the order of 50 or 100 eV, which is due to the discrete steps by which the accelerated particles lose their energy in the target, will be analyzed in what follows, in particular with respect to the displacement of the experimental resonance energy.

A few remarks about the history of this effect may be in order here. It was in 1956, when a measurement of the natural width of the well-known resonance $^{27}Al(p,\gamma)^{28}Si$ was made [1] in Zurich, that an "overshoot" in the thick target yield curve slightly above the resonance energy was repeatedly observed. However, prolonged runs with better statistics usually failed to confirm this anomaly. We nevertheless began to wonder how such an effect, if real, might be understood. A closer analysis soon revealed that the usual explanation of the thick target yield as the integral of the reaction cross section (folded with the experimental energy resolution) was actually based on the assumption that the protons were losing their energy continuously in the target. Clearly, this could only be a useful approximation, at best. For lack of a more detailed knowledge about the real loss spectrum (and the poor experimental evidence), I confined myself to evaluating the influence of discrete losses when all the steps are the same [2]. For this case (and the given experimental conditions), not only the amplitude of the wiggles in the yield curve was calculated as a function of the energy step, but a graph was also included showing the shift of the midpoint in the yield curve to lower energies, which turned out to be in this case half of the average energy loss per collision. For the case of a general loss spectrum, the method of the successive convolutions for determining the proton spectrum after a given number of collisions, which should prove to be so useful later, was already suggested in this first report on the straggling effect.

The first conclusive experimental evidence of the existence of a hump on yield curves - for the same reaction - seems to have been obtained in 1959 by A. del Callar [3]. One of his curves is reproduced in [11]. It was only two years later that similar observations (again with the Al resonance) were reported by a group at the University of Wisconsin [4]. Soon afterwards, Lewis [5] published a theoretical paper where the superposition of a damped oscillation on the normal yield curve was "predicted". Since some of the shortcomings of his mathematics have been corrected earlier [7], it may be sufficient to mention here that - as we shall see later - no oscillations occur actually (neither in experimental nor in theoretical yield curves) and that one of the principal contributions to the change in the shape (due to a δ function) has been neglected in his final formula for the effective proton spectrum. Nevertheless, the Wisconsin group has tried ever since to associate the name of Lewis with this effect. In view of the actual order of the events, however, this suggestion is somewhat difficult to justify and we prefer therefore the neutral designation "straggling effect". See also [6], [10], [14] and [15].

Today, its main interest lies undoubtedly in the influence on the apparent position of the resonance energy. In this contribution, we shall sketch a new method capable of evaluating this straggling effect with sufficient precision in a simple way.

2. The components of an experimental resonance yield

Generally speaking, the observed shape of a resonance yield Y as a function of energy is determined by the convolution [2]

$$Y(E_o - E^*) \sim \sigma(E - E^*) \circledast P(E - E_o) \quad , \tag{1}$$

where

σ is the resonance cross section for the reaction studied (e.g.

a Lorentzian, according to the Breit-Wigner one-level formula), and

P the effective energy spectrum of the accelerated particles (e.g. protons) which pass through the target and can then initiate a nuclear reaction.

For the convolution of two functions, the usual notation will be adopted, i.e.

$$f_1(x) * f_2(x) = \int f_1(x-\alpha) \cdot f_2(\alpha)\, d\alpha \qquad \text{and}$$

$$f_1(x) \circledast f_2(x) = \int f_1(x+\alpha) \cdot f_2(\alpha)\, d\alpha \ .$$

As is well known, these two types of convolutions correspond to the distribution of a sum or a difference of two random variables. Repeated use will also be made of the relation ([7], [17])

$$f_1(x) \circledast f_2(x) = f_2(-x) \circledast f_1(-x) = f_1(x) * f_2(-x) \ . \qquad (2)$$

The new variable $x \equiv E_o - E^*$, the difference between the (mean) energy of the incident particle and the (true) resonance energy, will be dropped as an argument in the following transformations whenever this is unambiguous. It is convenient to split the effective proton spectrum P into two parts by writing

$$P = F_+ \circledast R \ . \qquad (3)$$

Here

R includes all the broadening effects outside the target (finite geometrical energy resolution of the analyzer, limited field stabilization, etc.), whereas

F_+ describes the effective energy spectrum within the target for incident monoenergetic particles.

For practical reasons, the effect of Doppler broadening [2] is included in R. By means of (2), the yield can now be brought into the form (neglecting constant factors)

$$Y = \sigma \circledast (F_+ \circledast R) = F_- * \sigma * R = F_- * G \ , \qquad (4)$$

where $F_-(x) \equiv F_+(-x)$ and $G \equiv \sigma * R$.

We note that in $F_-(x)$ a positive argument stands for the energy loss and that the function G is made up of contributions which are independent of the energy loss mechanism in the target.

The crucial problem now lies clearly in the determination of the function F_-. The basic idea is to consider the energy loss process not as a function of time or penetration depth, as it is usual for diffusion processes, but rather as a function of the number of collisions a particle has suffered. Thereby we assume that the (unknown) probability density $f(x)$ for an energy loss x is independent of the initial energy and that the absorption of the particles can be neglected. In view of the small energy range involved, this seems to be justified. The total loss x after k collisions is then given by the k-fold convolution

$$f_k(x) = \left[f(x)\right]^{*k} , \qquad k = 1, 2, \ldots, \tag{5}$$

since the individual contributions are independent random variables. For a "thick" target, the total energy spectrum F_- is obtained by summing over all contributions, i.e.

$$F_-(x) = \sum_{k=0}^{\infty} f_k(x) , \tag{6}$$

where the first term $f_o(x) = \delta(x)$ corresponds to the monoenergetic beam at the target surface. The more complicated case of a "thin" or a "semithick" target will not be discussed here. We can therefore split F_- into

$$F_-(x) = \delta(x) + F(x), \qquad \text{with } F(x) = \sum_{1}^{\infty} f_k(x) . \tag{7}$$

By taking advantage of the Laplace transform

$$\tilde{f}(s) \equiv \int_0^{\infty} f(x) \cdot e^{-sx} \, dx ,$$

where convolutions correspond to simple multiplications, we obtain

$$\tilde{F}(s) = \sum_{1}^{\infty} \left[\tilde{f}(s)\right]^k = \frac{\tilde{f}(s)}{1 - \tilde{f}(s)} . \tag{8}$$

Since the transform of any normalized density can be written in the form

$$\tilde{f}(s) = E(e^{-sx}) = 1 - m_1 \cdot s + \frac{m_2}{2!} \cdot s^2 \mp \ldots ,$$

where $m_k = E(x^k)$ is the moment of order k, the Tauber theorem

$$\lim_{x \to \infty} g(x) = \lim_{s \to 0} s \cdot \tilde{g}(s)$$

gives for F the asymptotic result

$$\lim_{x \to \infty} F(x) = \lim_{s \to 0} \frac{s \cdot \tilde{f}(s)}{1 - \tilde{f}(s)} = \frac{1}{m_1} , \qquad (9)$$

which is independent of the special form adopted for f. Since (4) may now be written as

$$Y(x) = \left[\delta(x) + F(x) \right] * G(x) , \qquad (10)$$

the total yield can be resolved into three components (cf. Fig. 1)

$$Y(x) = Y_1(x) + Y_2(x) + Y_3(x) , \qquad (11)$$

where $Y_1(x) = \delta(x) * G(x) = G(x)$,

$$Y_2(x) = U(x) * G(x) = \int_{-\infty}^{x} G(x)\, dx ,$$

$$Y_3(x) = \left[F(x) - U(x) \right] * G(x)$$

and U(x) is the unit step function. Thereby, all the energies are measured in units of the average loss m_1. The first two components have a simple significance: Y_1 corresponds to the (normalized) "geometrical" resolution function, including the natural line width, and Y_2 is its integral. The third contribution, Y_3 , is of a more complicated nature. Its form as a difference might give rise to the hope that it could be taken as a small correction term. Although this is not true in general, Y_3 actually happens to be quite small in many cases of practical interest.

We note in passing that for continuous energy loss, F would be a simple unit step function, and therefore $Y = Y_2$. Any deviation of the yield curve from the usual shape is thus due to Y_1 and Y_3.

Previous attempts to determine F nearly always started from the special model for f where it is assumed that

$$f(E) \sim \begin{cases} E^{-2} & \text{for } E_{min} < E < E_{max} \\ 0 & \text{otherwise.} \end{cases} \qquad (12)$$

The upper limit E_{max} corresponds to the energy transferred to an electron in a head-on collision, whereas E_{min} is e.g. the first excitation energy of the target atom. Occasionally, a more refined collision spectrum has been used which includes the binding effects of the different electron shells. According to (5) and (6), this function f is then numerically folded a large number of times and summed up, either by a Monte Carlo method [9] or by performing some 50 self-convolutions of f on a computer [16]. Both solutions are clearly not very elegant, and in particular, for any change in the parameters, a new calculation must be begun from scratch.

In principle, a general method could be based on (8) by a partial fraction expansion of $\tilde{F}(s)$. By confining ourselves to the complex solution with the smallest negative real part, a good approximation might be expected since all other contributions die out more rapidly. Such an approach, which seems to explain the observed wiggles in a nice and simple way, has indeed been suggested [13] as a possible solution to the general problem. Unfortunately, this is not very useful, firstly because damped oscillations do not exist under normal experimental conditions, and secondly because the roots - if they can be found - depend strongly on the model adopted for f. The only general solution, namely $s=0$, gives rise to a step function, which we know already as the limiting behaviour of F. Another general attempt to find F, where only the moments of f are used as the parameters [7], did not prove very successful either. Additional assumptions would have to be made about the higher moments, which can clearly not be neglected.

3. The gamma model

In view of these various obstacles, a less ambitious solution of the problem was looked for. It is based on a model distribution for f and - although "general" in a restricted sense only - should be flexible enough to be adapted to specific experimental conditions and mathematically convenient. A simple model is readily offered by the family of gamma distributions, usually written in the form

$$f(t) = U(t) \cdot \frac{\rho(\rho t)^{\alpha-1}}{\Gamma(\alpha)} \cdot e^{-\rho t}, \tag{13}$$

with $\rho > 0$ and $\alpha > 0$.

The moment of order k for the variable t is given by

$$E(t^k) = \frac{\Gamma(k+\alpha)}{\rho^k \cdot \Gamma(\alpha)}$$

and the function is self-reproducing for convolutions (apart from a parameter). For the special application we have in mind it is advantageous to use instead of (13) the more general form

$$f(x) = U(z) \cdot \frac{\alpha \cdot z^{\alpha-1}}{(1-\Delta) \cdot \Gamma(\alpha)} \cdot e^{-z}, \tag{14}$$

where Δ is the shift of the gamma distribution (in units of the average energy loss) and $z = \alpha(\frac{x-\Delta}{1-\Delta})$.

Now $E(x) = 1$ and $\sigma^2(x) = \frac{1}{\alpha}(1-\Delta)^2$.

For practical reasons, we choose as parameters which characterize the actual energy loss distribution f the two quantities

$$Q = \left(\frac{\text{mean energy loss}}{\text{standard deviation}}\right)^2 = \frac{\alpha}{(1-\Delta)^2}$$

$$\text{and} \quad \Delta = \frac{\text{energy "shift" } E_{min}}{\text{mean energy loss}} . \qquad (15)$$

When chosen according to (14), the self-convolution $f_k(x)$ is easily determined for any order k and the evaluation of the sum $F(x)$ is now a simple matter of programming, where the parameters Q and Δ enter as variables.

4. Results and conclusions

In order to obtain numerical estimates for the shift S of the resonance (cf. Fig. 1), i.e. the distance from the energy corresponding to the half-plateau yield to the resonance energy E^*, the yield curve Y has to be determined by folding the effective proton spectrum F according to (4) with the (total) resolution function G. Since the shape of G depends on the experimental conditions, we have performed the convolution of F with three different simple functions, namely a triangle, a Gaussian and a Lorentzian, which are in each case characterized by their full width Γ at half maximum. Fig. 2 gives some final results for the shift S as a function of the parameters Q, Δ, and Γ. Their respective ranges are chosen such that they should cover the values which are of practical interest.

These calculations show also why no damped oscillations in the yield curves can be expected: they only occur for $Q \gtrsim 3$, whereas the actual value of Q seems to be somewhere near 0.2. The value of Δ has to be chosen rather carefully. For $E_{min} = 12$ eV and a mean energy loss per collision of about 60 eV, Δ is close to 20% and it happens that for energies slightly below the resonance the contribution Y_3 is practically negligible. Fig. 1 shows such a situation where the parameters correspond closely to real experimental conditions [16]. For those special cases, the shift S is only a function of the total experimental width Γ (Fig. 3) and the calculation becomes very simple.

We can therefore conclude - at least in an approximate way - that the energy shift S is not very sensitive to the details of the energy loss process. The main contribution stems from the component Y_1, which is caused by the particles with the full energy, i.e. at the very surface of the target. This happens to be just the term which used to be neglected by previous authors and shows that the discussion following [12] mainly missed the point. It also explains why the effect is only observed with freshly prepared targets: any surface layer would greatly reduce the "overshoot" in the yield curve and shift the apparent position of the resonance to higher energies, eventually overcompensating thereby the negative shift S.

The close analogy that exists between the energy loss mechanism and the time distribution of random pulses in counting experiments, which are both renewal processes [8], will be discussed elsewhere. It goes without saying that the present phenomenological approach should be completed by methods which allow one to determine the values of the parameters used (e.g. the average energy loss per collision and in particular its variance) with sufficient precision for all the targets of interest (solid and gaseous), but this must be left for further study.

All the numerical calculations have been performed on the IBM 1130 computer of the BIPM. The program, written in Fortran IV, is available upon request.

It is a pleasure to acknowledge the kind and continued interest of Prof. A. Allisy in this work.

References

(in chronological order)

[1] F. Bumiller, J.W. Müller, H.H. Staub: "Eine direkte Bestimmung der Resonanzbreite von Al-27(p,γ)Si-28 bei 991 keV", Helv. Phys. Acta 29, 234 (1956)

[2] J.W. Müller: "Direkte Bestimmung der Halbwertsbreite von Al-27(p,γ)Si-28 bei 991 keV", Master Thesis, University of Zurich (April 1958)

[3] A. del Callar: "An Absolute Determination of an Al(p,γ)Si Resonance", Master Thesis, Catholic University of America, Washington, D.C. (1959); not available

[4] W.L. Walters, D.G. Costello, J.G. Skofronick, D.W. Palmer, W.E. Kane, R.G. Herb: "Lewis Effect in Resonance Yield Curves", Phys. Rev. Letters 7, 284 (1961)

[5] H.W. Lewis: "Straggling Effects on Resonance Yields", Phys. Rev. 125, 937 (1962)

[6] W.L. Walters, D.G. Costello, J.G. Skofronick, D.W. Palmer, W.E. Kane, R.G. Herb: "Anomalies in Yield Curves over the 992-keV Al-27(p,γ)Si-28 Resonance", Phys. Rev. 125, 2012 (1962)

[7] J.W. Müller: "Stragglingeffekte bei Resonanzkurven", University of Zurich (August 1962)

[8] D.R. Cox: "Renewal Theory" (Methuen, London, 1962)

[9] D.W. Palmer, J.G. Skofronick, D.G. Costello, A.L. Morsell, W.E. Kane, R.G. Herb: "Monte Carlo Calculation of Resonance Yield Curves and Resonance Energy Determination", Phys. Rev. 130, 1153 (1963)

[10] J.W. Müller: "Doppel- und Tripel-Korrelationsmessungen an einer Gammakaskade bei Al-27(p,γ)Si-28", Doctoral Thesis, University of Zurich (Juris, Zurich, 1963)

[11] R.G. Herb: "Atomic Effects on Nuclear Reaction Yield Curves", in "Nuclidic Masses" (ed. W.H. Johnson), 241 (Springer, Wien, 1964)

[12] H.H. Staub: "Nuclear Resonance Energies", in "Nuclidic Masses" (as above), 257

[13] R. Bloch: "Die Reaktion A-40(p,γ)K-41", Master Thesis, University of Zurich (February 1964)

[14] D.G. Costello, J.G. Skofronick, A.L. Morsell, D.W. Palmer, R.G. Herb: "Atomic effects on nuclear resonance reaction yield curves of aluminium and nickel", Nucl. Phys. 51, 113 (1964)

[15] J.G. Skofronick, J.A. Ferry, D.W. Palmer, G. Wendt, R.G. Herb: "Detailed Forms of Yield Curves from Semithick Aluminum Targets", Phys. Rev. 135, A1429 (1964)

[16] R. Bloch, R.E. Pixley, H.H. Staub, F. Waldner: "Resonances in the 40-A(p,γ)41-K Reaction", Helv. Phys. Acta 37, 722 (1964)

[17] J.W. Müller: "Traitement statistique des résultats de mesure", Rapport BIPM-108 (1969 ff)

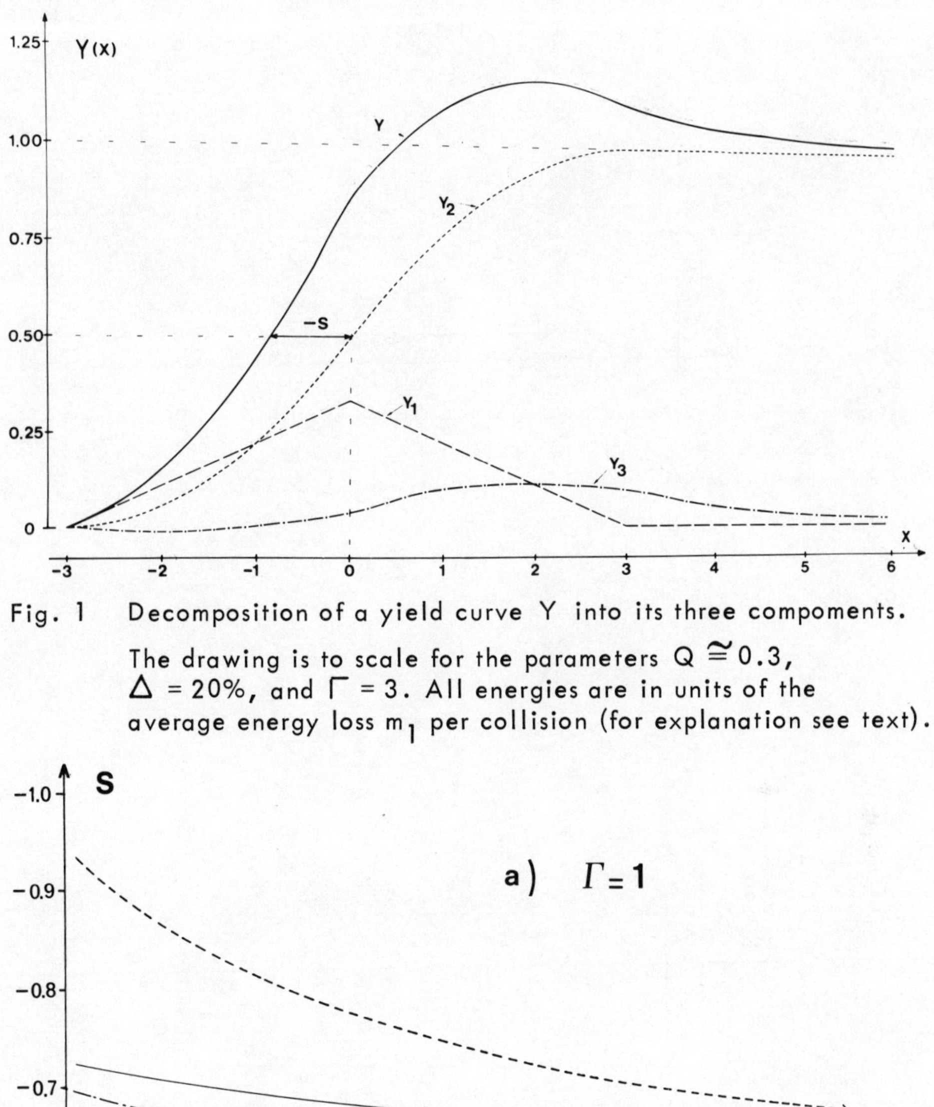

Fig. 1 Decomposition of a yield curve Y into its three compoments. The drawing is to scale for the parameters $Q \cong 0.3$, $\Delta = 20\%$, and $\Gamma = 3$. All energies are in units of the average energy loss m_1 per collision (for explanation see text).

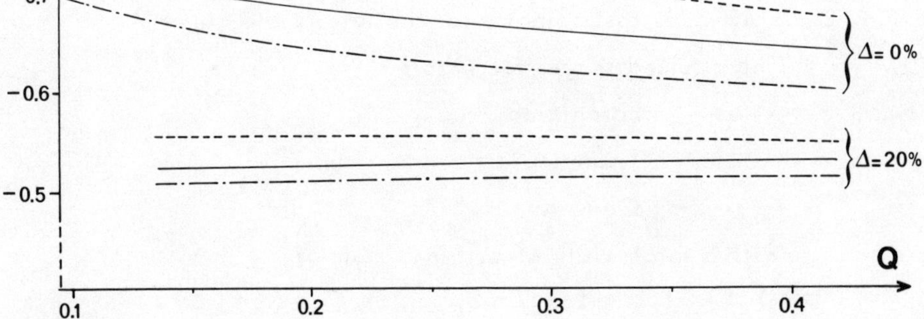

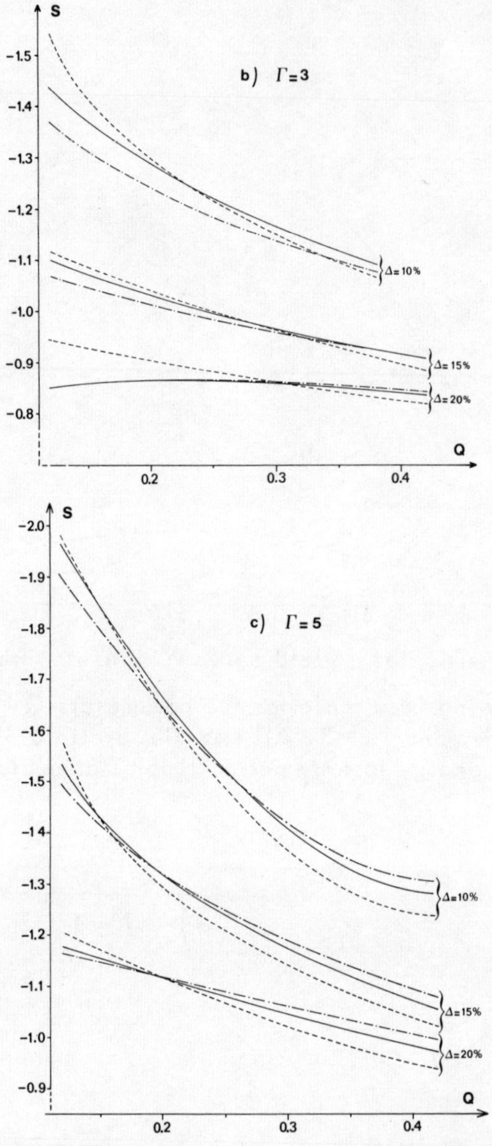

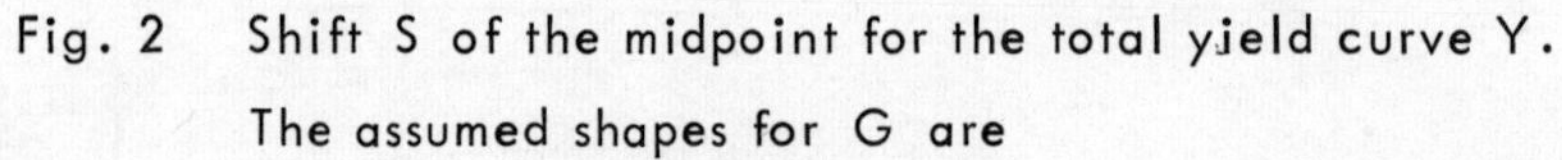
Fig. 2 Shift S of the midpoint for the total yield curve Y.

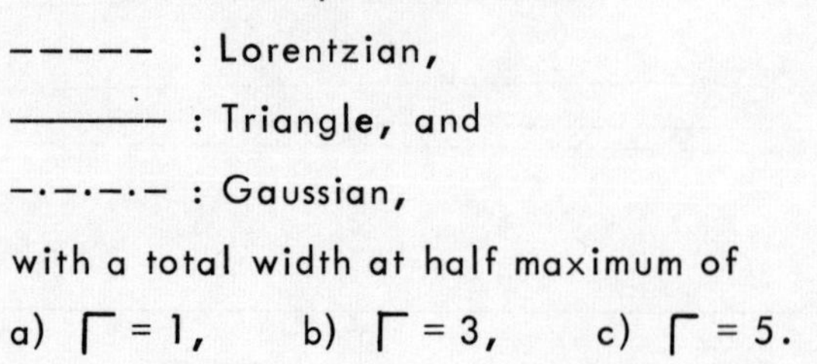
The assumed shapes for G are

----- : Lorentzian,

——— : Triangle, and

—·—·—·— : Gaussian,

with a total width at half maximum of

a) $\Gamma = 1$, b) $\Gamma = 3$, c) $\Gamma = 5$.

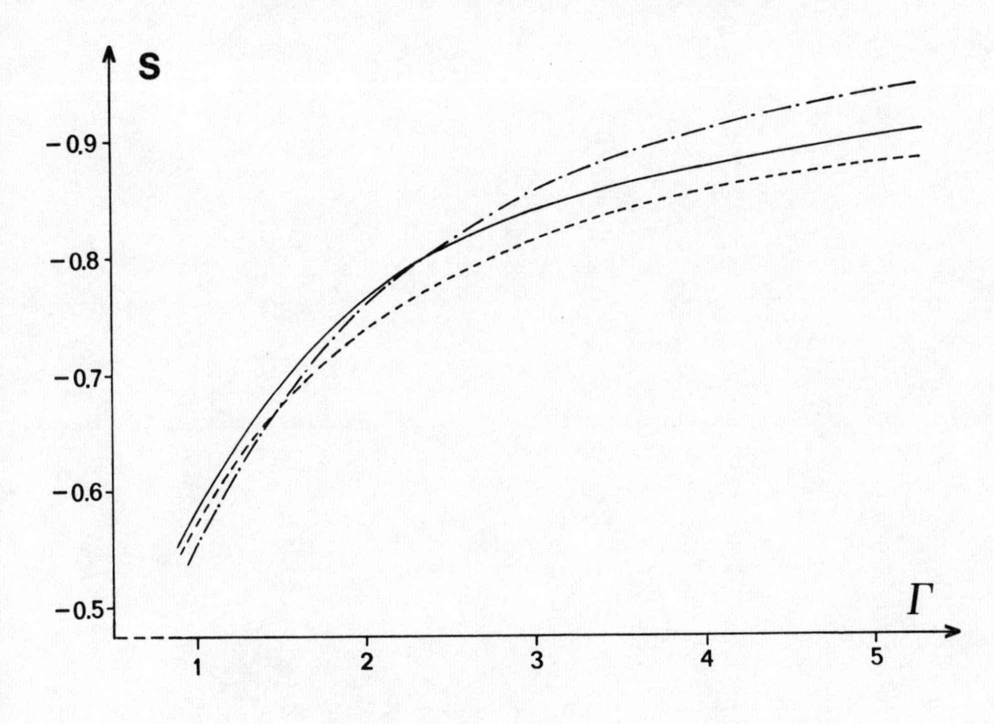

Fig. 3 Approximate values for the shift S according to the simplified model where Y_3 is neglected.

Part 2 Beta and Gamma Energies

Improvements in the Determination of β-spectra End-point Energies

Karl-Erik Bergkvist

Research Institute for Physics, and University of Stockholm, Stockholm 50, Sweden

1. Introduction

The subject 'Improvements in the determination of β-spectra end-point energies' has a broad content, covering in principle all changes in a desirable direction of the general circumstances for measurements of end-point energies. The present brief comments will be specialized to certain conditions and possibilities of relevance when the term improvement stands for an increase in absolute accuracy in an end-point energy determination. Improvements associated with e.g. the very production or isolation of the radioactive atoms will not be dealt with. By and large it will be understood that the parent atoms are available in any number but at a limited and fixed specific activity.

Many of the comments to follow are stimulated by the author's experience from a recent investigation of the tritium β-spectrum (1,2). Besides furnishing a measurement of the H^3-He^3 nuclear and atomic mass differences, a detailed study of particularly the end-point behaviour of this spectrum is informative for at least three other questions of present interest: the question of the electron-neutrino mass, the question about the presence of a universal degenerate sea of neutrinos, and, finally, the question about the magnitude of the axial-vector matrix element in the tritium decay, the latter being of interest for judging the applicability of PCAC in nuclei. Several new measurements of the tritium β-spectrum have recently appeared, focused on one or more of these three latter questions. For the third question it is the very magnitude of the end-point energy that has been of crucial and primary interest. The various conditions encountered when formulating generally the requirements for improved accuracy in an end-point energy measurement will be illustrated by means of these recent measurements on the tritium β-spectrum. For this conference, centred around the topic of atomic masses and fundamental constants, the measurement of the end-point energy of the tritium β-spectrum may perhaps stand out only as one single case among many where an atomic or nuclear mass difference is inferred from a measured end-point energy. In spite of this, because of the low end-point energy of the tritium spectrum and because of the number of efforts recently devoted to the experimental study of it, an inspection of certain features of these tritium experiments should offer a good means for illustrating, within a limited space, also certain more general features which will, at least ultimately, be involved when one tries to improve the accuracy of an end-point energy determination.

2. Two basic conditions

Consider a β-transition between two nuclear states A and B. How accurately can we infer the energy difference between A and B by measurements on the β-spectrum involved? Since the β-intensity goes to zero towards the upper end-point, some kind of extrapolation procedure must necessarily be relied upon for getting the energy difference considered. If the spectrum shape is not entirely known a priori, there will then inevitably be an ambiguity in the inferred value of the energy difference. There are primarily two uncertainties present in the spectrum shape in the general case: (i) an inherent uncertainty originating from contributions in the matrix element of forbidden or higher-order character and depending on the detailed structure of the states A and B; (ii) an experimental uncertainty stemming from distortion of the spectrum due to backscattering in the source. Both the effect (i) and the effect (ii) will go to zero with certainty only if more and more uppermost portions of the spectrum are employed for defining the extrapolated end-point energy. This suggests a first general condition for an accurate determination of an end-point energy.

Condition 1. Consider measurements on a β-spectrum with a resolution width w in an energy range $\Delta E_\beta >> w$, having the upper bound at the end-point energy $E_{\beta max}$. In order to avoid undue systematic uncertainties in the extrapolated energy $E_{\beta max}$, one must make the ratio $\Delta E_\beta / E_{\beta max}$ adequately small.

When, for a given β-decay, ΔE_β is reduced, the recorded intensity will decrease and the situation will eventually be reached where a statistical uncertainty dominates in the uncertainty in the extrapolated end-point. How rapidly will the intensity decrease when ΔE_β is reduced?

Let us assume that we keep the ratio $w/\Delta E_\beta$ roughly constant. To realize a sufficiently small resolution width w, a spectrometer based on the selection of particle orbits will, in general, have to be used. As conventionally employed, the permissible source width in such a spectrometer will be proportional to w. The permissible aperture will, in general, vary at least as $w^{1/2}$. Since the permissible thickness of the activity layer will also be proportional to w, we get primarily, for a given specific activity, a variation in the counting rate with w as $w^{2.5}$. A further factor w is introduced by the energy interval of the continuous β-spectrum selected in the spectrometer being proportional to w. In addition, when ΔE_β is reduced, the spectrum intensity will decrease primarily as $(\Delta E_\beta)^2$ in the vicinity of the end-point. All in all, we must expect the basic intensity to decrease with the employed energy interval ΔE_β at a rate like

$$\text{Intensity} \propto (\Delta E_\beta)^{5.5} . \qquad (1)$$

The relation (1) shows that the recorded intensity will decrease exceedingly rapidly when one tries to employ spectral regions closer to the end-point to define this point. This leads to a second condition:

Condition 2. Only if β-spectroscopic methods offering a dramatic improvement with respect to intensity can be devised, can significant progress in the determination of end-point energies along the lines defined by condition 1 be realized.

3. High-luminosity β-spectroscopic methods

Fig. 1 shows a schematic drawing of a combined electrostatic and magnetic β-spectrometer (3) offering an increase in luminosity over conventional spectrometers large enough to be of significance also when considering an intensity deterioration, as indicated by the relation (1). As illustrated in Fig. 1, the combined spectrometer consists basically of a $\pi \sqrt{2}$ magnetic spectrometer, but added to this are two electrostatic components, the Corrector and the source arrangement. The action of the corrector amounts to a cancellation, in principle, of the point source radial aberrations in the $\pi \sqrt{2}$ magnetic field. This is achieved by the deflecting action of two electrostatic fields in the corrector, formed between adjacent pairs of suitably shaped high-transmission grids. The electrostatic source arrangement allows an extended source area to be combined with a high resolving power. A non-equipotential source area introduces an energy change for the emitted particle which varies with the position of the point of emission of the particle from the source. By the combined action of the induced energy change and the dispersion in the magnetic field, an extended source can be imaged as a narrow line at the detector slit. For a point source there is no limitation by principle as to how completely the corrector can be made to reduce the radial aberrations of the magnetic component. Similarly, when a point aperture is used, there is no limitation with respect to the degree with which the effect of an extended source can be compensated for in the radial image width. When a finite aperture is combined with a finite source area, certain residual radial aberrations become unavoidable. These 'inherent aberrations' of the combined system are of third order and determine the degree to which the luminosity of the combined system can be made larger than that of the purely magnetic component. Increases in luminosity of between two and three orders of magnitude are possible at a level of resolution of around 0.1 % (3).

Fig. 2 shows a photo of an actual spectrometer along the principles illustrated by Fig. 1. The upper pole shoe of the $\pi \sqrt{2}$ magnet has been raised. The detailed set-up shown is from an experiment on the tritium β-spectrum (1,2). The tritium source, about $10 \cdot 20\ cm^2$ in area, can be seen to the left in the figure. The aperture of the electrostatic corrector, seen to the right in the figure, is around 0.6 % of 4π. At the selected instrumental resolution of 0.1 % in B_ρ, the figures stated imply an increase in luminosity, or intensity, of close to a factor of thousand over the basic $\pi \sqrt{2}$ component involved. A recording of the end-point region of the tritium β-spectrum, as obtained by the equipment shown in Fig. 2, is reproduced in Fig. 5(e). For further details see refs. (11) and (2).

4. The detailed end-point behaviour of the bare-nucleus spectrum.

Suppose we can approach, with adequate statistics, the end-point in such a way that any long-range curvature in the Kurie plot can be neglected in the extrapolation of the plot. Although all slowly varying features in the spectrum shape can be disregarded, there remains the possibility for a variation in the spectrum shape having a more singular behaviour at the end-point and becoming of importance as the end-point is more and more closely approached. In this section, we consider such effects in the spectrum shape which are present already in the purely nuclear β-spectrum, i.e. in the spectrum where the interplay with the atomic surrounding of the β-decaying nucleus is assumed to be absent.

(a) Ambiguity in extrapolated end-point energy due to residual lack of knowledge on neutrino mass. For a finite neutrino mass m_ν the spectrum shape is given by

$$N(E_\beta) \propto |M_\beta|^2 F(Z,E_\beta)\, p_\beta{}^2 (E_{\beta max}-E_\beta) \sqrt{(E_{\beta max}-E_\beta)^2 - m_\nu^2 c^4} \quad . \quad (2)$$

The Fermi function F varies slowly in the vicinity of the end-point. The dependence of the phase-space factor in eq. (2) on m_ν is illustrated by the thick dotted line in the Kurie plot in Fig. 3. The dependence of the extrapolated end-point caused by the uncertainty in the actual phase-space behaviour is decreasing as spectral parts more and more distant from the end-point are employed for defining the extrapolating straight line. Since difficulties of the kind considered in arriving at condition 1 will eventually enter, some kind of compromise will be required in the selection of the spectral portion giving the smallest uncertainty in the extrapolated end-point. With the presently established upper limits of 55 eV (1) on the neutrino mass, any ambiguity in an extrapolated end-point energy stemming from the uncertainty in the actual value of the neutrino mass should now not have to be larger than about ±5 eV (2).

Like the phase-space factor the so-called relativistic spinor term in the matrix element M_β exhibits a rapid variation with E_β in the vicinity of the end-point. The relativistic spinor term makes $|M_\beta|^2$

$$|M_\beta|^2 \propto 1 + a \frac{m_e\, m_\nu}{E_\beta\, E_\nu} \quad (3)$$

For a finite value on m_ν a variation of $|M_\beta|^2$ as given by eq. (3) essentially amounts to a parallel displacement of the Kurie plot by an amount proportional to the product $a \cdot m_\nu$. This feature is illustrated by the slim dotted line in Fig. 3. If only vector and axial-vector interaction contribute in β-decay, the value of the constant a in eq. (3) is zero (4). If small amplitudes other than V and A contribute, a small but non-zero value of the constant a may actually be present. With the above-mentioned upper limit of 55 eV on m_ν and with the small conceivable contributions other than V and A arrived at in a recent review of available evidence by Paul (5), any ambiguity in the significance of the extrapolated end-point due to the presence of a relativistic spinor term in $|M_\beta|^2$ should now not exceed 1 eV.

(b) Modification of spectrum shape due to radiative effects. Radiative effects in the β-decay process introduce a modification of the spectrum shape which exhibits a logarithmic divergence at the end-point and which hence may be potentially dangerous when the uppermost portion of the spectrum is employed for defining the extrapolated end-point energy. According to Sirlin (6) the uncorrected allowed spectrum $N_0(E_\beta)$ is modified by a factor $1 + g(E_\beta, E_{\beta max}, m_e)$ where the function g is given by

$$g(E_\beta, E_{\beta max}, m_e) = 3\ln\left(\frac{m_p}{m_e}\right) - \frac{3}{4} + 4\left[\frac{\tanh^{-1}\beta}{\beta} - 1\right]$$
$$\times\left[\frac{E_{\beta max}-E_\beta}{3E_\beta} - \frac{3}{2} + \ln\left\{\frac{2(E_{\beta max}-E_\beta)}{m_e}\right\}\right] + \frac{4}{\beta}\,L\left(\frac{2\beta}{1+\beta}\right)$$
$$+ \frac{1}{\beta}\tanh^{-1}\beta\left[2(1+\beta^2) + \frac{(E_{\beta max}-E_\beta)^2}{6E_\beta{}^2} - 4\tanh^{-1}\beta\right]. \qquad (4)$$

In eq. (4) m_e and m_p is the electron and proton mass respectively, $\beta = p_\beta/E_\beta$ and L in the Spence function. The third term on the right hand side of eq. (4) contains the logarithmic divergence. An inspection of the quantitative implications reveals that in the tritium spectrum, for example, ($E_{\beta max}$ = 18.6 keV) the logarithmic divergence amounts to a 2 % modification of the spectrum intensity at a distance of 2.5 eV below the end-point. This effect on the spectrum shape is too small to affect experimental measurements of presently conceivable accuracy.

(c) Modification of spectrum shape due to possible presence of neutrino degeneracy. It has been suggested by Weinberg (7) that there may exist throughout the universe a degenerate sea of neutrinos (or antineutrinos). If all antineutrino states are filled up to a certain Fermi energy $E_{F,\bar{\nu}}$, the intensity of a β^--spectrum will drop to zero at a β-particle energy which is an energy $E_{F,\bar{\nu}}$ smaller than the extrapolated end-point energy. This effect comes about because there is no phase-space available for the antineutrinos corresponding to β-particle energies in the normally uppermost spectrum portion. For a β^+-spectrum the presence of an antineutrino degeneracy will cause the Kurie plot to rise again above the extrapolated end-point since an antineutrino of finite energy can be absorbed by the β^+-unstable nucleus when emitting the β^+-particle. Present direct evidence from measurement on the tritium β-spectrum give upper limits of 55 - 60 eV (1) on the Fermi energies of a neutrino or antineutrino degeneracy. If appropriate portions of a β-spectrum are discarded in the definition of the extrapolating straight line, no ambiguity in the defined end-point energy will be caused by the possible exist-

of a degeneracy with a smaller Fermi energy than given by the above limit. This conclusion, of course, presumes that the Fermi surface is reasonably sharp.

5. Atomic effects affecting the interpretation of a measured end-point energy.

For the decay of a bare nucleus at rest, the end-point energy as discussed in the preceding section, after a minor correction for recoil, expresses just the energy difference between the nuclear states A and B. Clearly, however, it is difficult to conceive an experimental situation, suitable for a precision measurement of a β-spectrum, where it is the bare nucleus spectrum which is observed. We have to consider what meaning should be given to an observed end-point energy when the β-decaying nucleus is part of a more complex atomic or chemical system. We start by considering the β-decay of a free atom at rest.

Let M_i and M_f be the nuclear masses for the states A and B, and E_i and E_f the corresponding atomic energies involved. Conservation of energy implies

$$M_i c^2 + E_i = M_f c^2 + E_f + E_{rec} + E_\beta + E_\nu , \qquad (5)$$

where E_{rec} is the energy of the recoil atom (ion) and E_ν the energy of the neutrino. Let H_i be the initial state atomic Hamiltonian. The sudden change in nuclear charge in the β-decay involves a sudden change in the potential energy part of the atomic Hamiltonian, this changing to $H_f = H_i + \Delta V$. Let us assume that before the decay all atoms are in the eigenstate ψ_i of the atomic Hamiltonian H_i with the known eigenvalue E_i. Neglecting the recoil energy, we see from eq. (5) that in order to establish the relation between the energy sum $E_\beta + E_\nu$ and the nuclear mass difference $M_i - M_f$ we must know the final state atomic energy E_f. Consider an idealized measurement where both M_i and M_f are known and where both the β-particle energy and the neutrino energy are measured immediately after the decay. According to the relation (5) this actually involves a measurement of E_f. Immediately before the measurement of E_f the atomic system is in the state ψ_i ; the operator corresponding to the measured quantity E_f in H_f. From basic quantum mechanical principles this means that $\overline{E_f} = \langle \psi_i | H_f | \psi_i \rangle$ and $\overline{(\Delta E_f)^2} = \langle \psi_i | (H_f)^2 | \psi_i \rangle - (\langle \psi_i | H_f | \psi_i \rangle)^2$ Inserting $H_f = H_i + \Delta V$ we find from eq. (5) for the average value of the energy to be shared between the β-particle and the neutrino

$$\overline{E_\beta + E_\nu} = (M_i - M_f)c^2 - \langle \psi_i | \Delta V | \psi_i \rangle , \qquad (6)$$

and for the mean square spread $\overline{[\Delta(E_\beta + E_\nu)]^2}$ in this quantity.

$$\overline{[\Delta(E_\beta + E_\nu)]^2} = \langle \psi_i | (\Delta V)^2 | \psi_i \rangle - (\langle \psi_i | \Delta V | \psi_i \rangle)^2 . \qquad (7)$$

The operator ΔV amounts to e times the electrostatic potential at the nucleus caused by the presence of the atomic electrons. Since ψ_i is not an eigenfunction of ΔV, eq. (7) implies that there will be a spread in the energy available for the β-particle and the neutrino. Assuming the sharing of the energy between the two particles to be essentially given by the electron-neutrino phase-space volume, the result implies that there will be a spread in end-point energy in the β-decay of an ensemble of atoms. The average value of the observed end-point energy will be related to the nuclear mass difference involved in the way given by eq. (6). This latter relation is, of course, since long ago recognized and allowed for in the interpretation of end-point energies; it shows that the bare nucleus end-point energy in the atomic case is modified by an amount given by the change in potential energy of the β-particle at the nucleus due to the presence of the atomic electrons. The expectation value involved in eq. (6) can be obtained from various approximate atomic models.

That the β-decay of an atom leads to a distribution in atomic final states and that the β-spectrum should hence be expected to consist of a number of branches of slightly different end-point energies was recognized fairly long ago (8) but it does not seem as if this fact has until recently ever had to be taken into account in the analysis of a measured spectrum. Clearly each individual daughter atom will be in an eigenstate of the final state atomic Hamiltonian; the relation (7) expresses the width of the energy distribution of these states. There does not seem to be available any general evaluation of an expression like (7) for atoms of various Z. For the particular case of the tritium β-decay an evaluation (9) of the result of (7) reveals a mean square energy spread equal to $(27\ eV)^2$. This spread in end-point energy no doubt is astonishingly large recalling that the total atomic binding energy in tritium is only 13.5 eV. In the experimental investigation of the tritium spectrum (1,2) referred to in Fig. 2, the experimental resolution width is ~ 55 eV. The spread in end-point energy due to atomic effects is far from negligible in this case and has to be allowed for in the interpretation of the data. Fig. 4 shows how the response function of the measurement is modified by the fact that the tritium spectrum consists of a number of branches. 96.4 % of the total β-intensity goes to the three lowest s-states of the He^+ ion, with 70 % to the 1-s state (9). The distribution in end-point energy increases the effective resolution in the measurement from 55 eV to 70 eV. Little would clearly be gained by improving the experimental resolution width below 55 eV since the inherent spread would then start dominating.

When more and more uppermost portions of a β-spectrum are selected for defining the end-point energy (6), care has clearly eventually to be exercised in the interpretation of the observed behaviour in view of the presence of the spread (7). An evaluation of the expression (7) for various Z other than Z = 1 would be valuable as a guide for judging at what level of experimental accuracy the spread in end-point energy will have to be taken into account for decays of various Z. A thorough theoretical treatment of the interplay between a β-decaying nucleus and its atomic shell, although not centred around the particular question raised here, can be found in a paper by Bachall (10).

6. Effects associated with the β-active atom forming part of more complex chemical structure. The 'physics of the energy calibration' of a measurement.

For the β-decay of the free atom there should be no difficulty by principle in calculating the expectation value of the operator ΔV appearing in the relation (6) between the observed, average, end-point energy and the nuclear mass difference involved in the transition. For especially a low energy, high-Z β-transition some refinements of the relation (6) may be required, but even so, no ambiguity in the relation between the nuclear mass difference and the observed average end-point energy should be unavoidable.

In practice one does not, in general, measure the β-spectrum from the free atom. In all experiments suitable for precision measurements, the β-active atoms will form part of some more complex source structure. In lowest order, we should expect the very form of the relation (6) to remain valid also when the β-decaying nucleus is part of a molecule, provided that the operator ΔV then includes the effect of all electrons and of all other nuclei in the molecule. In the most general case, the operator ΔV in relation (6) will have to express the change in potential energy for the β-particle at the nucleus stemming from the presence of the atomic electrons of the decaying atom and from the nuclei and electrons from the whole remaining source material involved in the measurement. Calculations (9) of the influence of the atomic surrounding on the tritium spectrum for the case where the decaying tritium atom is part of a T - T molecule indicate that both the effect (6) on the average end-point energy and the spread in end-point energy (7) will not change drastically from the values for the case of the free tritium atom. With the admittedly very approximate molecular wave function employed, the results obtained do not differ by more than 10 - 15 % from the values 27 eV and $(27\ \text{eV})^2$ respectively, obtained for the free tritium atom. No results are available for decays of other Z, but it seems perhaps a fair guess that the expectation value of ΔV in eq. (6) should not vary by more than roughly the chemical binding energy of the decaying atom in the molecule. In the general case, the effect of the whole bulk of a source material will be to produce in the expectation value of ΔV a Madelung-term like contribution. It seems to me that an uncertainty of at least around ± 10 eV must be attributed to the precise meaning of a measured end-point energy due to the lack of precise knowledge of the expectation value occurring in relation (6). For a specific source with an accurately known structure, this uncertainty may of course at least in principle, be reduced.

In the discussion so far we have understood that the β-particle energy is defined as the kinetic energy (plus rest mass) at infinite distance from the β-decaying atom or assemblage of atoms. The electrostatic potential in the region where the β-particle kinetic energy is measured is assumed to be strictly the same as the potential at the site of the decaying atom or group of atoms. In a magnetic spectrometer it is the momentum of the β-particle that is measured. In this sense the measurement is in accordance with the above picture where the kinetic energy is assumed to be sensed. However, the electrostatic potential in the imaging domain of the spectrometer does not necessarily fulfil the condition of being entirely equal to the potential at the decaying atom in the sense required for the strict validity of the relation (6). The pertinent details of the relevant electrostatic potenti l are defined by the common Fermi level and the various work functions of the materials involved. In the general case at least three materials are involved: the material of the β-source, the material of the spectrometer and the material of the calibration source. Effective work functions are of the order of 4 - 5 eV and hence an error up to something like ± 15 eV could possibly be introduced by neglecting the effect of the work function. An analysis of the role of the work functions reveals as a somewhat astonishing circumstance that a large value of the effective work function of a β^--source will act so as to increase the observed end-point energy (2). When suitably allowed for, the various work functions involved should not have to produce any ambiguity in an end-point energy larger than a couple of eV.

Needless to say, the foregoing discussion is relevant if there is no accumulation of net charge in the source material. Appropriately accurate tests must of course be performed to ensure that this condition is fulfilled in an actual measurement (cf. e.g. ref. 2).

7. Discussion of some recent measurements on the tritium β-spectrum.

Fig. 5 shows end-point data from four recent measurements of the end-point energy of the tritium β-spectrum. For comparison, we have included also data from the well-known recording of the tritium spectrum by Langer and Moffat (11) from 1952. This latter recording, (a) in Fig. 5, was obtained with a magnetic spectrometer of a basically single-focusing type. A 2.5 • 0.6 cm^2 source of tritiated succinic acid was employed. The data illustrated in (b) are from a work by Daris and St-Pierre (12) from 1969. The measurement is basically similar to that of Langer and Moffat, a magnetic spectrometer again being employed. Details differ in so far as the spectrometer employed is of the double-focusing ($\pi\sqrt{2}$) type and in that an aluminium source with an adsorbed layer of tritium replaces the organic source material in the measurement of Langer and Moffat. The data shown in (c) are from a measurement by Salgo and Staub, also from 1969 (13), using an integral electrostatic β-spectrometer. A very large (5 cm radius) tritium source, consisting of an evaporated layer of T_2O ice, was employed in this work. In the measurement illustrated in (d) in Fig. 5, by Lewis (14) and from 1971, a litium drifted silicon detector was used, with the tritium activity implanted in the depletion layer. The recording shown in (e) is from the investigation reported in refs. (1) and (2) and obtained by means of the combined electrostatic-magnetic β-spectrometer illustrated in Fig. 2.

The recordings (a) - (d) in Fig. 5 are excerpts from actual figures in the respective publications, scaled to a common unit in E_β. The abscissa in (e) is magnified twice with respect to this unit. Except for the data in (c), which represent an integral spectrum, the recordings illustrated are Kurie plots of the data obtained.

Some relevant features of the recordings illustrated in Fig. 5 are collected in Table I.

Compared with the old measurement by Langer and Moffat, the measurement by Daris and St-Pierre represents improvements with respect to the resolution realized and with respect to the control of certain systematic effects. Although there are no statistical error bars included in the figure from Langer and Moffat's work (a) and

TABLE I

Key letter in Fig. 5	Author	Method	Instrumental resolution	Basic statistical accuracy in definition of end-point	Final assigned error in $E_{\beta max}$	$E_{\beta max}$
(a)	Langer and Moffat (1952)	Magnetic spectrometer, organic source.	260 eV	Not stated	± 100 eV	17.95 keV
(b)	Daris and St-Pierre (1969)	Magnetic spectrometer, T-activity adsorbed in Al.	90 eV	± 20 eV	± 75 eV	18.570 keV
(c)	Salgo and Staub (1969)	Electrostatic integral spectrometer, T_2O ice source.	Integral recording	± 60 eV	± 60 eV	18.70 keV
(d)	Lewis (1970)	Si(Li) detector, T-activity implanted in detector.	380 eV	± 25 eV	~ ± 60 eV	18.540 keV
(e)	Bergkvist (to be published)	Electrostatic magnetic spectrometer, T-activity in Al_2O_3 layer.	40 eV	± 2 eV	± 16 eV	18.610 keV

Some basic data for the measurements of the tritium β-spectrum illustrated in Fig. 5 (a) - (e).

hence a judgement of the statistical accuracy of the data points exhibited must be somewhat uncertain, it does not appear from a comparison between (a) and (b) in Fig. 5 that the measurement by Daris and St-Pierre represents any very large improvement with respect to the purely statistical accuracy achieved in the data points close to the end-point. Unfortunately there is a considerable systematic error present in the end-point energy derivable from Langer and Moffat's data.

The very large source permitted by the integral electrostatic spectrometer used by Salgo and Staub implies a favourable basic intensity situation. The philosophy of the measurement is to make the end-point energy measurement independent both of any assumptions about the source quality and of any a priori knowledge of the true spectrum behaviour in the vicinity of the end-point by sensing simply the highest β-particle energy present in the spectrum. To obtain this energy a least square fit to the end-point data is made of a fourth order polynomial, with all coefficients to be determined by the data points employed. A particular feature limiting the accuracy in the measurement by Salgo and Staub is the presence of a high background which seems difficult to reduce. A more inherent difficulty in the method adopted is, it seems to the present author, the very fact that the use of an integral spectrometer makes the recorded spectrum intensity go to zero at least with the third power of the distance to the end-point. If the finite thickness of the source is taken into account, the actual fourth order variation assumed by Salgo and Staub results. The accurate localization along the energy axis of such a flat minimum will inevitably, it seems to me, be relatively difficult.

It should be remarked here that a measurement sensitive to the highest β-particle energy present in a β-spectrum does not measure the end-point energy as defined by the relation (6) above, a difference equal to the mean excitation energy of the daughter ion being involved between the two end-point energies. Also, obviously there will be an ambiguity in the meaning of the end-point energy equal to the full magnitude of the possible mass of the neutrino.

In the measurement by Lewis, (d), the possibility of the presence of a distortion in the spectrum due to backscattering effects is essentially avoided by implanting the tritium activity in the sensitive volume of an Si(Li) detector. In the low-energy tritium spectrum there should be no significant inherent departures from the allowed shape, and in the measurement by Lewis the use of a fairly extended portion of the spectrum for defining the end-point should hence not be objectionable. Because of this the basic statistical definition of the end-point energy of ± 25 eV quoted in Table I should be significant in spite of the fairly crude data in the immediate vicinity of the end-point as appearing from Fig. 5 (d).

In the recording illustrated in (e) in Fig. 5, obtained with the electrostatic-magnetic high-luminosity spectrometer shown in Fig. 2, a significant improvement in the basic statistical accuracy of the end-point data has been achieved, while at the same time sustaining an experimental resolution width not larger than 55 eV. As appears from Table I, the basic statistical error in the definition of the end-point is a factor of ten smaller than in the measurements (b) and (d) in Fig. 5, and a factor of thirty smaller than in the measurement (c). In the over-all errors quoted in Table I the relative variation is smaller due to the relatively larger importance of uncertainties of other origin in the measurement (e). In the error ± 16 eV in this measurement an uncertainty in the momentum ratio between the extrapolated end-point and the calibration line (from a Tm^{170} activity) contributes by ± 7 eV. Uncertainties associated with the lack of knowledge of the precise chemical structure of the source material (cf. section 6) are estimated at ± 10 eV. A third significant contribution in the over-all error stems from the error in the absolute magnitude of the energy of the calibration lines. Clearly the presence of the chemical uncertainty of ± 10 eV makes it not particularly worthwhile to try to reduce the other uncertainties involved. A degree of accuracy is achieved, the further improvement of which will require a detailed knowledge of the chemical structure of the source and a more thorough analysis of how this specific structure influences the observed end-point energy.

From the result $E_{\beta,max}$ = 18.610 ± 0.016 keV for the tritium β-spectrum and the value of - 27 eV for the expectation value of ΔV **in the relation (6), a value of 18.651 ± 0.016 keV can be deduced for the H^3 - He^3 atomic mass difference. A correction of 3 eV for recoil is induced in the stated mass difference.**

The measurements (a), (b), (c) and (e) illustrate, it seems to me, fairly well the requirements formulated in Section 2. The measurements in (a), (b) and (e) are basically similar. The initial, very statistical accuracy does not differ much between (a) and (b). Only when simultaneously an adequate improvement in resolution and a substantial, i.e. orders of magnitude, improvement in intensity are realized as in (e), can the end-point region be recorded in such a way that a significant improvement in the definition of the end-point becomes possible. Although the method employed in (c) is favourable from the intensity point of view, it violates the requirement of adequate resolution in the sense that a narrow energy window is replaced by an integral recording.

The recording (d) to a certain extent circumvents two basic points of Section 2: the use of an implanted source in the Si(Li) detector essentially eliminates the possibility of distortion of the spectrum due to backscattering and the inherent spectrum shape in the particular case of the tritium decay should not significantly depart from the allowed shape. From these points of view fairly intense spectral regions not too close to the end-point should be possible to employ without introducing undue systematic ambiguities in the extrapolated end-point. The statistics in (d) is primarily limited by the necessity of avoiding damage of the Si(Li) detector by the implantation of the tritium.

The investigation of the end-point region of the tritium β-spectrum has attracted many experimental efforts because if offers a suitable case for gaining information about several general questions of current interest. From the point of view of establishing nuclear or atomic mass differences, the measurements of the tritium spectrum represents just one particular case. As to the possibilities of applying the methods examplified in Fig. 5 and Table I to other β-spectra - and this essentially means β-spectra of higher end-point energies - the measurements (a) and (b) by and large represent the standard way of measuring an end-point energy. As to the method (c), apart from the flat minimum inherent in the integral method employed, it seems to me that the purely electrostatic method will inevitably present problems at higher energies due to discharge effects. As to the general method of using a solid state detector, as illustrated by (d), such a detector has of course an immense appeal from the point of view of versatility and efficiency. It seems to me, however, that in the general case the conditions 1 and 2 formulated in Section 2 will be difficult to meet with such a device. With higher end-point energies it will be difficult to avoid distortion of the spectrum and the statistical accuracy in the uppermost portion of a spectrum will inevitably have to be severely limited by the over-all counting rate of the β-spectrum in the detector.

Concerning the possibilities of applying the combined electrostatic and magnetic β-spectroscopic methods used in the recording (e) the most severe general limitation is set by the risk for discharge effects on the non-equipotential source area inherent in these methods. Experience with the spectrometer shown in Fig. 2 indicates that no serious difficulties should be encountered below 1 MeV. Up to this energy a number of β-spectra should be possible to measure under considerably improved conditions with respect to intensity versus resolution. As to the actual accuracy to be expected in the end-point energies, general statements are of course very difficult and dangerous. Assuming as a rough guide that the purely statistical accuracy attainable will be proportional to the magnitude of the end-point energy, it does not seem impossible that the uncertainty of ± 10 eV assigned to the influence from the chemical structure of the source will remain a substantial part of the over-all error for at least a number of β-spectra.

References

1. K.-E. Bergkvist, to be published. Preliminary reports are given in Proc. Top. Conf. Weak Inter., p. 91, CERN 1969, and in AFI Annual Report 1969, p. 135.

2. K.-E. Bergkvist, to be published. Preliminary reports are given in AFI Annual Report 1969, p. 137 and AFI Annual Report 1970, p. 126.

3. K.-E. Bergkvist, Arkiv Fysik 27, 383; 27, 439 (1964).

4. J.J. Sakurai, Phys. Rev. Lett. 1, 40 (1958).

5. H. Paul, Nucl. Phys. A154, 160 (1970).

6. A. Sirlin, Phys. Rev. 164, 1767 (1967).

7. S. Weinberg, Phys. Rev. 128, 1457 (1962).

8. M.S. Fredman, F. Wagner and D.W. Engelkemeir, Phys. Rev. 88, 1155 (1952);
R. Serber and H.S. Snyder, Phys. Rev. 87, 152 (1952).

9. K.-E. Bergkvist, to be published.

10. J.N. Bachall, Phys. Rev. 129, 2683 (1963).

11. L.M. Langer and R.J.D. Moffat, Phys. Rev. 88, 689 (1952).

12. R. Daris and C. St-Pierre, Nucl. Phys. A138, 545 (1969).

13. R.C. Salgo and H.H. Staub, Nucl. Phys. A138, 417 (1969).

14. V.E. Lewis, Nucl. Phys. A151, 120 (1970).

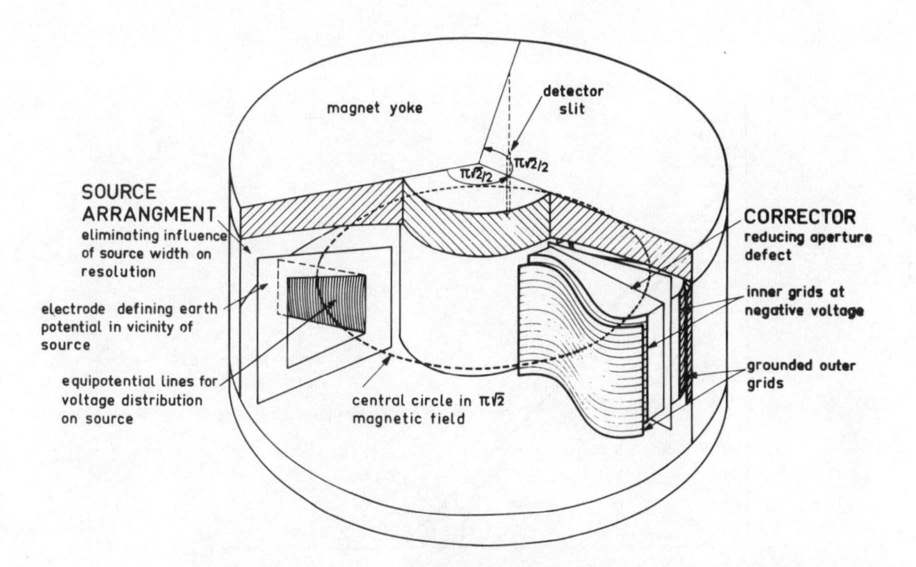

Fig. 1. Schematic drawing of high-luminosity β-spectrometer according to Bergkvist (3), based on a combination of a magnetic $\pi\sqrt{2}$ spectrometer and two electrostatic components.

Fig. 2. Actual execution of the high-luminosity principles illustrated in Fig. 1 as involved in an investigation of the end-point behaviour of the tritium β-spectrum (1,2).

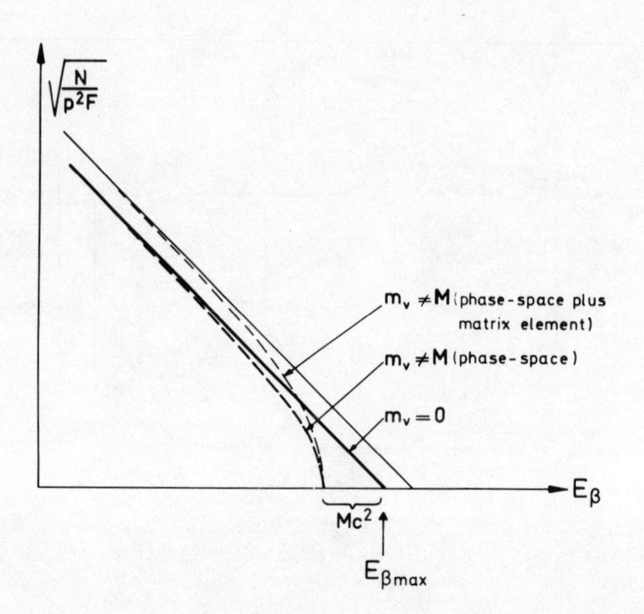

Fig. 3. To illustrate influence of neutrino mass on the end-point behaviour of a β-spectrum.

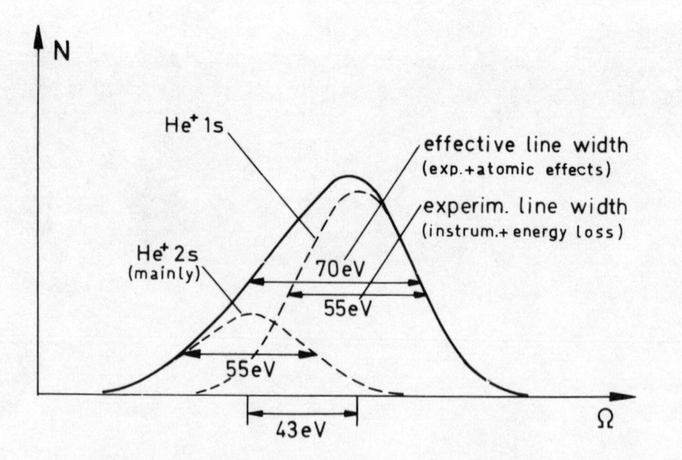

Fig. 4. Modification of actual response function in measurement of tritium spectrum due to the distribution in atomic daughter states after the β-decay. The resulting spread in end-point energy broadens the purely experimental resolution of 55 eV illustrated to an effective over-all resolution width of about 70 eV.

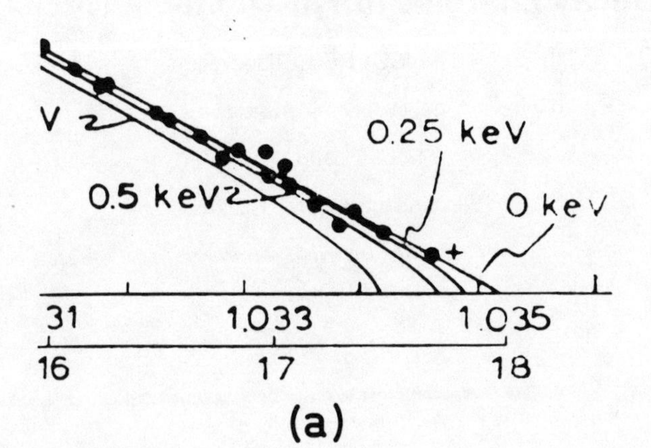

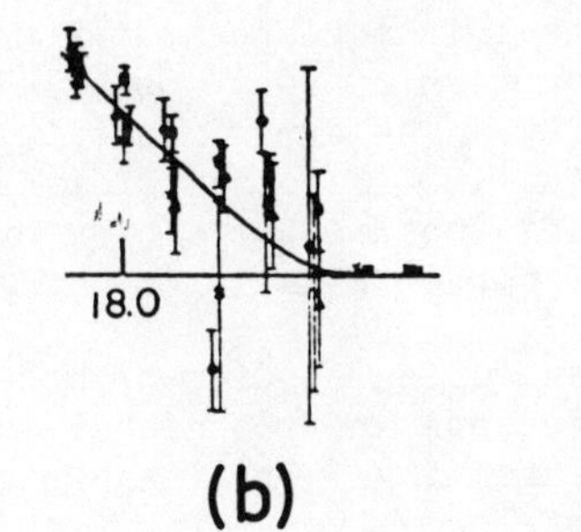

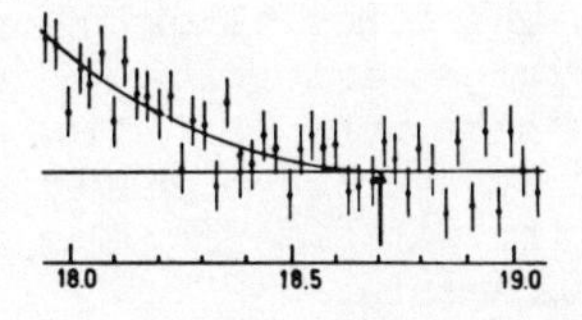

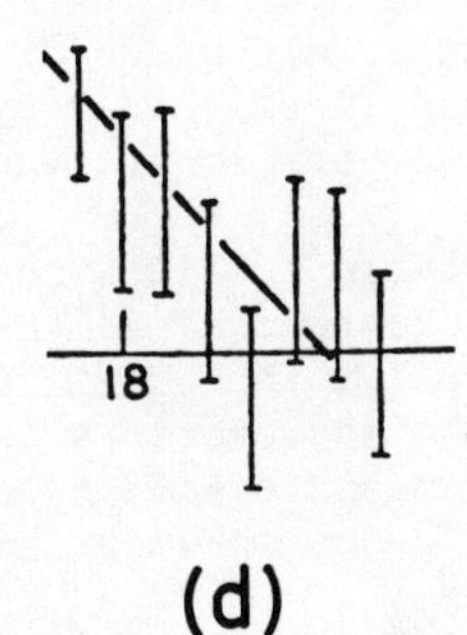

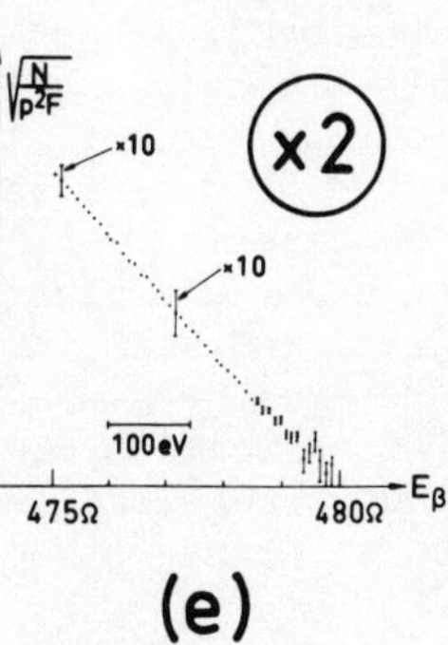

Fig. 5. Comparison of end-point data on the tritium β-spectrum as obtained in five different experimental investigations. Only the very uppermost portion of the spectra obtained are exhibited. The original drawings have been scaled to a common unit in E_β, except (e) which is enlarged twice, relative to this. Apart from (c), which shows an integral spectrum, the data represent Kurie plots of the measurements. Some pertinent data for the recordings illustrated are collected in Table I.

Decay Energies of Far Unstable Nuclei

The ISOLDE Collaboration

CERN, Geneva, Switzerland

and

L. Westgaard and J. Zylicz*

CERN, Geneve, Switzerland

and

O. B. Nielsen

The Niels Bohr Institute, Copenhagen, Denmark

1. Introduction

Studies of decay energies far from stability are some of the results due to the success of several isotope separator on-line projects, which now allow the production of very unstable nuclides with sufficient intensity and purity. The experiments to be discussed here were performed at the ISOLDE facility at CERN (1), which supply sources of neutron-deficient isotopes of many elements of a quality satisfactory for most conventional types of radioactivity measurements.

We still report only on low-resolution studies of modest accuracy, not because precision measurements are technically unfeasible, but rather because they are time-consuming. On-line experiments differ from conventional studies of long-lived activities in the need for disposing of a high-energy accelerator during measurements, and the machine is only available for limited periods of time. Under these circumstances, it was decided first to survey a reasonably broad range of nuclei rather than to perform precise measurements on a few cases.

The large decay energies of far unstable nuclides result in short half-lives and also in very complex spectra. As the isotope separators prepare the sources continuously, the short life-times are no great complication, but the complexity presents serious problems, since determinations of total disintegration energies always demand some understanding of the decay schemes.

2. Experimental methods

Figure 1 illustrates a typical decay mode of a far unstable nucleus. A major part of the beta transitions populates the daughter in an energy region of such a level density that the excitations can only be described in a statistical way, practically by a function expressing the population per unit energy. This so-called beta strength function has already been determined in many cases (2) and is usually found to vary only slowly at high excitation energies. As a consequence, the population is approximately described by the energy dependence of the beta transitions. The upper part of the continuum receives then a fairly high rate of excitation by electron capture, depending on the energy in only second power.

*) Permanent address: Institute of Nuclear Research, Swierk near Warsaw, Poland

In the decay of the most neutron-deficient nuclides, which can be produced by the ISOLDE facility, excitation occurs at such an energy that proton emission through the Coulomb barrier is not only energetically possible, but also able to compete with gamma-decay (Figure 1). Studies of the continuous proton spectra offer a new possibility for obtaining nuclear mass differences, as the endpoint energy corresponds to $Q_{EC} - B_p$, B_p being the proton binding energy. The values are of particular interest, because they establish a connection between even- and odd-mass numbers, a result which is neither obtained in alpha- nor in beta-decay.

Figure 2 shows the proton spectra following the beta-decay of ^{115}Xe and ^{117}Xe. Just as in the case of beta-ray spectra, the determination of the endpoints can be based upon an analysis of the spectral shape that is fairly well understood in terms of the statistical model (3). The first spectrum is an exception from this rule, as the upper part has the character of a weak tail, obviously unsuited for an endpoint determination. It represents in fact the only case for which we have to assume a strong energy dependence of the beta strength function. However, $Q_{EC} - B_p$ could be deduced from the ratio between electron capture and positron transitions to the proton emitting levels. This ratio is a sensitive function of the available energy in a certain region, and it was found by measuring the yield of annihilation radiation in coincidence with protons. The results for both mass numbers are compared with the extrapolated values of Gove and Wapstra (4) in figure 3.

In most cases, measurements of positron spectra represent the best possibility for estimating decay energies of the neutron-deficient nuclides. For the on-line experiments we have used the arrangement shown in figure 4. The positrons are detected by a plastic scintillator, and gamma-rays by a Ge-Li crystal in measurements, during which gamma-ray spectra coincident with positrons as well as positron spectra in coincidence with selected parts of the gamma-ray spectrum are obtained. These coincident gamma-ray spectra are usually much simpler than the single spectra as one of the effects of the 5th power dependence for positron transitions, due to which the population of the lowest levels in the daughters is favoured. Frequently one can identify positron transitions of considerable strength to well-established low-energy levels. The determination of decay energies is then based upon the recording of such positron groups in coincidence with an appropriate part of the gamma-ray spectrum. In other cases, the understanding of the decay scheme is still insufficient, and new investigations may result in reevaluation of some results.

3. Treatment of data

Figure 5 shows a coincident positron spectrum as measured with the low resolution detector. Instead of correcting the data by means of the response functions of the scintillator and subsequently performing a Fermi analysis, we have adapted a simpler treatment by transforming the spectrum according to the formula given on the graph. It represents an attempt to utilize all information on energy contained in the data by representing each point by the mean energy of all events recorded in higher channel numbers.

The method is illustrated in figure 6 by some spectra of different end-point energy after transformation. The plots are obtained by calculation, assuming allowed theoretical shape and Z = 0. Energy calibration can be performed along any straight line through the origin, and is easily verified to be almost linear. The distorsion due to the low resolution of the detector is also shown for one of the spectra. It is responsible for some deviation from linearity in the experimental calibration curves.

Some of the complications met in actual cases are discussed briefly in connection with figure 7.

4. Results

The results of our decay energy measurements on neutron-deficient cadmium isotopes and their daughters are summarized in Table I. The agreement with earlier measurements, partly performed with more accurate instruments, is consistent with the estimated accuracy of 1oo-2oo keV on the values based on the positron spectra. No large discrepancies with the decay energies calculated from mass formulae are observed.

The experiments have been extended to the largest distance from the stability line in the osmium-mercury region. This is a result of the possibility of obtaining short-lived mercury isotopes from the ISOLDE plant with high yield. ^{181}Hg, which is 18 mass units lighter than the most stable mercury isotope, has indeed been produced with an intensity sufficient for coincidence experiments. The positron branches from these very neutron-deficient isotopes of heavy elements are still intense enough for the determination of decay energies, but a major difficulty in obtaining nuclear masses from the Q-values arises due to the absence of positron transitions in the region close to stability.

TABLE I

Experimental and theoretical Q-values of neutron-deficient cadmium and silver isotopes, MeV

Nuclide	Experimental		Calculated			
	This work	Ref. 5, 6	Ref. 7	Ref. 8	Ref. 9	Ref. 10
^{101}Cd	5.35	5.53	5.o1	6.14	5.3o	5.15
^{101}Ag	4.1o	3.92	4.25	4.35	3.6o	4.11
^{102}Cd	2.5o ± 0.1o 2.58 ± 0.o7 *)		2.73	3.23	2.44	2.49
^{102}Ag	5.35	5.3	5.39	5.81	5.15	5.52
^{103}Cd	4.25	4.2o	3.88	4.69	4.o2	3.9o
^{104}Ag	4.35	4.27	4.15	4.61	3.99	4.29
^{105}Cd	2.6o	2.8	2.64	3.13	3.o9	2.67

*) This value is determined from measurements of positron to electron capture ratios

The experimental uncertainties are estimated as 1oo-2oo keV

TABLE II

Experimental Q-values for the 182 and 186 isobars, MeV

Nuclide	Q-value	Nuclide	Q-value
^{182}Os	1.1 est		
^{182}Ir	5.70		
^{182}Pt	2.90		
^{182}Au	6.85	^{186}Au	5.95
^{182}Hg	4.95	^{186}Hg	3.25

Only for the A = 182 chain we have been able to establish connection to the masses of stable nuclei, though the Q-value for the decay $^{182}Os \rightarrow {}^{182}Re$ is still based on an estmate rather than a measurement. The adopted energy, 1.1 MeV, corresponds to probable values of log ft for electron capture transitions to low energy states in the daughter, and is in agreement with the extrapolated values of Gove and Wapstra (4). However, figure 8 shows that this link is sufficient to create an experimental basis for the masses also for a number of nuclides outside the mass 182 isobar due to additional interconnections from known alpha- and delayed proton-decay energies (5, 11, 12).

Table II contains our experimental beta decay energies. The Q-values in the closed loop involving the masses 182 and 186 fit together within 300 keV. This is consistent with the assumed uncertainty of 100-200 keV for the energies, though some results are still preliminary due to the ambiguities attached to the underlying assumptions on the decay schemes.

Experimental masses for the 182 isobar deduced from the Q-values of Table II and the decay energies from ref. (5, 11, 12) are in figure 9 compared to the results of some recent mass calculations. We ascribe an uncertainty 500 keV to the experimental masses. To this should be added a possible systematic contribution of a few hundred keV from an error in the estimate of the $^{182}Os \rightarrow {}^{182}Re$ mass difference. However, the conclusion drawn in the following do not depend critically on the assigned uncertainties.

It is a remarkable feature that the table of Garvey et al. (7) reproduce the experimental masses much closer than any other calculation in the region close to stability, but that it deviates strongly from the experimental results for the far unstable nuclei. Three other calculations, which are all based upon the liquid-drop model with corrections related to nuclear structure, follow the main trend in the masses over the whole region without giving the observed local structures. Further discussions of the mass calculations are outside the scope of this contribution.

The situation is similar for mass 186 (figure 10), except that the data refer to nuclei closer to stability. The peaking of about 1 MeV relative to the calculations, which was observed around ^{182}Ir, has moved to higher values of Z and A. We have indications that it represents a sort of ridge, which can be followed through several Z-values almost parallel to the line of stability, in the sense that electron capture energies at the distant side are systematically lower than the tables of, say, Myers and Swiatecki (8), and higher on the nearer side. Thus, the decay energies for all odd-odd iridium isotopes from mass 182 to mass 190 are from 500 to 900 keV above the calculated values.

Figure 11 surveys the experimental masses of the mercury isotopes from mass 182 to 190 in relation to the tables of Myers and Swiatecki. The general agreement through 25 neutron numbers is remarkable, though also in this representation the local irregularities are not reproduced.

Acknowledgement: We are indebted to Mr. Peer Tidemand-Petersen for performing the calculations on theoretical beta spectra, to Mrs. Bodil Poulsen for careful treatment of data, and to numerous members of the ISOLDE group for continuous help.

REFERENCES

1. The ISOLDE Collaboration (A. Kjelberg and G. Rudstam ed.) CERN Report No. 70-3 (1970)

2. DUKE, C. L., HANSEN, P. G., NIELSEN, O. B. and RUDSTAM, G., Nucl. Phys. A151, 609 (1970)

3. HORNSHØJ, P., WILSKY, K., HANSEN, P. G., JONSON, B. and NIELSEN, O. B., to be published

4. WAPSTRA, A. H. and GOVE, N. B., Nuclear Data Tables, in press.

5. LEDERER, C. M., HOLLANDER, I. M. and PERLMAN, I., Table of isotopes, 6th ed. (Wiley, New York, 1967)

6. BECK, E., Proc. International Conference on the Properties of Nuclei Far from the Region of Beta Stability, Leysin August 31-September 4, 1970, CERN Report No. 70-30 p. 353

7. GARVEY, G. T., GERACE, W. J., YAFFE, R. L., TALMI, I. and KELSON, I., Revs. Modern Phys. 41 no. 4 (Pt. II) S1 (1969)

8. MYERS, W. D. and SWIATECKI, W. J., Univ. of Calif. Lawrence Radiation Lab. UCRL-11980 (1965)

9. SEEGER, P. A. and PERISKO, R. C., Los Alamos Scientific Laboratory LA-3751 (1967)

10. ZELDES, N., GRILL, A. and SIMIEVIC, A., Mat.Fys.Skr.Dan.Vid. Selsk. 3 no. 5 (1967)

11. HANSEN, P. G., NIELSEN, H. L., WILSKY, K., ALPSTEN, M., FINGER, M., LINDAHL, A., NAUMANN, R. A. and NIELSEN, O. B., Nucl. Phys. A148, 249 (1970)

12. LINDAHL, A., NIELSEN, O. B. and RASMUSSEN, I. L., Proc. International Conference on the Properties of Nuclei Far from the Region of Beta Stability, Leysin August 31-September 4, 1970, CERN Report No. 70-30 p. 331

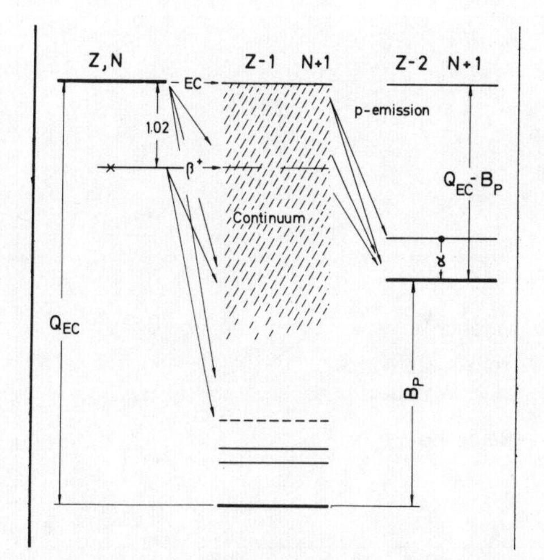

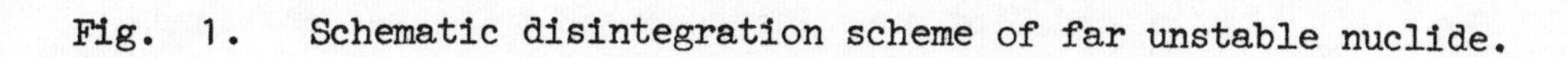
Fig. 1. Schematic disintegration scheme of far unstable nuclide.

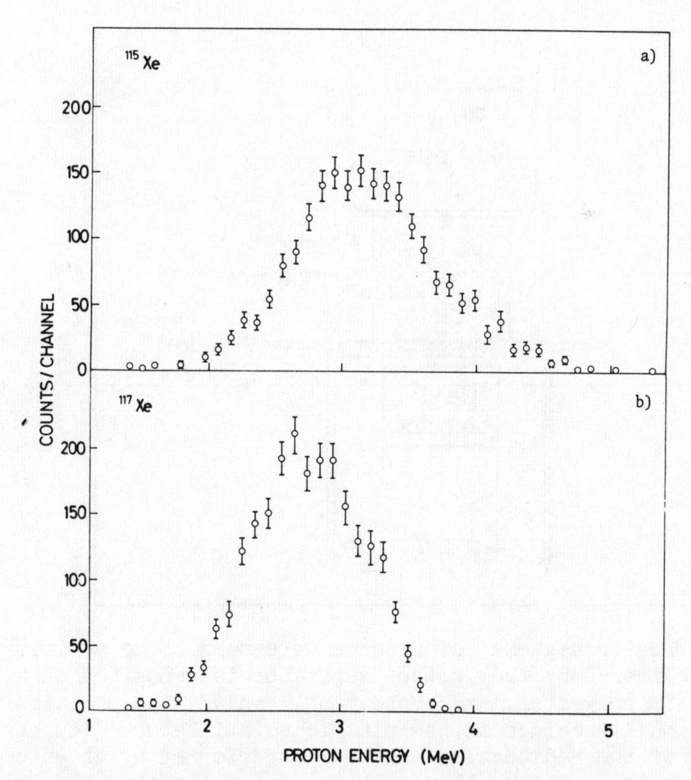

Fig. 2. Spectra of delayed protons following the beta dacay of ^{115}Xe and ^{117}Xe.

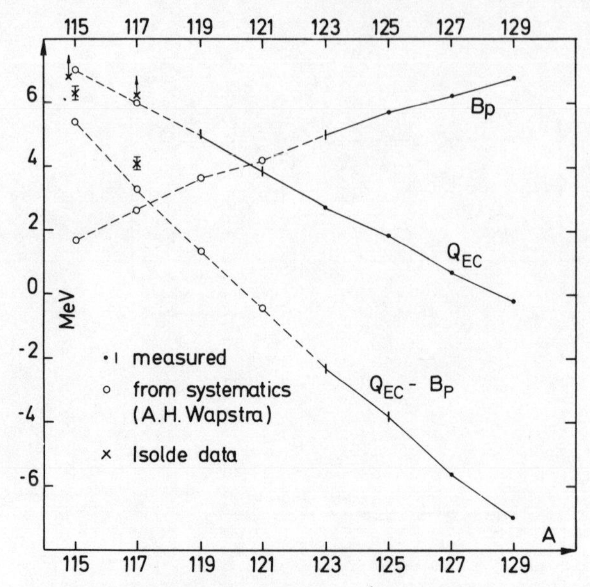

Fig. 3. Proton endpoint energies compared to values obtained by extrapolation (4).

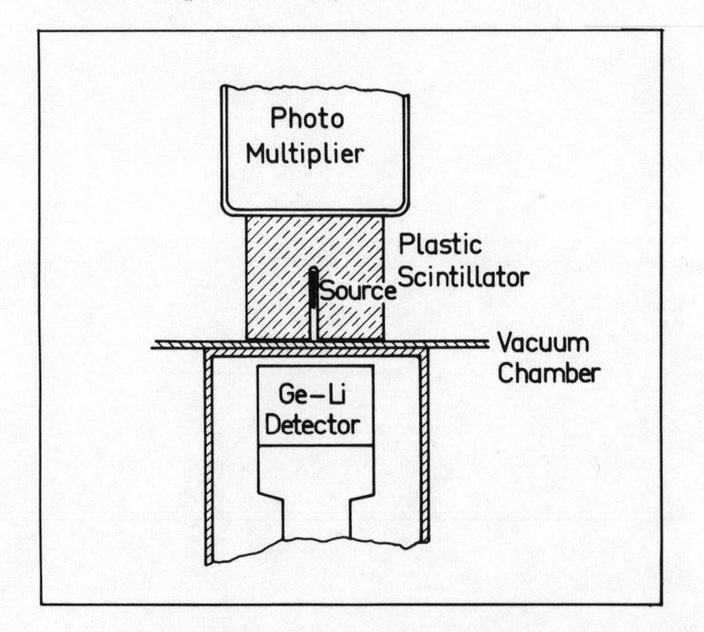

Fig. 4. Positron-gamma coincidence detectors. The activity delivered by the isotope separator is deposited on a tape of polyester, and subsequently moved into position in a slit machined in the plastic scintillator. The handling of the radioactive sources is performed by an automatic system, which allows to optimize collection and measuring periods for the mother or any longer-lived daughter activity.

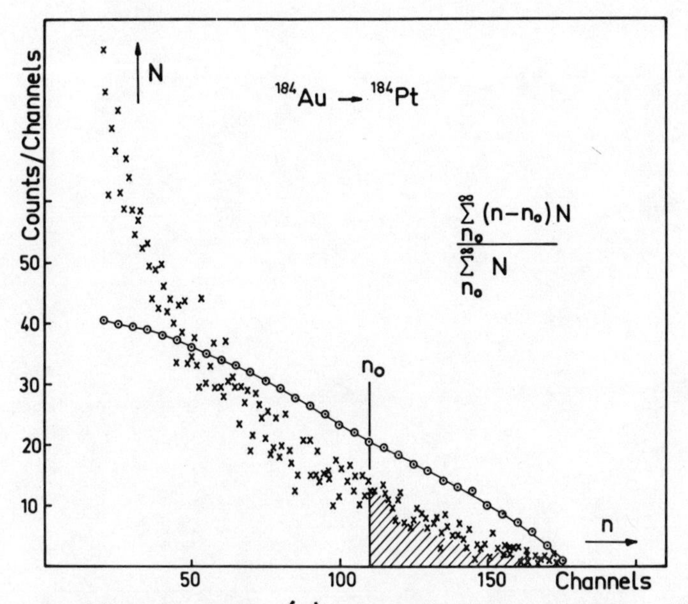

Fig. 5. Positron spectrum (x) measured in coincidence with a single gamma ray. The transformation into the curve (o) is defined in the figure.

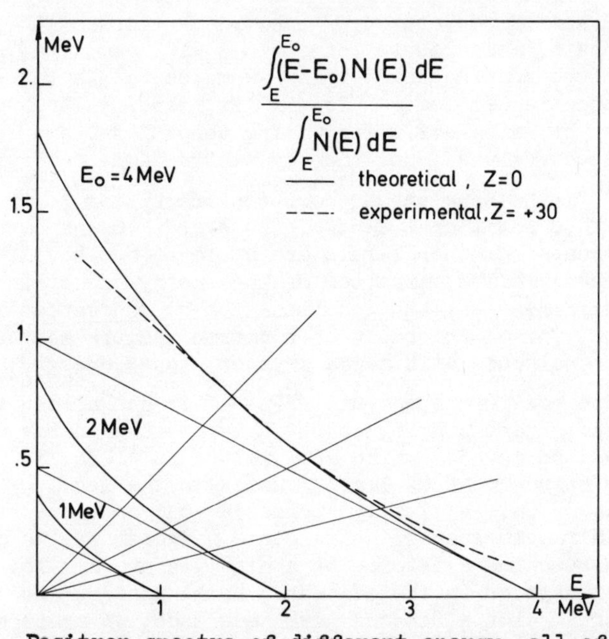

Fig. 6. Positron spectra of different energy, all of allowed theoretical shape, transformed according to the formula given in the figure. Note that the tangents to the spectra in the endpoints have the slope 1:2. This result of the transformation is independent of Z.

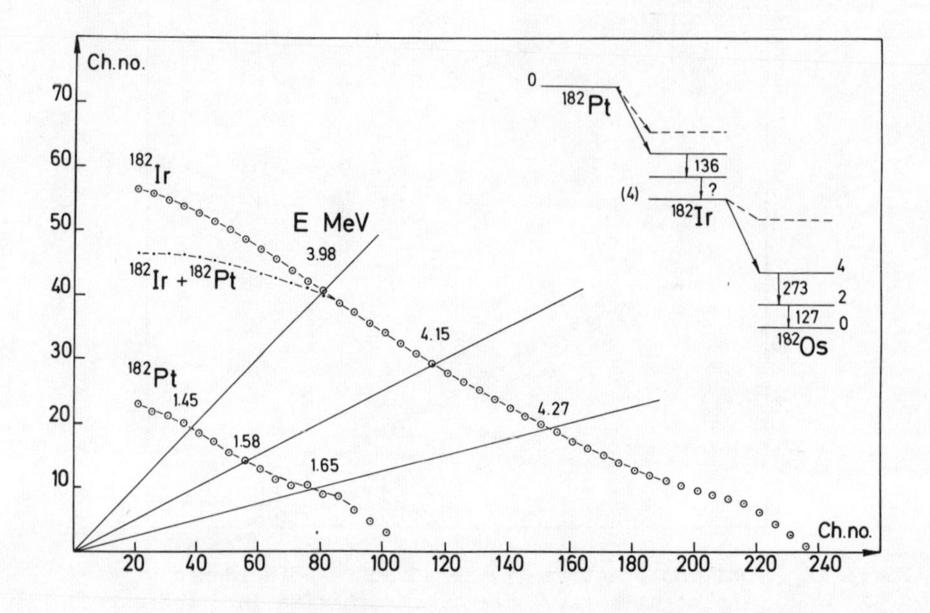

Fig. 7. Transformed version of positron spectra of the mass 182 chain, measured in coincidence with gamma ray pulses corresponding to energies from 1oo to 52o keV. The two spectra denoted ^{182}Ir + ^{182}Pt, and ^{182}Ir, were obtained by using different timing conditions, and the ^{182}Pt spectrum was found then by subtraction.

^{182}Ir has one strong positron transition to the ground-state rotational band of the even ^{182}Os, and positron groups to other levels are at least 8oo keV lower in energy. The variation of the energy as obtained along different calibration lines can be understood and corrected for on the basis of positron spectra measured in coincidence with gamma rays of higher energy.

The positron spectrum ^{182}Pt → ^{182}Ir has almost the character of one single group. However, the energy of the level or levels, which are excited in ^{182}Ir, is uncertain. A spin gap of at least 3 units to the level, which decay to ^{182}Os, is clearly suggested, but only one strong gamma ray is seen in coincidence. Therefore, we have to assume the existence of a highly converted low-energy transition in the cascade. Until measurements on the conversion electrons have been made, we arbitrarily estimate the energy as 1oo keV, as a much higher energy would imply an observable gamma ray.

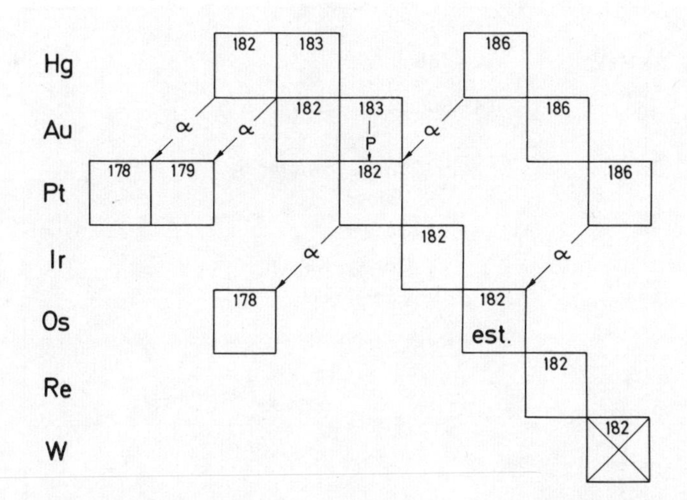

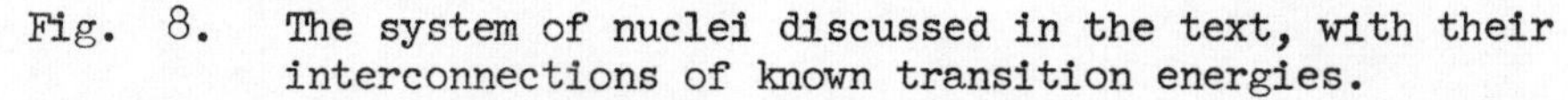

Fig. 8. The system of nuclei discussed in the text, with their interconnections of known transition energies.

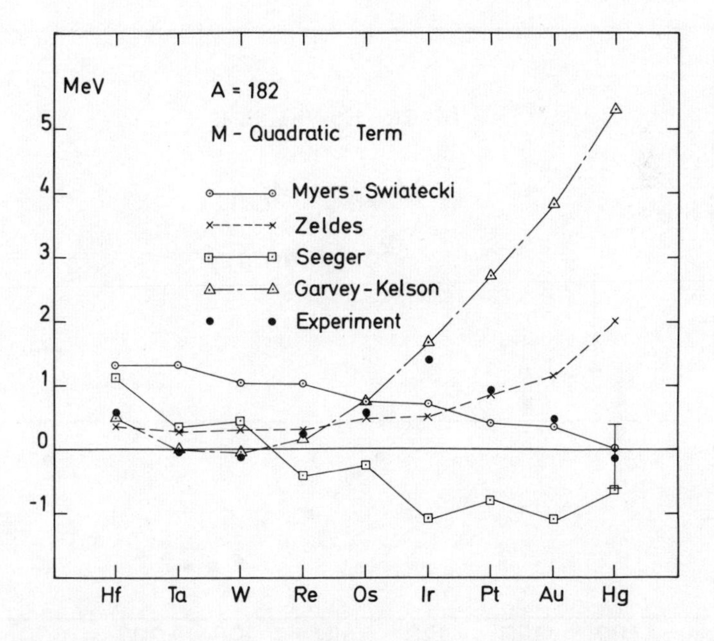

Fig. 9. Experimental and calculated masses for the 182 isobar after subtraction of a quadratic term. The mass tables are refs. (7-10).

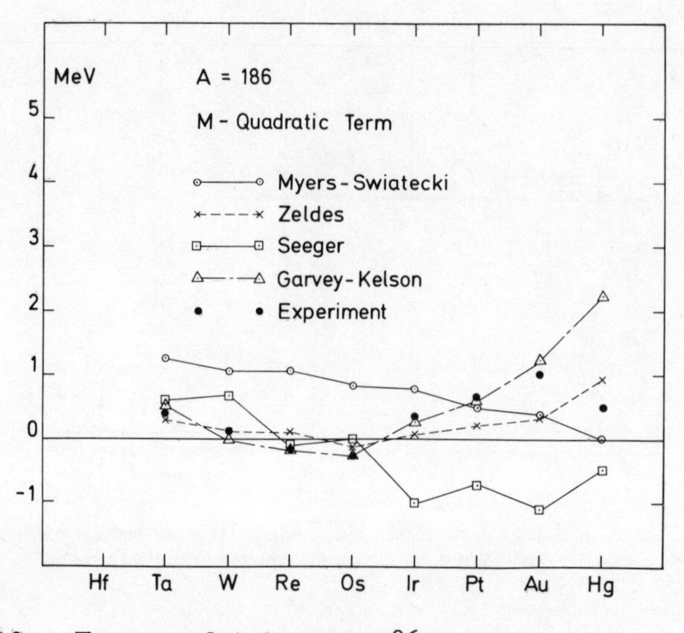

Fig. 10. The same plot for mass 186.

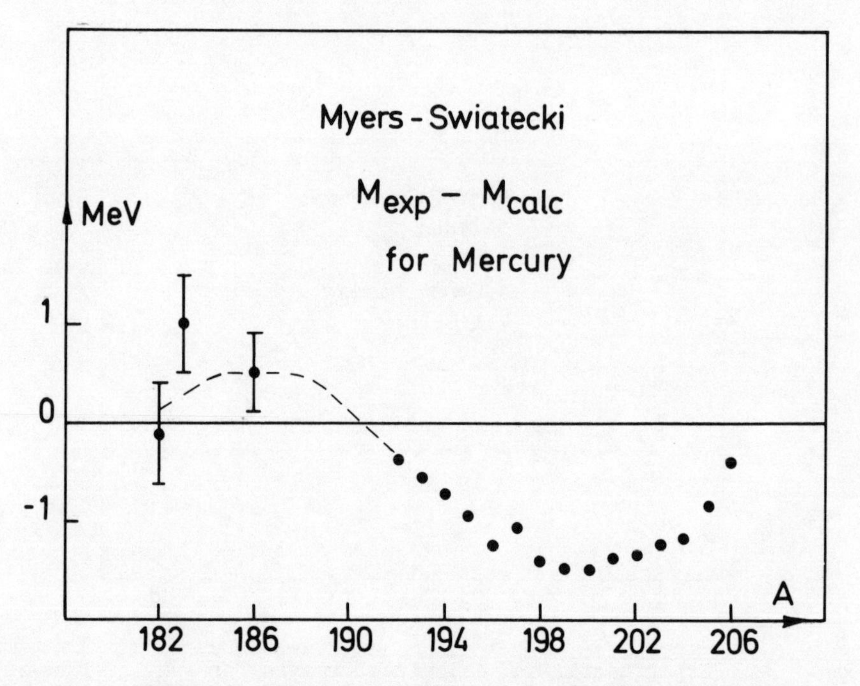

Fig. 11. Difference between the experimental masses and the Myers-Swiatecki calculation (8) for the mercury isotopes.

Accurate Fermi Beta Decay Measurements and the Magnitude of the Weak Interaction Vector Coupling Constant

J. M. Freeman and G. J. Clark

Atomic Energy Research Establishment, Harwell, Berks.

J. S. Ryder

Nuclear Physics Laboratory, Oxford

W. E. Burcham and G. T. A. Squier

Physics Department, University of Birmingham

J. E. Draper

AERE, Harwell and University of California, Davis

1. Introduction

This paper is concerned mainly with the magnitudes of the coupling constants in weak interactions. Special consideration will be given to the vector coupling constant in beta decay, about which there are significant uncertainties. We begin with a brief general survey of the weak interaction processes; for a more complete account and a comprehensive reference list see ref. (1).

2. Weak interaction Processes

A weak interaction takes place between two pairs of fermions which may be leptons (electron or muon and their associated neutrinos) and/or hadrons (baryons and mesons). The process is regarded as a very short range interaction between leptonic and hadronic currents. The propagator may be a massive vector boson (W) in the same way that the electromagnetic interaction is propagated by the photon, but the existence of the intermediate vector boson has not yet been established, and the process is normally treated as a contact interaction.

Weak interactions are categorized as purely leptonic, semi-leptonic, or non-leptonic. We shall concern ourselves here with the first two classes:

(i) <u>pure leptonic decay</u>, for example the muon decay

$$\mu^- \longrightarrow e^- + \bar{\nu}_e + \nu_\mu$$

This is the simplest type of weak interaction since leptons are not subject to the effects of strong forces. It is represented as the interaction of two leptonic currents j_λ^μ and j_λ^e (where the λ's represent the four components in the Dirac notation). The Hamiltonian for the interaction can be written

$$H_\mu = \frac{G}{\sqrt{2}} \left[(j_\lambda{}^\mu)^\dagger (j_\lambda{}^e) + \text{hermitean conjugate} \right] \quad (1)$$

G being the basic weak interaction coupling constant.

(ii) Semi-leptonic decays, involving the interaction of a leptonic current and a hadronic current ($\mathcal{J}_\lambda$) can be classified as:

(a) strangeness conserving decays ($\Delta S=0$), for example the beta decay

$$n \longrightarrow p + e^- + \bar{\nu}_e$$

(b) strangeness non-conserving decays ($|\Delta S| = 1$), for example

$$\Lambda \rightarrow p + e^- + \bar{\nu}_e$$

$$\text{or } K^+ \longrightarrow \Pi^0 + e^+ + \nu_e$$

Strong forces are also involved in these decays and might be expected to modify the strength of the interactions, i.e. to renormalise the coupling constants.

According to the Cabibbo theory (2) the hadronic current has two components, J_λ (strangeness conserving) and S_λ (strangeness non-conserving), the sum of the squares of the coefficients requiring to be unity. We thus write

$$\mathcal{J}_\lambda = \cos \Theta_c \, J_\lambda + \sin \Theta_c \, S_\lambda \, , \tag{2}$$

where the parameter Θ_c is known as the Cabibbo angle. Thus for beta decay the interaction Hamiltonian can be written as

$$H_\beta = \frac{G'}{\sqrt{2}} \cos \Theta_c \left[(J_\lambda)^\dagger (j_\lambda{}^e) + \text{h.c.} \right] \tag{3}$$

and for the strangeness-non conserving decays

$$H_S = \frac{G'}{\sqrt{2}} \sin \Theta_c \left[(S_\lambda)^\dagger (j_\lambda{}^e) + \text{h.c.} \right] \tag{4}$$

The prime on the basic coupling constant G represents the effect of a possible renormalisation.

Now in beta decay we have to distinguish two modes of interaction, namely vector (V) and axial vector (A), which correspond to the allowed Fermi and Gamow-Teller transitions respectively. It is known, from comparing pure Fermi with pure G-T decays, that the axial-vector coupling constant $G_\beta{}^A$ is not equal to the vector coupling constant $G_\beta{}^V$, and that there is in fact some renormalization in the axial-vector case, due to strong forces. For the vector interaction, the conserved vector current (CVC) hypothesis (1) postulates no renormalization and thus we expect the coupling constants:

for β decay, $G_\beta^V = G \cos \Theta_c$; (5)

for $|\Delta S| = 1$ decay, $G_S^V = G \sin \Theta_c$; (6)

with muon decay: $G_\mu = G$ (7)

3. Muon decay

The muon coupling constant is well established (3) in terms of the accurately known mass m_μ and mean life τ_μ of the muon:

$$G_\mu^2 = \frac{192 \Pi^3 \hbar^7}{c^4} \cdot \frac{(1 + 8(m_e/m_\mu)^2)}{(m_\mu)^5 (\tau_\mu)(1 - \delta_R)} \tag{8}$$

Here δ_R is a small electromagnetic radiative correction due to the emission of real photons and the exchange of virtual photons by the interacting particles. It has been calculated (3) to be + 0.12%, with an uncertainty of the order 0.01%.

Using the following values (3,4,5):

$m_\mu/m_e = 206.7683 \pm 0.0006$

$\tau_\mu = (2.1983 \pm 0.0008) \times 10^{-6}$ s.

$m_e = (9.109558 \pm 0.000054) \times 10^{-28}$ g.

$\hbar = (1.0545919 \pm 0.0000080) \times 10^{-27}$ erg. s.

$c = (2.9979250 \pm 0.0000010) \times 10^{10}$ cm. s^{-1}

one obtains

$$G_\mu = (1.4354 \pm 0.0003) \times 10^{-49} \text{ erg. cm}^3 \tag{9}$$

4. Superallowed Fermi beta decay

For an allowed beta decay the ft value is related to the vector and axial vector coupling constants by the equation

$$ft\,(1 + \delta_R) = \frac{K}{(G_\beta^V)^2 |M_V|^2 + (G_\beta^A)^2 |M_A|^2} \tag{10}$$

where $K = 2\Pi^3 (\ln 2)\hbar^7/m_e^5 c^4 = 1.230627 \times 10^{-94}$ c.g.s. units; $|M_V|$ and $|M_A|$ are the vector and axial vector nuclear matrix elements; δ_R is an electro-magnetic radiative correction (6); and the ft value is calculated from the beta end-point energy and half life as discussed in references (7,8). (We assume here that ft is the quantity $\widetilde{ft}\ \overline{C(E)}$ of ref. (8)).

The radiative correction δ_R, which is not negligible, comprises two parts Δ_R and δ'_R termed by Wilkinson and Macefield(9) the "inner" and "outer" radiative corrections. The outer correction δ'_R, dependent on the β end-point energy, is model-independent and therefore exactly calculable to order α (ref. 9). The inner correction Δ_R is not energy dependent but does depend on the details of the strong interactions involved and on the structure of the decay process (e.g. whether or not it is mediated by a vector boson). Thus Δ_R is not very reliably estimated, but it is constant for all the decays under discussion. Normally the radiative corrections δ_R of Källén (10) are adopted; in these Δ_R = +0.7%. An uncertainty of about ± 0.2% in this value is usually assumed.

For $0^+ \rightarrow 0^+$ superallowed Fermi transitions between T=1 states, $|M_A|^2 = 0$, and assuming CVC theory, $|M_V|^2$ can be written

$$|M_V|^2 = 2(1 - \delta_c) \tag{11}$$

where the correction factor δ_c represents the effect of isospin impurities and dynamical distortion in the initial and final states of the β decay; theoretical calculations indicate δ_c to be small ($\leq$few tenths per cent) and positive (1).

Thus for the $0^+ \rightarrow 0^+$ superallowed decays eqn. (10) becomes

$$ft(1 + \delta_R) = \frac{K}{2(G_\beta^V)^2(1 - \delta_c)} \tag{12}$$

Assuming CVC theory we expect the values $ft(1 + \delta_R)$ for all these decays to be the same apart from the positive factors δ_c which cannot be reliably estimated. If these factors are not negligible we expect the $ft(1 + \delta_R)$ value to be least for the decay in which δ_c is least, and this would then be the best result to adopt in calculating G_β^V.

5. Experimental results

Current experimental ft values, including the Källén radiative corrections, are shown in Fig. 1. (11,12) for a number of pure Fermi transitions. The best studied cases are those of ^{14}O and $^{26}Al^m$ decay.

The result for $^{26}Al^m$ is based on three independent Q-value measurements from which the β end-point energy was deduced, and a careful measurement of the half-life. The Q-value results for the reaction $^{26}Mg(p,n)^{26}Al^m$ were:

-5012.2 ± 2.2 keV (13) [(p,n) thresholds in ^7Li, ^{19}F]
-5013.5 ± 1.6 keV (11) [ThC α-particle energies]
-5015.58 ± 0.57 keV (14) [γ-ray energies]

where the respective brackets ( ) and [ ] give references and sources of the energy calibrations. The agreement is satisfactory; the weighted mean leads to the result 3210.6 ± 1.0 keV for the $^{26}Al^m$ beta end point energy. The half life for $^{26}Al^m$ was obtained (11), from a large number of runs, using a procedure (15) for data analysis equivalent to the maximum likelihood method. It was shown clearly in this work that the analysis of decay curves by the commonly used method of least squares fitting to the logarithm of the yield can lead to an over-estimate in the deduced half life of the order 0.5%, depending on the initial rate.

The ft value for $^{26}Al^m$ obtained from this work has for some time appeared anomalously low. However a recent measurement of the ^{14}O half life in our laboratory (12), following the analysis procedure mentioned above, has given the result 70.580 ± 0.035 s which, combined with our measurement (16) of the beta end-point via the $^{14}N(p,n)^{14}O$ reaction, leads to an ft value for ^{14}O consistent with the $^{26}Al^m$ result, as seen in fig. 1. This consistency strongly supports the choice of the $^{26}Al^m$ ft value for the calculation of G_β^V, using the criterion discussed above, i.e. choice of the lowest result. The ft values for some of the higher Z cases given in Fig. 1, notably ^{34}Cl, appear to be significantly higher, with the implication that the charge-dependent factor δ_c may not be negligible. Further measurements and analyses for these cases are clearly required to resolve this question, and we are currently undertaking some of these.

6. The magnitude of G_β^V

If we assume $\delta_c = 0$ for $^{26}Al^m$ decay then the calculation of G_β^V from eqn. (12) gives

$$G_\beta^V = (1.4102 \pm 0.0012) \times 10^{-49} \text{ erg. cm}^3 \qquad (13)$$

excluding the uncertainty in the Källén radiative correction. Thus, from eqn. (5) and the value of G_μ, $\theta_\beta^V = 0.188 \pm 0.004$.

However, from the experimental results (17,4) for the strangeness non-conserving decay $K^+ \rightarrow \Pi^0 \; e^+ \nu$ (i.e. Ke3 decay), an independent result $\theta_K^V = 0.223 \pm 0.004$ is obtained (6), neglecting radiative and SU3 symmetry breaking corrections. This result for θ^V implies a value for G_β^V 0.7% lower than that quoted in eqn. (13). The discrepancy would be resolved if the true inner radiative correction Δ_R in the beta decay were in fact at least three times as great as the Källén value used above. Such values of Δ_R are predicted theoretically (18) on the assumption of the existence of an intermediate vector boson, provided its mass is sufficiently high. It can be shown (6) that the $^{26}Al^m$ result becomes in this way compatible with the Ke3 decay data if the boson has a mass M_W of at least 20 proton masses. In these circumstances the magnitude of the vector coupling constant would be as implied by the Cabibbo angle obtained from the Ke3 decay data, i.e.

$$G_\beta^V = (1.3999 \pm 0.0014) \times 10^{-49} \text{ erg. cm}^3 \qquad (14)$$

However, it has recently been pointed out by Fishbach et al.(19) that an alternative theoretical treatment can be given of the form factors required in deducing Θ_K^V from the Ke3 data. This leads to a value for $\Theta_K^V = 0.192 \pm 0.016$, compatible with the $0^+ \rightarrow 0^+$ result based on Källén radiative corrections, and therefore supporting the magnitude of G_β^V of eqn. (13). Fishbach et al. (19) suggest that the question raised by this alternative theory may be resolved experimentally.

7. The ratio G^A/G^V for beta decay

For transitions such as mirror decays, where axial vector as well as vector interactions are allowed, equation (10) applies but the axial vector nuclear matrix element $|M_A|$ is not in general model-independent. For the neutron decay however, we know that $|M_V| = 1$ and $|M_A| = \sqrt{3}$. Thus the ratio of the ft value for the neutron to that for the $0^+ \rightarrow 0^+$ decay ($^{26}Al^m$) provides a value for the ratio G^A/G^V. The main uncertainty in this value lies at present in the uncertainty in the neutron half life. The present best value for this is 10.61 ± 0.16 min (Christensen et al. (20)), and the corresponding value for G^A/G^V is

$$G^A/G^V = 1.239 \pm 0.011$$

8. Summary

The fundamental weak interaction coupling constant G, calculated from muon decay data, is well defined, but there still remains some uncertainty in the magnitudes of the beta decay coupling constants G_β^V and G_β^A. The magnitude of G_β^V is likely to lie between the value of eqn. (13), derived from Fermi beta decay data with conventional radiative corrections, and the result of eqn. (14) derived from Ke3 data. The true value depends on theoretical interpretations and the question of the intermediate vector boson. The magnitude of G_β^A depends largely at present on the value of the neutron half life.

References

1. BLIN-STOYLE, R.J., Isospin in nuclear physics, ed. D.H. Wilkinson (North-Holland Publ. Co. 1969) p.115.
2. CABIBBO, N., Phys. Rev. Letters 10 531 (1963).
3. ROOS, M., and SIRLIN, A. CERN preprint No. TH1294 (1971).
4. Particle Data Group, Phys. Letters 33B, 1 (1970).
5. TAYLOR, B.N., PARKER, W.H. and LANGENBERG, D.N. Rev. Mod. Phys. 41, 375 (1960).
6. BLIN-STOYLE, R.J. and FREEMAN, J.M. Nucl. Phys. A150, 369 (1970).
7. BEHRENS, H. and BÜHRING, W. Nucl. Phys. A106, 433 (1968).
8. BLIN-STOYLE, R.J. Phys. Lett. 29B, 12 (1969).
9. WILKINSON, D.H. and MACEFIELD, B.E.F. Nucl. Phys. A158 110 (1970).
10. KÄLLÉN, G. Nucl. Phys. B1, 225 (1967).

11. FREEMAN, J.M., JENKIN, J.G., MURRAY, G., ROBINSON, D.C. and BURCHAM, W.E. Nucl. Phys. A132, 593 (1969).
12. CLARK, G.J., FREEMAN, J.M., ROBINSON, D.C., RYDER, J.S., BURCHAM, W.E. and SQUIER, G.T.A. Phys. Letters 35B, 503 (1971).
13. FREEMAN, J.M., MONTAGUE, J.H., WEST, D. and WHITE, R.E. Phys. Letters 3, 136 (1962).
14. de WIT, P. and van der LEUN, C. Phys. Letters 30B, 639 (1969).
15. ROBINSON, D.C. Nucl. Instruments and Methods, 79, 65 (1970)
16. FREEMAN, J.M., JENKIN, J.G., ROBINSON, D.C., MURRAY, G. and BURCHAM, W.E. Phys. Letters 27B, 156 (1968).
17. BOTTERILL, D.R. et. al. Phys. Rev. 174, 1661 (1968).
18. BRENE, N., ROOS, M. and SIRLIN, A. Nucl. Phys. B6, 255 (1968).
19. FISHBACH, E., NIETS, M.M., PRIMAKOFF, H., SCOTT, C.K. and SMITH, J. preprint (1971).
20. CHRISTENSEN, C.J., NIELSEN, A., BAHSEN, A., BROWN, W.K. and RUSTAD, B.M. Risö Report No. 226 (1971).

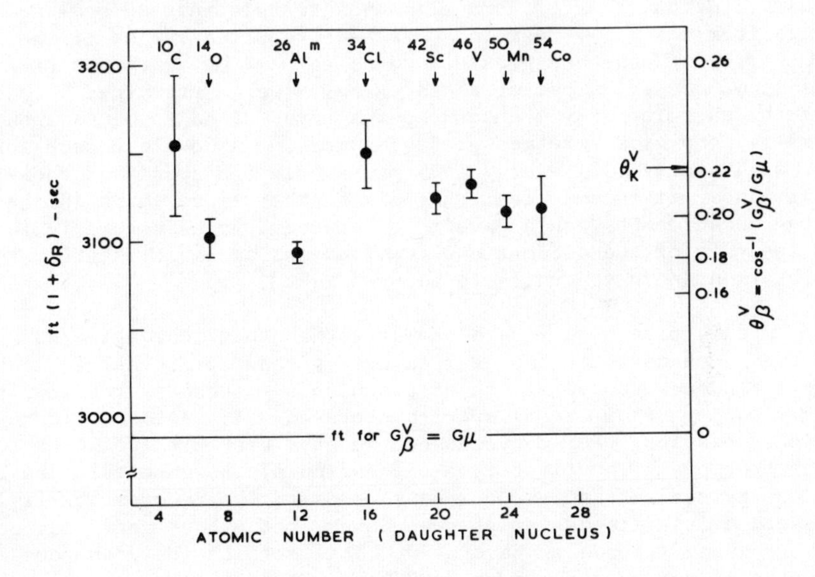

Fig. 1 ft values (including Källén radiative correction δ_R) for superallowed Fermi decays. Right hand ordinate scale gives corresponding Cabibbo angles.

Gamma-ray Energies Measured with Ge(Li) Spectrometer

R. G. Helmer, R. C. Greenwood and R. J. Gehrke

Aerojet Nuclear Company, National Reactor Testing Station, Idaho Falls, Idaho 83401

1. Introduction

In this paper, we discuss the capability of Ge(Li) spectrometers for the precise determination of gamma-ray energies. Of special interest here is the consideration of the problems, and the precision attainable, for gamma rays with energies of a few MeV. Previously, the energies of some gamma-ray transitions and nuclear levels from the (n,γ) reaction have been reported by one of the authors (1). These energies extend up to $\sim$ 11 MeV with uncertainties of $\sim$ 0.5 keV. Some of these energies were determined by summing the energies of the cascade gamma rays. Also in many cases the gamma-ray energies were measured by comparing a double-escape (DE) peak with nearby full-energy (FE) peaks. In the course of these experiments, it became clear that any further improvement in these energy measurements would have to be preceded by a systematic review and analysis of the assumptions involved and an improvement in the calibration energies available below 3 MeV.

In this paper, we give a summary of a study to measure and compare gamma-ray energies over an energy range of 50 keV to 11 MeV. This includes a review of the data analysis procedures that can be used with data from Ge(Li) spectrometers as well as some of our preliminary results. For convenience, we have broken the paper into three sections. The first is a discussion of the uncertainties in the energy scales that are commonly used and their compatibility. The second is a discussion of techniques; both those generally used and the specific ones we have used. This section includes some examples of our results up to 3500 keV. Steps that may lead to significant reduction in the uncertainties in the gamma-ray energies are noted. The third and last section includes comments on the problems in the use of DE peaks and the precision that we feel can be obtained in the few MeV region.

2. Energy Scales

In gamma-ray spectroscopy, energies have been commonly measured on two energy scales. The first scale has used the energies, or wavelengths, of the atomic x rays for calibration and reference. This "x-unit" scale was used initially to make the best use of the high precision that was obtained with crystal diffraction spectrometers. Currently these measurements are often based on the wavelength scale suggested by Bearden (2). In this work, the previous intercomparisons of the x-ray wavelengths were re-analyzed in order to obtain the most consistent and accurate set of energies possible. For this energy scale, Bearden has chosen to use the W $K\alpha_1$ line as the reference; he quotes a value of 59.318 24 keV for this line. From the adjustment of the fundamental constants by Taylor et al. (3), we obtain an adjusted W $K\alpha_1$ x-ray energy of 59.319 18 keV (±5.9 ppm). It might be noted that one problem with the use of x-ray lines for calibration is their intrinsic width. For example, the W $K\alpha_1$ line has a width of ∿ 45 eV, compared to the order of 10^{-8} eV for a typical gamma ray of this energy.

The second energy scale that is in common use is based on the energy corresponding to the rest mass of the electron, m_oc^2. The electron mass is derived in the adjustments of the fundamental constants, and in the latest adjustment (3) its value is quoted as (511.0041 ± 0.0016) keV (or ± 3.3 ppm). Although this unit of energy is, in principle, directly available for calibration in any process involving the annihilation of a positron-electron pair, this annihilation line is generally not satisfactory for energy measurements. Since most positrons annihilate in matter with bound electrons, the requirement that both momentum and energy be conserved in the reaction leads to a line that is several keV wide. Also, the transfer of energy to the bound electron and to the atom shifts the observed photon energy to a somewhat lower and uncertain value. One method of overcoming this problem is to use the narrow (but very weak) component of the annihilation line that results from the annihilation of positronium. Such measurements must be carried out with a very high resolution spectrometer and with the annihilation in a material that maximizes the positronium production. The only measurement of this type is that of Murray et al. (4) in which the energy of the 411-keV gamma ray from the decay of ^{198}Au was measured with respect to this narrow component by means of external-conversion-electron spectroscopy. Since all subsequent measurements on this energy scale have been made with respect to the energy of this 411-keV line, the effective uncertainty in this scale is the total error of the 411 energy, which we take to be 19 ppm.

The question that arises next is whether these energy scales are in agreement. The results of three previous comparisons of these scales have been summarized in our first energy measurement paper (5). These results have also been corrected to correspond to the latest values of the fundamental constants (3). A more indirect comparison made by us, and reported in ref. 5, gives the inconsistency between the energy scales as 0 ± 28 ppm, in agreement with two of the other three results. The most direct and most precise scale comparison (6) involves the measurement of the ratio of the energies of the annihilation and W $K\alpha_1$ lines from the Ta(n,γ) reaction. Although this comparison indicates a discrepancy of 56 ± 14 ppm (re-evaluated according to the Ref. 3 information), subsequent measurements by Van Assche (7) have shown that in the Ta(n,γ) reaction several gamma-ray lines are superimposed on this wide annihilation line with an especially strong line with an energy of 509.9 keV.

Therefore, it is felt that the energy scales are consistent, at least within the accuracy of a Ge(Li) spectrometer.

3. Gamma-Ray Energy Measurements

A. General Methods

Our gamma-ray energy measurements have been made on several different Ge(Li) spectrometers involving planar, true coaxial, and five-sided coaxial detectors with volumes from 2 to 50 cm^3. All of these detectors have preamplifiers with field-effect transistors in the first stage; some have cooled and others have room-temperature first stages. Each system uses a high-quality linear amplifier and a 4096-channel analog-to-digital converter. All systems have good energy resolution; that is, a full-width-at-half-maximum of ≤ 1.2 keV at 100 keV and < 2 keV at 1 MeV. This good resolution is of value primarily so that we can measure the energy differences between lines that are just a few keV apart. For the counting times used (generally less than 16 hours), the stability of each system was such as to show no observable effects (e.g., degradation of resolution). In all cases, the deviation from linearity was less than $\pm$ 2 channels from channel 200 to 4000. In the following discussion, certain assumptions will be made about the shapes of the observed peaks. These assumptions imply certain restrictions on the quality of the Ge(Li) detector itself as well as the ability of the electronics to handle the counting rates used without distortion of the peaks.

In Fig. 1 is shown a typical spectrum for our energy measurements. The presence of a large number of calibration lines spread over most of the spectrum should be noted. The essential steps in the analysis of such a spectrum are as follows:

1. the determination of the location of the peaks of interest,

2. the correction of the peak positions for the non-linearity of the system,

3. the determination of the energy vs. channel function E(x) from the energies and positions of the calibration peaks, and

4. the computation of the gamma-ray energies and uncertainties; or the calculation of the difference in energy of specific pairs of peaks.

Our methods of carrying out these operations are outlined in the following.

In order to determine the position (and area) of a peak it is usually fit with either a simple Gaussian (plus some function to represent the background) or a more complex function. Since we are primarily interested in the peak positions, we have chosen to use a simple Gaussian function. The background is taken as a straight line (generally of zero slope) matched to the average count in the background on the high-energy side of the peak. The computer program currently used for this analysis includes a routine which automatically locates the peaks. In cases where a sloping background is needed (e.g., for a peak on a Compton edge) or other special cases, this information can be supplied manually to the program and the automatic peak location routine is by-passed.

The precision of the energy measurements is limited in part by the errors in the peak locations. For the spectrum in Fig. 1, the computed errors in the peak positions of interest are 0.006 - 0.05 channels with a mean of 0.018 channels. (This is less than 1/100th of the peak width.) Since only 5 to 9 channels are used in these peak fits, the extremely small peak location errors (i.e., < 0.02 channels) may be fortuitous. However, if they are, this fact should show up when the average value is computed for several measurements.

As noted above, the second step in our computation of the gamma-ray energies is the correction of the peak positions for the nonlinearity of the system. For each spectrometer, this correction is obtained from spectra of several sources that have gamma rays whose energies are well known. From the peak positions and energies of two of these lines, a straight line is determined that would represent the energy if the system were linear through these two points. Then for all of the peaks the differences between the energies computed from this line and the actual values, when converted to channels, give the deviation from linearity. By making these measurements with different gamma rays (i.e., different isotopes) and at different amplifier gains, the effects of the errors in the gamma-ray energies and of any very local fluctuations in the nonlinearity are averaged out. The linearity correction used is taken from a smooth curve drawn through the experimental points.

The question of the accuracy of the linearity correction is a complex one. Even at the time of the linearity measurements, this curve is probably not defined to better than 0.05 channels. Since this value is larger than the uncertainty in the peak positions, this is of some concern. Also since a smooth curve is used to represent the nonlinearity correction, any local fluctuations in the nonlinearity will not be taken into account.

In our measurements the following steps have been taken to minimize the effects of errors in the nonlinearity correction. Most of our measurements are of energy differences so that the long-ranged errors have only a very small effect. Second, each energy determination involves measurements taken as several different gains, in order to average out any effects from local fluctuations in the nonlinearity. Also, most of the results involve measurements taken from two to four different spectrometers. The agreement of the results from the different systems is a verification of the quality of the linearity corrections.

The third step in the determination of the gamma-ray energies is the computation of the energy vs. channel function E(x) which is taken to be either linear $(a + b \cdot x)$ or quadratic $(a + b \cdot x + c \cdot x^2)$. The coefficients a, b and c are computed from a weighted linear least-squares fit to the energies of the calibration lines. Since both the calibration energies and channels have uncertainties, the most correct procedure would be to do a bi-variate fit; however, we have chosen a less exact but faster method. If the peak positions were exact, the weights for the E vs. x fit would be computed from the calibration line errors as simply $1/\sigma^2(E)$. The error in the peak position is taken into account in an approximate way by converting it to energy units and adding it in quadrature to $\sigma(E)$; that is, the weights are $1/[\sigma^2(E) + \sigma^2(x)]$.

The matrix that is computed and inverted in order to obtain the parameters a, b and c also contains the information on the associated variance and covariance terms. Finally, these parameters and associated error terms as well as the peak positions and uncertainties are used to determine the energies and uncertainties of the gamma rays.

B. Specific Methods and Results

Our first group of energy measurements (5) were concerned with the energy region below 400 keV. In this region there have been many precise measurements, both with crystal diffraction and electron spectrometers. Therefore, our major effort was to compare the existing energies to determine their compatibility. We found the energies compatible to the precision that we could measure. Therefore, we have used these values as a basis for work at higher energies.

Our second group of gamma-ray energy measurements (8) cover the energy range from 400 to 1300 keV. In this region the number of lines available from long-lived radioactive sources whose energies have been reported with an accuracy of 35 ppm or better was not sufficient for the accurate calibration of Ge(Li) spectrometers. Therefore, we determined the energies of a number of new calibration lines in this region.

Since Ge(Li) spectrometers are inherently nonlinear, we did not consider it satisfactory to rely on interpolation over a large energy, or channel, range even with the use of a nonlinearity correction. Two techniques were used in order to avoid such interpolations. First, we used cascade-crossover combinations where we were able to determine accurately the energies of two of the three transitions (usually the cascade transitions), so that the third energy could be computed. The second technique involves the measurement of small energy differences (less than 3% of the energy, or 30 keV at 1 MeV).

The accuracy of such energy differences can be estimated. Suppose that two peaks at 1 MeV are observed at about channel 3000 with position errors of ± 0.02 channels (i.e., the average for the peaks in Fig. 1). One energy difference measurement then gives an uncertainty of ∿ 10 eV; and the average of 10 measurements gives ∿ 3 eV (or 3 ppm of the gamma-ray energy). At this point, the errors in the energy scale and the calibration line (of the pair) are the major contributions to the total error.

Figure 2 shows a diagram of two chains of energies that have been built up to ∿ 1300 keV by the use of these techniques. The cascade-crossover combinations and the energy differences measured are indicated. One chain is based on the transitions from ^{192}Ir with ^{160}Tb and the other on those of ^{198}Au with ^{182}Ta and ^{59}Fe. Since the ^{192}Ir energies have been determined primarily on the m_oc^2 (i.e., ^{198}Au 411 keV) scale, both of these chains are on this scale. In addition to providing two independent routes up to 1300 keV, these chains provide a method of accurately testing for the accumulation of systematic errors in the energies of the chains. One such test is illustrated in Fig. 3 where data for the energy difference 1177 (^{160}Tb) - 1189 (^{182}Ta) measurement is shown. From three such comparisons of the energies of the two chains, it was determined that there was a discrepancy in the preliminary energies of 17 ± 4 eV. (See Ref. 8 for details of this calculation.) Our final reported energies have been adjusted to remove this discrepancy.

As an example of the precision obtained in these energy differences and the consistency of the data, Table I gives the results for the two energy differences shown in Fig. 3. In both cases, the data from the different spectra and the different detectors are completely consistent. That is, the values of ε^2 $\left[\text{i.e., } \frac{1}{N-1}\sum_{i=1}^{N}\left(\frac{E_i-\bar{E}}{\sigma_i(E)}\right)^2\right]$ for the 14 and 29 measurements are both ∿ 1.0

TABLE I

Energy Difference Measurements

Transitions	Detector	Number of Measurements	Energy Difference (eV)	Average Difference ε^2	Average Difference Value (eV)
1189(^{182}Ta) - 1177(^{160}Tb)	P-40	6	11082 ± 4	0.6	11085 ± 2
	P-60	6	11088 ± 3		
	ND-2	2	11086 ± 9		
1221(^{182}Ta)- 1231(^{182}Ta)	P-31	5	9612 ± 4	1.1	9606 ± 2
	P-40	6	9608 ± 5		
	P-60	16	9603 ± 2		
	ND-2	2	9603 ± 9		

(namely, 0.6 and 1.1). It should also be noted that the computed errors in these differences are both ± 2 eV. (This value does not include any systematic error that would be common to all of the detector systems.)

From the energies of the transitions in the two chains shown in Fig. 2, other line energies were determined from small differences and cascade-crossover combinations. Of the ∿ 55 energy differences measured, most have uncertainties of $\leq$ 5 eV. In computing the total error in a new gamma-ray energy, this measurement uncertainty must be combined with the measurement error in the reference line (of the pair) as well as the uncertainty in the energy scale. This process has resulted in energies for ∿ 60 transitions in the 400 - 1300 keV region. The associated errors are 21 - 31 ppm with the dominant contribution from the 19 ppm error in the energy scale. This fact suggests that a significant reduction in these errors would be forthcoming if the energy of the ^{198}Au 411-keV line were improved. However, an even more dramatic improvement could be accomplished if two lines, such as the 411- and 675-keV lines from ^{198}Au, were accurately determined (to say 1 ppm). Then our energy differences could be used to generate a whole set of calibration lines up to 1300 keV with uncertainties of ∿ 5 ppm. (The work of R. D. Deslattes (9) at the National Bureau of Standards may provide such calibration lines.)

By use of the same methods (i.e., cascade-crossover combinations and measurement of small energy differences), we are in the process of determining gamma-ray energies from radioactive decay from 1300 to 3500 keV. Table II gives some of our preliminary results in this region. The associated errors are ≈ 22 ppm (i.e., 22 eV/MeV). The isotopes for which we have measured energies so far in this region include ^{58}Co, ^{82}Br, ^{84}Rb and ^{132}Cs in addition to those in Table II.

4. Double-escape Peaks

In all of the measurements described so far only full-energy peaks have been used. As we go to higher energies (by means of the prompt transitions from the neutron-capture gamma-ray process), it is clear that DE peaks will have to be used. This will introduce two problems. First, it is well-known (10) that the primary electrons from gamma-ray interactions are accelerated by the electric field in the detector. As shown in Table III, the magnitude of this acceleration has been measured for some sample energies for three detector configurations. These values are based on the assumption that there is no acceleration if the gamma rays enter a planar detector perpendicular to the electric field. As shown in the Table, the observed shift is ≈ 100 ppm for these planar and five-sided coaxial detectors for FE peaks and about a factor of 10 smaller (and in the opposite direction) for DE peaks. It is, however, presumed that one can compare DE and FE peaks by having the gamma rays enter perpendicular to the electric field for a planar detector or a true coaxial detector. But for very precise measurements one must be very careful about the angle at which the gamma rays enter and about the orientation of the electric field.

TABLE II

Preliminary gamma-ray energies determined above 1300 keV

Parent Isotope	Method		Energy (keV) Sum	Average
^{110m}Ag	706+677	SUM	1384.265	
	763+620	SUM	1384.265	
	937+446	SUM	1384.270	1384.267 ± 0.029
	818+657	SUM	----	1475.757 ± 0.034
	818+687	SUM	1505.010	
	884+620	SUM	1505.004	1505.006 ± 0.032
	818+744	SUM	1562.266	
	884+677	SUM	1562.262	1562.264 ± 0.033
^{124}Sb	645+1045	SUM	1690.950	1690.948
	722+968	SUM	1690.947	
^{144}Ce	696+1489	SUM	----	2185.618
^{56}Co	1037+1175+1238	SUM		3451.071

TABLE III

Field Effect - Preliminary Results

Values assume no field effect for gamma rays entering detector perpendicular to electric field.

Detector type	Detector voltage	Gamma energy (keV)	FE peak shift	Field effect (eV) DE peak shift	Field effect (eV) FE-DE peak shift
planar	1900	2600	256	-24	280 ± 20
planar	1900	2180			255 ± 30
planar	1900	1180	75		
five-sided	2000	2180			300 ± 40
coaxial	2500	2180			25 ± 30

The second question that arises with respect to the use of DE peaks for energy measurements is whether the DE-FE peak separation for one transition is exactly $2m_oc^2$. In the pair production process a certain fraction of the positrons will annihilate with bound electrons. This means that some of the $2m_oc^2$ energy will be left in the detector as the two annihilation photons escape. To the extent that this energy produces ion pairs (or holes), the energy of the DE peak will be shifted to a somewhat higher energy. Therefore, it would seem very desirable to measure the FE-DE peak energy difference. Previous measurements (11) have been carried out; they allow this difference to be as much as 50 eV less than $2m_oc^2$. We have made one comparison involving three transitions from ^{124}Sb (namely, 645, 1045 and 1691 keV). Measurements were made of the 1045-keV energy and the difference between the 645-keV FE peak and the 669-keV DE peak (from the 1691-keV transition). The latter measurement was made with the gamma rays entering a planar detector perpendicular to the electric field. From the preliminary results of this experiment, we find that the FE-DE peak difference is less than $2m_oc^2$ by 15 ± 25 eV. If this preliminary result proves to be correct, we can use the DE peaks to extend our energy calibration set to higher energies with an uncertainty of ≈ 25 eV/MeV. (This assumes that the field effect has been eliminated.) This would be quite satisfactory for the present, since our m_oc^2 energy scale (via ^{198}Au 411 keV) has a basic uncertainty of 19 eV/MeV.

Therefore, it would seem that with the present energy scales we can expect to establish a set of calibration energies with uncertainties of ≈ 25 ppm. It is our plan to extend our measurements from the 3500 keV that we can reach with radioactive sources to ≈ 11 MeV by means of some transitions and cascades from the (n,γ) reaction.

1. GREENWOOD, R. C., Phys. Lett. 23, 482 (1966) and Phys. Lett. 27B, 274 (1968).

2. BEARDEN, J. A., Rev. Mod. Phys. 39, 78 (1967).

3. TAYLOR, B. N., PARKER, W. H. and LANGENBERG, D. N., Rev. Mod. Phys. 41, 375 (1969).

4. MURRAY, G., GRAHAM, R. L. and GEIGER, J. S., Nucl. Phys. 45, 177 (1963) and ibid. 63, 353 (1965).

5. GREENWOOD, R. C., HELMER, R. G. and GEHRKE, R. J., Nucl. Instr. and Meth. 77, 141 (1970).

6. KNOWLES, J. W., in Proc. 2nd Intern. Conf. Nuclidic Masses, Vienna, 1963 (ed. W. Johnson; Springer-Verlag, Vienna, 1964) p. 113.

7. VAN ASSCHE, P., Intern. Conf. on Precision Measurement and Fundamental Constants, NBS, 1970.

8. HELMER, R. G., GREENWOOD, R. C. and GEHRKE, R. J., Nucl. Instr. and Meth. to be published.

9. DESLATTES, R. D., Intern. Conf. on Precision Measurement and Fundamental Constants, NBS, 1970 and ibid. SAUDER, W. C.

10. HEATH, R. L., Proc. 1968 Intern. Conf. on Modern Trends in Activation Analysis, NBS Special Publication 312 (1969) Vol. II; and GUNNICK, R., MEYER, R. A., NIDAY, J. B. and ANDERSON, R. P., Nucl. Instr. and Meth. 65, 26 (1968).

11. WHITE, D. H., GROVES, D. J. and BIRKETT, R. E., Nucl. Instr. and Meth. 66, 70 (1968).

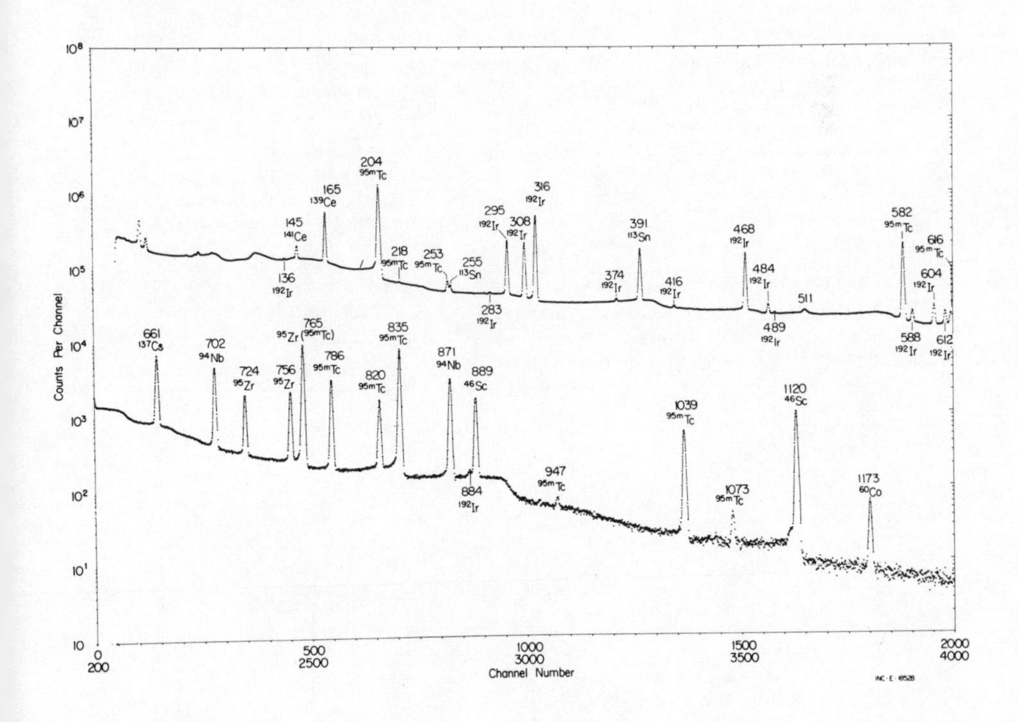

1. Typical energy calibration spectrum for determination of the energies of the lines from ^{94}Nb and ^{95}Zr. Energies are in keV.

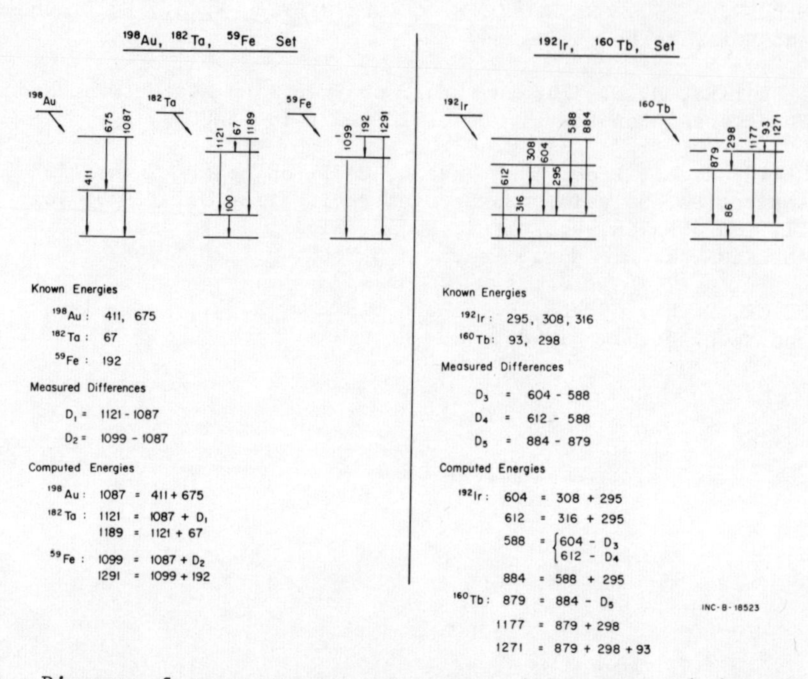

2. Diagram of gamma-ray transitions used in energy chains up to 1300 keV. The measured energy differences and energy values determined are shown.

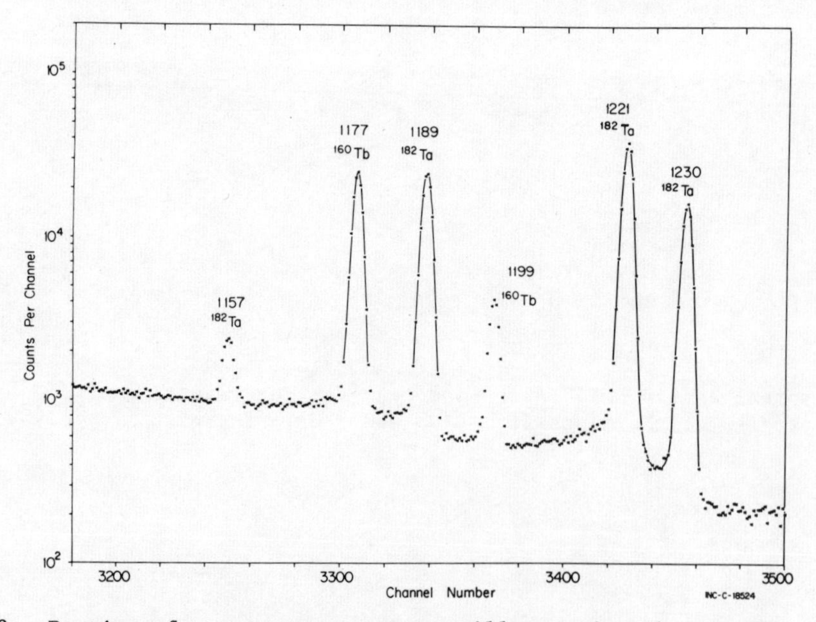

3. Portion of a gamma-ray spectrum illustrating the energy difference measurements near 1200 keV.

Measurement of (n,γ)-lines with a High-resolution Dumond-type Diffractometer

O. W. B. Schult, H. R. Koch, H. A. Baader, and D. Breitig

Physics-Department, Technical University, Munich, Germany

and

Research Establishment, Risø, Denmark

1. Introduction

The high resolving power of X-ray diffractometers has led DuMond and his co-workers to investigate the applicability of such a device for the detailed study of low-energy γ-spectra. Previous attempts have suffered from the extremely low luminosity of flat-crystal spectrometers. Therefore, DuMond (1) has constructed a diffractometer with a cylindrically bent, focussing crystal, the efficiency of which was about three orders of magnitude larger than that of a flat crystal.

DuMond's collaborators in Pasadena have used such spectrometers for numerous, extended studies of the γ-spectra from radioactive decay (2). Similar diffractometers have been built and utilized at different laboratories.

Only few of them have been applied extensively for studies of (n,γ)-spectra which are well known to exhibit a very large complexity, so that high resolution is the primary requirement to be met by (n,γ)-spectrometers. A diffractometer with a very large crystal has been installed at the ANL CP-5 reactor (3), and it has been improved and used with great success by Smither (4). The Munich group has set up a relatively small system at the DR-3 reactor at Risø, where many (n,γ)-spectra have been measured in the period from 1962-197o.

2. Diffractometer

With a DuMond type diffractometer, γ-spectra are measured point by point. Gamma rays from a very narrow source on the focal circle are reflected in Laue transmission and detected with a suitable counter. The γ-ray energies are derived from a precise determination of the Bragg angle ϑ which is given by the orientation of the crystal relative to the source. Measurements with crystal spectrometers require high specific activities of the small sources which generally weigh between 1 and 100 mg.

For (n,γ)-spectroscopy these sources must be located close to the reactor core. They have to be mounted on a light support for keeping the background radiation sufficiently low. Thermal and other effects introduce small and irregular movements of the source holder, so that φ cannot be determined precisely, unless the source migrations are controlled. This can be achieved by cumbersome manual measurements of the reflection of a given γ-line on both sides of $\varphi=0$, or with the help of a monitor crystal which permanently determines the location of the source through a suitable measurement of the reflection of a strong line.

Measurements of γ-spectra with many hundreds of lines, (n,γ)-studies of isotopes which give rise to double neutron capture, and slow-neutron capture γ-ray spectroscopy of nuclei with very high cross sections, which burn out in a relatively short time, require an automatized diffractometer.

The automatization of the curved crystal spectrometer at Risø was accomplished through the installation of a "sine screw" for the movement of the main spectrometer crystal (5). The monitor crystal was coupled to a feed back system in such a way (see Fig. 1) that a correction equal to the angular movement of the source in the beam hole of the reactor was superimposed on both, the orientation of the main crystal and the monitor crystal. The detector, a 2" dia x2" scintillation counter, and its shield, 18 cm lead plus 10 cm paraffin, weighing about 1.2 tons, was moved with a second screw, which was driven together with the sine screw. The electronical signals from the first, second, third, fourth, and fifth orders of reflection are recorded in parallel.

3. Characteristics of the Diffractometer

The simultaneous observation of the spectra reflected from the different orders, and the great number of lines in one spectrum have allowed the elimination of systematical errors. In this way it was possible to measure the γ-spectrum with 30 keV$\lessapprox$E$\lessapprox$2 MeV with an angular accuracy of 0.18 seconds of arc (6). Through the automatized measurement lines with medium and high intensity are measured in all five orders of reflection, weak lines appear in the lowest three orders of reflection and only very weak lines in the first and second order. Therefore, the accuracy of the energy determination of (n,γ)-lines has been improved considerably through the automatization of the spectrometer which had been used for manual measurements during its first period of operation. The relative energy error δE(=standard error) of a strong γ-line measured in the order n of reflection is given by

$$\delta E = 0.35 \times E^2/(nGeV). \qquad (1)$$

For weak transitions the energy errors are impaired by the statistical uncertainty of the determination of the angle of reflection. The absolute energy errors include the calibration error which in general amounts to $2 \times 10^{-5} \times E$ because of the uncertainties of the standards, (usually the $K\alpha_1$ line of the isotope formed through slow neutron capture).

Besides the reduction of the energy errors, efforts have been made to fabricate sources with more perfect shapes. This work was very successful and has made possible the measurement of (n,γ)-spectra with a resolution ΔE (=FWHM) as low as

$$\Delta E = 2.4 \times E^2/(nGeV) \qquad (2)$$

if only the central half (about 7cm^2) of the area of the 4 mm thick quartz crystal reflecting from the 110 planes was utilized.

The efficiency of the spectrometer is quite low. It reaches a value of about 3×10^{-7} between 50 keV and 120 keV for the second order of reflection and decreases proportional to E^{-2} above 200 keV, where the fifth order reflection has approximately 5% of the strength of the second order reflection. Absorption effects, especially within the source, strongly reduce the number of low-energy quanta which can be detected. The low efficiency is counterbalanced, however, by the high activity of the sources, which has an upper limit of about 1000 Curies at the DR-3 reactor. The source strength becomes too small for the (n,γ)-study with the Risø spectrometer if the neutron absorption cross section is less than 1 barn, although the minimum partial cross section for a γ-line in the region of the maximum sensitivity is only a few millibarns.

4. Comparison with Ge(Li)-and Si(Li)-spectrometers

The high efficiency of the solid state spectrometers and their good energy resolution make these detectors well suited for coincidence measurements, for a very fast survey of (n,γ)-spectra and for all kinds of studies with low activities. The crystal spectrometers cannot be used for any of such experiments. The diffractometer yields better data than Ge(Li)- or Si(Li)-detectors for one dimensional low-energy γ-ray spectroscopy from neutron absorption in a target with a sufficiently large capture cross section.

The line width ΔE of the diffractometer is smaller than that of the solid state counter and it is not affected by high counting rates.

The reflections appear on a very smooth background. Compton edges and escape peaks, which influence and complicate the Ge(Li)-pulse height spectrum, do not exist in measurements with crystal spectrometers.

The dynamic range of the diffractometer is very large in the lower energy region ($E \lesssim 400$ keV), which is the domain of such a device (see Fig. 2).

In Gd158, Baader (7) has observed a 150.5 keV line with an intensity of 1.5×10^{-6} per neutron capture at an intensity limit of only 4×10^{-7} per neutron capture, in which case the counting time was 1000 sec per point. Such weak lines can be searched for, if they are of sufficient interest for the nuclear level scheme under study. The total measuring time (20 - 40 days for one isotope) is not increased remarkably by a few of such runs. This kind of measurements is, however, far beyond the capability of a Ge(Li)- or Si(Li)-spectrometer, even if very good anti-Compton spectrometers (8) are used.

Very precise low-energy γ-energy measurements have been performed with Ge(Li)-detectors in exceptional cases (9), where the density of lines was so low that peaks in the pulse height spectra could be considered as singuletts. Meaningful and precise transition energies cannot be extracted from complex spectra, unless the resolving power of the spectrometer is appropriate. The basic difficulty constitutes a limitation for the usefullness and reliability of all γ-spectroscopic data. It is a much more severe problem for Ge(Li)- and Si(Li)-data than for low-energy (n,γ)-data taken with a curved crystal spectrometer and with an intense source. Occasionally even the best diffractometers are still not sufficient for good (n,γ)-studies even at a few hundred keV. Mühlbauer's measurement (6) of the γ-spectrum from slow-neutron capture in Eu151,where more than 2500 lines have been resolved with energies between 30 and 500 keV, is probably the best example in support of the above statement.

5. Results

With the automatized diffractometer, the low-energy γ-spectra have been measured from slow-neutron capture in Ag107, Ag109, La139, Sm150, Eu151, Eu152, Gd155, Gd157, Tb159, Yb168, Yb171, Yb174, Hf178, Ta181, Ta182, Au197, Au198, Hg199, and Th232. The study of the reactions Eu152(n,γ), Ta182(n,γ) and Au198(n,γ) was possible because of the long half-lives and the high cross sections of these isotopes (double neutron capture). The neutron capture experiments on Hf178, Ta181, Ta182, and Sm150 were performed in collaboration with the group of Van Assche, Mol, Belgium.

The evaluation and nuclear structure interpretation of several of the obtained spectra has already been completed. The remaining part of the data is being analysed at the present time. The high quality of the curved crystal spectrometer data has allowed to locate many of the measured γ-lines in the low-energy part of the nuclear level diagrams. The decay schemes of Gd158 (7) and Yb175 (10) are representatives of the results of such studies.

The main interest of these (n,γ)-studies has been the investigation of the structure of the observed levels for tests and refinements of the present nuclear models. Relevant to accurate determinations of differences of atomic masses is the high precision of the energies of low lying nuclear states as obtained from the γ-ray energies measured with the diffractometer. Absolute level energies can be determined with accuracies of 2- 5 $\times 10^{-5}$ (1eV - 100eV) for excitation energies below 2 MeV. The combination of these data with accuratcly measured primary (n,γ)-transitions, which can be measured with good single Ge(Li)-detectors or Ge(Li)-NaI(Tl)-pair spectrometers, allows the determination of neutron separation energies with uncertainties between 0.5 and 1 keV. The neutron binding energy cannot be obtained from the primary (n,γ)-data without a sufficiently detailed knowledge of the excitation energies of final states, i.e. the low-energy level scheme.

The widths of the high-energy (n,γ)-lines measured with good germanium diodes (FWHM $\approx$ 6 keV) should permit the determination of the absolute transition energies in favourable cases to better than 1/30 of the FWHM. Absolute energies of low-energy transitions have been measured to about 1/100 of their FWHM (9), and the statistics of the high-energy (n,γ)-data does frequently yield fitting errors of 0.2 keV or less (13).

A possible occasional multiplett structure of primary (n,γ)-lines does not reduce the quality of very precisely measured high-energy spectra, since the knowledge of a sufficiently complete low-energy level scheme aids for the identification of such complex structures.

Therefore, it is of interest to considerably increase the precision of the neutron binding energies through more accurate measurements of primary (n,γ)-spectra. The progress in the development of electronical equipment should make such measurements possible.

6. Conclusions

Although level energies with an accuracy of few electron volts are far beyond the precision of the theoretical predictions and what is needed for atomic mass formula at the time being, it is desirable to improve the energy accuracy dE considerably beyond what has been achieved with the Risø spectrometer. It should be emphasized that the error δE is relatively large in comparison with the line width ΔE at this device. This is due to the extreme difficulty to mechanically measure small angles with higher accuracy. The application of the energy combination principle, a very powerful tool for the construction of complicated nuclear level schemes, does already benefit a lot, if dE is reduced by a factor of only two. This can in principle be achieved by angular measurements with a translation-invariant interferometer, which yields an accuracy of better than 0.1 seconds of arc. Such a system has been built by Marzolf (11). A much simpler and less expensive interferometer of that kind is being tested at the present time in Munich (12). The future will tell us, how far we can proceed improving and applying an old method for high resolution studies of γ-spectra and precision measurements of γ-ray energies, in particular, after the new high flux reactor has started its operation in Grenoble, where the measurements performed at Risø should be continued.

Acknowledgement

The authors wish to thank Prof. H. Maier-Leibnitz, Prof. Kofoed-Hansen, Prof. T. Bjerge, Prof. K.O. Nielsen, Dr. Fl. Juul and Civ. Ing. Th. Friis-Sørensen, for the continuous support and sponsorship of this work and the staff of the DR-3 reactor and the computer groups at Risø and the Bavarian Academy of Sciences for the excellent co-operation.

References

1. DUMOND, J.W.M., Rev.Sci.Instr., 18, 626(1947).
2. LEDERER, C.M., HOLLANDER, J.M. and PERLMAN, I.,Table of Isotopes, Sixth Edition,John Wiley & Sons,Inc., New York 1968.
3. ROSE, D., OSTRANDER, H. and HAMERMESH, B.,Rev.Sci. Instr., 28, 233(1957).
4. SMITHER, R.K. and BUSS, D.J., Neutron Capture Gamma-Ray Spectroscopy, p.55, IAEA, Vienna 1969.
5. KOCH, H.R., BAADER, H.A., BREITIG, D., MÜHLBAUER, K., GRUBER, U., MAIER, B.P.K., and SCHULT, O.W.B., Neutron Capture Gamma-Ray Spectroscopy, p.65,IAEA,Vienna, 1969.
6. MÜHLBAUER, K., Thesis, Technical University, Munich, 1969.
7. BAADER, H.A., Thesis, Technical University, Munich, 197o.
8. MICHAELIS, W., and KÜPFER, H., Nucl.Instr. & Meth. 56, 181(1967).

9. GREENWOOD, R.C., HELMER, R.G., and GEHRKE, R.J., Nucl.Instr. & Meth., 77, 141(1970).
10. BREITIG, D., Z. Naturforschung, 26a, 371(1971).
11. MARZOLF, J., Rev.Sci.Instr., 35, 1212(1964).
12. BORCHERT, G., Diploma-Thesis, Physics-Department, Technical University, Munich, 1969.
13. SHERA, E.B., private communication of the primary Eu151(n,γ)-data.

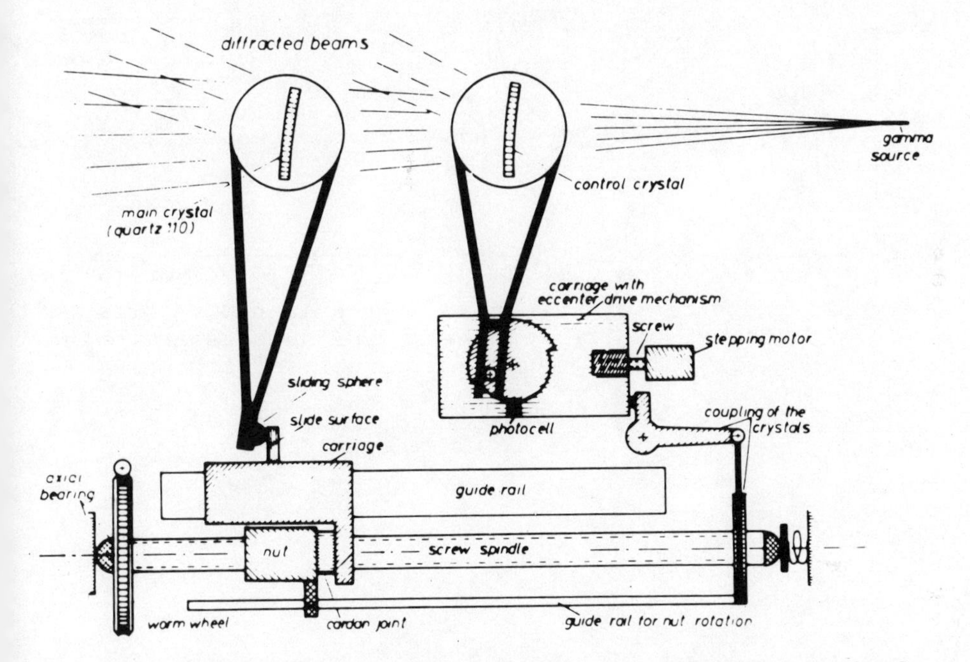

Fig. 1. Schematic view of the driving mechanism and of the control system for the compensation of the source movement.

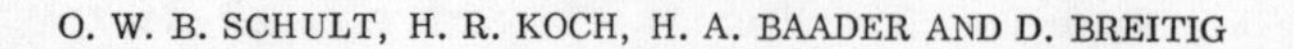

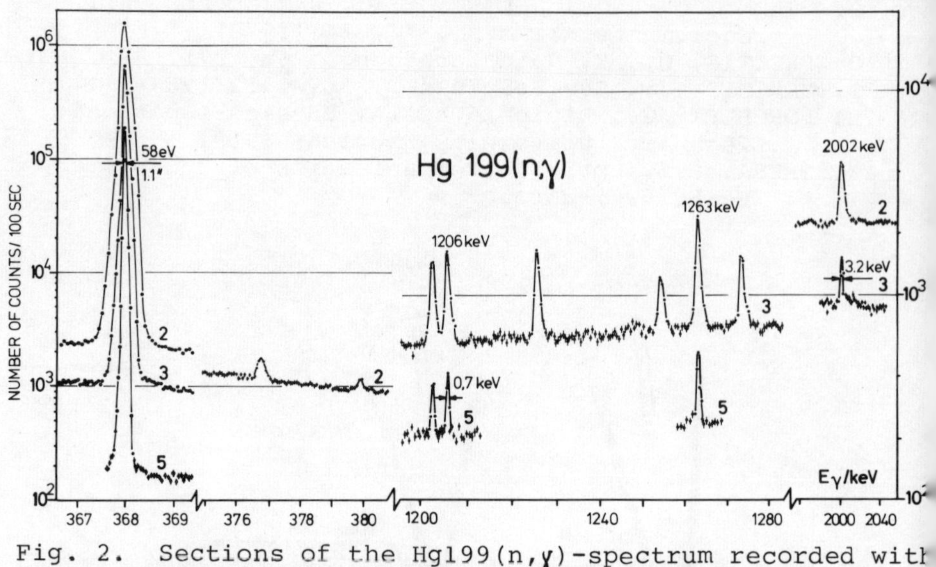

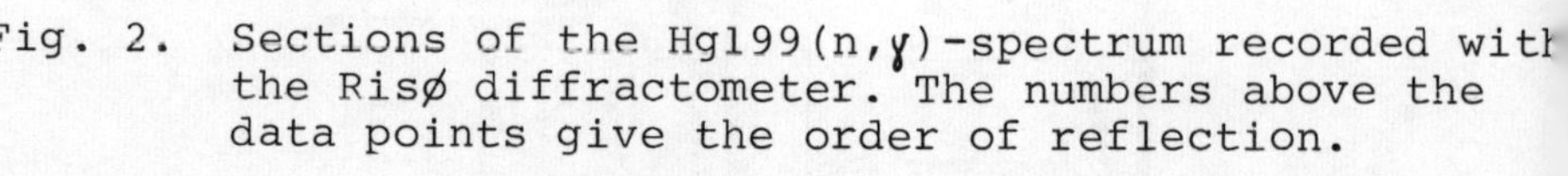
Fig. 2. Sections of the Hg199(n,γ)-spectrum recorded with the Risø diffractometer. The numbers above the data points give the order of reflection.

Capture Reaction Q-values with Sub-keV Accuracy

C. van der Leun and P. de Wit

Fysisch Laboratorium, Rijksuniversiteit, Utrecht, The Netherlands

1. Introduction; some problems

A set of accurate Q-values has proven to be of some use in many respects. The question whether further reduction of the errors into the sub-keV range warrants some extra efforts can better be discussed after a stock-taking of the technical possibilities and a discussion of some applications.

A measurement of the Q-value of a charged-particle capture reaction consists of two parts:

(i) The resonance energy is measured relative to one or more of the thoroughly studied (1) calibration resonances, of which the energy is known with a precision of the order of 0.1 keV. As long as the analyzing magnet does not approach its saturation range, the magnetic field -usually measured with a NMR fluxmeter- is considered to be a reliable measure for the particle energy. Many (p, γ) resonance energies, e.g. for sd-shell nuclei, have been published with a precision of 0.1-0.2 keV (2). There seems to be no problem here. Recent measurements, however, cast some doubt on this optimistic point of view; see section 5.

(ii) The γ-ray energies are measured relative to that of a series of well-known calibration γ-lines, mainly from radioactive sources (3). For capture reactions with low Q-value -and thus low-energy γ-rays of say $E_\gamma \leq 3$ MeV- there is no serious problem. The difficulties arise at the more typical γ-ray energies of $E_\gamma = 4$-10 MeV. The most frequently used calibration line in this range is the 6.13 MeV $^{19}F(p, \alpha\gamma)^{16}O$ transition. The popularity of this calibration line is due to the old-time nuisance of (p, γ) studies, the ubiquitous ^{19}F contamination of targets. Though the energy of this common background line is accurately known, a word of caution is justified. An asymmetry of this peak, that may invalidate its use as a calibration line, has been observed in many experiments. Fig. 1 demonstrates the cause of this asymmetry. If the target, as usual, makes an angle of ≈ 45° with the beam, and if the detector, as usual in precision energy measurements, is placed at 90°, the shape of the 6.13 MeV line wil depend on the position of the detector, or -as illustrated in fig. 1- on the orientation of the target..In the upper part of the fig., the $^{16}O^*$ nucleus recoiling in the direction of the detector will slow down in the target before the 6.13 MeV γ-ray is emitted ($\tau_m = 24 \pm 2$ ps). Recoil in the opposite direction -in vacuum- will lead to observation of Doppler-shifted γ-rays of lower energies. This causes the bump at the low-energy side of the 6.13 MeV peak. If the target is turned over 180°, as in the lower part of fig. 1, the situation is reversed and the Doppler shift bump appears at the high-energy side of the peak. In practice, the bumps will of course be less outspoken than in these spectra measured with a target with a deliberately contaminated surface. It is, however, precisely the less clear-cut asymmetry of actual measurements that -when overlooked in a (computer)determination of the centre of gravity- may invalidate the results of γ-ray energy determinations of sub-keV precision.

2. An alternative γ-ray source

Gamma-rays from thermal neutron capture in many different materials can in principle be used as convenient calibration lines. It has been shown that the thermal-neutron flux from a moderator-surrounded 2.6 x 10^6 n/s ^{241}Am-Be source is sufficiently high to produce capture γ-ray intensities that are comparable with the γ-ray intensities produced in typical proton and α-particle capture reactions (4). A sketch of the set-up is given in fig. 2. By varying the distance L_1 and/or L_2 the γ-ray strength of the source is adjusted to that of the (p, γ) or (α, γ) spectrum studied. Only for the strongest (p, γ) resonances the distance L_1 has to be minimized. The source-detector distance then is 25 cm. For the calibration of (α, γ)-spectra, this distance typically is about 1 m. An example of an (α, γ) spectrum (5) with Fe(n, γ) calibration lines is given in fig. 3; in this experiment the source-detector distance was 50 cm, and could have been still larger.

The capturing materials most frequently used up to now are
(i) Fe, with the 7.63-7.64 MeV doublet from ^{56}Fe(n, γ)^{57}Fe transitions to the ground state and first excited state at 14.41 keV; this doublet is a well-known testcase for the quality of Ge(Li) spectrometers, and
(ii) Cl, not only since the thermal-neutron capture cross section is high (33b), but also since it has a γ-ray spectrum with rather well-isolated peaks over a broad energy range (0.52-8.58 MeV). The Fe-cap of fig. 2 is then replaced by a 5 mm thick vessel with CCl_4 in front of the detector.

Of course, these neutron capture lines can only be fully used in the sub-keV precision range, when the energies have been determined with adequate accuracy. Before discussing these measurements (sect. 4), it may be useful to point to some applications in which knowledge of the precise energies is not required.

3. Precision Q-values, without precise calibration energies

An example (6) is our measurement of the ^{26}Al^m-^{26}Mg mass difference -relevant to the problem of the weak-coupling constants discussed this morning by Miss Freeman. The principle of the experiment is demonstrated in fig. 4. Apart from the measurement of a few low-energy γ-rays and a low-energy resonance energy, the problem is reduced to a measurement of the energy-difference $E_{\gamma 2n}-E_{\gamma 0p}$. The ideal experiment would be a simultaneous measurement of the ^{25}Mg(n, γ)^{26}Mg and ^{25}Mg(p, γ)^{26}Al^m spectra. The ^{25}Mg mass then drops out. This experiment, however, requires a proton accelerator at a high-flux reactor, or -in the set-up described above- a dangerously strong α-Be source near the accelerator, since the capture cross section of ^{25}Mg (with 10% natural abundance) is low.

Therefore, the experiment has been performed in two steps:
(i) a mixed Mg-Cl sample was exposed to the thermal-neutron beam of a high-flux reactor giving a combined ^{25}Mg(n, γ) and Cl(n, γ) spectrum;
(ii) a ^{25}Mg(p, γ) plus Cl(n, γ) spectrum was measured in the set-up described above.

The Cl(n, γ) lines, of which the energy was not yet accurately known at the time, served only as a common reference standard.

In principle this method can, of course, be applied to many more difference-measurements. The mass-differences of mirror nuclei (and other members of isobaric multiplets), relevant to the discussion of Coulomb energies, can in this way be measured with an accuracy of a few tenths of a keV.

4. Energies of thermal-neutron capture lines

Precision-values for the energies of several promising thermal-neutron calibration lines have already been published; some of them with 0.1-0.2 keV errors. In principle, such a precision can be reached. In several cases, however, inspection of the papers reveals that systematical errors have not been included, or treated as statistical errors. It is not excluded that some of these values even slipped into recognized mass-tables. This may be one of the problems to be discussed next Friday.

Measurements of γ-ray energies with the accuracy quoted above are certainly not of the routine type. Even in the keV accuracy range discrepancies occur frequently. An example is the above mentioned reaction ^{25}Mg(n, γ) with published Q-values of 11 098.1 ± 1.0 keV (7), 11 096.0 ± 0.5 keV (8) and 11 0 93.4 ± 1.2 keV (9).

In view of discrepancies of this size, it seemed worthwhile to measure the energies of a few thermal-neutron capture γ-rays independently. Not since in principle (p, γ)-groups are expected to produce more reliable data than (n, γ)-groups, but since it is always advisable to perform precision measurements in different ways.

The obvious advantages of (n, γ) experiments,
(i) a high flux of thermal-neutrons and (ii) the exactly known energy of the incoming particle, are compensated by the fact that the one and only (n, γ) capturing state for each target nucleus is replaced by say 100 resonances in (p, γ). This abundance of initial states makes it possible to select resonance levels which decay strongly both in a cascade of low-energy γ-rays ($E_\gamma \lesssim 2.5$ MeV) and directly with a high-energy γ-ray to one of the lowest bound states. At these resonances it is possible to measure the excitation energy of the resonance level (typically 5-10 MeV in the sd-shell) with a precision of typically 0.2 keV (sum of the errors of three of four $\lesssim$ 2 MeV γ-rays). The deexciting high-energy γ-rays are unsuitable as standard calibration lines, but they can be used to determine a precise value for a few convenient (n, γ) lines.

In these measurements the E_p = 316, 390, 592, 724, 1044, 1106 and 1165 keV resonances of the reaction ^{25}Mg(p, γ)^{26}Al and the E_p = 860 keV resonance of ^{35}Cl(p, γ)^{36}Ar were used. It should be noted that for the present discussion the precise proton-energies of these resonances are irrelevant. A relatively large number of resonances was used to reduce the risk of a misinterpretation of a decay-scheme.

Table 1

Energies (in keV) of neutron-capture γ-rays from (p, γ) experiments

Fe(n, γ)	Cl(n, γ)
9 299.5 ± 0.4	7 791.0 ± 0.4
7 646.63 ± 0.16*)	7 414.5 ± 0.3
7 632.22 ± 0.16*)	6 640.4 ± 0.4
7 279.9 ± 0.2	6 111.3 ± 0.2
6 019.4 ± 0.2	5 716.2 ± 0.5
5 921.4 ± 0.2	

*) The 14.41 keV difference was used a an input-datum.

The preliminary values given in table 1 are essentially based on the E_γ < 2.75 MeV energies listed in ref. (3), with assumedly uncorrelated uncertainties. Since it is difficult to trace the degree of correlation, a reanalysis of the data with the consistent set of low-energy values discussed by Helmer may be more profitable.

In experiments of this precision care has to be taken that
(i) all measurements of spectra an calibration lines are taken simultaneously;
(ii) the relative counting rates of the (p, γ) and (n, γ) reactions do not change too much (e.g. through target deterioration);
(iii) the spectral and calibration γ-rays reach the detector from the same side.
To reduce the risk of the influence of possible inhomogeneties of one detector it may also be advisable to use different Ge(Li) spectrometers.

The values listed in table 1 agree with the (unpublished) less precise energies deduced from (n, γ) experiments summarized in ref. (10).

5. Proton- and α-particle energies

Precision Q-value measurements of capture reactions can profitably be performed with low-energy accelerators, since at low energies the E_p's can be determined with small absolute errors.

So far analyzing magnets have been used to connect resonance energies to calibration resonances; saturation of the magnet is usually supposed to be negligible. Published resonance energies for E_p < 1 MeV have a typical precision of a few tenths of a keV (2).

The linear relation between the proton energies and deexcitation γ-ray energies in one reaction can be used to check the relative proton energies. Calibration of the high-energy γ-rays with a Fe(n, γ) line (of which the exact energy again is irrelevant) allows the calculation of E_p-differences with a precision of a few tenths of a keV. A preliminary check on the published resonance energies for the reaction ^{25}Mg(p, γ)^{26}Al, reveals systematic discrepancies of up to about 2 keV. It is not yet clear whether this disturbing situation occurs only for ^{25}Mg(p, γ). Further experiments are in progress to clarify the situation.

Along the lines described above, it is possible to determine a set of accurate secondary calibration standards for resonance energies. The bombarding particle energy of a resonance may be connected to that of a well-known resonance in the same nucleus through precision γ-ray energy measurements.

In this way, the recent progress in γ-ray spectroscopy may be used to define some more accurate charged particle energies. For many years the situation has been the other way round.

6. Perspectives

The high precision obtained in Q-value measurements in capture reactions opens a way to measure the mass-differences of almost all the light nuclei with good accuracy. Through co-operation of groups working on (p, γ) and (n, γ) reactions, it should be possible to construct a network of capture reaction Q-values. For the stable nuclei in the middle of the sd-shell such a network is illustrated in fig. 5. The system is overdetermined if also (α, γ) reactions are included. A least-squares analysis should lead to a complete set of Q-values with an accuracy of the order of at most 1 keV and with some extra effort of the order of 0.1 keV.

What, however, is the use of such a precision if theoretical calculations reproduce the nuclidic masses and excitation energies of the levels with errors that are at least a few orders of magnitude larger than those discussed here?

At present such a precision is indeed not required for general theoretical reasons, but mainly for more specialized applications like Coulomb displacement energies and the vector coupling constant of weak interactions mentioned above.

Furthermore, the measurements of Q-values and excitation energies are so closely interrelated, that precise Q-value determinations will also lead to a large number of accurate excitation energies with the ensuing separation of so far unresolved doublets, that may play a crucial role in nuclear theories. In fact, many of the published Q-values are just a by-product of measurements of this type.

Another application of rather general character should be mentioned in connection with the resonance reactions discussed above. Many unbound states can be excited in different resonance reactions. Identification of e.g. a ^{24}Mg(α, γ) resonance with a ^{27}Al(p, γ) or ^{27}Al(p, α) resonance, which to a large extent depends on precision energy determinations, can be an important step in spectroscopic experiments. Similar arguments apply to the identification of levels excited in stripping, pick-up and other direct reactions on the one hand, and proton or α-particle capture reactions on the other.

The flow of information from these spectroscopic studies certainly warrants some extra efforts in the measurement of precision E_x- and Q-values, also from the point of view of more fundamental nuclear physics.

1. MARION, J.B., Revs. Mod. Phys., 38, 660 (1966).
2. ENDT, P.M. and VAN DER LEUN, C., Nucl. Phys., A105, 1 (1967).
3. MARION, J.B., Nucl. Data, A4, 301 (1968).
4. VAN DER LEUN, C. and DE WIT, P., Physics Lett., 30B, 406 (1969).
5. DE VOIGT, M.J.A. et al., Nucl. Phys., to be published.
6. DE WIT, P. and VAN DER LEUN, C., Physics Lett., 30B, 639 (1969)
7. JACKSON, H.E. et al., Physics Lett., 17, 324 (1965).
8. SPILLING, P. et al., Nucl. Phys., A102, 209 (1967).
9. SELIN, E. and HARDELL, R., Nucl. Phys., A139, 375 (1969).
10. RAPAPORT, J., Nucl. Data, B3, No. 3-4, 103 (1970).

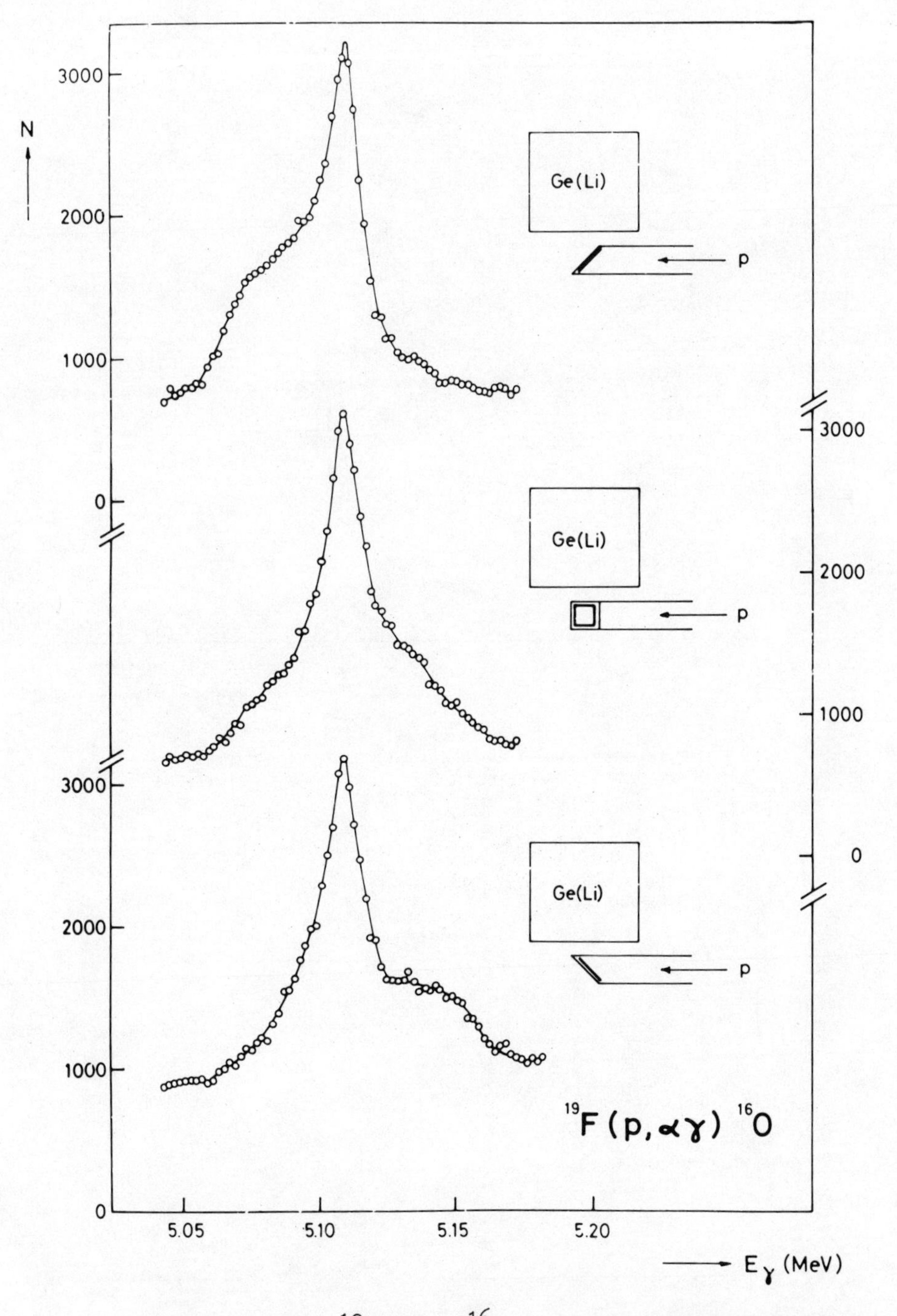

Fig. 1. The 6.13 MeV, $^{19}F(p, \alpha\gamma)^{16}O$ γ-ray peak measured with target and Ge(Li) detector in different relative positions.

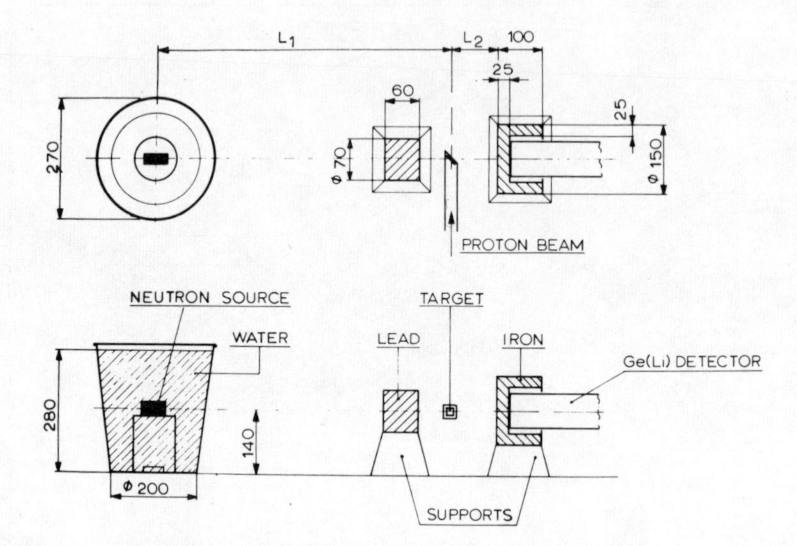

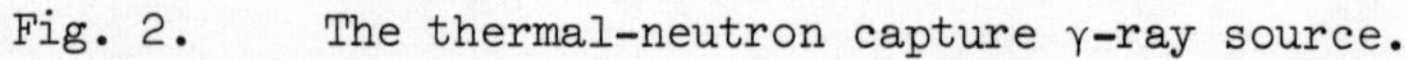

Fig. 2. The thermal-neutron capture γ-ray source.

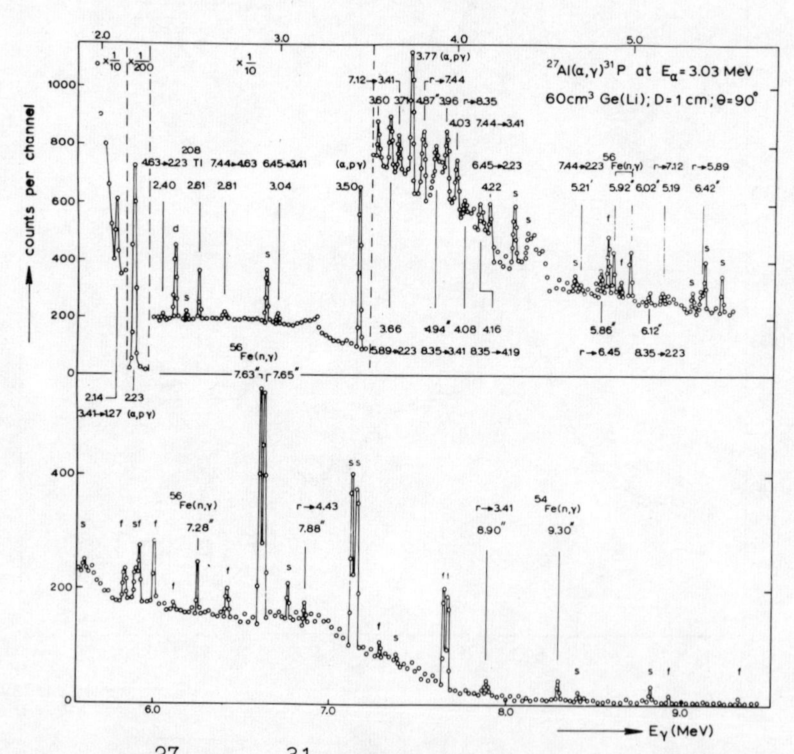

Fig. 3. An $^{27}Al(\alpha, \gamma)^{31}P$ spectrum with Fe(n, γ) calibration lines. In the peaks the average value of two consecutive channels is plotted; in between the average of ten channels.

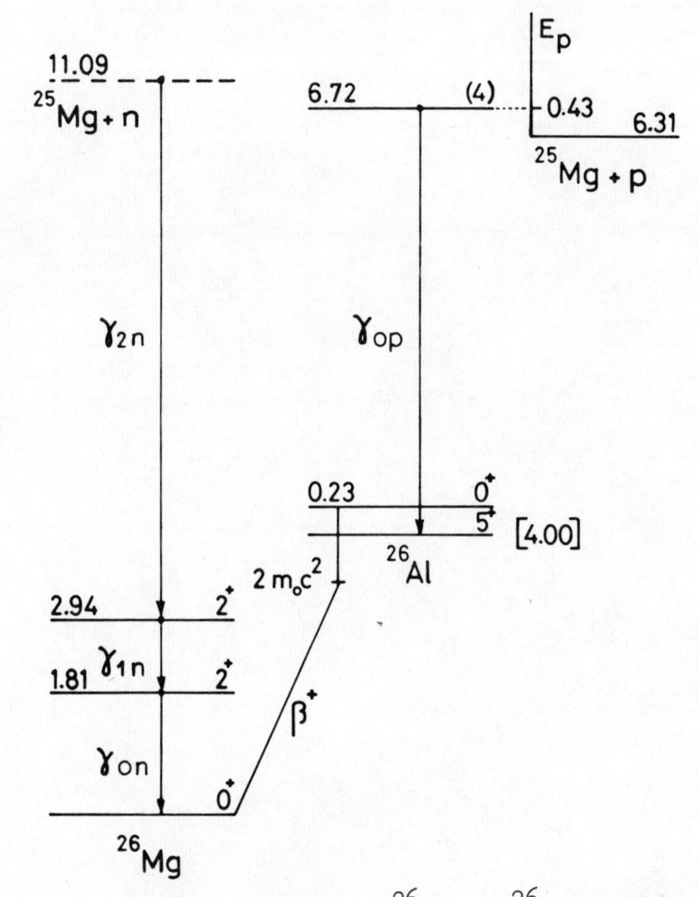

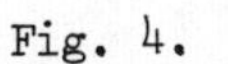

Fig. 4. Partial decay schemes of ^{26}Mg and ^{26}Al levels.

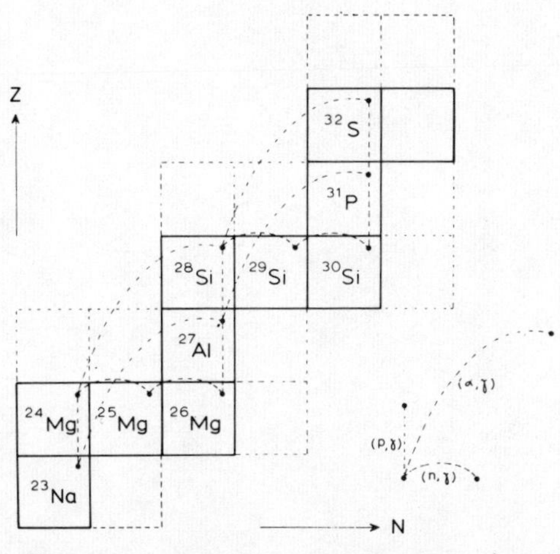

Fig. 5. Possible network of capture-reaction Q-value determinations.

Part 3 Mass Spectroscopy

Recent Determinations of Atomic Mass Differences at the University of Manitoba

R. C. Barber, J. W. Barnard, R. L. Bishop, D. A. Burrell
H. E. Duckworth, J. O. Meredith, F. C. G. Southon, and P. Williams

Department of Physics, University of Manitoba, Canada *

1. Introduction

For some time a systematic study of atomic mass differences between naturally-occurring nuclides in the region above Pr has been in progress in this laboratory. This region has been of special interest inasmuch as the mass differences and masses have, until recently, been least well known here. Moreover, it is in this region that strong nuclear deformation makes its appearance.

A substantial body of data on which this study has been based was obtained using the 2.7 m. radius Dempster type mass spectrometer and has appeared in a series of three papers [1, 2, 3]. More recently, a new mass spectrometer, constructed at the University of Manitoba has been used to improve the precision of some data, and also to include in the study values for doublets not previously studied, particularly those involving isotopes of low (< 0.2%) relative abundance.

2. The Mass Spectrometer

The instrument was constructed according to the second-order focusing theory of Hintenberger and Konig [4] and employs a 94.65° cylindrical electrostatic analyser of radius 1.00 m followed by a 90° uniform magnetic field. It was normally operated with a resolving power ($\Delta M/M$) of 100,000 to 150,000 measured at the base of the peaks. The reader is referred to detailed descriptions of the instrument in its early [5] and present [6] forms which have appeared elsewhere.

The determination of the mass difference between members of a doublet depends on an exact theorem described by Bleakney [7]. On alternate sweeps of the display oscilloscope, the potential (V) applied across the electrostatic analyser, and the source potential (V_a) are changed by the same fractional amounts to $V \pm \Delta V$ and $V_a \pm \Delta V_a$. For the proper values of ΔV and ΔV_a the trajectory of the second member of a doublet is displaced to the identical path previously described by the first member (before the voltage increments were applied) and thus appears at the same position on the oscilloscope screen. Adjusting the value of ΔV to obtain coincidence is the process known as "matching". When the members of the doublet are so matched, the mass difference, ΔM, is given by

$$\Delta M = M\frac{\Delta V}{V}$$

where M is the mass of the displaced ion.

*Work supported by the National Research Council of Canada.

In this instrument, the determination of the matched condition has been carried out by either of two techniques, both of which make use of a 1024-channel signal averager. The first of these is the "visual null method" of Benson and Johnson [8] as described in reference 6. Fig. 1 is

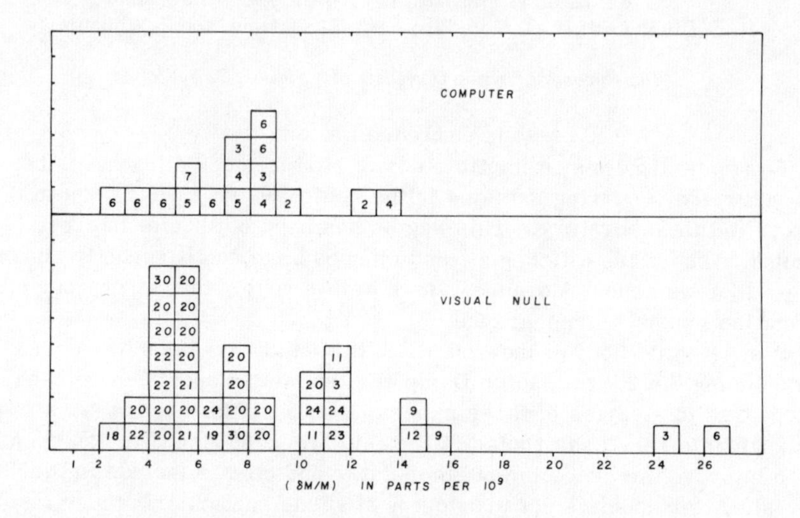

Fig. 1. Histogram of the precision obtained by the "visual null method." δM is the error on the doublet and M is the mass at which the difference was determined.

a histogram showing the precision ($\delta M/M$) obtained by this technique for the various doublets studied to date. As is evident, the typical precision is $\sim 5/10^9$.

A second technique of peak matching has recently been introduced on our instrument [9]. In this arrangement the 1024-channel memory of the signal averager is divided into four sections of 256-channels each. The sawtooth sweep is adjusted to coincide with the scan through a 256-channel quadrant.

As indicated in Fig. 2, $\Delta V = 0$ during the first scan and so the reference peak is stored in the first quadrant. During the second, third and fourth scans, voltages ΔV_1, ΔV_2 and ΔV_3 respectively are added to the electrostatic analyser voltage, V, and the other peak of the doublet is stored in the remaining three quadrants at three positions which bracket the matched condition. The whole cycle is repeated many times to improve the signal-to-noise ratio.

Usually ΔV_2 is very near the expected value while the ΔV_1 and ΔV_3 differ from it by ~ .1% to 1% (the displacements D_1 and D_3 are much exaggerated here). As shown here the ΔV_1, ΔV_2 ΔV_3 are in order of increasing size.

This latter order may be reversed, the direction of sweeping the ion beam across the collector slit may be changed and either of the doublet

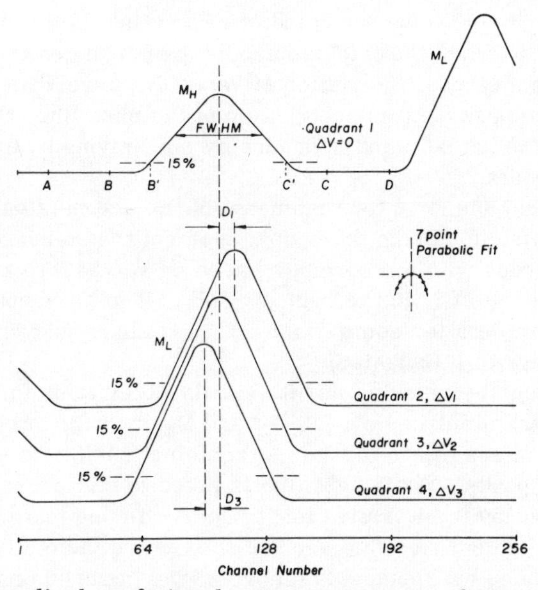

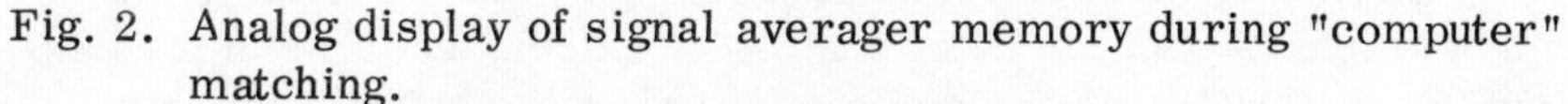

Fig. 2. Analog display of signal averager memory during "computer" matching.

members may be regarded as the reference peak (i.e., ΔV may be added to or subtracted from V). These arrangements are permuted to give 8 different matching configurations and the weighted average of the eight values so obtained is the value for one run.

In order to make maximum use of the information available, the peak of interest is adjusted to occupy as many channels as is convenient in the center of a quadrant, leaving about 30 channels at each end to establish a base line. The memory contents are read onto magnetic tape, then the matching configuration is changed, and the process repeated. As many complete runs as desired can be recorded with minor refocusing taking place between runs.

The analysis of the raw data is carried out off-line with the university IBM 360/65 computer as follows. The spectra are plotted by the line printer for visual inspection, and the points A, B, C, D are identified by eye and read into the computer via punched cards. The base line is calculated between A and B and between C and D for each quadrant and is subtracted from the peak between B and C.

The overlapping of tails from the peaks would have the effect of reducing the calculated separation. Moreover it would be desired to exclude from the calculations ions which have been significantly scattered. We have therefore followed the general lead of Stevens and Moreland [10] and have used only the part of the peak lying 15% or more above the base line.

As Campbell and Halliday [11] have shown, the fundamental limit on the precision of locating a peak is determined by the number of ions in the peak. The idealized shape of the peak here is triangular and the weighted mean, or centroid, is the best method of estimating the location. For a large number of ions, N, the triangle can be approximated by a normal distribution with a standard deviation of $W/\sqrt{24\,N}$ where W is the full width at the base of the peak. After subtraction of the base line, the peak height is converted to the actual number of ions by multiplying by a previously determined constant.

The centroid and first four moments of the distribution between B^1 and C^1 are calculated. From the second moment the standard deviation is derived; this agrees with the estimate based on $W/\sqrt{24\,N}$ to within a few percent when the cutoff is in the region ~ 15%. The third and fourth moments allow more detailed comparisons of peak shape (skewness and sharpness relative to a Gaussian).

After finding the centroid and its standard deviation for the peak in each quadrant, the displacement D_1, D_2, and D_3 are calculated. The errors in the D_i are d_i where d_i are the r.m.s. combinations of the errors in locating the two peaks. A straight line is fitted to the three pairs of points ($D_i \pm d_i$, $\Delta V_i \pm v_i$) by the iterative least squares fitting procedure of Williamson [12] which takes into account the errors in both coordinates.

As prescribed by Bleakney's theorem, the matched condition is given by the intercept of the fitted line. The calculated error in the intercept agrees very well with the error expected on the basis of the total number of ions in the match.

A series of experiments were carried out to determine if the linear fitting was justified. In each of the eight possible matching configurations a series of ten matches were carried out where a peak was matched to itself. That is, $\Delta V = 0$ and the amounts that the peaks in quadrants 2 and 4 were displaced from the reference peak were varied over a range much larger than that ever used in matching.

A straight line, then second and third order polynomials were fitted to the resulting data. For each of the eight configurations it was found that the quadratic term was three orders of magnitude smaller than the linear term and has the same sign for both positive and negative slopes. The cubic coefficient was 4 orders of magnitude smaller than the quadratic and had essentially no effect on the magnitude of the lower order coefficients.

In matching a peak to itself it was found that the peak in quadrant 1 was displaced by 0.18 ± 2 channels with respect to the other three quadrants. If uncorrected, this would leave the mean for a run unaffected but would result in an unnecessarily large error for that value. Accordingly, a correction is now applied before the fitting procedure is carried out.

Five well known doublets in the spectrum of $CdCl_2$ and $NdCl_2$ were studied to test the operation and precision of the system. In Table I we present the results for a typical doublet, giving the working resolving power (FWHM as measured from the plotted contents of the signal averager's memory), value of ΔM for the given run, the "internal" and "external" standard deviations as defined by Birge [13]. The "internal" standard deviation is the error expected on the basis of the errors calculated

TABLE I

$^{114}Cd^{35}Cl-^{112}Cd^{37}Cl$ $M/\Delta M \approx 42{,}000$

All Masses in Micro Units

Date	R.P. (FWHM)	ΔM	σ_{int}	σ_{ext}
March 17/71	163,000	3553.8	0.54	2.1
March 17	152,000	3547.6	0.25	1.2
March 18	156,000	3546.3	0.28	1.5
March 18	150,000	3545.3	0.27	2.8
March 22	145,000	3551.6	1.13	2.6
March 22	139,000	3549.9	0.71	2.1
March 25	176,000	3547.9	0.57	2.0
	Weighted Mean	3548.5 ±	1.0	cf. 3547.9 ± .9

for each match. The "external" standard deviation is often a factor of ~4 larger than the internal standard deviation. This means that differences between results from the eight configurations are larger than would be expected on the basis of the precision associated with each match. The internal error (~0.5 μu at M ~ 200 u) is the lower limit attainable with the number of ions collected, and the resolution used. The larger spread of the eight matches is not completely understood, but is probably due to changes in the spectrometer operation from match to match. At present there does not appear to be any systematic variation of the result with matching configuration.

In calculating the final weighted mean, σ_{ext} has been used and the final error quoted is the larger of the resulting σ_{ext} or σ_{int}. Both are about the same size, so that the assignment of error to a run appears realistic and the reproducibility of the spectrometer from run to run and from day to day is as good as the stability from match to match within a given run.

This technique is now being used routinely in our study of atomic mass differences in the rare earths and in the region immediately above the rare earths.

3. Determination of Relatively Wider Mass Doublets

The atomic mass differences which have been determined by this group over the past several years have all involved doublets for which the spacings were less than ~ 1/20,000. Inasmuch as almost all of these doublets have involved the $^{37}Cl - ^{35}Cl$ mass difference, several determinations of this difference have been undertaken since the 1964 Mass Table was published [8, 14, 15, 16]. We felt it would be desirable to contribute an additional value via the

$$^{12}C\,^{35}Cl\,D_2 - ^{12}C\,^{37}Cl\,H_2$$

doublet which is obtained from methylene chloride and has a spacing, $\Delta M/M \sim 1/3{,}300$.

In preparation for this determination, the $^{37}Cl - {}^{35}Cl$ difference was studied as a doublet of width 2 u at various mass numbers in the region 100 to 250 u. This work revealed the presence of a systematic error which was consistently negative and which behaved in the same general fashion as that encountered by Stevens and Moreland [10]. In our instrument the error appears to arise as a result of surface potentials on the electrostatic analyser plates. The formation and decay of such potentials as a result of low intensity ion or electron bombardment have been studied by Petit-Clerc and Carette [17] at Laval University and appear to be consistent with the nature of the error observed by us.

The electrostatic analyser was cleaned with ether, absolute alcohol and distilled water and, in the early stage of pumpdown, with a Tesla discharge in Ar. Subsequently the error was found to be ~ 0 ppm; thereafter it slowly increased to ~ 100 ppm over a period of a few weeks.

Accordingly, the results obtained for the narrow chlorine doublet mentioned earlier were corrected by a factor determined by a 1 u calibration doublet (the ^{1}H mass) which was measured immediately before or after the narrow doublet.

In the upper part of Table II are given calculated values (where possible) for the doublet which was measured and the new value from our laboratory. The high degree of precision associated with the last two values and the agreement between them is noteworthy, particularly in the light of the fundamental differences between the two instruments used.

In the lower half of the table we give the values of $^{37}Cl - {}^{35}Cl$ mass difference which, in the first four cases, is calculated from the measured mass of ^{37}Cl and ^{35}Cl. The value from this work is calculated from the doublet separation and from the weighted averages of all experimental values for the D and H masses.

4. Rare Earth Mass Differences

The rare earth mass differences which have been determined by this research group both at McMaster University and the University of Manitoba are of the types

$$^{A}X - {}^{A}Y = \Delta_1 \qquad (1)$$

$$^{A+2}X^{35}Cl - {}^{A}Y^{37}Cl = \Delta_2 \qquad (2)$$

$$^{A+2}X^{35}Cl_2 - {}^{A}Y^{37}Cl_2 = \Delta_3 \qquad (3)$$

where X and Y may or may not be the same element. Thus mass connections are obtained between nuclides differing in A by 0, 2 or 4 units. In Fig. 3 the new mass connections are indicated by solid lines with the number indicating the size of the error in keV. Values obtained with the 2.7 m radius instrument at McMaster are indicated by broken lines. (All naturally-occurring nuclides are shown.) As is evident from the figure the new data are generally more precise (0.7 keV to 4.2 keV) than those obtained with the older instrument and provide many connections not previously

TABLE II

$^{12}C\,^{35}Cl\,^{2}D_2 - {}^{12}C\,^{37}Cl\,H_2$ (μu)			Reference
1964 Mass Table	15 506.66	± 140	18
Benson and Johnson (1966, 1968)	15 504.30	± 60	8, 19
Stevens and Moreland (1968, 1970)	15 503.59	± 13	10, 15
Smith (1971)	15 503.774	± 65	16
This work (1971)	15 503.796	± 91	
$^{37}Cl - {}^{35}Cl$			
1964 Mass Table	1.997 047 4	± 14	18
Benson and Johnson (1966)	1.997 049 7	± 6	8
Dewdney and Bainbridge (1965)	1.997 048 89	± 59	14
Stevens and Moreland (1970)	1.997 050 59	± 42	15
Smith (1971)	1.997 049 711	± 63	16
This work*	1.997 049 730	± 98	

*using experimental values quoted in ref. 16 to get the following weighted averages:

H = 1.007 825 034 ± 12 u D = 2.014 101 797 ± 28 u

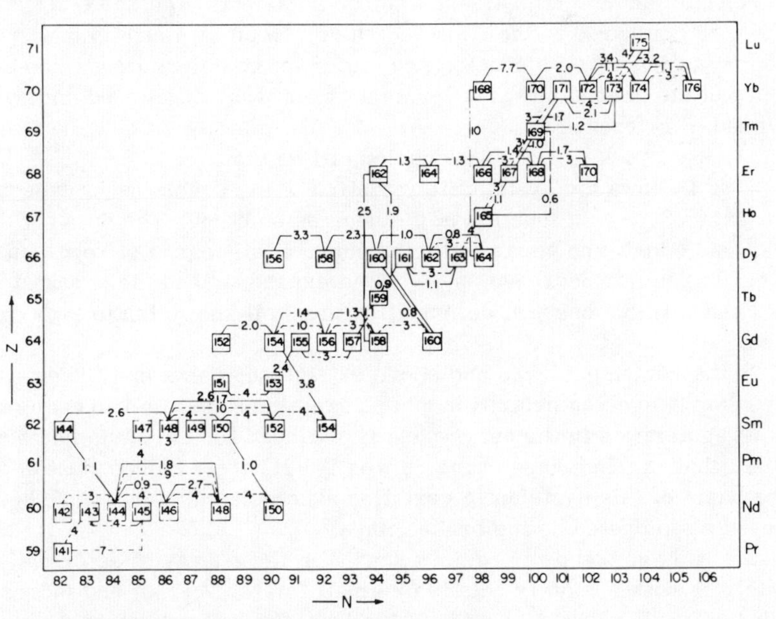

Fig. 3. Mass connections provided by the data tabulated in references 1, 3, and 20.

determined. The actual values for the mass differences (except for Yb and Lu) are presented in a paper which has been submitted to the Canadian Journal of Physics [20] along with a detailed comparison with other mass spectroscopic data and the 1964 Mass Table.

As there is a substantial, but incomplete, body of nuclear decay and reaction Q-values in this region it is desirable to combine the two kinds of data to obtain "best" values for the mass differences throughout this region. The only other mass spectroscopic data with sufficiently significant weights are those from the University of Minnesota [8] involving the Nd masses. In order best to combine these data, we have performed a least squares adjustment [21] following the general procedure outlined by Mattauch [22] and his coworkers. The region in which the mass spectroscopic values lie, has been divided arbitrarily into two regions for the adjustment; the first covers the region $59 \leq Z \leq 69$ and the second $67 \leq Z \leq 72$. The details of these calculations, including the input data, and a discussion of the results will appear in the near future [21, 23]. We wish, however, to draw attention to certain features of the work at this time.

In the first adjustment, one Q-value was eliminated from the beginning on the basis of gross inconsistency with related data, namely the ^{153}Gd (e.c.)^{153}Eu decay Q-value which was low by a factor of ~ 2. Of the 167 nuclides in the region, 83 form an overdetermined set in which there are 143 mass differences subject to 53 closed loop constraints. When this smaller problem was solved, it was found that the ^{148}Nd (d, p) ^{149}Nd and ^{150}Nd (d, t) ^{149}Nd reaction Q-values made major contributions to χ^2. As was previously done in the Nuclear Data Group adjustment, these values were rejected and the adjustment was recalculated. The value of $\sqrt{\chi^2/f} = .95$ compared to the expected value, 1.000 ± 75 and so it was unnecessary to introduce a consistency factor in the calculation. The consistency of the second and smaller adjustment was not as good and will be discussed in reference 23.

4. Discussion of Results

Fig. 4 shows the systematic variation of the double neutron separation energy, S_{2n}, as a function of the neutron number. The data for even-N and odd-N are separated to emphasize the degree of regularity that exists. The major decrease as N = 82 is exceeded, and the break at N = 88 associated with the onset of deformation are well known features of these curves.

In the region below 88 and above 92, the segments of adjacent curves between any two given neutron numbers are almost parallel to each other. Thus irregularities in the curves are reproduced for the same neutron number, that is, the shape of the curves is relatively independent of Z. On the basis of this systematic variation extrapolated curves have been proposed as indicated by the broken lines.

As we have noted in previous work [24] the energy of deformation for nuclides with N = 90 is relatively small for Nd (Z = 60) and increases in going to Sm (Z = 62) and Gd (Z = 64). This figure further suggests that at N = 88, Dy (Z = 66) may already have a ground state deformation inasmuch

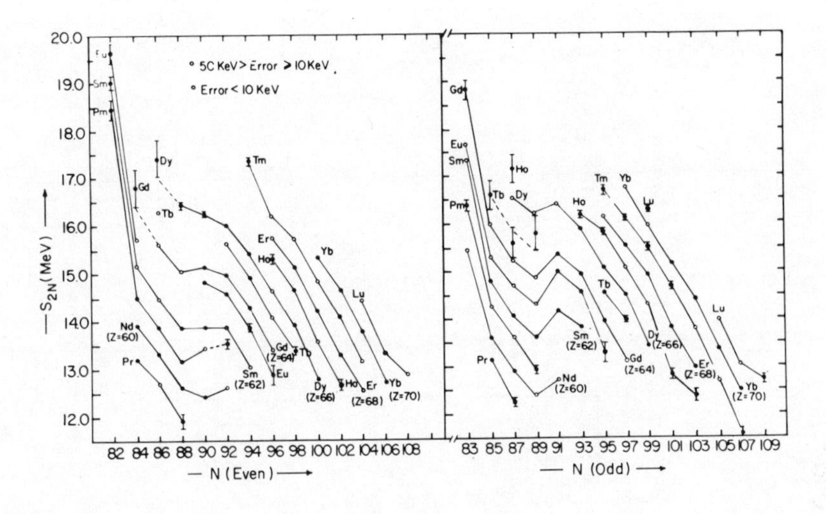

Fig. 4. Systematic variation of S_{2n} vs. N.

as that point lies somewhat higher than would otherwise be expected. That is, as Z increases, the effect due to nuclear distortion occurs somewhat earlier. Some support for this comes from the change in the slope of the N = 89-91 segments of the odd-N curves with increasing Z. For both sets of curves it appears that when N ~ 92 is reached the nuclides have approximately the same energy of deformation.

The single-neutron separation energies, S_n, are shown as a function of N in Fig. 5. The nature of the discontinuity in the region of deformation (N ~ 90) has previously been commented on [1] for both even- and odd-Z, as well as the pairing effect, as manifested in the displacement of the odd-Z curves in each of the plots (i.e., for even-N and odd-N). It should be noted that, for odd-odd nuclides, the neutron-proton pairing appears to occur in the region $N \geq 85$. For even-N nuclides which are distorted, this does not appear to be significant although it does occur for the region below N = 88.

The S_n plots again suggest that in odd-N nuclides the distortion may not be as large as in even-N nuclides, but continues to increase beyond N = 92. Again, the even-N plot suggests that at N = 88, Dy has already acquired some distortion.

The plots of the pairing energy, P_n, and the proton separation energies, S_{2p} and S_p, will be presented elsewhere [21, 23]. For the latter two quantities this work confirms the lack of dramatic structure [1, 3].

Although not presented here, a plot of P_p vs. Z does not exhibit any systematic behaviour. However, when we plot P_p vs. N a systematic change with the addition of neutrons is evident. P_p has a relatively large value which increases somewhat until the addition of neutrons precipitates deformation. Thereafter there is a drastic decline until N = 92 at which

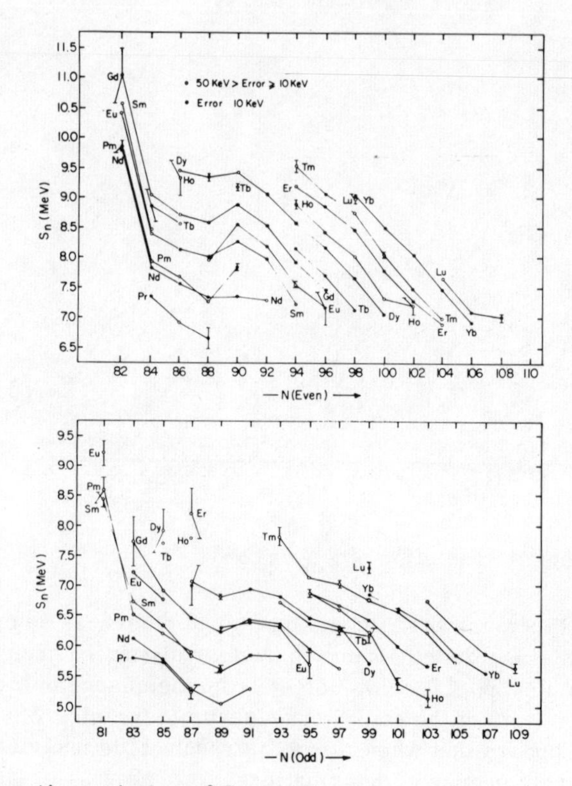

Fig. 5. Systematic variation of S_n vs. N.

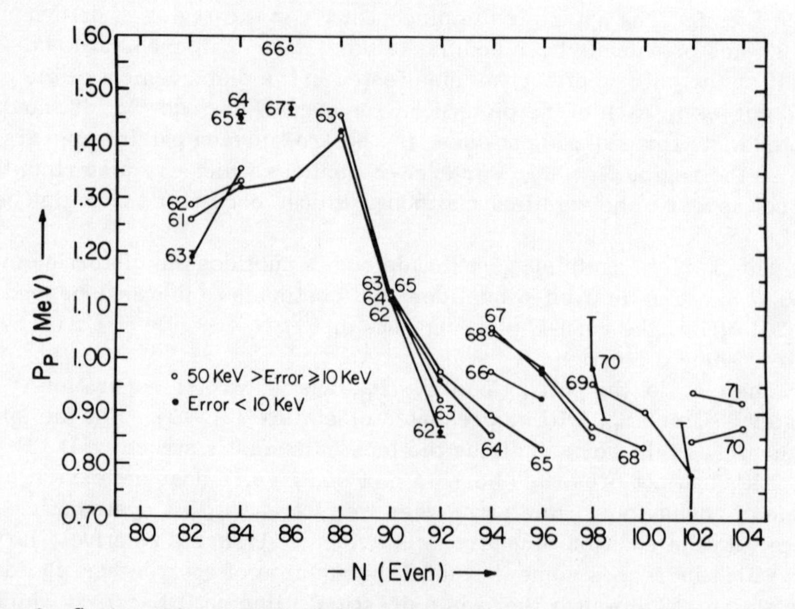

Fig. 6. Systematic variation of P_p vs. N.

point some dependence of Z begins to reappear, although the values all lie much lower than those for the undeformed nuclides.

References

1. MACDOUGALL, J. D., MCLATCHIE, W., WHINERAY, S. and DUCKWORTH, H. E., Nucl. Phys. A145, 223 (1970).
2. MCLATCHIE, W., WHINERAY, S., MACDOUGALL, J. D. and DUCKWORTH, H. E., Nucl. Phys. A145, 244 (1970).
3. WHINERAY, S., MACDOUGALL, J. D., MCLATCHIE, W., and DUCKWORTH, H. E., Nucl. Phys. A151, 377 (1970).
4. HINTENBERGER, H. and KONIG, L. A., Advances in Mass Spectroscopy (Pergamon Press, London, 1960), p. 16.
5. BARBER, R. C., MEREDITH, J. O., BISHOP, R. L., DUCKWORTH, H. E., KETTNER, M. E. and VAN ROOKHUYZEN, P., Proceedings of the Third International Conference on Atomic Masses (University of Manitoba Press, Winnipeg, 1968), p. 717.
6. BARBER, R. C., BISHOP, R. L., DUCKWORTH, H. E., MEREDITH, J. O., SOUTHON, F. C. G., VAN ROOKHUYZEN, P., and WILLIAMS, P., Rev. Sci. Instrum. 42, 1 (1971).
7. BLEAKNEY, W., Amer. Phys. Teacher, 4, 12 (1936).
8. BENSON, J. L. and JOHNSON, W. H., JR., Phys. Rev. 141, 1112 (1966).
9. MEREDITH, J. O., Ph. D. Thesis, University of Manitoba (unpublished, 1971).
10. STEVENS, C. M. and MORELAND, P. E., Proceedings of the Third International Conference on Atomic Masses (University of Manitoba Press, Winnipeg, 1968), p. 673.
11. CAMPBELL, A. J. and HALLIDAY, J. S., 13th Annual Conference on Spectroscopy and Allied Topics (1965).
12. WILLIAMSON, J. H., Can. J. Phys. 46, 1845 (1968).
13. BIRGE, R. T., Phys. Rev. 40, 207 (1932).
14. DEWDNEY, J. W. and BAINBRIDGE, K. T., Phys. Rev. 138, 540 (1965).
15. STEVENS, C. M. and MORELAND, P. E., Recent Developments in Mass Spectroscopy (University Park Press, Tokyo, 1970), p. 1296.
16. SMITH, L. G., Phys. Rev. C4, 22 (1971).
17. PETIT-CLERC, Y., CARETTE, J. D., Vacuum 18, 7 (1968).
18. MATTAUCH, J. H. E., THIELE, W., and WAPSTRA, A. H., Nucl. Phys. 67, 1 (1965).
19. JOHNSON, W. H., HUDSON, M. C., BRITTEN, R. A., and KAYSER, D. C., Proceedings of the Third International Conference on Atomic Masses (University of Manitoba Press, Winnipeg, 1968), p, 793.
20. BARBER, R. C., BISHOP, R. L., MEREDITH, J. O., SOUTHON, F. C. G., WILLIAMS, P., DUCKWORTH, H. E., and VAN ROOKHUYZEN, P. to be published.
21. MEREDITH, J. O., BARBER, R. C., and DUCKWORTH, H. E., to be published.

22. MATTAUCH, J. H. E., Proceedings of the International Conference on Nuclidic Masses (University of Toronto Press, Toronto, 1960), p. 3.
23. MEREDITH, J. O., SOUTHON, F. C. G., BURREL, D. A., BARBER, R. C., DUCKWORTH, H. E., and WILLIAMS, P., to be published.
24. DUCKWORTH, H. E., BARBER, R. C., VAN ROOKHUYZEN, P., MACDOUGALL, J. D., MCLATCHIE, W., WHINERAY, S., BISHOP, R. L., MEREDITH, J. O., WILLIAMS, P., SOUTHON, G., WONG, W., HOGG, B. G., and KETTNER, M. E., Phys. Rev. Letters, 23 592 (1969).

Recent Determination of the Atomic Masses of H, D, ^{35}Cl and ^{37}Cl at Osaka University

I. Katakuse and K. Ogata

Department of Physics, Faculty of Science, Osaka University, Toyonaka, Japan

At the days of the Winnipeg Conference in 1967, we had some difficulties in the focussing characteristics of our reconstructed large mass spectrograph. The cause of the difficulties had not been found out as long as the machine was operated under the graphic mode. In changing from graphic to metric, it became clear that the main cause of this difficulty was the stray a.c. magnetic field filling in the room [1]. Since then, we have been operating the machine in the mass-spectrometric mode, and for the measurement of doublet mass differences the peak-matching techniques have been adopted.

After finishing the focuss-adjustment in 1968, the reliability of the measuring procedures was checked using the doublets at mass 28. First, serious discrepancies were found between mass differences measured under different conditions. They were overcome by moving the matching-circuit panel as far as possible from the ion-path. The details of this consistency check were already reported at the 1969 Kyoto Conference. In this case, the orthodox visual peakmatching techniques were used, and the mass differences ($^{12}C_2H_4$-$^{12}C^{16}O$), ($^{12}C_2H_4$-$^{14}N_2$) and ($^{14}N_2$-$^{12}C^{16}O$) were determined[2].

A program was then started for the determination of the masses of the important mass-substandards H, D and ^{35,37}Cl. As an introduction, the effects of the peak-shapes of the doublet components on the measured mass differences were checked. If the peak-shapes are asymmetric and different, the measured mass differences depend on the definition of the position of the peak.

The following methods were used in this check since visual matching then can not be applied. Digital pulse-counting techniques were used with a 400-channels memory-oscilloscope, and each peak of the doublets to be measured, for a set value ΔR, was accumulated into a set of 200 channels. The doublet scanning speed was about 3 cycles/sec., and 10 to 40 cycles were used depending on the intensity of ion current.

The peak-shape can be expressed by the step function $f(x)$, being the pulse numbers accumulated in each channel x during the total scanning. The positions of representative points of each peak (expressed by channel number) were calculated according to the following five relations:

(I) $G_1 = \frac{\Sigma x f(x)}{\Sigma f(x)}$,

(II) $G_2 = \frac{\Sigma x f^2(x)}{\Sigma f^2(x)}$,

(III) $G_3 = \frac{\Sigma x f^3(x)}{\Sigma f^3(x)}$,

(IV) S : the channel number corresponding to the half-area of a peak (median),

(V) G : the channel number corresponding to the tope of Gaussian curve fitted to a peak (Gaussian method).

The mismatch between two peaks for a certain value of the matching-resistance ΔR is given by the difference between the corresponding representative points of each peak; for instance, $\Delta G_1 = G_1(M) - G_1(M+\Delta M)$, $\Delta G_2 = G_2(M) - G_2(M+\Delta M)$, and so on.

A method different from the above methods using the representative points, here indicated "parabolic method" starts by defining a function $A(\Delta i)$ as follows:

$$A(\Delta i) = \Sigma\{f_1(x) - f_2(x - \Delta I + \Delta i)\}^2$$

Here, $f_1(x)$ and $f_2(x)$ are the peak-shape functions corresponding to the doublet components of the masses M and $M+\Delta M$ respectively, ΔI is the mismatch and Δi is a small variable chosen arbitrarily for finding out the matching grade. If the mismatch is small, the above function $A(\Delta i)$ can be reduced to the following form by neglecting the terms of the higher order of Δi:

$$A(\Delta i) = \Sigma\{f_1(x) - f_2(x)\}^2 + \Sigma f'^2_2(x) \cdot (\Delta i - \Delta I)^2.$$

Thus, $A(\Delta i)$ is a parabolic function of (Δi) with a minimum for Δi equal to the mismatch ΔI at the chosen ΔR-value.

Of the above six methods used for calculating the mismatches of the doublet components, the first four methods belong to the category of "peak-top matching", and the Gaussian- and parabolic-method of the "whole-peak matching".

Fig. 1 shows that in the present case the mismatches as calculated in these six ways are in rather good agreement with each other, and the peak-shape asymmetry seems not to have so serious effect[3].

In order to find out the matching-resistance (ΔR) for the doublet components to be measured, the mismatches are plotted as a function of ΔR (fig. 2a). The matching resistance, corresponding to mismatch zero, is determined in a least squares calculation. The differences as obtained with different methods are shown in fig. 2b, whereas the small difference between the actual curves and Gaussian shapes is illustrated in fig. 2c.

Table I

Measured Mass Differences

Doublet	Mass No.	No. of measurement	Mass Difference (μu)	
$C_3-H^{35}Cl$	36	13	23,322.328±0.325	
$C_2D_6-H^{35}Cl$	36	11	107,933.422±0.538	
$C_3-C_2D_6$	36	10	84,611.730±0.411	
$C_3H_2-H^{37}Cl$	38	14	41,922.176±0.305	
$^{37}Cl-D^{35}Cl$	37	14	17,051.816±0.185	
$C_2H_4-C_2D_2$	28	16	3,096.444±0.127	
$C_3H_8-C_3D_4$	44	8	6,192.705±0.265	
$C_3-H^{35}Cl$	36	6	23,321.830±0.630	ref. (3)
$C_3-C_2D_6$	36	4	84,611.600±0.340	ref. (3)

Table II

Zero-cycle check at Mass 36

Doublet	Mass difference (μu)	Closure Error	
$C_2D_6-C_3$	84,611.60±0.34		ref. (3)
$C_2D_6-H^{35}Cl$	107,934.90±0.54		ref. (3)
$C_3-H^{35}Cl$	23,321.83±0.63	1.47μu	ref. (3)
$C_2D_6-C_3$	84,611.730±0.411		
$C_2D_6-H^{35}Cl$	107,933.422±0.538		
$C_3-H^{35}Cl$	23,322.328±0.325	-0.636μu	

The above checks show that the peak-shapes of doublet components are nearly similar and symmetric in the present case. It therefore suffices to take the median (S) of each peak as the representative points of the peaks. As the results of these consistency checks for the measuring procedures, the doublet mass difference measurements have been carried out by the "median method" in the present measurements.

For the purpose of determining the masses of H,D and $^{35,37}Cl$, the following seven mass differences were measured: (C_3-$H^{35}Cl$), (C_2D_6-$H^{35}Cl$) and (C_2D_6-C_3) at mass 36; ($D^{35}Cl$-^{37}Cl) at 37; (C_3H_2-$H^{37}Cl$) at 38; (C_2H_4-C_2D_2) at mass 28; (C_3H_8-C_3D_4) at mass 44.

During these measurements, ($\frac{\Delta R}{R}$)-value calibrations were carried out using one hydrogen intervals, especially when samples were changed. The measurements have been done with the half-height resolution of about 400,000 to 500,000. The widths of the object- and image-slits are both about 10μ, and those of the α and β slits are both about 0.1mm. The heights of the object- and image-slits are 0.3mm and 0.5mm respectively. The image broadning caused by second order aberrations is roughly estimated to be at most about 0.3μu. This value is completely within the statistical deviations of the measured mass differences in each run. So in the present case, the image aberration effect can be neglected.

The mass differences of nine doublets were determined. The results obtained are tabulated in the Table I. The given mass differences are the weighted means calculated from the doublet mass differences determined in some ten individual runs. As to the weights, the statistical errors for each run were used. In this table, the mass difference of (C_2D_6-$H^{35}Cl$) doublet measured in last year is rejected since judged to be in error by the zero-cycle check shown in the Table II[3].

The daily consistency has been checked for each doublet, an example is shown in Fig. 3.

Masses of H, D and $^{35,37}Cl$ were calculated from these doublet mass differences in a least square treatment, using the errors attached to each doublet for calculating their weights. The masses thus determined are shown in the Table III. The errors in the external-column are those obtained in the least square calculations, and the errors in internal-column are those calculated by multiplying the inverse ratio of $R_e/R_i = \{\frac{\Sigma(pvv)}{N-n}\}^{\frac{1}{2}} = (0.254)^{\frac{1}{2}}$ to each external error.

Table III

Mass Excesses of H, D, ^{35}Cl and ^{37}Cl

	Mass Excess (μu)	External Error	Internal Error
H	7,825.065	0.015	0.030
D	14,101.930	0.020	0.040
^{35}Cl	-31,147.21	0.10	0.20
^{37}Cl	-34,097.10	0.11	0.22

The accuracy of the doublet mass differences is estimated between a few times 10^{-5} and several times 10^{-6}, depending on the doublet separations. The accuracy of the atomic masses determined is estimated to be 1 - 4 x 10^{-8}.

Using these masses, the masses of ^{14}N and ^{16}O were calculated from the doublet mass differences of C_2H_4-$^{14}N_2$= 25,152.44 ± 0.19 and C_2H_4-$C^{16}O$ = 36,386.01 ± 0.24 which were previously reported [2]. The results are: ^{14}N : 3,073.91 ± 0.14 and ^{16}O : -5,085.75 ± 0.25μu.

Fig. 4 shows a comparison between the present results those previously reported by others, including the 1964 table values.

The agreement for separate nuclides is not excellent, but the mass difference values for H_2-D and for the chlorine isotopes seem to be in rather good agreement with each other. It may be necessary to repeat these measurements, together with the investigations of systematic errors which might still be included in the atomic mass determinations.

1. OGATA, K., MATSUMOTO, S., NAKABUSHI, H., KATAKUSE, I., Proc. 3rd Intern. Conf. on Atomic Masses, (Ed. Barber, Univ. Toronto Press), p748(1967); Mass Spectroscopy (Japan), 16, 379(1968).

2. NAKABUSHI, H., KATAKUSE, I., OGATA, K., Recent Devel. in Mass Spectroscopy (Ed, Ogata, Hayakawa, Univ. Tokyo Press), p482(1970); Mass Spectroscopy (Japan), 17, 705(1969).

3. KATAKUSE, I., NAKABUSHI, H. OGATA, K., Mass Spectroscopy (Japan), 18, 1276(1970).

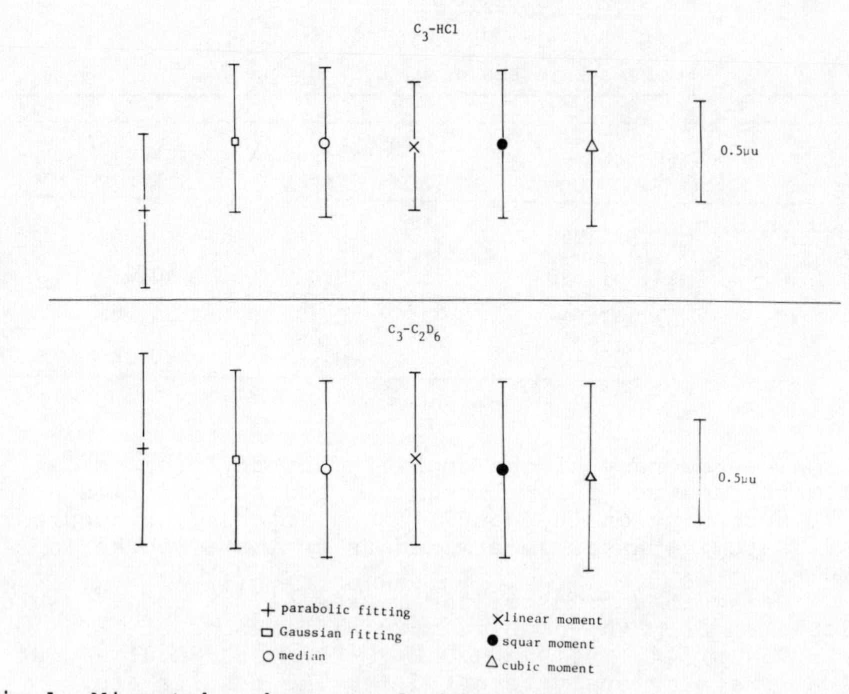

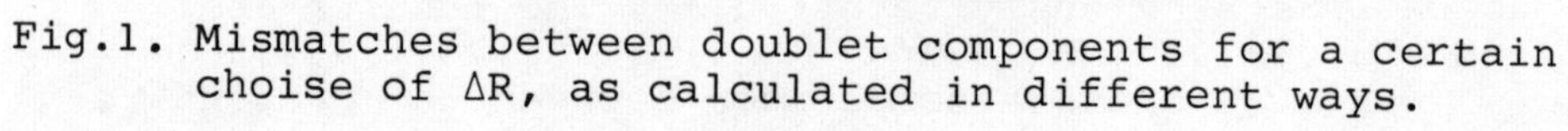
Fig.1. Mismatches between doublet components for a certain choise of ΔR, as calculated in different ways.

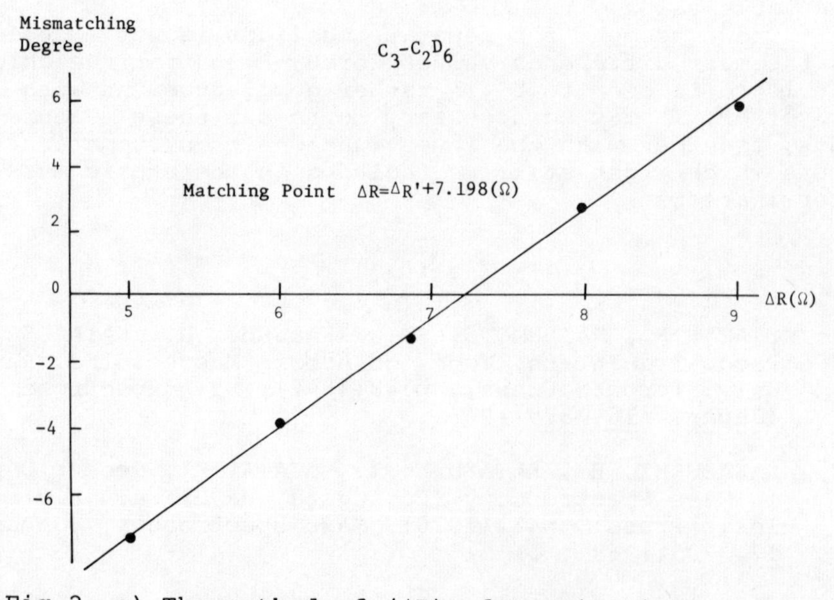

Fig.2. a) The method of (ΔR). determination.

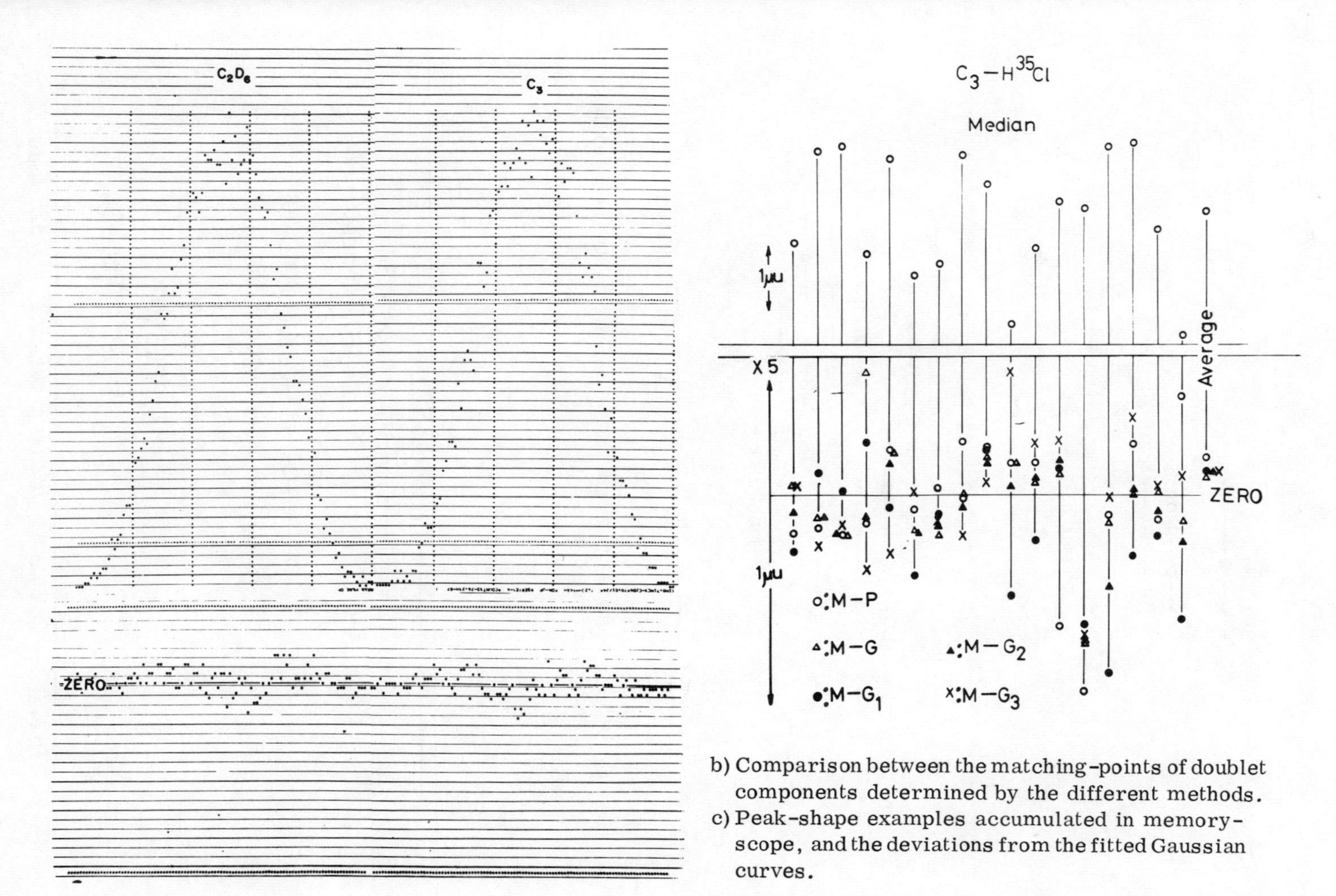

b) Comparison between the matching-points of doublet components determined by the different methods.

c) Peak-shape examples accumulated in memory-scope, and the deviations from the fitted Gaussian curves.

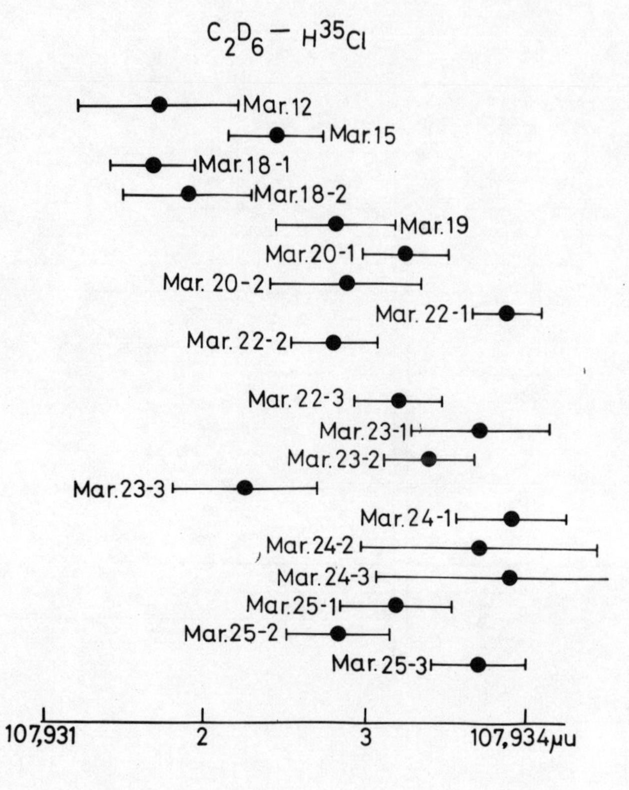

Fig.3. Daily consistensies between the measured doublet mass differences.

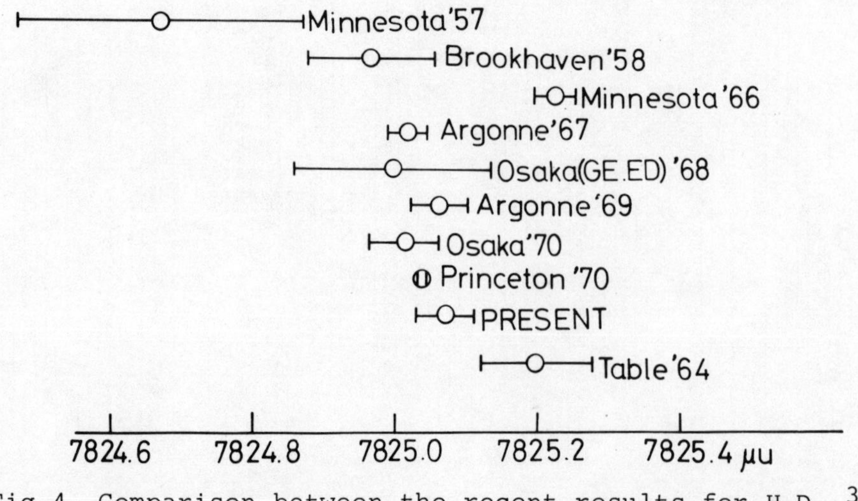

Fig.4. Comparison between the recent results for H,D, ^{35}Cl and ^{37}Cl masses.

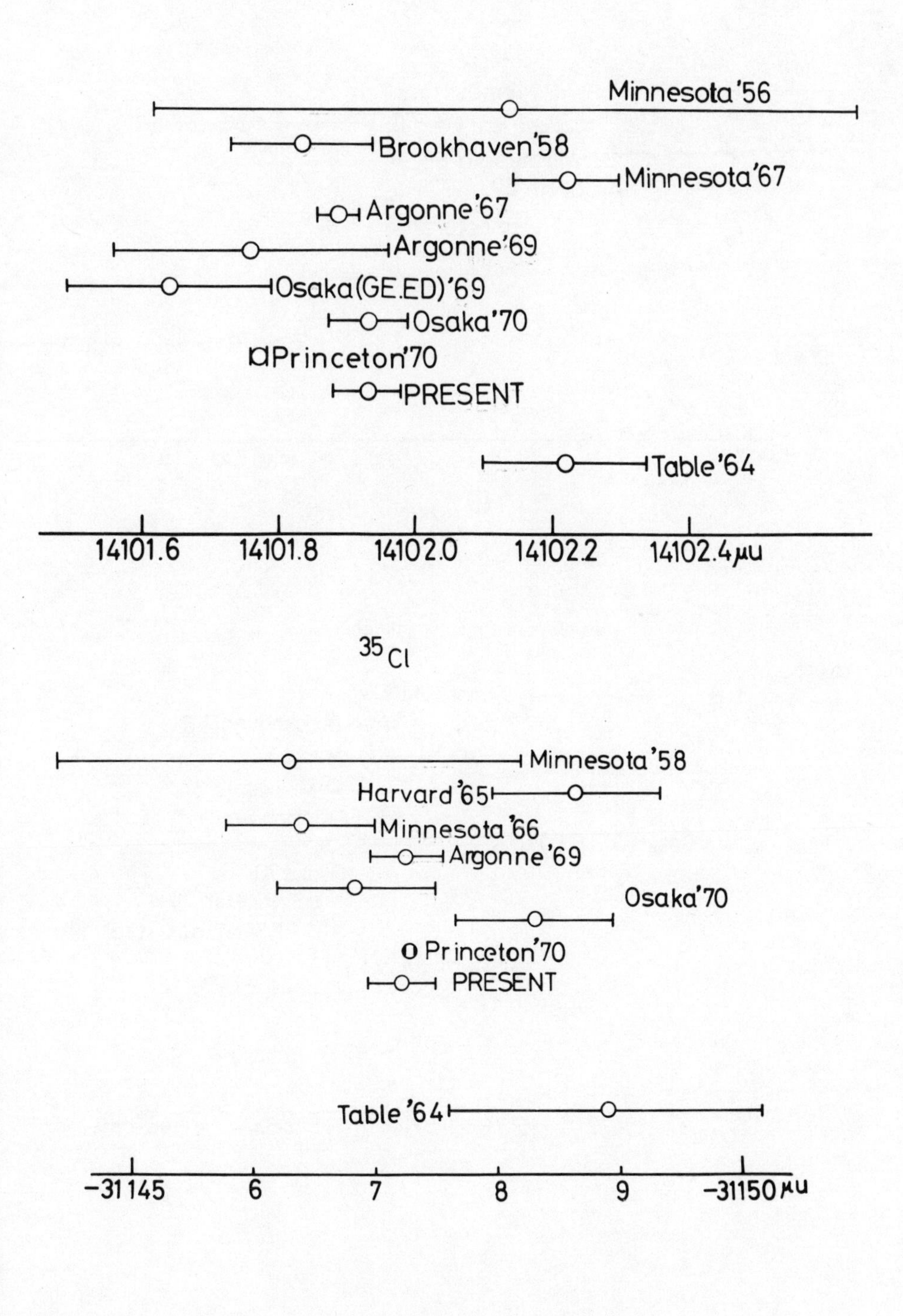
D
Minnesota'56
Brookhaven'58
Minnesota'67
Argonne'67
Argonne'69
Osaka(GE.ED)'69
Osaka'70
Princeton'70
PRESENT
Table'64
14101.6
14101.8
14102.0
14102.2
14102.4 μu
35 Cl
Minnesota'58
Harvard'65
Minnesota'66
Argonne'69
Osaka'70
Princeton'70
PRESENT
Table'64
−31145
6
7
8
9
−31150 μu

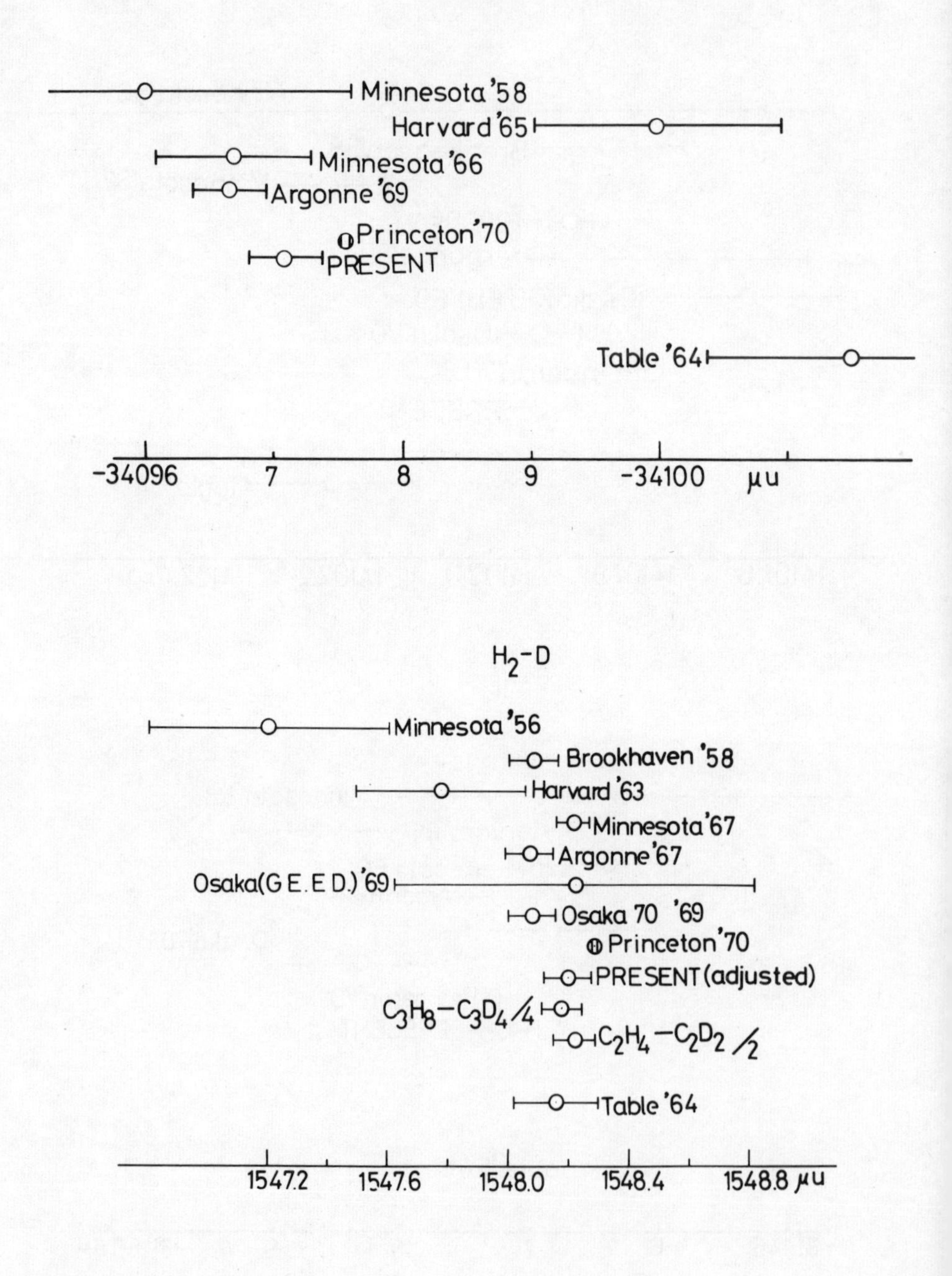

^{37}Cl
Minnesota '58
Harvard '65
Minnesota '66
Argonne '69
Princeton '70
PRESENT
Table '64
-34096
7
8
9
-34100
μu
H_2-D
Minnesota '56
Brookhaven '58
Harvard '63
Minnesota '67
Argonne '67
Osaka(G E. E D.) '69
Osaka 70 '69
Princeton '70
PRESENT (adjusted)
$C_3H_8-C_3D_4/4$
$C_2H_4-C_2D_2/2$
Table '64
1547.2
1547.6
1548.0
1548.4
1548.8 μu

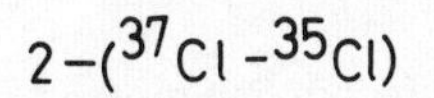

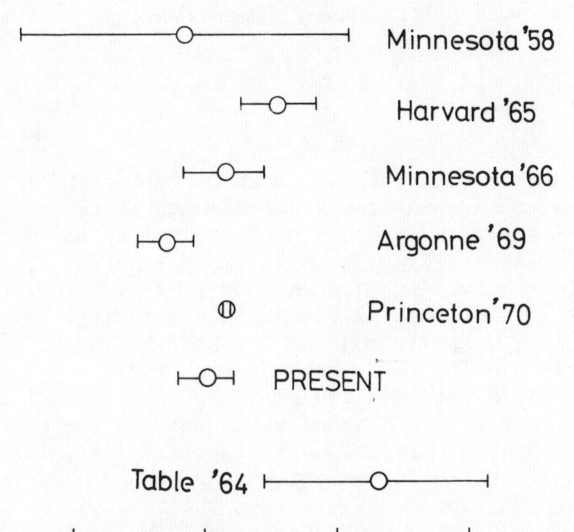

Minnesota'56: Phys. Rev. 102, 1071 (1956); Minnesota'57: Phys. Rev. 107, 1664 (1957); Brookhaven'58: Phys. Rev. 111, 1606 (1958): Minnesota'58: Phys. Rev. 110, 712 (1958); Harvard'65: Phys. Rev. 138, 540 (1965); Minnesota'66: Phys. Rev. 141, 1112 (1966); Argonne'67: Proc. Int. Conf. "Atomic Masses" p673 Winnipeg (1967); Minnesota'67: ibid. p793; Osaka (GE. ED.)'68: J. Phys. Soc. Japan 25, 950 (1968); Argonne'69: Proc. Int. Conf. "Mass Spectroscopy" p1296 Kyoto (1969); Osaka (Ge. ED.)'69 : ibid. p477; Osaka'70 Mass Spectroscopt (Tokyo) 18, 1276 (1970); Princeton'70: Private communication; Table'64: Nuclear Physics 67, 1 (1965)

Recent Precision Mass Measurements at Princeton

Lincoln G. Smith

Joseph Henry Laboratories, Princeton University

1. Introduction

A description of the principles of operation and essential elements of the rf spectrometer was presented in 1967(1). Further details of procedures, a discussion of precision, and measurements of sixteen doublets yielding the mass excesses of H, D, ^{14}N, ^{16}O, ^{35}Cl and ^{37}Cl were given in a recent paper (2). It has been shown: (a) that, with the technique of visual peak matching, the standard "setting" error of a single reading is about 1/2500 of the half-width of a peak; and (b) that the frequency ratio (f_2/f_1) may easily be determined within a few parts in 10^{10} so that no calibration uncertainty arises from equating this to the mass ratio (M_1/M_2) for the highest precision attainable. Thus, in the absence of other, greater sources of error, masses should be determinable for resolutions achieved ($2\text{-}4 \times 10^5$) from the mean of ten readings with an accuracy between 3 and 6 parts in 10^{10}. After elimination of a number of purely instrumental sources of error, the primary known remaining sources that limit precision are small electric fields that do not shift with the applied voltages in inverse proportion to the two masses being compared. Their presence is readily shown by the study of wide "calibration" doublets for which the differences in mass numbers are one or two units. Considerable progress in neutralizing the effects of these fields has been made by applying small adjustable d.c. voltages in series with the high voltage applied to the ion source and with the lesser voltages applied to one of each pair of external deflectors and to each internal deflector as explained in reference (2).

It is the purpose of this paper to present the results of further studies of these electric field effects as well as measurements of four doublets at mass numbers 3 and 4 whence have been obtained the mass excesses of ^4He, ^3He and T.

2. Instrument Modifications

Figs. 1 and 2, reproduced from reference (1), show the complete ion orbit from source to detector and, in somewhat more detail, the orbit in the magnetic field. Most of the essential elements of the instrument are shown. Not shown are axial focusing lenses traversed by the beam just before entering the modulator the first time and just after leaving it the second. Each is enclosed in a shield box along with a pair of deflecting plates to allow axial (vertical) bending of the beam of adjustable amount. Installed prior to the taking of the data of reference (2), these

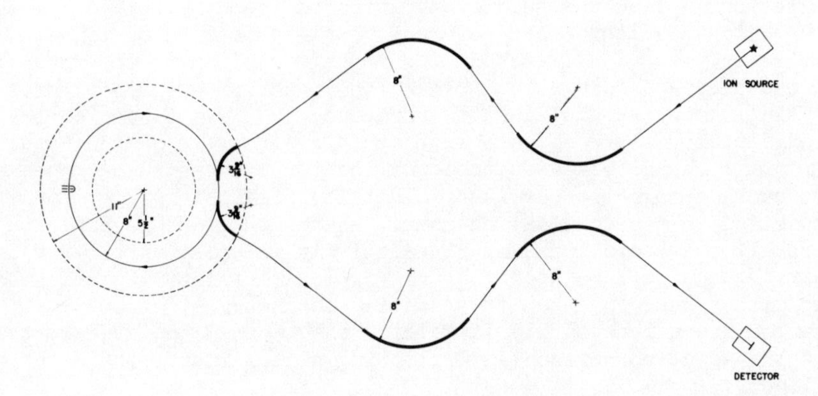

Fig. 1 Plan view of entire orbit

lenses and deflectors have enhanced transmission appreciably and resolution slightly.

A pair of vertical deflecting plates which allow adjustment of the pitch of the beam as it enters the magnet is shown in Fig. 2. A similar pair of plates is located inside the inner cylinder of the main accelerating lens, its midpoint being 1-1/16 inches from the output aperture of the source (presently a hole 0.040 inch in diameter). Its purpose is to allow vertical centering of the beam on the slit whereon it is focused by the main lens 8 inches from the source. Because of the two-directional focusing of the 8-inch spherical lenses, the effects on the pitch of the beam in the magnet of both these vertical deflectors are quite similar. A very large non-switched voltage on one plate of the pair near the magnet

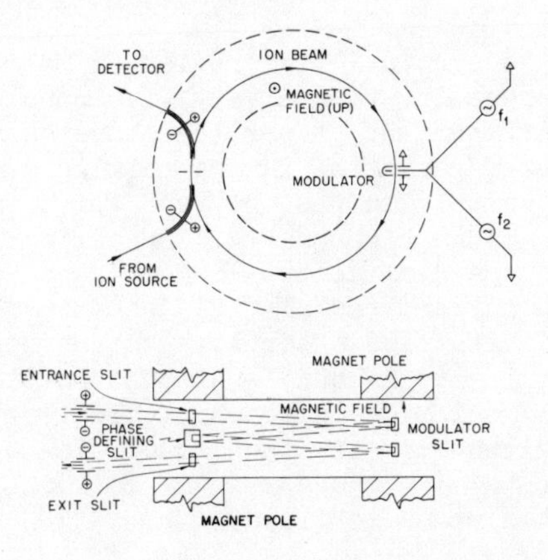

Fig. 2 Plan and elevation views of orbit in magnet

gap was often required to make peaks of a doublet vary in the same way as one varied the switched voltage on the other plate. This non-switched voltage changed considerably from time to time. As it was varied, the non-switched d.c. voltage applied to the inner jaw of the phase-defining slit to make the peaks of a wide calibration doublet match (as explained in reference (2)) was found to vary considerably. It would appear that this effect is caused by differential change of the pitches of the two ion beams and hence of the average magnetic (or stray electric) fields to which each is subjected in the toroidal chamber. These observations indicated strongly that some ions in the relatively intense, unresolved beam near the source were striking the vertical deflecting plates there and either charging them or producing secondary electrons that charged the unnecessarily large mica sheets used to insulate them and the adjacent horizontal deflecting plates inside the main lens cylinder.

Since the results of reference (2) were obtained, the mica has been replaced with small alumina washers presenting very little surface to the beam and a metal baffle containing a hole 0.100 inch in diameter has been placed in front of each pair of deflecting plates near the source so as to prevent ions from striking either. This change has virtually eliminated the need for any non-switched voltages on one vertical deflecting plate at the magnet gap and has considerably improved the matchings of peaks of wide doublets.

Another change made after the results of reference (2) were obtained was the introduction of two baffles in the modulator housing, each with a horizontal slot 0.050 inches high, so as to limit the accepted beam height to that amount just before it enters the modulator each time. These baffles should materially reduce the chance of any ions striking the housings of the injector or ejector lenses at glancing incidence. The reduction of intensity has not proven severe and resolution has been slightly increased by this change.

For making measurements in the region of masses 3 and 4, it is necessary to use for calibration a very wide doublet involving ions of masses 3 and 4 or 4 and 5. After the above changes were made, a considerably greater variation in the setting of the PD voltage (see reference (2)) with time was observed for such wide doublets than had been observed at high masses. Specifically it was found that the setting of the dial controlling the applied PD field decreased (i.e. the outward radial field became less positive and sometimes negative) with time after turn on approaching an assymptotic value in from 1/4 to 1 hour. This assymptotic setting was lower and apparently the time to reach it greater the greater the ion current. For given current, the dropoff in PD dial setting was greater the smaller the lower mass of the calibration doublet and hence the smaller the magnetic field.

After the data on doublets at masses 3 and 4 reported below were obtained, it was discovered that, if the rf voltage is kept below the critical value such that ions just reach the jaws of the phase-defining slit, no dropoff at all is observed in the PD dial setting for any wide doublet. A determination of the settings for the doublets with lower masses 3(^{4}He-HD), 4(DT-HT) and 18(CD_4-CD_3) was found to yield the same value as nearly as it could be determined, regardless of current, so long as no ions struck the phase-defining slit. This is regarded as a most promising result.

The cause of this effect, which is obviously unique with this instrument, is as yet unknown. If the surfaces of the slit jaws are charged by the beam striking them, the surface charges must be negative, which seems unreasonable on the grounds that the ions are positive and eject negative electrons and that the time required is long. It seems more likely that a few secondary electrons are migrating the 1 inch across the field to the insulators supporting the jaws thereby building up a radially outward field at a slow rate and to an extent that decreases as the magnetic field increases. These insulators will be thoroughly screened from the beam as soon as possible.

3. Measurements At Masses 3 And 4

In column 3 of Table 1 are given the results of measurements of one doublet at mass 3 and three at mass 4. All results have been corrected for differences in kinetic energy, chemical binding energy and ionization potential as were the results of reference (2). From 17 to 20 measurements, nearly all at 25 kv energy, were made of each doublet in at least four different runs on three or more days. A calibration run on D_2-HD for at least 1/2 hour was made before each run on HD-^{3}He at roughly the same conditions. Both D_2-HD and DT-D_2 were observed to determine the PD dial setting for the doublet D_2-HT. For the two wide doublets involving ^{4}He, either

TABLE I

Measured and adjusted (underlined) values in nu of doublets at masses 3 and 4

Mass No.	Doublet	Adjusted and Measured Values (nu)	Error Estimates (nu) σ_m	σ_{sm}	σ_e	σ_r
4	D_2-^{4}He	25,600,328±8 (underlined)				7.84 (underlined)
4	D_2-^{4}He	25,600,315±14	2.30	1.68	14.26	14.44
4	HT-^{4}He	21,271,070±8 (underlined)				7.76 (underlined)
4	HT-^{4}He	21,271,075±12	2.17	1.82	11.85	12.05
4	D_2-HT	4,329,257±3 (underlined)				2.73 (underlined)
4	D_2-HT	4,329,257±3	2.32	1.68	2.41	3.35
3	HD-^{3}HE	5,897,512±5	3.96	1.68	3.29	5.14

or both of the doublets DT-D_2 and DT-HT were used for calibration and particular care was taken to use the same current and other conditions for the narrower as for the calibration doublets.

For the triplet at mass 4, the closing error (17nu) of the three experimental doublet values, is large in comparison with the standard errors of the means (σ_m) of the 17-20 measurements. A least squares adjustment was performed to yield the three underlined values. It will be noted that the errors σ_m are only slightly greater than the "setting" errors (σ_{sm}) while the errors (σ_e) ascribed to uncertainties in the PD dial setting are considerably larger for the wide doublets involving ^{4}He. Each error σ_e corresponds to an uncertainty of 6.5 divisions in the PD dial setting or 0.033 volts in the potential applied to the inner jaw. This is the amount required to give the χ^2 of the adjustment its expected value of unity. It is 30% larger than the uncertainty in the PD setting assumed in reference (2). The same uncertainty of 6.5 divisions is assumed in determining σ_e for HD-^{3}He. Each standard error (σ_r) of a measured value is the resultant of σ_m and σ_e.

The doublet values are compared with previously measured values in Fig. 3. We note poor agreement with the mass synchrometer values of the two wide doublets D_2-^{4}He and HT-^{4}He. This indicates the presence of electric field effects in the older results obtained at much lower energy.

Combining the new results with values in reference (2), for

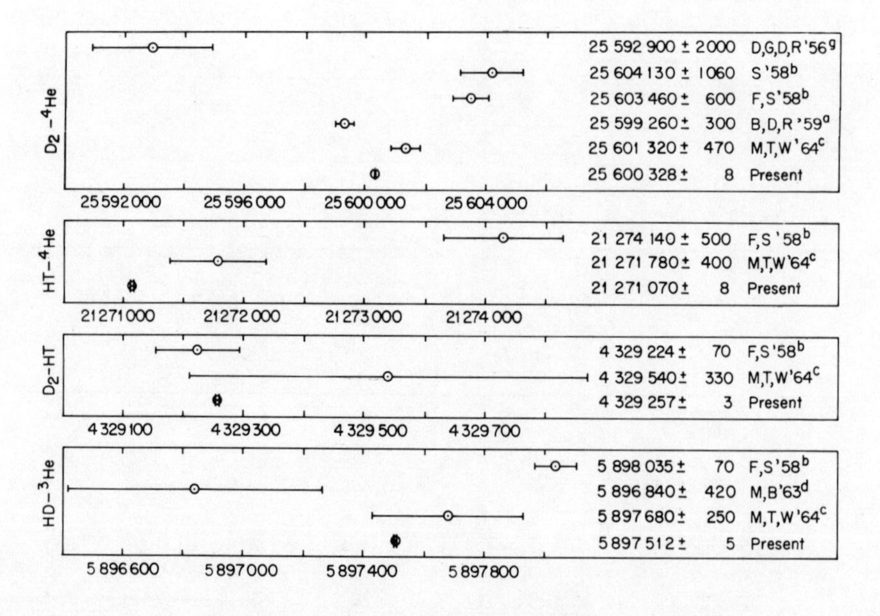

Fig. 3 Present adjusted and previously measured values of four doublets in nu

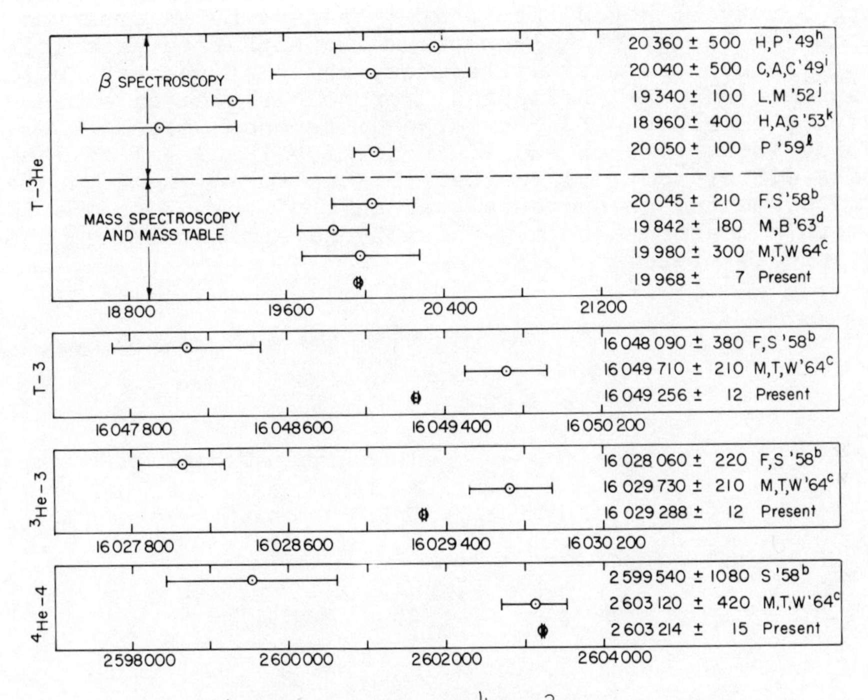

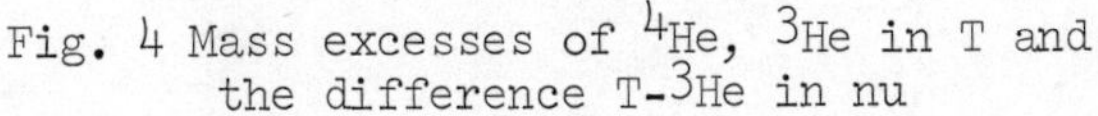
Fig. 4 Mass excesses of ^{4}He, ^{3}He in T and the difference T-^{3}He in nu

H-1, D-2 and H_2-D, one obtains the mass excesses shown in Fig. 4 which are also compared with previously measured values. Here we note large discrepancies with the Brookhaven results for all these mass excesses. However, for the small difference T-^{3}He, the Brookhaven results are in good agreement with the present one, while the directly measured value of Moreland and Bainbridge agrees somewhat less well. Both these other results have been corrected (in this case only) for differences in molecular binding energies and ionization potentials.

Values of T-^{3}He in nu derived from measurements of the extrapolated value(E_{extr}) of the β spectrum of tritium are shown at the top of Fig. 4. To convert these measured values to the mass difference of neutral atoms, the difference in energies required to strip all electrons from the two atoms has been subtracted from each value of E_{extr} (as it should be in all comparisons between atomic masses and Q-values). This difference in complete ionization potential is -65.4 volts or -70.2 nu. If one adds this amount to the present value of T-^{3}He and converts to volts, he obtains E_{extr}=18,534±7 volts.

Note added by editor.
Dr. Bergkvist noted that what is sensed by an observed endpoint energy of a β-spectrum is neither the atomic, nor the nuclear mass difference, and that the particular case of the ^3H - ^3He mass difference is treated extensively in his paper. The large difference between his value 18651 ± 16 eV and Smith's value 18609 ± 7 eV (atomic mass difference) can scarcely be due to an error in his measurements; he suggests that part of the $^3He^+$ ions in Smith's measurements may possibly have been in the metastable 2s state.

4. References

1. SMITH, L.G. in "Proceedings of the Third International Conference on Nuclidic Masses", R.C. Barber, Ed., Univ. of Manitoba Press, 1967, p. 811.
2. SMITH, L. G., Phys. Rev., C 4, 22 (1971).

(The following references are for Figs. 3 and 4.)

B, D, R '59[a] BENSON, J. L., et al., Phys. Rev., 113, 1105 (1959).

S '58[b] SMITH, L. G., Phys. Rev., 111, 1606 (1958).

F, S '58[b] FRIEDMAN, L. and SMITH, L. G., Phys. Rev., 109, 2214 (1958).

M, T, W '64[c] MATTAUCH, J. H. E., et al., Nucl. Phys., 67, 1 (1965).

M, B '63[d] MORELAND, P. E. and BAINBRIDGE, K. T., in "Nuclidic Masses, Proceedings of the Second International Conference", W. H. Johnson, Jr., Ed., Springer-Verlag, Vienna, 1964, p. 425.

D, G, D, R '56[g] DEMIRKHANOV, et al., Atomnaya Energ., 2, 21 (1956).

H, P '49[h] HANNA, G. L. and PONTECORVO, B., Phys. Rev., 75, 983 (1949).

C, A, C '49[i] CURRAN, S. C., et al., Phys. Rev., 76, 853 (1949).

L, M '52[j] LANGER, L. M. and MOFFAT, R. J. D., Phys. Rev., 88, 689 (1952).

H, A, G '53[k] HAMILTON, D. R., et al., Phys. Rev. 92, 1521 (1953).

P '59[l] PORTER, F. T., Phys. Rev., 115, 450 (1959).

Recent Minnesota Mass Results*

D. C. Kayser, R. A. Britten and W. H. Johnson

School of Physics and Astronomy,
University of Minnesota,
Minneapolis, Minnesota, U.S.A.

I wish to report on several features of the work of the Mass Measuring group at the University of Minnesota that have occurred since the 1967 Winnipeg conference. The main efforts recently have been in three areas; (a) the measurements of U^{235}, U^{238}, and Th^{232}, (b) a study of the feasibility of mass measurements of short half-life radioactive isotopes, and (c) development of a new technique for peak matching which may be used with only partially resolved doublets.

Because our present instrument was described in some detail in the 1967 conference proceedings, only a brief description will be given [1-2]. Our mass spectrometer is made up of a 16 inch radius, 60° magnetic analyzer and a 19.75 inch radius, 90° cylindrical electrostatic analyzer combined to produce a first and second order angle focusing and a first order energy focusing [3]. Mass doublets are measured using peak matching techniques. For some years we have used a signal averaging device to improve the signal-to-noise ratio in the measurement. The instrument operates routinely at a full width at half maximum resolution of 100,000 and in certain instances as high as 200,000.

The first part of this discussion will concern measurements made by Rodney Britten of the masses of U^{235}, U^{238}, and Th^{232}. Measurements of these masses were begun because there were no modern measurements available. During the course of these measurements, however, a paper by Kerr and Bainbridge describing mass differences involving isotopes of uranium and lead was published [4]. Before making comparisons with the measurements of Kerr and Bainbridge, let us first consider the doublets measured in this work. For each isotope, two separate doublets were measured. These are illustrated in Table I together with the preliminary doublet results. In order to check for proportional errors similar to those which appeared in earlier rare earth measurements, a set of doublets were measured which result in a series of closed loops. These are illustrated in Table II. The first three doublets check the closure of metal –metal doublets. As can be seen the closure as shown by the sum of the first two minus the third, yields a satisfactory result. The $Cl^{37}-Cl^{35}$ mass difference determined from this over-determined set is about 20

*Supported in part by Contract N00014-67-A-0113-0018 with the Office of Naval Research and a grant from the Graduate School Research Fund.

TABLE I

Preliminary values of measured doublets

Doublet	Δm in u
$C_9H_8 - \frac{1}{2}Th^{232}$	0.043 570 ± 1
$C_{12}H_8 - \frac{1}{2}Th^{232}Cl^{37}Cl^{35}$	0.076 195 ± 1
$\frac{1}{2}U^{235} - C_9H_9$	0.451 534 ± 1
$C_9H_{10} - \frac{1}{2}U^{235}$	0.556 293 ± 1
$C_9H_{11} - \frac{1}{2}U^{238}$	0.060 683 ± 1
$C_{12}H_{10} - \frac{1}{2}U^{238}Cl_2^{35}$	0.084 005 ± 1

TABLE II

Closed loops

Doublet	Δm in u
$\frac{1}{2}U^{238}Cl_2^{37} - \frac{1}{2}U\ Cl^{37}Cl^{35}$	0.998 541 ± 4
$\frac{1}{2}U^{238}Cl^{37}Cl^{35} - \frac{1}{2}U^{238}Cl_2^{35}$	0.998 545 ± 4
$\frac{1}{2}U^{238}Cl_2^{37} - \frac{1}{2}U^{238}Cl_2^{35}$	1.997 091 ± 4
Closure	−5 ± 7
Resultant $Cl^{37} - Cl^{35}$	1.997 089 ± 3
Doublet is 39 μu higher or 19.5 ppm*.	
$C_9H_{10} - \frac{1}{2}U^{235}$	0.556 303 ± 2
$\frac{1}{2}U^{235} - C_9H_9$	0.451 543 ± 2
H =	1.007 846 ± 3
Doublet is 21 ppm high	

*Minnesota value for $Cl^{37} - Cl^{35}$ of 1.997 050 u was employed.

parts per million high. A similar conclusion can be reached from the two hydrocarbon-metal doublets, doublets 4 and 5. The sum of these two doublets result in a hydrogen mass that is 21 ppm high. This conclusion is also reached by direct hydrocarbon-hydrocarbon doublets which yield a mass difference of one hydrogen. More than 30 measurements of this sort resulted in a mass difference that was 20 ppm high. Whether dealing with metal-metal, metal-hydrocarbon or hydrocarbon-hydrocarbon doublets, a result that is high by about 20 ppm was obtained. A proportional correction of 20 ± 3 ppm was therefore made to all doublets. The doublet results listed in Table I include a correction for this proportional error. The value of the mass of a particular isotope was then determined by forming the unweighted average of the individual masses determined from each doublet. The preliminary value for each isotope measured is:

Th^{232}	232.038 060 ± 2 u
U^{235}	235.043 921 ± 3 u
U^{238}	238.050 790 ± 2 u

For this calculation and subsequent calculations, the masses of H^1, S^{32} and Pb^{206} were taken from the 1964 mass table [5].

Except for the U^{238}-U^{235} doublet measured by Kerr and Bainbridge, no direct comparison between the present results and those of Kerr and Bainbridge may be made. If one assumes masses for H^1, S^{32}, and Pb^{206},

TABLE III

Comparison of results

U^{238}-U^{235}	
Present Result	3.006 869 ± 3
Kerr and Bainbridge	3.006 859 ± 8
1967 Mass Table	3.006 876 ± 17
Nuclear Reactions	3.006 873 ± 4
U^{235}-Th^{232}	
Present Result	3.005 866 ± 3
1967 Mass Table	3.005 864 ± 15
Nuclear Reaction	3.005 868 ± 4
U^{235}	
Present Result	235.043 921 ± 2
Kerr and Bainbridge	235.043 936 ± 12
1967 Mass Table	235.043 943 ± 12
U^{238}	
Present Result	238.050 790 ± 2
Kerr and Bainbridge	238.050 795 ± 12
1967 Mass Table	238.050 819 ± 12
Th^{232}	
Present Result	232.038 060 ± 2
1967 Mass Table	232.030 079 ± 11

then one can calculate the masses of U^{235} and U^{238} and an indirect comparison may be made. Table III shows the comparisons that may be made. We have added for further comparison the 1967 least square masses and mass differences from Wapstra [6], and a nuclear reaction difference for U^{235}-Th^{232} and U^{238}-U^{235} which resulted from a least squares fit of available nuclear reactions in this region.

With the new atomic mass results of Kerr and Bainbridge and the present results available, it was clear that one no longer needed to rely as heavily on the reaction chains to tie uranium and thorium to the lead masses. For this reason it appeared worthwhile to make a revision of the atomic mass table for the elements above A = 206. A least squares adjustment similar to that described by Mattauch [7] was programmed on our

TABLE IV

Results of least squares fit

Th^{232}	
Present Result	232.038 060 ± 2
1967 Mass Table	232.038 079 ± 11
1971 Minn. L.S.F.	232.038 062 ± 2
U^{235}	
Present Result	235.043 921 ± 3
1967 Mass Table	235.043 943 ± 12
1971 Minn. L.S.F.	235.043 924 ± 2
U^{238}	
Present Result	238.050 790 ± 2
1967 Mass Table	238.050 819 ± 12
1971 Minn. L.S.F.	238.050 790 ± 2

CDC 6600 computer. Input data previous to 1967 were taken directly from the input data for Wapstra's 1967 table [6].

Twenty-one nuclear reactions that have been published since 1967 were added together with the new mass measurements. The final mass table will be presented as part of a publication describing these measurements to the Physical Review. Of interest in the present discussion are the masses of U^{238}, U^{235}, and Th^{232}. These are listed in Table IV.

This comparison is indicative of the general comparison of our least squares fit and the 1967 mass table, that is, that the present table is about 20 μu lower than the 1967 mass table.

Next I will discuss our study of the feasibility of mass measurement of radioactive atoms. The success of the Bernas group [8] in making low resolution mass spectrometric measurements of short half-life radioactive atoms suggested to us that mass measurements using a higher resolution instrument might be possible. We have built a small double-focusing instrument in order to begin studies in this area. One of the first considerations that was made in this work was a study of the precision of mass measurement when only a small number of ions are collected. This problem has been considered by A. J. Campbell and J. S. Halliday and reported at an ASTM Committee E-14 meeting [9]. Because this problem may have some general interest, I will discuss it in some detail. In order to study this problem, we can resort to an interpretation in which the shape of an ion peak is just the statistical distribution function which will give the probability of location of ions of a particular mass M in the spectrum. A typical peak shape is shown in Fig. 1. Under the assumption that this peak

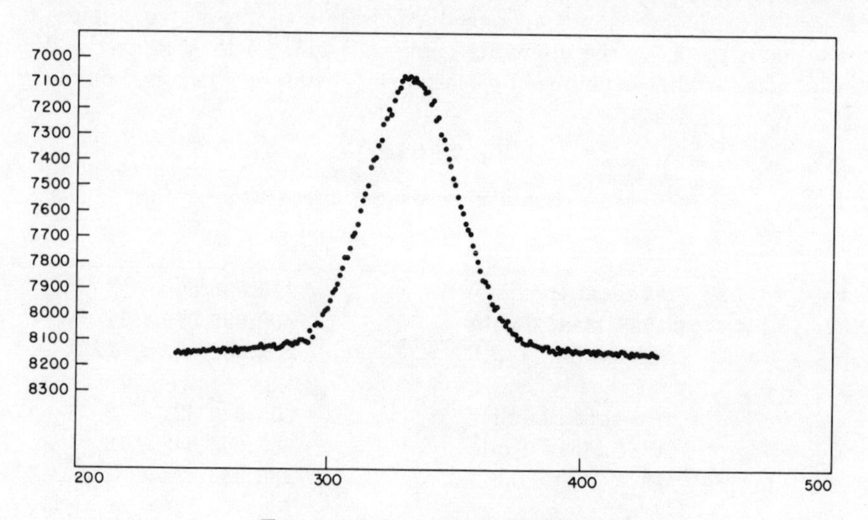

Fig. 1. A typical peak shape.

shape represents the distribution function, a standard deviation may be calculated in the usual manner. The standard deviation of the particular distribution shown was determined to be 0.51 times the full width at half maximum of the ion peak. Suppose, for example, that we wish to determine the mass of a single ion of mass 90 in the spectrum of an instrument with a full width at half maximum of 3000. The Δm corresponding to the full width at half maximum is 90000 mu/3000 or 30 mu. The standard deviation associated with the measurement of a single ion in this spectrum is therefore 15 mu. If instead an ion peak made up of N ions is considered, the usual method for compounding variances yields a standard deviation for this measurement of $\sigma_N = \sigma/\sqrt{N}$. A calculation of the standard deviation for this circumstances as a function of N, the number of ions in the ion peak, is given in Table V.

In order to check the accuracy of this standard deviation formula, we recorded a series of 100 ion peaks with an average number of ions of 55. The location of the center of each of these peaks was determined by a simple centroid calculation. A standard deviation of a single measurement was determined from these 100 means. This result was ± 3.6 channels. This result may be compared with the result using the previously discussed formula for standard deviation.

The full width at half maximum for the peak shape while the 100 runs were made was 59 ± 4 channels. Applying the formula $\sigma_N = .5\ (\mathrm{FWHM})/\sqrt{N}$ yields $\sigma_N = 3.9 \pm .3$ channels. I believe that this is adequate agreement to allow the use of the standard deviation formula.

TABLE V

Standard deviation of a single determination of the center of an ion peak containing N ions

N	in mu	σ_N/FWHM
1	15	.5 or 1 in 2
100	1.5	.05 or 1 in 20
10^4	.15	.005 or 1 in 200
10^6	.015	.0005 or 1 in 2,000

This summer we are beginning preliminary measurements of this sort by studying some long-lived cesium isotopes formed by (p, fission) reactions with protons from our High Voltage Corp. Model MP Tandem Van de Graaff. The samples will then be run off-line in our laboratory. If these preliminary measurements are successful, we will mount a mass spectrometer on-line at this accelerator.

Before moving to the third area of discussion, it is worthwhile to point out that this standard deviation formula may be employed to calculate the maximum precision obtainable using a peak matching technique as a function of the number of ions collected. Table V also lists the ratio σ_N divided by the full width at half maximum as a function of the number of ions in the peak.

I would now like to outline some research whose primary purpose was to develop ways of handling unresolved mass doublets. The resulting mathematics are as interesting as the experimental techniques because they generalize the concept of an error signal as employed in our peak matching technique, and show that unresolved doublets are mathematically no different than resolved ones.

To simplify the discussion let us consider peak type functions, that is, continuous ones that are arbitrarily near zero except in a limited region. Then if f(t) is such a function, consider a spectrum $g_N(t)$ defined by

$$(1)\quad g_N(t) \equiv f(t) + \sum_{j=1}^{N} A_j f_j(t-b_j),\ b_1 \neq b_2 \neq \ldots . b_N \neq 0$$

The Laplace transform, f(s), of f(t) is well defined, and hence so is $g_N(s)$:

$$(2)\quad g_N(s) = (1 + \sum_{j=1}^{N} A_j e^{-sb_j})f(s).$$

The transform of our data yields $g_N(s)$, so that if we knew the A_j and b_j we would divide out this factor, obtaining f(s) and then f(t). Let B_j approximate A_j and β_j approximate b_j. Define $h_N(s)$ as

$$\text{(3)} \quad h_N(s) \equiv \left(1 + \sum_{j=1}^{N} A_j e^{-sb_j}\right)\left(1 + \sum_{j=1}^{N} B_j e^{-s\beta_j}\right)^{-1} f(s).$$

Then $h_N(t)$ is obtained by taking the inverse transform of (3):

$$\text{(4)} \quad h_N(t) = \sum_{n=0}^{\infty} (-1)^n \left[\sum_{n_1+n_2+\ldots n_N=n} \frac{n!}{n_1!n_2!\ldots n_N!} \left(\prod_{j=1}^{N} B_j^{\,n_j} \right) g(t-\vec{\beta}\cdot\vec{n}) \right]$$

where $\vec{\beta}\cdot\vec{n} = \beta_1 n_1 + \beta_2 n_2 + \ldots + \beta_N n_N$.

$h_N(t)$ is an N+1 function generalized error signal of $g_N(t)$. It is possible to show from (4) that when $B_j = A_j$ and $\beta_j = b_j$ for all j that

$$h_N(t) = f(t) + \rho_N(t)$$

where $\rho_N(t)$ represents a noise and baseline contribution. $h_N(t)$ has a periodic behavior which reduces infinite summations to finite ones and allows a practical computer evaluation.

In the case where N = 1 we obtain a generalized doublet error signal $h_1(t)$ from (4):

$$\text{(5)} \quad h_1(t) = \sum_{n=0}^{\infty} (-1)^n B^n [f(t-\beta n) + A f(t-b-\beta n)],$$

with the actual summation on n running to n = 1 for resolved doublets and perhaps n = 4 for badly unresolved ones. The numerical implementation of (5) involves finding a minimum for the sum of squares of $h_1(t)$ over one period. An alternative implementation involves the construction of a multi-increment system for the spectrometer so that a generalized error signal may be formed directly.

A simple example of such a system provides understanding that (4) never could. The system cycles through three equal voltage increments and then resets to the nominal voltages, with inversion and multiplication of the signal according to the sequence 1, $-B$, B^2, $-B^3$. In Fig. 2, triangular peaks separated by half a peak width yield a resolved peak usually requiring twice the resolution with a null between the resolved members. Such a system on our spectrometer requires extensive changes, so for now we are using the numerical approach.

REFERENCES

1. JOHNSON, W. H., HUDSON, M. C., BRITTEN, R. A. and KAYSER, D. C., Proceedings of the Third International Conference on Atomic Masses (R. Barber, Ed.) University of Manitoba Press, Winnipeg (1967), p. 793.
2. BENSON, J. L. and JOHNSON, W. H., Phys. Rev. 141, 1112 (1966).
3. JOHNSON, E. G. and NIER, A. O., Phys. Rev. 91, 10 (1953).
4. KERR, D. P. and BAINBRIDGE, K. T., Can. Jour. Phys. 49, 756 (1971).
5. MATTAUCH, J. H. E., THIELE, W. and WAPSTRA, A. H., Nuc. Phys. 67, 1 (1965).
6. WAPSTRA, A. H., Proceedings of the Third International Conference on Atomic Masses (R. Barber, Ed.) University of Manitoba Press, Winnipeg (1967), p. 153.
7. MATTAUCH, J. H. E., Proceedings of the International Conference on Nuclidic Masses (H. E. Duckworth, Ed.) University of Toronto Press, Toronto (1960), p. 3.
8. BERNAS, R., Recent Developments in Mass Spectroscopy (K. Ogata and T. Hayakawa, Eds.) University of Tokyo Press, Tokyo (1970), p. 535.
9. CAMPBELL, A. J. and HALLIDAY, J. S., Thirteenth Annual Conference on Mass Spectroscopy and Allied Topics, ASTM Committee E-14, May 16-21, 1965, unpublished.

Fig. 2. Example of a generalized error signal.

On-line Mass Spectrometric Analysis of Far Unstable Light Nuclei Produced in High Energy Nuclear Reactions. Present Status and the Prospect of Direct Mass Measurements

R. Klapisch and C. Thibault

Centre de Spectrométrie Nucléaire et de Spectrométrie de Masse du C.N.R.S.–91 Orsay–France

In the last five years or so, a large number of very neutron rich isotopes of the light elements have been identified. Fig. 1 shows the changes brought in this part of the chart of the nuclides through the work done by groups at Berkeley, Princeton, Orsay and Doubna (1).

What do we know about these nuclei ? not much at present. The first thing we usually learn is that they exist a time long enough to be registered by an instrument (that is 10^{-9} or 10^{-3} second depending on the technique) and that tells us already that they are stable against particle decay. In terms of masses, one has e. g. :

$$M\,(^{17}C) \leqslant M\,(^{16}C + n)$$

(or else ^{17}C would decay in 10^{-21} s)

In some other cases, one knows the β half-life in addition. We have measured this quantity using a mass spectrometer (2) in the cases of ^{11}Li 9.0 ms and five new sodium isotopes ^{27}Na (288 ms) to ^{31}Na (16 ms) as will be explained later in this report.

Obviously, the knowledge of the ground state masses would be an important piece of information if one wanted to study in more detail the nuclear properties in this region. At some distance from stability the current theoretical predictions begin to diverge significantly. We shall give 2 examples to illustrate this fact (3).

1) - ^{11}Li was not expected to be bound according to the early Garvey-Kelson calculation. Since then other calculations, based on various models have been done but they still vary significantly as is shown in Table I.

TABLE I

Different calculated values for the mass of ^{11}Li (in MeV)

Author	M ^{11}Li	E_n ^{11}Li	E_{2n} ^{11}Li
Vinogradov	41.1		0
Janecke	40.2 ± 2.2	+ 3.4	+ 0.9 ± 2.2
Garvey et al	41.7 ± 0.6	+ 0.5 ± 0.6	- 0.6 ± 0.6
Vorobiev	40.7	+ 1.3	+ 0.4
Cohen and Kurath	42.	+ 0.1	- 0.9
Goldhammer	40.9	+ 1.3	+ 0.2

Except Janecke (but with a high uncertainty) and the very empirical evaluation of Vorobiev, all the others fail to account for a bound ^{11}Li. The evaluation by Garvey et al being marginal. The two last lines refer to calculations by C. Thibault following the prescriptions of the authors here quoted.

Clearly here a measurement with a precision of 0.5 MeV would be very valuable.

2) - The second example is the case of the sodium isotopes and is shown on Fig. 2. It is really dramatic to see how after 4 or 5 masses numbers from stability, the scatter between different predictions can be of a few MeV. This is of course a reflection of the fact that all mass formulas derive their parameters from experimental data measured near the valley of stability.

The conclusion of this brief survey is that, at some distance from stability, mass measurements with an accuracy of 1 MeV or a little better would be valuable at this stage.

Taking these considerations into account we decided in 1970 to investigate the possibility of measuring directly these masses with a precision of 200 keV for ^{11}Li and around 600 keV for ^{30}Na. This means a mass measurement with an accuracy of 2.10^{-5} i.e. 3 to 4 orders of magnitude less than what is usually achieved in the field of high precision mass measurements. There are of course specific difficulties associated with this new field of research and this is what we would now like to explain.

a) These exotic nuclei are made by the reaction of high energy particles on a heavy target e.g. uranium. Fig. 3 shows the excitation function for the production of ^{24}Na. One sees that it is really necessary to be well over one GeV to have an appreciable yield and this is the reason why these experiments can be done only at the few laboratories in the world where there are high energy accelerators. Our group has built at Orsay an instrument designed to be moved to and installed in a beam of the CERN proton synchrotron.

b) Another requirement is apparent from Fig. 4 showing the cross section that we have measured for the different isotopes of sodium. As one would expect the cross sections decrease as one departs from stability. This points to the necessity of a high sensitivity and one will always be limited by statistics as one aims at reaching the limit of particle stability.

c) With such a high excitation energy one makes essentially all nuclei that are particle-stable and lighter than the target nucleus. Because of this, it is necessary to use very selective technique to isolate a particular species. In contrast to other groups who have used electronic techniques to identify fragments in flight we use a mass spectrometer to analyze stopped reaction products.

d) Finally, far unstable light nuclei have short half-lives (milliseconds to seconds). Thus, not only have the measurements to be made on-line but also, all delays (chemical separation etc...) should be kept to a minimum.

We now come to describing how we meet these requirements.

The experiment is based on the fact that certain chemical elements (the alkalies) have a very fast diffusion in heated graphite and can also be selectively ionized by surface ionization on a hot metallic surface.

Fig. 5 shows a schematic of the experiment. The target consists of a sandwich of thin (70 μm) graphite slabs on which a layer of e.g. uranium carbide has been coated. The assembly is wrapped in a Re or Ta foil heated to 2000°. The energetic recoils from the reactions will be brought to rest in the graphite. The alkalies will diffuse out quickly and will be ionized by surface ionisation on the Re surface. This assembly then acts as an ion source for a mass spectrometer which is a single stage Nier-type magnetic sector. Detection of ions is achieved by an electron multiplier counting single pulses. Fig. 6 shows the installation in an external beam at the CERN P.S..

We said that the diffusion is fast. This is readily measured and shown in Fig. 7. If the instrument is set on a particular mass then one will record the decrease of current as a function of time following a burst of protons from the synchrotron. One sees that very short lived isotopes can be recorded even in the range of milliseconds

The sensitivity is limited by the signal-to-background-ratio with the background being strongly time dependent as is shown in Fig. 8. This shows how ^{11}Li could be seen over the background and its half-life (9.0 ms) measured by comparing its diffusion to the diffusion of stable ^{6}Li.

This way, it was possible to measure the half life not only of ^{11}Li but also of 5 new sodium isotopes Na (27-31) and also of some new Rb and Cs (4).

This was the status at the beginning of 1970 when we began investigating the possibility of direct mass measurements. As we said earlier, a precision of 2.10^{-5} is modest compared to the standards of high precision mass measurements. However, even that seemed discouraging considering the poor statistics of our 1969 run.

It thus appeared to us that a prerequisite of mass measurements was a substantial improvement in the counting rate. This was in any case necessary to extend our half-life measurements beyond ^{31}Na as far as ^{34}Na or ^{35}Na.

On the other hand, while mass measurements are usually performed with double focusing instruments, it did not seem to us that the improvement in resolution that we could expect was worth the increased effort and complexity. For this reason, the instrument that we have built and will take to CERN by the end of next month is a single stage machine.

It is a magnetic sector ($\Phi = 90°$ $r = 35$ cm) with a wide gap (50 mm), slightly inhomogeneous ($n = 0,23$) and it gives a resolving power of 450 with wide slits (4/10 mm).

We hope to increase the production by ~ 50 due to : longer target, better transmission, higher proton beam intensity and to decrease the background by 2 orders of magnitude through careful shielding.

To measure an isotopic doublet we will accumulate counts on two isotopes following two alternate P. S. pulses (i. e. every 10 seconds) and these will be stored in two different subgroups of a multiscaler memory. To jump from one member of doublet (M_1) to the other (M_2) we will add a voltage V_2 from a second regulated power supply to the accelerating voltage V_1 corresponding to mass M_1 of the doublet. V_1 and V_2 are measured continually by a digital voltmeter accurate to 10^{-6}.

The contents of the multiscaler memory will then be transferred to a small on-line computer that calculates the centroid of the peaks. Accordingly one varies V_2 until the peaks are exactly matched.

Since the measurements on M_1 and M_2 will alternate every 10 seconds the problems resulting from the drifts of V_1, V_2, B and the measuring instruments should be minimised and the adequate precision of 2.10^{-5} should be reached. This has been tested off-line by generating an artificial doublet and the mismatch for a 0.2 volts difference in a 10 kV accelerating voltage could be clearly seen.

To reach the precision of 2.10^{-5} with a resolving power of 500, it is necessary to locate the centroid of a peak with an accuracy of 1 %. This situation has been discussed by Professor W. H. Johnson (5). In our case, however, we have to take into account the presence of background. The calculation is quite straightforward and we shall only give the results here.

Given a triangular peak shape with height R, base 2 s and a constant background B, N = R s is the total number of counts.

A centroid can be defined in many different ways. We give the uncertainty in two cases of interest :

a) the median

$$\frac{\Delta M}{s} = \frac{1}{\sqrt{N}} \sqrt{\frac{1}{4} + \frac{1}{2}\,\frac{B}{R}}$$

b) the mean

$$\frac{\Delta M}{s} = \frac{1}{\sqrt{N}} \sqrt{\frac{1}{6} + \frac{2}{3}\,\frac{B}{R}}$$

Depending on the background one sees that one can sometimes be more advantageous than the other. Taking advantage of our on-line computer we will continually calculate both.

References :

1 - See KLAPISCH, R., Proceedings of the International Conference on the Properties of Nuclei far from the Region of Beta Stability ; Leysin (1970) ; Rpt. CERN 70-30, p 21 , for references therein.

2 - KLAPISCH, R., THIBAULT-PHILIPPE, C., DETRAZ, C., CHAUMONT, J., BERNAS, R., BECK, E., Phys. Rev. Letters, 23, C 652 (1969).

3 - THIBAULT, C., Thesis, Orsay (1971).

4 - CHAUMONT, J., ROECKL, E., NIR-EL, Y., THIBAULT-PHILIPPE, C., KLAPISCH, R., BERNAS, R. ; Phys. Letters 29 B, 652 (1969).; and TRACY, B. L., CHAUMONT, J., KLAPISCH, R., NITSCHKE, J. M., POSKANZER, A. M., ROECKL, E., THIBAULT, C., Phys. Letters, 34 B, 277, (1971).

5 - JOHNSON, W. H., Jr., Proceedings of the International Conference on the Properties of Nuclei far from the Region of Beta Stability ; Leysin (1970) ; Rpt. CERN 70-30, p 307 ; and these proceedings.

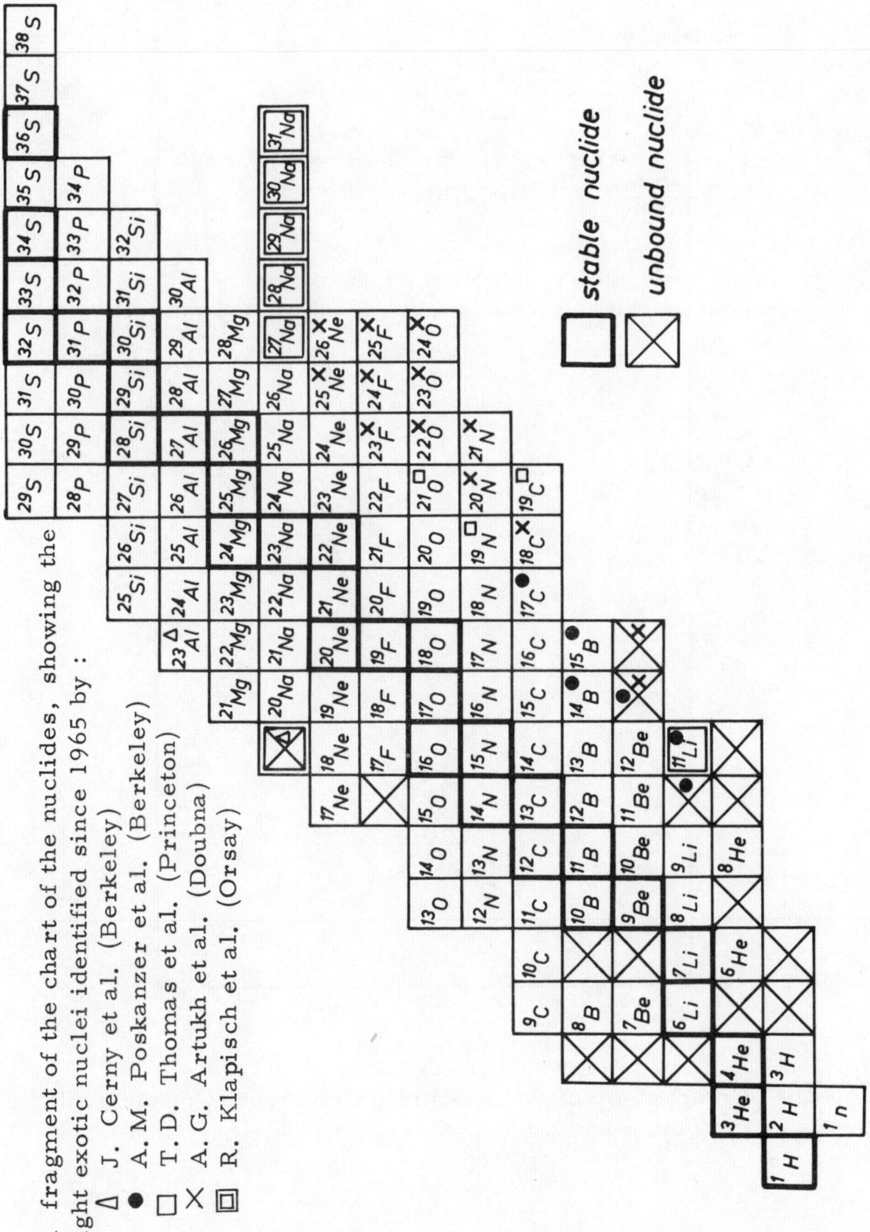

Fig. 1. A fragment of the chart of the nuclides, showing the light exotic nuclei identified since 1965 by :

- Δ J. Cerny et al. (Berkeley)
- ● A. M. Poskanzer et al. (Berkeley)
- □ T. D. Thomas et al. (Princeton)
- × A. G. Artukh et al. (Doubna)
- ⧈ R. Klapisch et al. (Orsay)

stable nuclide

unbound nuclide

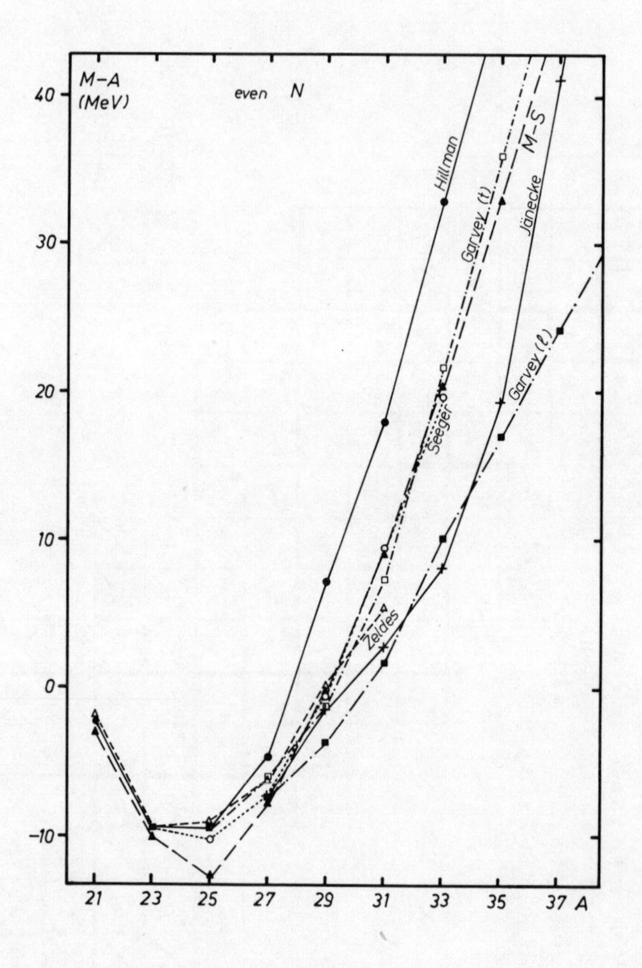

Fig. 2. Comparison of various predicted masses of the sodium isotopes.

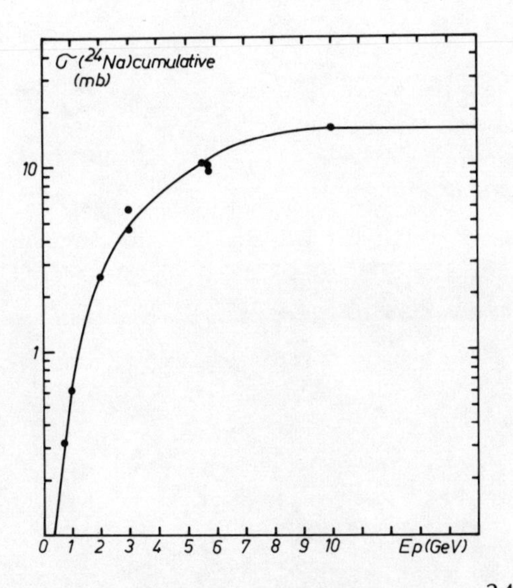

Fig. 3. Excitation function for the production of ^{24}Na from proton bombardment of uranium.

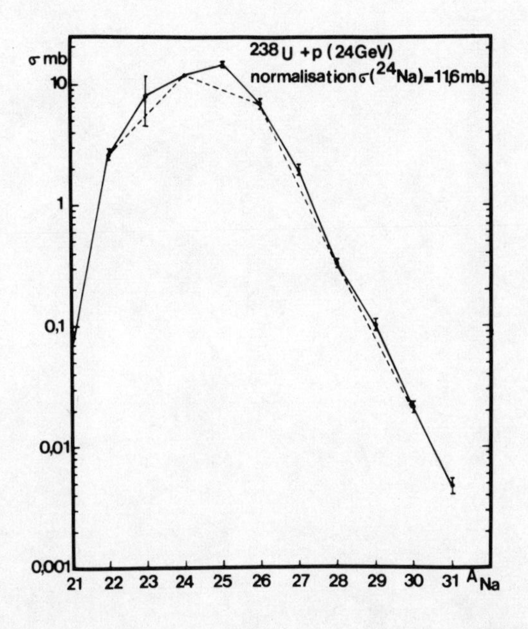

Fig. 4. Isotopic distribution of the cross sections of production of sodium from proton bombardment of uranium.

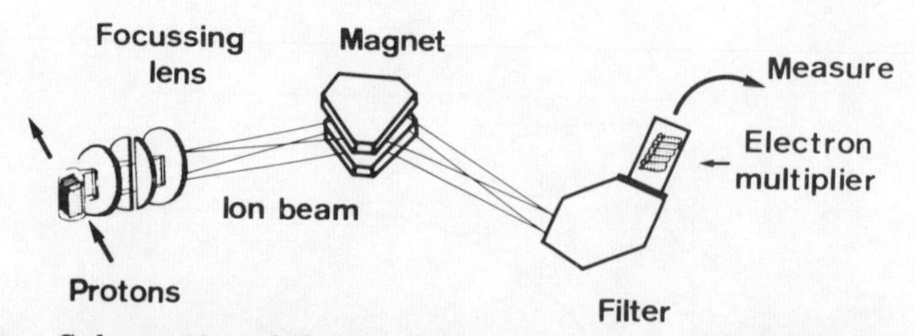

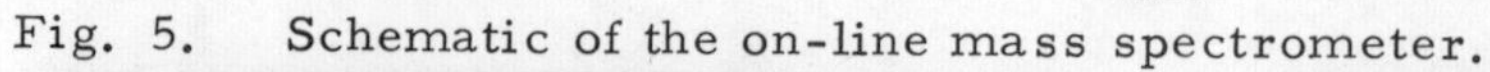
Fig. 5. Schematic of the on-line mass spectrometer.

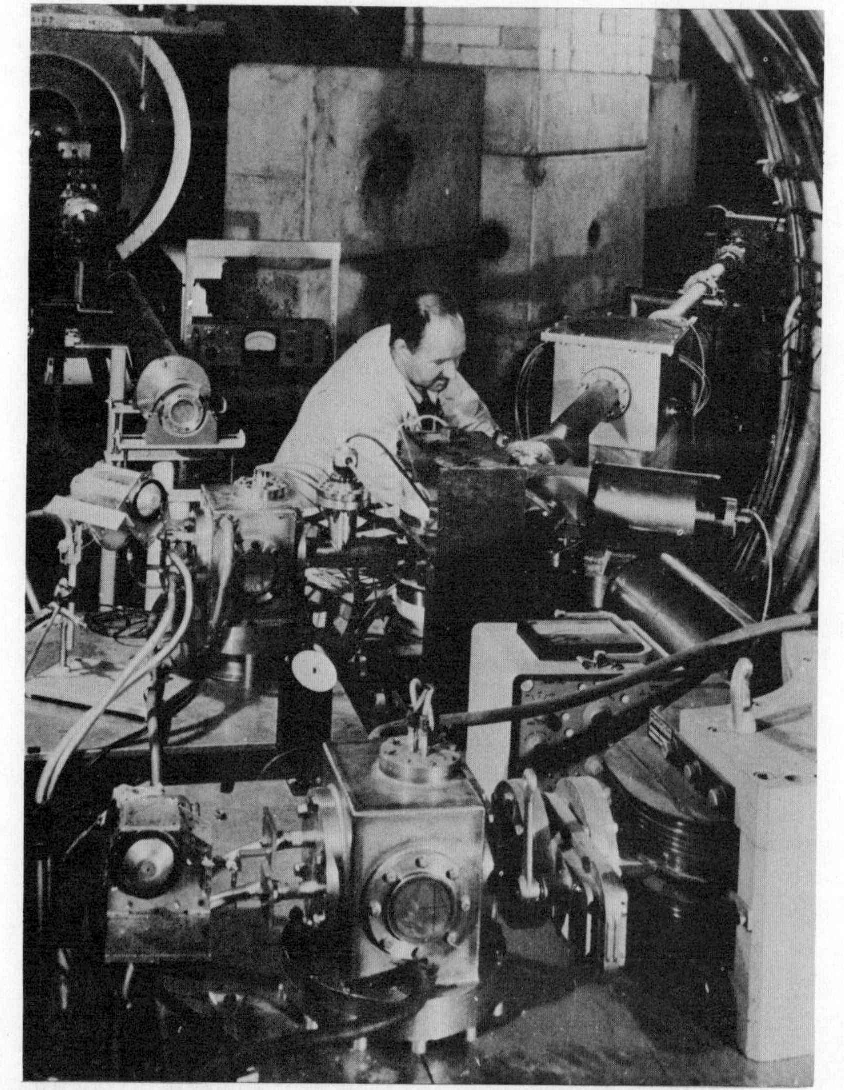

Fig. 6. Installation of two on-line mass spectrometers in an external proton beam at the CERN P. S.

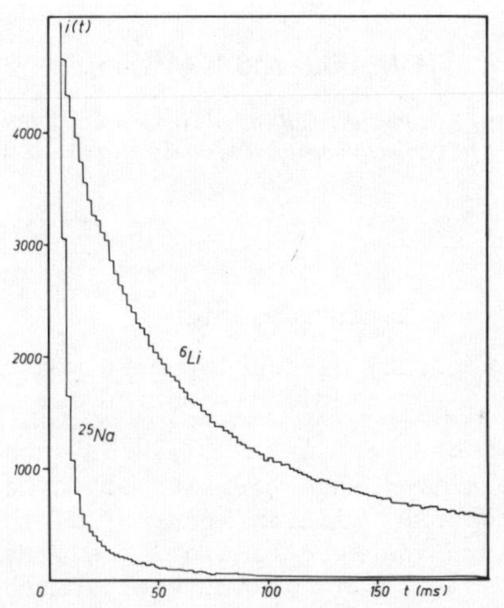

Fig. 7. Time dependance of ion current due to diffusion following the proton burst.

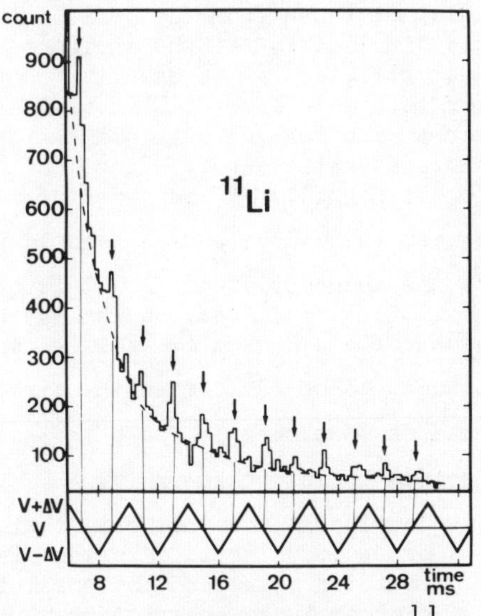

Fig. 8. Diffusion and radioactive decay of ^{11}Li. Following a proton pulse, a triangular modulation of the accelerating potential and the counting on a multiscaler are triggered simultaneously. The ^{11}Li peaks appear over the background (dashed line). The change to their intensity with time is due to both diffusion and radioactive decay of ^{11}Li. From a comparison with the peaks of ^{6}Li produced in the same conditions, the half life of ^{11}Li is found to be 9.0 ± 0.5 ms. The spectrum shown is the result of a 4h-run at the CERN P. S.

Image Aberrations of Double Focusing Mass Spectrometers

H. Matsuda and T. Matsuo

Institute of Physics, College of General Education, Osaka University, Toyonaka, Japan

1. Introduction

Recently, the accuracy of doublet mass measurement has been greatly increased and the atomic masses are determined with an accuracy as small as one part in 10^8(1). In such a high accuracy measurement, the relative image position of the components of mass doublet should be measured with an accuracy of the same order. For a double focusing mass spectrometer, the final position of an ion coming from an object slit of extremely narrow width is given in a second order approximation as follows:

$$x = A_\alpha \alpha_o + A_\delta \delta + A_{\alpha\alpha} \alpha_o^2 + A_{\alpha\delta} x_o \delta + A_{\delta\delta} \delta^2 + A_{yy} y_o^2 + A_{y\beta} y_o \beta_o + A_{\beta\beta} \beta_o^2 \quad , \tag{1}$$

where x is the radial distance from the optic axis, α_o and β_o the radial and the axial inclination of the incident beam, y_o the half height of the object slit, and δ the relative energy deviation. If the values of x for the two different kind of ions are not exactly equal to each other at the peak matching time, the measured mass value may have some systematic error. Since the mass dispersion of a mass spectrometer is normally of the order of magnet radius a_m, the difference between two values of x/a_m should be less than 10^{-8} in order to assure the accuracy of 10^{-8}. It is quite likely that two different kind of ions of a mass doublet may have slightly different mean values of α_o, β_o and δ. For instance, if the differences are assumed to be 10^{-4}, then the first order coefficients A_α/a_m , A_δ/a_m should be less than 10^{-4} and the second order coefficients A_{ij} should be less than unity.

From the above considerations, not only the exact first order but also the second order focusing is important in the mass measurement work. About ten years ago the possibilities of correcting the second order image aberrations were investigated by Hintenberger and König (2). The calculations, however, were carried out neglecting the influence of the fringing field. The influence of the fringing field on the trajectories of charged particles has been calculated in a third order approximation for a magnetic field by Matsuda and Wollnik (3) and for an electric field by Matsuda (4). In this report the possibilities of the second order focusing are investigated considering the influence of the fringing field. The instruments here investigated consists of a toroidal electric field and a homogeneous magnetic field.

2. Method of Calculation

Image aberrations of a double-focusing mass spectrometer are calculated most conveniently by using transfer matrices which transform a second order vector from one profile plane to another. Fig.1 shows the arrangement of the fields and the necessary profile planes, which are indicated by numbers 0-9. The second order vector to be transformed has 12 elements and is expressed as:

$$k^{(2)} = x,\ \alpha,\ \delta,\ x^2,\ x\alpha,\ x\delta,\ \alpha^2,\ \alpha\delta,\ \delta^2,\ y^2,\ y\beta,\ \beta^2 \quad , \qquad (2)$$

where x, α, and y, β are, respectively, the radial and the axial displacement and inclination of the trajectory relative to the main path, and δ is the relative energy deviation. The vector at different profile planes is distinguished by the subscript $i (i = 0,\cdots,9)$.

At each boundary of the field, we consider two profile planes which correspond to the entrance and the exit side of the boundary, and all the influences due to the fringing field, including the effects of an oblique and curved field boundary, are contained in the transfer matrix between them. The field between the boundaries is assumed to be an ideal field. There is a small parallel shift between the main path inside the field and the path outside the field.

The transformation between the 0-th vector (source) and the 9-th vector (image) is obtained by calculating the product of nine matrices which give the transformation between (0-1),(1-2),···,(8-9). respectively. If the matrix elements of the first row of the product matrix are calculated and denoted by A_i , $(i = x,\alpha,\delta,\cdots,\beta\beta)$, then we have the relationship

$$x_9 = A_x x_o + A_\alpha \alpha_o + A_\delta \delta + A_{xx} x_o^2 + A_{x\alpha} x_o \alpha_o + A_{x\delta} x_o \delta + A_{\alpha\alpha}\alpha_o^2 + A_{\alpha\delta}\alpha_o \delta + A_{\delta\delta}\delta^2 + A_{yy} y_o^2 + A_{y\beta} y_o \beta_o + A_{\beta\beta}\beta_o^2 \quad , \qquad (3)$$

where A_x is the total magnification of the image and the other A_i are the coefficients of the first and second order image aberrations.

The transfer matrices of the free space, the ideal electric field and the ideal magnetic field have been described by many authors (5,6,7), and therefore we do not repeat them here. We will. however, explain the fringing field matrices briefly.

3. Influence of Fringing Field

The real particle trajectory in the fringing field can be calculated by solving the equation of motion if we know a suitable mathematical expression of the fringing field. From a different point of view, we can assume that the same particle moves without any deflection up to the ideal boundary, experiences a small shift and bend there and moves in the ideal field after having passed through the boundary. If the shift and bend at the boundary are properly given, the trajectory sufficiently inside the field would coincide completely with the real trajectory. Therefore the influence of the fringing field can be reduced to a transformation of the vector given by eq.(2).

3.1. Toroidal electric field

From the third order calculation described in ref.(4), the following results are obtained. The shift of the main path at the boundary is given by

$$\Delta x_e = r_e I_{1a} \quad . \tag{4}$$

At the entrance boundary the transformation between the profile planes 1 and 2 is given by

$$x_2 = x_1[1+cI_{1a}-2I_{1b}] + r_e I_{1a} + x_1^2 r_e/2 \quad , \tag{5a}$$

$$\alpha_2 = x_1[-2I_{4a}/r_e] + \alpha_1[1-cI_{1a}+2I_{1b}] + x_1^2[2(1+2c)I_{4a} - 6I_{4b} - 4I_{1a}I_5+4I_7]/r_e^2 + x_1\delta[4I_{4a}/r_e] \quad , \tag{5b}$$

$$y_2 = y_1 \quad , \qquad \beta_2 = \beta_1 + y_1\beta_1/r_e \quad , \tag{5c}$$

where $c = r_e/R_e$ and R_e is the vertical radius of curvature of the equipotential surface.

At the exit boundary the transformation between the profile planes 3 and 4 is given by

$$x_4 = x_3[1-cI_{1a}+2I_{1b}] - r_e I_{1a} - x_3^2 r_e/2 \quad , \tag{6a}$$

$$\alpha_4 = x_3[-2I_{4a}/r_e] + \alpha_3[1+cI_{1a}-2I_{1b}] + x_3^2[(4c+3)I_{4a}-6I_{4b}-4I_{1a}I_5+4I_7]/r_e^2 + x_3\delta[4I_{4a}/r_e] \quad , \tag{6b}$$

$$y_4 = y_3 \quad , \qquad \beta_4 = \beta_3 - y_3\beta_3/r_e \quad , \tag{6c}$$

In the above equations, the quantities denoted by I_{1a}, ... etc. are given by the following definite integrals:

$$I_{1a} = E_o^{-1}\int_a^b\int E\, d\eta d\eta - \tfrac{1}{2}\eta_b^2 \quad , \qquad I_{1b} = E_o^{-2}\int_a^b\int E^2 d\eta d\eta - I_{4a}\eta_b - \tfrac{1}{2}\eta_b^2 \quad ,$$

$$I_{4a} = E_o^{-2}\int_a^b E^2 d\eta - \eta_b \quad , \qquad I_{4b} = E_o^{-3}\int_a^b E^3 d\eta - \eta_b \quad ,$$

$$I_5 = E_o^{-2}\int_a^b (E')^2 d\eta \quad , \qquad I_7 = E_o^{-3}\int_a^b (E')^2 (\iint E\, d\eta d\eta) d\eta \quad , \tag{7}$$

where $E/E_o = E(\eta)/E_o$ is a distribution function of the electric fringing field along the main path and $r_e\eta$ is the distance from the ideal field boundary. These integrals can be evaluated if we know the distribution function $E(\eta)$, which has been given by Herzog(8) for any kind of electrode structure. As an example, assume that the gap width between two condenser electrodes is $0.062\, r_e$ and a thin Herzog plate with a slit of $0.021\, r_e$ is placed at both entrance and exit, the distance between theHerzog plate and the end of electrodes being $0.015\, r_e$. In this case the ideal field boundary coincides with the end of the condenser electrodes and the necessary integrals are evaluated to be

$I_{1a} = 0.000093$, $I_{1b} = 0.000114$, $I_{4a} = -0.00746$,
$I_{4b} = -0.01172$, $I_5 = 22.75$, $I_7 = 0.00130$.

The transfer matrices for the electric fringing field are constructed from the transformations (5) and (6).

3.2. Homogeneous magnetic field

We treat here the magnetic field with straight field boundaries. From the third order calculation of ref.(3), we obtain the following results. The shift of the main path at the boundary is given by

$$\Delta x_m = r_m I_1 / c^2 \quad , \tag{8}$$

where c stands for cos ε' at the entrance and cos ε" at the exit, ε' and ε" being the incident and the exit angle respectively. The transformation between the profile planes 5 and 6 is given by

$$x_6 = x_5 + \alpha_5[-2r_m(t'/c'^2)I_1] + \delta[\tfrac{1}{2}r_m I_1/c'^2] + x_5^2[-\tfrac{1}{2}t'^2/r_m] + y_5^2[\tfrac{1}{2}/c'^2 + (\tfrac{1}{2}t'/c')(5+6t'^2)I_4]/r_m \quad , \tag{9a}$$

$$\alpha_6 = x_5[t'/r_m] + \alpha_5[1 - 2(t'/c')^2 I_1] + \delta[\tfrac{1}{2}(t'/c'^2)I_1] + x_5\alpha_5[t'^2/r_m] + x_5\delta[-\tfrac{1}{2}t'/r_m] + y_5^2[\tfrac{1}{2}t'(1+2t'^2) + \tfrac{1}{2}(t'^2/c')(7+10t'^2)I_4]/r_m^2 + y_5\beta_5[-t'^2 - (t'/c')(1+2t'^2)I_4]/r_m \quad , \tag{9b}$$

$$y_6 = y_5 \quad , \tag{9c}$$

$$\beta_6 = \beta_5 + y_5[-t' - (1+2t'^2)I_4/c']/r_m \quad , \tag{9d}$$

where c', t' stand for cos ε', tan ε'. The transformation between the profile planes 7 and 8 is given by

$$x_8 = x_7[1-2(t''/c'')^2 I_1] + \alpha_7[-2r_m(t''/c''^2)I_1] + \delta[-\tfrac{1}{2}r_m I_1/c''^2] + x_7^2[\tfrac{1}{2}t''^2/r_m] + y_7^2[-\{\tfrac{1}{2}/c''^2 + (\tfrac{1}{2}t''/c'')(5+6t''^2)I_4\}/r_m] \quad , \tag{10a}$$

$$\alpha_8 = x_7[t''/r_m] + \alpha_7 + x_7^2[-\tfrac{1}{2}t''^3/r_m^2] + x_7\alpha_7[-t''^2/r_m] + x_7\delta[-\tfrac{1}{2}t''/r_m] + y_7^2[-\tfrac{1}{2}t''^3 - t''^2(1+2t''^2)I_4/c'']/r_m^2 + y_7\beta_7[t''^2 + (t''/c'')(1+2t''^2)I_4]/r_m \quad , \tag{10b}$$

$$y_8 = y_7 \quad , \tag{10c}$$

$$\beta_8 = \beta_7 + y_7[t'' + (1+2t''^2)I_4/c'']/r_m \quad , \tag{10d}$$

where c", t" stand for cos ε", tan ε". The definite integrals which depend on the fringing field distribution now are

$$I_1 = B_o^{-1}\int_a^b \int B\, d\eta d\eta - \tfrac{1}{2}\eta_b^2 \quad , \quad I_4 = B_o^{-2}\int_a^b B^2 d\eta - \eta_b \quad , \tag{11}$$

where $B/B_o = B(\eta)/B_o$ is a distribution function along the direction perpendicular to the pole boundary and $r_m\eta$ is the distance from the ideal field boundary. As an example, if we assume that the gap width of magnet pole is $0.05\,r_m$, then we have the following values:

$$I_1 = 0.00175 \quad , \quad I_4 = -0.0260 \quad .$$

TABLE I. Values of field parameters and the coefficients of image aberrations. The upper part shows the comparison among the following three cases: a) results of Hintenberger and König(2) in which the influence of fringing field is neglected. b) results of present calculation for the same parameters. c) results for which the field parameters are adjusted so as to satisfy the first order double focusing condition. In the lower part three examples of second order double focusing mass spectrometers are given.

	ϕ_m	ε'	ε''	ϕ_e	r_e/r_m	ℓ_e'/r_m		d/r_m	ℓ_m''/r_m	A_α/r_m	A_δ/r_m	$A_{\alpha\alpha}/r_m$	$A_{\alpha\delta}/r_m$	$A_{\delta\delta}/r_m$	$A_{\beta\beta}/r_m$
							a	3.4364	1.4239	0.000	0.000	-0.000	0.000	-0.000	0.00
1	55	0	0	81.510	1.2762	0.4768	b	"	"	-0.005	-0.011	0.184	0.546	0.684	-33.40
							c	3.4835	1.4109	0.000	0.000	0.115	0.745	0.517	-33.72
							a	2.9930	1.2145	0.000	0.000	-0.000	-0.000	-0.000	0.00
2	60	0	- 3	90.999	1.2203	0.2835	b	"	"	-0.003	-0.011	0.107	0.511	0.867	-28.90
							c	3.0271	1.2047	0.000	0.000	0.060	0.632	0.767	-29.12
							a	4.5202	0.8377	0.000	0.000	-0.000	0.001	-0.001	0.00
3	60	0	-15	56.569	1.8143	1.8204	b	"	"	-0.010	-0.015	0.625	0.337	0.160	-59.21
							c	4.6347	0.8277	0.000	0.000	0.350	1.026	-0.309	-60.43
							a	4.9390	0.9116	0.000	0.000	-0.000	-0.000	0.001	0.00
4	70	15	-15	50.036	2.6663	4.1875	b	"	"	-0.024	-0.015	1.558	0.470	0.146	-72.18
							c	5.0764	0.9019	0.000	0.000	0.943	1.528	-0.351	-73.14
							a	2.8459	0.6637	0.000	0.000	0.000	-0.001	-0.000	0.00
5	60	-15	-15	97.849	1.0081	0.0879	b	"	"	0.001	-0.013	0.059	0.346	0.636	-35.70
							c	2.8801	0.6566	0.000	0.000	0.011	0.464	0.535	-36.07
	74.5	0	0	81.51	1.496	1.485		1.9255	1.0202	0.000	0.000	-0.002	0.003	0.002	-36.03
	76.1	0	0	84.8	1.449	1.278		1.8380	0.9924	0.000	0.000	0.002	0.001	0.003	-33.28
	79.1	0	0	90.0	1.384	1.018		1.6997	0.9397	0.000	0.000	-0.000	0.002	0.002	-29.90

4. Image Aberrations

In order to see the magnitude of the effect due to the influence of the fringing field, calculations of image aberrations are done for five different examples of the second order double focusing mass spectrometers which have been presented by Hintenberger and König(2). In this case the electric field is cylindrical. For reasons of practical application, we investigated only the instruments with deflection in the same sense in the electric and the magnetic field. In these calculations, we used the above mentioned values of integrals which are given as an example.

The results are shown in TABLE I, where three different cases are given for each instrument. Row (a) shows the results of Hintenberger and König in which the influence of fringing field is neglected. Row (b) shows the results of the present calculation using the same parameters as (a). On account of the presence of the fringing field, small changes are observed in the first order coefficients A_α and A_δ. To satisfy the first order double focusing conditions $A_\alpha=A_\delta=0$, the values of d/r_m and ℓ''_m/r_m were adjusted slightly. In row (c) the values of d/r_m and ℓ''_m/r_m thus obtained are given together with the coefficients of second order image aberrations. As can be seen in the table, the influence of the fringing field is important if we want to correct image aberrations of second order.

The calculation to find suitable field parameters for the second order double focusing are also done in the case of normal incidence and exit to the magnetic field. Three examples are given in the lower part of TABLE I. It is seen that the parameters are considerably different from those given in the example 1. For instance, in the case of $\phi_e = 81.51°$, ϕ_m should be $74.5°$ (fringing field considered) instead of $55°$ (fringing field neglected).

In the case of a mass spectrometer with a cylindrical electric field and a homogeneous magnetic field, the coefficients of image aberrations for the off median plane particles, that is, A_{yy}, $A_{y\beta}$, $A_{\beta\beta}$ become very large because of the presence of the fringing field, though it is possible to find suitable parameters for the second order double focusing on the median plane. These coefficients, however, can be reduced very much by using a toroidal electric field

TABLE II. Values of field parameters and the coefficients of image aberrations for off median plane. On the median plane $A_\alpha=A_\delta=A_{\alpha\alpha}=A_{\alpha\delta}=A_{\delta\delta}=0$.

$\phi_e=90°$, $R'_e=dR_e/dr_e=1/c$, $\varepsilon'=\varepsilon''=0$.

c	ϕ_m	r_e/r_m	ℓ'_e/r_m	d/r_m	ℓ''_m/r_m	A_{yy}/r_m	$A_{y\beta}/r_m$	$A_{\beta\beta}/r_m$
0	79.1	1.384	1.018	1.700	0.940	-0.903	-10.306	-29.768
0.1	77.1	1.315	1.021	1.778	0.985	-0.324	- 4.988	-21.148
0.2	74.8	1.247	1.021	1.873	1.036	-0.081	- 0.970	-14.018
0.3	72.4	1.180	1.021	1.978	1.090	-0.221	1.684	- 8.340
0.39	70.1	1.121	1.023	2.085	1.141	-0.727	2.807	- 4.473
0.5	67.0	1.050	1.024	2.240	1.212	-1.947	2.351	- 1.408

instead of a cylindrical electric field. Fig.2 shows an example of the behaviour of the coefficients A_{yy}, $A_{y\beta}$, $A_{\beta\beta}$ when the characteristic constant $c = r_e/R_e$ is changed. In this case the conditions for the second order focusing on the median plane, $A_\alpha = A_\delta = A_{\alpha\alpha} = A_{\alpha\delta} = A_{\delta\delta} = 0$, are of cource satisfied. In the case $\phi_e = 90°$, a optimum condition is obtained at c=0.48 and the value of $A_{\beta\beta}$ is reduced to about 1/20 of the value for cylindrical case c=0. The values of field parameters as well as the coefficients are given in TABLE II.

The values of suitable field parameters for the second order double focusing mass spectrometers with toroidal electric field are given in Fig.3.

From the calculations described above, it has been shown that the influence of the fringing fields on the image aberrations is relatively large. In order to design a mass spectrometer with second order double focusing, it is important to consider the fringing fields. A small correction is also necessary even in first order focusing.

REFERENCES

1. DUCKWORTH, H. E., Proc. Int.Conf. on Mass Spectroscopy (ed. K. Ogata, T. Hayakawa) Univ. of Tokyo Press, Tokyo 1970, p.26.
2. HINTENBERGER, H. and KÖNIG, L. A., Z. Naturforsch., 12a,773(1957).
3. MATSUDA, H. and WOLLNIK, H., Nucl. Instr. Methods, 77,40,283(1970).
4. MATSUDA, H., Nucl. Instr. Methods, 91, 637 (1971).
5. BROWN, K. L., BELBEOCH, R. and BOUNIN, P., Rev. Sci. Instr., 35, 481 (1964).
6. TAKESHITA, I., Z. Naturforsch., 21a, 9 (1966).
7. WOLLNIK, H., Nucl. Instr. Methods, 52, 250 (1967).
8. HERZOG, R., Phys. Z. 41, 23 (1940).

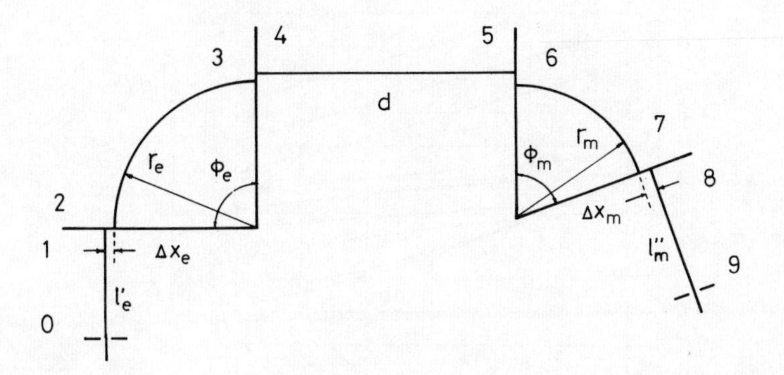

Fig.1. Arrangement of the fields of a double focusing mass spectrometer. The necessary profile planes are indicated by numbers 0–9.

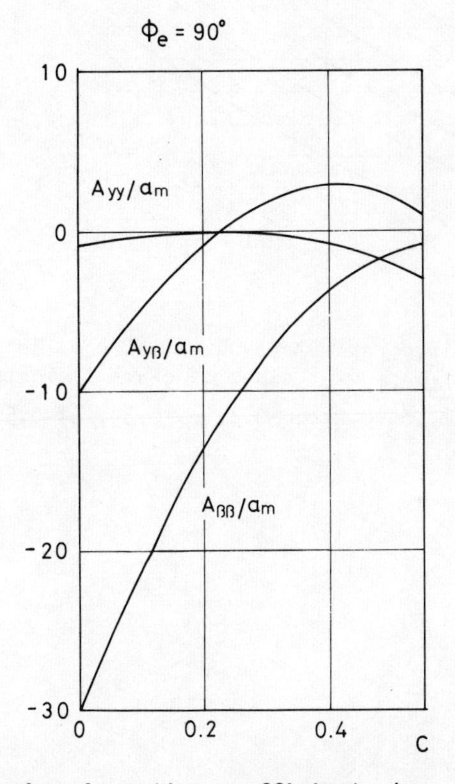

Fig.2. Second order aberration coefficients A_{yy}, $A_{y\beta}$, $A_{\beta\beta}$ as functions of $c = r_e/R_e$. Second order double focusing conditions on the median plane, $A_\alpha = A_\delta = A_{\alpha\alpha} = A_{\alpha\delta} = A_{\delta\delta} = 0$ are satisfied.

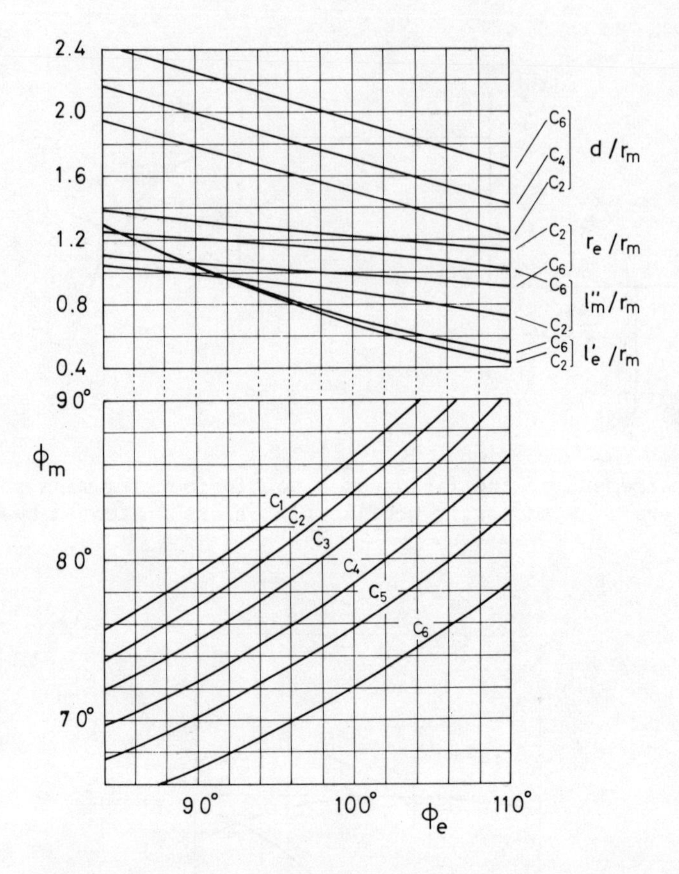

Fig.3. Values of field parameters which satisfy the conditions $A_{\alpha}=A_{\delta}=A_{\alpha\alpha}=A_{\alpha\delta}=A_{\delta\delta}=0$. Values of c are indicated by c_1-c_6: $c_1=0$, $c_2=0.1$, $c_3=0.2$, $c_4=0.3$, $c_5=0.39$, $c_6=0.5$.

Fast On-line Analysis of Fission Products and the Products of Heavy Ion Reactions

H. Ewald

II. Physikalisches Institut, Giessen, Germany

1. Introduction

Fission products which emerge with high kinetic energies of about 50 to 100 MeV from thin sheets of Uranium are highly ionized (1). Their ionic charge numbers are distributed between about 18 and 26. Rays of these particles therefore can be deflected in magnetic and electric fields of reasonable small size. The radii of curvature of their trajectories in the fields are of the order of 1 to 3 m.

Keeping this in mind we constructed about 10 years ago a fission particle mass spectrograph at the research reactor Garching near Munich (2,3,4). Fig.1 shows a scheme of the arrangement which is of a modified Mattauch-Herzog type using a toroidal condenser in combination with a homogeneous magnetic field. The thin sheet of fissionable material is located in an evacuated beam tube near the core of the reactor. With the neutron flux available at Garching ($3 \cdot 10^{13}$ n/s cm^2) we have more than 10^9 fissions per second in the sheet.

A small part (10^{-5}) of the fission products emitted in the direction of the beam tube can pass the entrance slit and the deflecting fields of the spectrometer. The particles are separated into mass lines at the exit boundary of the magnetic field according to their M/e values (M = mass number, e = ionic charge number). They arrive there about 10^{-6} sec after the fission processes. It is worthy to note that a deflection voltage as high as 300 kV is needed for the condenser.

In Fig.2 a spectrum is shown of the heavy fission particle group measured with definite values of the field strengths of the spectrograph. It was collected on nuclear emulsion plates and counted in narrow strips under the microscope. The periodic intensity distribution of the lines originates from the fact that the particles occur in the different ionic charge states which are shown above the periods together with the energies of the selected particles. Below the lines their mass numbers are written. The intensity of the spectrum is such that about 25 particles per second arrive in the most abundant lines.

The separated fission particles of known mass numbers, ionic charge numbers and kinetic energies were used for a number of different studies (4,5). For a part of these studies the relatively low intensity of the separated particles was quite sufficient, for instance for certain examinations of their interactions with matter. For others, especially for studies of the nuclear structure of the particles by coincidence measurements of their short-lived ß- and γ-radiations much higher intensities are desired.

2. The fission particle parabola spectrograph at Grenoble.

For that reason a more powerful fission particle mass spectrograph is now under construction at the French-German high-flux reactor at Grenoble. This is done in a cooperation of members of the Institut Max von Laue - Paul Langevin at Grenoble, of the Institut für Neutronenphysik der Kernforschungsanlage at Jülich and of our institute at Giessen (6). The new separating system is a stigmatic focusing parabola spectrograph (7), which consists of a homogeneous magnetic sector field and a cylinder condenser (Fig.3). The mean planes of deflection of both fields are perpendicular to each other. The system has a size of about 20 meters and will be completed about the end of this year. I should mention here that a focusing parabola spectrograph in a size of about 2 meters is working since a number of years in our institute at Giessen. It is used successfully in atomic collision works with mass and energy resolutions of a few thousands (8).

The focusing fission particle parabola spectrograph has the advantage that it uses nearly the whole energy width of more than 10% of the fission particles of one mass number. The Mattauch-Herzog system at Garching on the other hand uses only about 1% energy width. The image errors of focusing parabola spectrographs have been calculated by H. Wollnik and coworkers (9). In this manner the dimensions of the new spectrograph could be chosen such as to give a maximum of intensity. Thus about 10^4 to 10^5 separated particles per second are expected in the most abundant lines. This is an intensity sufficient to study effectively the radiations and nuclear structures of the members of the separated decay chains.

The dimensions with which the spectrograph is built can be seen from Fig.3. The upper part of this figure shows the projections of the trajectories into the mean plane of deflection of the electric field, the lower part correspondingly into that of the magnetic field. In Fig.4 views are given of the whole arrangement, seen from the side and from above.

The main components of the arrangement are the following:

1. the beam tube, containing the fissionable source with

adjustable diaphragms defining the entrance slit, a collimator for the neutrons and gammas, a fast shutter and a sophisticated system for a remote controlled changing of the source.

2. the C-shaped magnet with a horizontal mean plane of deflection and 25 cm gap between the pole pieces.
3. the deflecting condenser with electrodes 30 cm apart in a vacuum of 10^{-6} mm Hg and voltages up to 600 kV. The cathode on -300 kV is of aluminium, the anode on +300 kV of stainless steel. In a preliminary experimental setup 600 kV could be kept between the electrodes during several days without breakdown.
4. an oblique exit diaphragm in the focusing plane of 72 cm length for the particles of one selected mass number corresponding to 11% energy width of the particles and a window behind the diaphragm. Beyond this exit window the fission products traverse some millimeters of air before they are trapped in a collector. Thus the measuring systems can work outside of the vacuum offering in this manner a large versatility. In order not to loose the good accessibility by a shielding against the background of the experimental hall, this part of the spectrometer is located inside of a big shielded measuring room which can serve also as a dark room for the detectors.

The used pieces of the parabolas of 72 cm length are straight lines in a good approximation. The term parabola is used here only for historical reasons. According to the calculations the apparatus shall yield a mass resolution of 800 with a source width of 1.6 mm, a source length of 40 mm, a horizontal aperture of 2.3° and a vertical aperture of 0.23°. This is enough to resolve most of the parabolas of the different mass numbers and ionic charge numbers from each other.

If a certain parabola is brought to the exit diaphragm particles of the whole decay chain of that mass number are selected, i.e. particles of 4 or 5 neighbouring atomic numbers. It raises the question whether it is possible with the arrangement at Grenoble to resolve the particles of these chains into the corresponding mass parabola multiplets. For such a purpose a resolution of about 20000 to 30000 would be necessary. This is a resolution 10 times as high as realized with our small parabola spectrograph at Giessen. But there seems to be no reason why it should not be possible to reach this higher resolution. As to our experience the parabola spectrograph is easily to be adjusted. The adjustment is achieved by turning semi-circular insertions in the edges of the pole-shoes of the magnetic field. In this manner the radial and axial images are made to move against each other to coincide to deliver a stigmatic focusing. This holds for fairly long pieces of the parabolas.

In order to obtain the high resolution it will be necessary to reduce the source width to about 0.1 mm, its length to about 1 cm and the radial and axial apertures by factors 3 or 5. Then only a small number say of the order of 10 particles per second will reach the focused parabola multiplet lines which have a usable length of 72 cm. Nuclear emulsion plates or solid state detectors can be used for the detection of these particles. Then it is clear that exposure times of minutes or hours are needed to accumulate enough particles in the multiplet components, say 100 per cm of their length.

The places of impact of the particles in the emulsion or in the solid detector can be determined with the microscope. Thus the multiplet structure could be observed. The intensity distribution along the single component would give the fission yield distribution Y(M,Z,E) for the particles of that component which thus could be measured in the most direct way. E is the kinetic energy of the particles.

We think that there is a certain chance to realize such a program. Therefore all parts of the spectrograph have been constructed very carefully. The same holds for a number of provisions which have been made for the adjustment of the apparatus. But probably additional efforts will have to be made. Possibly difficulties will come from mechanical vibrations of the beam tube of the reactor and therefore also of the source. Such vibrations may result from the vehement streaming of water for cooling purposes in the reactor. There is a proposal of H. Wollnik (10) to compensate for these vibrations by the use of small correction coils which deflect the particles opposite to the mechanical vibrations of the source. The vibrations are recorded by a laser beam reflected from a mirror which is fixed next to the source.

3. The analysis of the products of heavy ions reactions

Now I shall reproduce some preliminary considerations on a task which is analogous to the analysis of fission product rays but even more difficult. This is the analysis of the energetic products of heavy ion reactions emerging from a thin target. We are interested in this problem, because a heavy ion accelerator is in construction at Darmstadt at a distance of about 80 km from Gießen. This accelerator is intended to accelerate particles of all possible atomic numbers to energies of about 8 MeV per nucleon.

The products of the heavy ion reactions may be a small number of compound nuclei formed from the projectile and target nuclei, a larger number of fission products formed from very short-living compound nuclei and a fairly large number of products from other reactions.

Of advantage for the analysis of this mixture of particles by a spectrograph is the fact that all these particles are peaked heavily in the forward direction. The compound nuclei are within $\pm 1^{\circ}$ of the direction of the primary particles, the other particles within say $\pm 10^{\circ}$ of this direction.

The complication for the analysis of the beam of the product particles is that it is mixed with the beam of the primary particles, the projectiles. The number of the expected compound nuclei will be smaller by 12 to 16 orders than the number of the projectiles. This is quite an extreme ratio. Therefore the separation of the compound nuclei probably cannot be done in a one-stage separator, two stages will be necessary. This is a consequence of the severe scattering of the projectiles from the edges of diaphragms, from the surfaces of electrodes and pole-shoes and from rest gas molecules. The scattering can be reduced substantially by the use of materials of low atomic number for the diaphragms and the mentioned surfaces. This is known in detail from experimental (11,12) and theoretical (13) work on the scattering of fission products from surfaces.

In discussion is in our institute (14) at this time a separating system for the products of heavy ion reactions the first stage of which is shown in Fig.5. This stage consists of a Wien filter in combination with an achromatic deflection system. The second stage of the system in principle shall be a focusing parabola spectrograph. Because most particles of interest from the reactions have velocities which are considerably smaller than the velocity of the primary particles a velocity filter is best suited for the elimination of the primary particles from the beam. The field strengths in the filter have to be adjusted to such values that the primary particles are deflected most effectively to the side while the particles of interest of a certain velocity range undergo only small deflections. Typical values of the field strengths are 50 kV/cm and 2 kG. A modified type of the Wien filter with separated fields shall be used in order to avoid the difficulties of the technical realization of overlapped fields (15).

A velocity filter is achromatic in the sense that it selects particles of a certain velocity range to first order independently from the ionic charge numbers of the particles. This is of advantage from intensity reasons. Such a velocity filter on the other hand has nearly no focusing action. This action has to be delivered by some additional device, for instance by an achromatic deflection system consisting of three magnets as shown in Fig.5. Such a system has stigmatic focusing to first order. It can be designed to deliver achromatic focusing in the radial direction to higher order. This is of importance because particles with velocity and charge number ranges of 10% and more shall be focused. These achromatic systems deliver both these focusings. Thus all particles which enter the achromatic system through a relatively wide diaphragm are focused at the same place where they can pass an intermediate slit only a few millimeters wide. This holds with the restriction that by the action of the Wien filter we have a small velocity dispersion at the place of this slit. This would be a disadvantage. This velocity dispersion can be compensated by the arrangement of a second Wien filter behind the magnets of the achromatic system (shown by dotted lines in Fig. 5).

The intermediate slit is the entrance slit of the second stage of the separator arrangement, the parabola spectrograph. It is expected that the background of the scattered primary particles which passes the intermediate slit is smaller by 8 to 10 orders than the initial number of the primary particles (10^{13} per second). We try to verify this expectation by Monte-Carlo calculations of the trajectories of scattered particles through the arrangement. For these calculations we use the mentioned experimental and theoretical knowledge (11-13) of the scattering of heavy ions from surfaces.

The particles which pass the intermediate slit can undergo then mass and energy separation by the parabola spectrograph in principle in the same manner as it is done with the spectrograph at Grenoble for fission products. We think that a mass resolution of 500 to 1000 could be achieved. It is hoped that the number of scattered primary particles expected to reach the place of the single parabolas would be smaller by more than 15 orders compared to the initial number of the primary particles. The parabola spectrograph probably could be designed somewhat smaller than the arrangement at Grenoble. But its ion optics would have to be changed to some extent compared to that of the Grenoble instrument. This is because the particles from the heavy ion reactions partly need a mean radius of curvature in the electric field which is larger by about a factor 2.

The whole separator system outlined here is possibly quite ideal but also quite expensive. We examine therefore possibilities to simplify the arrangement. If our calculations should show that the background of scattered primary particles at the place of the final parabolas is not too critical we can firstly omit the second Wien filter and secondly open the intermediate slit say to about 10 cm width. Because of the velocity dispersion produced by the first Wien filter a full parabola spectrograph is not required then in the second stage of the system. A single magnetic or electrostatic field as a second stage together with the Wien filter in the first stage will act similar to a parabola spectrograph. We calculate now in detail the ion optical properties of such field combinations. It has to be mentioned here that a one-stage separator is in construction at the MIT by Enge and Betz (16) for use at the heavy ion accelerator at Berkeley. It shall consist of a Wien filter and a split pole magnet and resembles in the effect to a parabola spectrograph similar to our proposal.

----

1. Lassen, N.O., Kgl. Danske Videnskab. Selskab, Mat.Fys. Medd. 26, Nr. 5 (1951).
2. Ewald, H., Konecny, E., Opower, H., Rösler, H., Z. Naturf. 19a, 194 (1964).
3. Konecny, E., Opower, H., Ewald, H., Z. Naturf. 19a, 200 (1964).
4. Ewald, H., Mass Spectroscopy and Nuclear Fission, in Advances in Mass Spectrometry, Vol. 4, pages 899-918, London 1968
5. Ewald, H., Recoil Particle Spectrometers, in Proc. Int. Conf. on Electromagnetic Isotope Separators, Marburg 1970, Bundesministerium f. Bildung u. Wissenschaft, Forschungsbericht K70-28, pages 225-240.
6. Moll, E., Ewald, H., Wollnik, H., Armbruster, P., Fiebig, G., Lawin, H., A Mass Spectrometer for the Investigation of Fission Products, in Proc. Int. Conf. on Electromagnetic Isotope Separators, Marburg 1970, pages 241-254.
7. Neumann, S. u. Ewald, H., Z. Phys. 169, 224 (1962); see also Ewald, H., Parabola Spectrographs with 2 and 3 Fields, in Recent Developments in Mass Spectroscopy, University Park Press, 1970, pages 88-97.
8. Vogler, M. and Seibt, W., Z. Phys. 210, 337 (1968); see also Ewald, H. and Seibt, W., Mass Spectroscopic Researches on Atomic and Molecular Collisions, in Recent Developments in Mass Spectroscopy, University Park Press 1970, pages 39-59 and 939-943.
9. Fiebig, G., Lawin, H., Wollnik, H., Die ionenoptischen Eigenschaften von fokussierenden Parabelspektrographen, Bericht der Kernforschungsanlage Jülich - Nr.713- NP, 1970.
10. Wollnik, H., internal proposal made to the Institut Laue-Langevin.
11. Engelkemeir, D. and Walton, G.N., United Kingd. Atom. Energ. Auth. Rep. No. AERE-R 4716 (1964); Engelkemeir, D., Phys. Rev. 146,304 (1966).
12. Albrecht, J. and Ewald, H., Z. Naturf. 26a, 1296 (1971).
13. Güttner, K., Z. Naturf. 26a, 1290 (1971).
14. Members of the institute participating in this work: Eichler, W., Ewald, H., Güttner, K., Hinckel, P., Kumar, D., Münzenberg, G., Wollnik, H.
15. Wollnik, H., Nucl. Instr. and Meth., in print; Ewald, H., unpublished.
16. Betz, H.D., private communication.

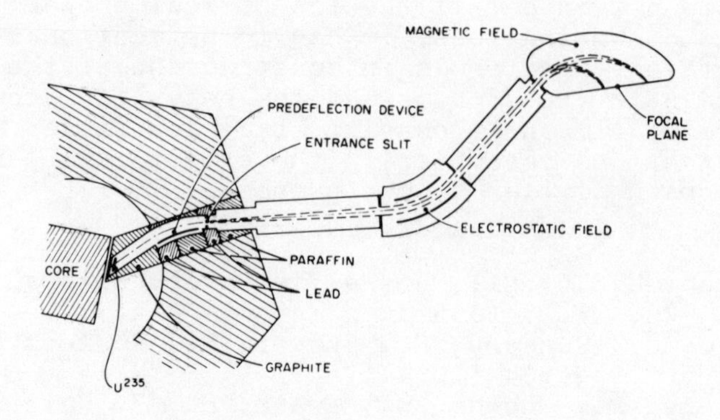

Fig. 1. Scheme of the fission particle mass spectrograph at Garching

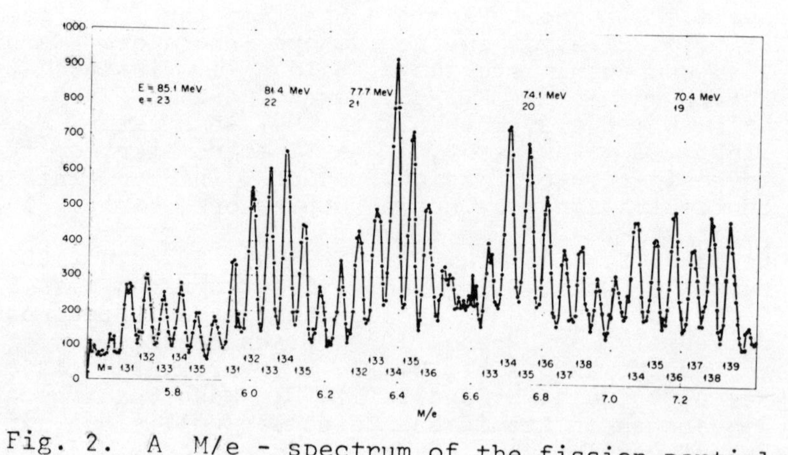

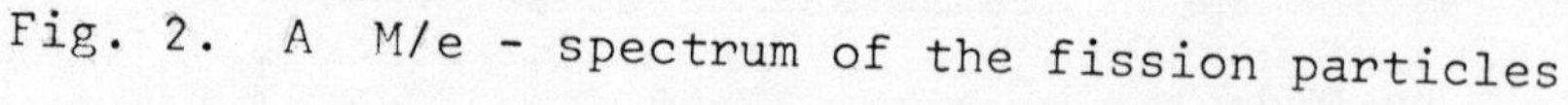

Fig. 2. A M/e - spectrum of the fission particles

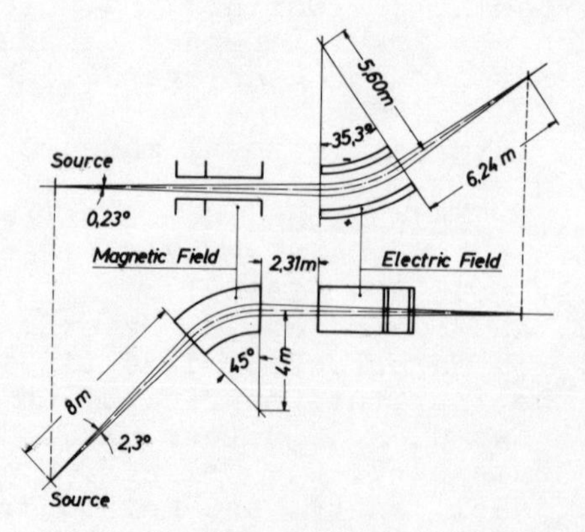

Fig. 3. Scheme of the parabola spectrograph

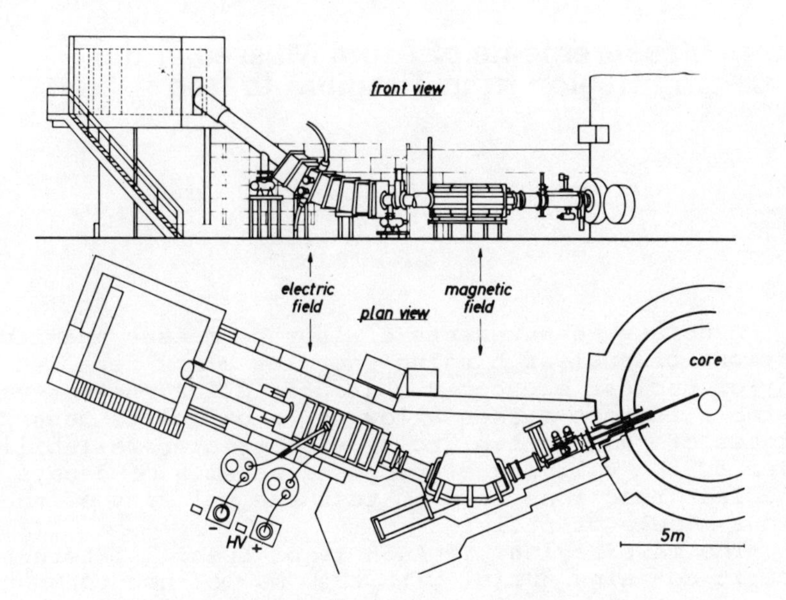

Fig. 4. The fission particle parabola spectrograph at Grenoble

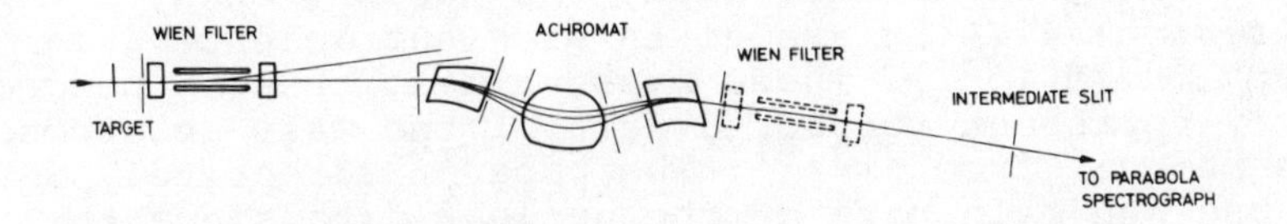

Fig. 5. The first stage of the proposed separator for the products of heavy ion reactions

Measurements of Atom Masses in the Region from Titanium to Iron

R. A. Demirkhanov, V. V. Dorokhov, M. I. Dzkuya

Physical Technical Institute of the Georgian Academy of Sciences, Sukhumi 2, USSR

Accurate measurements of isotope masses and investigations of nuclear binding energies allow detailed study of nuclear energetic characteristics. Better values of stable isotope masses allow us to calculate masses and energies of radioactive nuclei with greater reliability, to verify the validity of proposed schemes of decays and of nuclear reactions, and to test the validity of the various models of nuclei.

The mass region: $45 \leq A \leq 58$ is of special interest since it contains nuclei with the "magic" neutron number N=28. The accurate definition of the quantitative energetic parameters of nuclei on approaching the "magic" number, at the "magic" number and after filling the shell by the neutron is of great interest in solving a number of problems related to the theory of the nuclei formation and their structure.

In the recent table of masses[1)] made using all of the newest experimental data, the nuclear mass values in the region: $45 \leq A < 58$ are given with sufficiently high accuracy ($\Delta M/M \sim 10^{-7}$). These mass values, though, predominantly represent corrected data of the mass-spectroscopic measurements from [2,3,4)]. The paper[1)] effectively does not use the main part of the nuclear data since the errors of Q-values in the nuclear data are higher by approximately an order of magnitude than those in mass-spectroscopic data. Despite the fact that a great number of nuclear and some mass-spectroscopic data was rejected, the arbitrary "consistency factor: 2,65" for the obtained experimental errors was used and some other "agreement" rules were applied, we could not obtain a good agreement for various experimental data. This conclusion applies both to the comparison of nuclear and mass-spectroscopic experimental data and to that of various mass-spectroscopic values between each other. Also, the number of mass spectroscopic data is limited: besides references[2,3,4)] only the papers[5,6,7,8)] are available, which moreover should be considered. The measurements of isotope masses in[2,3,4)] are given with high accuracy (2 to 10 μu). However, in a number of cases, on measuring ΔM of the same doublet with various devices there arise very serious divergences which are far out of boundaries of the measurements total error. Thus, for the doublet $Ti^{46}\ Cl^{35}Cl^{37}-Ti^{48}Cl_2^{35}$ the divergence in the values is 9μu with the total error being equal to 4μu [4]. The values for the doublet $Ti^{47}\ Cl_2^{37}-V^{51}\ Cl_2^{35}$ are equal to 24±7μu, respectively. Thus, the divergences in the doublet values exceed the error by factor 2 to 3. It should be emphasized that the divergences in the values for the same doublet obtained from the nuclear data are very gross too. As an example we give numbers for the doublet $Ti^{49}\ Cl^{37}-V^{51}Cl^{35}$. The value $\delta\ \Delta M$ for the mass-spectroscopic data from [2] and [4] is equal to 18±4μu. The same value in comparison with the nuclear data is equal to:

from paper [2] 48 ± 25 μu
from paper [4] 30 ± 25 μu

It must be noted that the nuclear data on Q-values considerably differ from each other too. Thus, for example, the above doublet $Ti^{49}\ Cl^{37} - V^{51}\ Cl^{35}$ value, calculated with the use of different reactions, has a divergence of 61 ± 40 μu.

In [2,3,4] no attempt is made to obtain redundant results for the obtained mass values. Each mass value is measured with a single doublet; thus, existence is not excluded of unobserved systematic errors which may exceed statistic ones several times. Comparison of data in [2] and [4] shows that such a situation is quite possible.

The negligible errors in the isotope mass values over the range being studied, given in [1], are in addition due to rejection of a great number of nuclear and mass-spectroscopic data, which did not "agree" with the results of the measurements in [2,3,4] within the given accuracy. As the analysis of the values of some doublets shows the rejection of a considerable amount of the experimental data given in [1] to increase the accuracy of the given atom mass values is not sufficiently justified.

We therefore consider it desirable that nuclear mass values in the range considered are remeasured with additional control measurements of high accuracy, independent from each other measuring each mass several times and checking the obtained results for consistency in each case in order to exclude possible systematic errors.

The mass measurements have been performed with a doublefocusing mass-spectrograph [9]. The resolution is 70000 to 90000 allowing complete separation from fragments containing the isotopes of poor abundance C^{13} and N^{15}. Thus, any organic compound could be used as a substandard. Where C^{13} and N^{15} containing fragments had sufficient intensity, they were used as references too.

The mass of each isotope has been calculated from several doublets of various organic compositions. In a number of cases, fragments of the same mass but obtained from various organic compounds have been used as reference lines in the doublet. The result of each measuring cycle were treated individually and averaged afterwards. The doublet value given here represents the average value over all cycles, including different reference line origins. These origins could be: molecular ones - introduced into the ion source; fragmentary ones - isolated from the main mass of an atom or two and more atoms of hydrogen or methyl groups; and associative ones - generated in an arc of the ion source as a result of ion-molecule reactions. The ions of the materials investigated are obtained by the metal evaporation in the crucibles of special constructions. The impurity content in the materials studied does not exceed 0.1%.

The use of several doublets on measuring the mass of each isotope allowed to verify the "inner" consistency of the results. Such methods give the possibility of disclosing systematic errors which can take place within the individual measuring cycles. The probability for systematic errors is especially large in using ions of associative origin.

The measurement of "isotopic doublets" is of special interest in verifying the absence of systematic errors. The experimental errors in measuring "isotopic" doublets become rather large (because of inaccuracy on calculating the dispersion), but the basic value of an "isotopic" doublet allowes us to give firm judgements about the presence (or absence) of systematic errors within the isotope pleiad of the element being studied. The data control with "isotopic" doublets is carried on with titanium, chromium and iron. In the case of titanium the "isotopic" doublets were used both with titanium (masses between 46 and 50) and titanium oxide (62 to 66). The final values of isotope masses were calculated weighting inversely proportional to the square of the statistical errors. Each mass value neither directly nor indirectly depends on the measurements of other isotopic masses.

The apparatus dispersion calculation has been carried out individually for each doublet measurement. Consequently, any change in the doublet value because of dispersion was taken into consideration automatically during the spectrum treatment. The dispersion was determined from fragments of organic compounds only different by one or two hydrogen atoms. Such a procedure insures the relative error of the dispersion calculation for the given range of masses ~ 10^{-5}. The average relative error of measurements obtained in the present paper considering the "inner" consistency of the results is $\Delta M/M \simeq 2.10^{-7}$ which corresponds to ~ 10 keV. In most cases, the errors of individual cycles and the isotope mass values obtained using all the measured doublets were equal to 2 to 5 μu. In such cases we adopted an error of 8 μu corresponding to the instrumental error of the comparator with the given width of the mass-spectral line. The errors, given in the values of doublets, masses and average magnitudes of masses represent the average statistical errors calculated in the usual way [10].

The mass values of H^1, C^{13}, N^{14}, O^{16}, S^{32}, Cl^{35} and Cl^{37} recommended in [1)] are used as substandards. The designations H, C, N and O always stand for H^1, C^{12}, N^{14} and O^{16}. All data are given in the scale C^{12} = 12.000000 u.

A check the reliability of the dispersion calculations and the correctness of the mass-spectrograph ion-optical system adjustment is obtained by comparing differences between reference lines at each mass number with well-known values, mostly taken from ref. [1)]. A number of such comparisons is given in table I, they form part of a large set.

The values of the calculated control doublets agree well with the similar data from [1)]. No trend is displayed in the differences, over the whole range of the measured masses. All these facts indicate the high quality of the mass-spectrograph adjustment and the reliability of the apparatus dispersion calculation during the measurements.

Table II gives the measured values of 82 basic doublets and the resulting masses for 16 stable isotopes of Ti, V, Cr, Mn and Fe. The error values given, refer to the last significant digit.

The Mass Measurements

1. Titanium. Titanium was evaporated in tungsten crucibles. Every isotope of titanium has been measured using organic doublets of various contents and origins. In addition, some isotopes were related to S^{32}, Cl^{35} and to copper isotopes. Since titanium isotopes are related to each other by "isotopic" doublets, the whole pleiad of titanium isotopes is related to the elements listed above.

We used the following organic compounds and their fragments: ethyl alcohol (m 46, C_2H_6O), methyl ether of formic acid (m 60, $C_2H_4O_2$), benzol (m 73, C_6N_6) and cyclohexane (m 82, C_6H_{16}). The comparison with cyclohexane is collected for the masses: 62 to 66, using titanium oxide (TiO).

Since the low-voltage arc plasma source mainly operated with helium (~ 95%) to produce the ions of the elements investigated the arc contained a great number of ions of several compounds which were used as reference lines in some cases.

2. Vanadium. Vanadium ions have been produced by evaporating metallic vanadium in a tungsten crucible. The vapour of the organic compounds used as references have been introduced into the arc of the ion source. Alongside with the fragments which directly take part in the doublet formation as reference lines, the mass spectrum contained lines corresponding to $\pm$ H, $\pm$ 2H and higher. These lines have been used in the present paper to calculate the dispersion constants of the mass-spectrograph.

On measuring vanadium isotopes it turned out

possible to obtain simultaneously a qualitative spectrum of two isotopes (with the resolution of ~ 80.000) though the abundance of the isotope: V^{50} is only 0.25%. The statistics for both isotopes involving the rest for the "inner" consistence have been carried out separately.

The organic compounds benzol (m 78, C_6H_6), vinyl cyanide (m 53, C_3H_3N), cyclohexane (m 82, C_6H_{10}) and n-amyl alcohol (m 88, $C_5H_{12}O$) were used as references. Cyclohexane and n-amyl alcohol have been used as references for forming fragments for measuring vanadium oxides (m 66 and m 67).

3. Chromium. Chromium was evaporated in a tantalum crucible with the organic compounds vinyl cyanide (m 53, C_3H_3N), cyclohexane (m 82, C_6H_{10}) and n-amyl alcohol (m 88, $C_5H_{12}O$) and their fragments. All of these compounds give the fragments of a good intensity for chromium isotope masses.

4. Manganese. Manganese ions have been produced by evaporating the metallic manganese in a tantalum crucible. The compound vapour n-amyl alcohol (m 88, $C_5H_{12}O$) has been introduced into the ion source arc by means of a special valve. We measured the mass of an isotope Mn^{55} using the n-amyl alcohol fragment C_3H_3O, m 55 and the fragments of C_3H_3N and C_4H_7 which are formed in the ion plasma source arc. In order to test the consistency of the results on measuring the Mn^{55} mass we used the fragments of cyclohexane (m 82, C_6H_{10}) and isopropyl alcohol (m 60, C_3H_8O) as references.

5. Iron. Iron ions were produced from metallic iron with a minimum content of carbon. The following organic compounds and their fragments have used as references: acetone (m 58, C_3H_6O), cyclohexane (m 82, C_6H_{10}) and acetamide (m 59, C_2H_5ON). The mass of Fe^{58} isotope could not be obtained simultaneously with the other iron isotopes because of its poor abundance (~ 0,26%). For this isotope we had to collect individual statistics. For this reason we had not measured the "isotopic" doublet involving Fe^{58}.

Discussion of the Results Obtained.

Table III gives the comparison of the masses of titanium, vanadium, chromium, marganese and iron isotopes, obtained in the present paper with the data of earlier mass-spectroscopic measurements. Most of the results do not agree within the given total errors. The agreement is best with data from [2)] (see Ti^{49}, V^{50} et al.) Yet even here several serious discrepancies exist that can not be explained by statistical errors (Ti^{47}, Cr^{53}, Mn^{55} et al.). Yet, the agreement with these data is far better than that with the earlier data from [8)], as shown in table III.

The analysis of the differences of the present data with the results from [2,3] and [8] shows that they are of a statistic character. We are convinced that they are at least partly due to absence of tests for systematic errors, i.e., the tests for the "inner" consistence of the results. This fact explains both the discrepancies between values of the same doublets obtained in various laboratories (see [2] and [4]) and in the same laboratory (see [8] and [2]) at different times and with various devices. The authors do not give satisfactory treatments for the divergences which frequently exceed the total measurements errors several times.

The absence of any trend in the deviations considerable complicates the procedure of finding the reasons for these errors. The situation available must be studied on the basis of a wider range of the experimental data obtained both by mass-spectroscopic and nuclear ways. One of the most promising techniques for finding and removing systematic errors is the analysis of the variation of nuclear energetic parameters over the range studied.

The existence of these discrepancies exceeding the measurement total errors two or three times indicated the presence of systematic errors in the measurements that cannot be decreased by increasing statistics. The most effective way to exclude them is to measure the same mass using doublets of various organic compositions obtained from various compounds. It is desirable that the reference masses should be both of molecular and fragmentary origin.

The comparison of the masses measured in the present paper with the nuclear data for mass differences shows that there is a considerable number of Q-values rejected in [1] that confirm the validity of our data. However, it should be noted that between the various nuclear data there exist divergences which often exceed 100-150 μ at units. Consequently, such a comparison can not give a well-defined indication of the reliability of any data.

The authors express their acknowledtements to E.E. Baroni and K.A. Kovirzina for affording the necessary organic compounds and to G.A. Dorokhova for the practical help during the work.

1. MATTAUCH, J.H.E., THIELE, W., WAPSTRA, A.H., Nucl. Phys. 67, 1 (1965).

2. GIESE, C.F., BENSON, J.L., Phys. Rev. 110, N3, 721 (1958).

3. QUISENBERRY, K.S., SCOLMAN, T.T., NIER, A.O.. Phys. Rev. 104, N2, 461 (1956).

4. BARBER, R.C., Mc.LATCHIE, W. et all., Can. J. Phys. 42, 591 (1964).

5. OKUDA, T., OGATA, K., Phys. Rev. 60, 690 (1941).

6. OGATA, K., Phys. Rev. 75, 200 (1949).

7. DUCKWORTH, H.R., JOHNSON, H.A., Phys. Ref. 78. 179 (1952).

8. COLLINS, T.L., NIER, A.O., JOHNSON, W.H., Phys. Rev. 86, 408 (1952).

9. DEMIRKHANOV, R.A., GUTKIN, T.I., DOROKHOV, V.V., RUDENKO, A.D., Atomiç Energy 2, 21 (1956).

10. DEMIRKHANOV, R.A., DOROKHOV, V.V., DZKUYA, M.I., Izv. AN USSR series of Physics, 27, N10, 1338 (1963).

11. JOHNSON, W.H., Phys. Rev. 87, N1, 166 (1952).

* Paper received by editor after the conference and accepted because of its importance.

TABLE 1

Isotope	Doublet	ΔM value obtained in the present paper, m at units (1971)	ΔM value calculated from [1]) m at units	$\delta = \Delta M - \Delta M^1$ μ at units
1	2	3	4	5
Ti^{46}	$C_2H_8N-CC^{13}H_5O$ ($CH_3N-C^{13}O$)	28,272 ± 16	28,280 ± 5	−8 ± 21
	$C_2H_8N-CH_4NO$ (CH_4-O)	36,399 ± 12	36,385 ± 3	14 ± 15
	$C_2H_8N-CN_2O_2$ (CH_6N-O_2)	60,190 ± 16	60,195 ± 5	−6 ± 21
V^{51}	$C_4H_3-C_3NH$ (CH_2-N)	12,583 ± 12	12,576 ± 3	7 ± 15
	$C_5H_7-C_4H_5N$ (CH_2-N)	12,587 ± 17	12,576 ± 3	11 ± 20
Cr^{54}	$C_4H_6-C_3H_4N$ (CH_2-N)	12,573 ± 22	12,576 ± 3	−3 ± 25
	$C_4H_6-C_2NO$ (C_2H_6-NO)	48,961 ± 39	48,962 ± 4	−1 ± 34
	$C_3H_4N-C_2NO$ (CH_4-O)	36,388 ± 25	36,385 ± 3	3 ± 28
Fe^{58}	$C_3H_8N-C_3H_6O$ (NH_2-O)	23,806 ± 18	23,810 ± 3	−4 ± 21
	$C_3H_8N-C_2H_4NO$ (CH_4-O)	36,383 ± 18	36,385 ± 3	−2 ± 21
	$C_3H_6O-C_2H_4NO$ (CH_2-N)	12,577 ± 19	12,576 ± 3	1 ± 22

TABLE 2

Mass Charge	Doublet	ΔM value in at units	Isotope mass value, at units	Average mass value, at units
$_{22}Ti^{46}$	$C_2H_8N-Ti^{46}$	113,071 ± 7	45,952605 ± 8	45,952602 ± 8
	$CC^{13}H_5O-Ti^{46}$	84,799 ± 13	45,952596 ± 13	
	CH_4NO-Ti^{46}	76,672 ± 8	45,952618 ± 9	
	$C_5H_2-Ti^{46}O$	68,145 ± 15	45,952590 ± 15	
	$CH_2O_2-Ti^{46}$	52,881 ± 14	45,952599 ± 14	
	$C^{13}HO_2-Ti^{46}$	48,423 ± 9	45,952587 ± 10	

Mass Charge	Doublet	ΔM value in at units	Isotope mass value, at units	Average mass value, at units
	$CC^{13}H_8N-Ti^{47}$	117,329 ± 14	46,951701 ± 14	
	$C_2H_7O-Ti^{47}$	98,012 ± 7	46,951679 ± 8	
$_{22}Ti^{47}$	$C_5H_3-Ti^{47}O$	76,869 ± 10	46,951692 =10	46,951691 ± 8
	$CH_3O_2-Ti^{47}$	61,608 ± 10	46,951698 ± 10	
	Ti^{47} $O-Cu^{63*}$	17,036 ± 23	46,951713 ± 24	
	$Ti^{47}-Ti^{46}$	999,071 ± 41	46,951673 ± 42	
	$C_5H_4-Ti^{48}O$	88,492 ± 24	47,947894 ± 24	
	$C_5H_5-Ti^{48}OH$	88,494 ± 27	47,947892 ± 27	
$_{22}Ti^{48}$	$C_4H_2N-Ti^{48}O$	75,935 ± 17	47,947875 ± 17	47,947900 ± 8
	C_4-Ti^{48}	52,109 ± 19	47,947891 ± 19	
	$Ti^{48}H-Ti^{49}$	7,876 ± 7	47,947915 ± 11	
	$Ti^{48}OH-Ti^{49}O$	7,874 ± 27	47,947913 ± 29	
	$Ti^{48}-Ti^{47}$	996,209 ± 48	47,947900 ± 49	
	$C_5H_5-Ti^{49}O$	96,348 ± 19	48,947863 ± 19	
	$CH_5S^{32}-Ti^{49}$	63,365 ± 14	48,947834 ± 14	
$_{22}Ti^{49}$	C_4H-Ti^{49}	59,967 ± 10	48,947858 ± 10	48,947864 ± 8
	$CH_2Cl^{35}-Ti^{49}$	36,637 ± 13	48,947864 ± 14	
	$Ti^{49}O-Cu^{65*}$	15,030 ± 10	48,947901 ± 13	
	$Ti^{49}-Ti^{48}$	999,957 ± 36	48,947857 ± 37	
	$C_5H_6-Ti^{50}O$	107,253 ± 18	49,944783 ± 18	
	$C_4H_2-Ti^{50}$	70,860 ± 8	49,944790 ± 9	
$_{22}Ti^{50}$	$C_3C^{13}H-Ti^{50}$	66,401 ± 21	49,944779 ± 21	49,944787 ± 8
	C_3N-Ti^{50}	58,279 ± 43	49,944795 ± 43	
	$CH_3Cl^{35}-Ti^{50}$	47,550 ± 23	49,944777 ± 23	
	$Ti^{50}-Ti^{49}$	996,925 ± 38	49,944789 ± 39	
	$C_4H_2-V^{50}$	68,485 ± 14	49,947165 ± 14	
$_{23}V^{50}$	C_3N-V^{50}	55,903 ± 23	49,947171 ± 21	49,947167 ± 8
	$CH_3Cl^{35}-V^{50}$	45,158 ± 17	49,947169 ± 17	
	$C_5H_7-V^{51}O$	115,921 ± 13	50,943940 ± 13	
$_{23}V^{51}$	$C_4H_5N-V^{51}O$	103,334 ± 13	50,943951 ± 13	50,943951 ± 8
	$C_4H_3-V^{51}$	79,526 ± 9	50,943950 ± 10	
	C_3NH-V^{51}	66,943 ± 7	50,943957 ± 8	
	$C_4H_2-Cr^{50}$	69,608 ± 8	49,946042 ± 9	
$_{24}Cr^{50}$	C_3N-Cr^{50}	57,051 ± 7	49,946023 ± 8	49,946032 ± 8
	$CH_3Cl^{35}-Cr^{50}$	46,290 ± 14	49,946037 ± 15	
	$C_4H_4-Cr^{52}$	90,826 ± 9	51,940475 ± 10	
$_{24}Cr^{52}$	$C_3C^{13}H_3-Cr^{52}$	86,373 ± 18	51,940457 ± 18	
	$C_3H_2N-Cr^{52}$	78,253 ± 6	51,940472 ± 7	51,940472 ± 8
	$Cr^{52}-Cr^{50}$	1994,434 ± 41	51,940466 ± 42	

Mass Charge	Doublet	ΔM value in at units	Isotope mass value, at units	Average mass value, at units
	$C_4H_5-Cr^{53}$	98,529 ± 8	52,940579 ± 9	
	$C_3H_3N-Cr^{53}$	85,958 ± 10	52,940592 ± 10	
$_{24}Cr^{53}$	$C_2C^{13}H_2N-Cr^{53}$	81,507 ± 27	52,940572 ± 27	52,940593 ± 8
	C_3HO-Cr^{53}	62,152 ± 14	52,940588 ± 14	
	$Cr^{53}-Cr^{52}$	1000,115 ± 46	52,940587 ± 47	
	$C_4H_6-Cr^{54}$	108,018 ± 17	53,938933 ± 17	
	$C_3C^{13}H_5-Cr^{54}$	103,569 ± 15	53,938911 ± 15	
	$C_3H_4N-Cr^{54}$	95,445 ± 13	53,938930 ± 13	
$_{24}Cr^{54}$	$C_2C^{13}H_3N-Cr^{54}$	90,960 ± 24	53,938944 ± 24	52,938927 ± 8
	C_2NO-Cr^{54}	59,057 ± 26	53,938932 ± 26	
	$Cr^{54}-Cr^{53}$	998,338 ± 48	53,938931 ± 49	
	$C_4H_7-Mn^{55}$	116,757 ± 8	54,938019 ± 8	
	$C_3C^{13}H_6-Mn^{55}$	112,281 =25	54,938025 ± 25	
$_{25}Mn^{55}$	$C_3H_5N-Mn^{55}$	104,202 ± 10	54,937998 ± 10	54,938013 ± 8
	$C_2H_3N_2-Mn^{55}$	91,618 ± 28	54,938006 ± 28	
	$C_3H_3O-Mn^{55}$	80,372 ± 10	54,938019 ± 10	
	$C_4H_6-Fe^{54}$	107,368 ± 11	53,939583 ± 11	
	$C_3C^{13}H_5-Fe^{54}$	102,908 ± 48	53,939572 ± 48	
$_{26}Fe^{54}$	$C_3H_4N-Fe^{54}$	94,791 ± 8	53,939584 ± 8	53,939582 ± 8
	C_2NO-Fe^{54}	58,411 ± 8	53,939578 ± 8	
	$Fe^{56}-Fe^{54}$	1995,245 ± 47	53,939600 ± 48	
	$C_4H_8-Fe^{56}$	127,754 ± 10	55,934848 ± 10	
	$C_3C^{13}H_7-Fe^{56}$	123,300 ± 47	55,934831 ± 47	
$_{26}Fe^{56}$	$C_3H_6N-Fe^{56}$	115,171 ± 13	55,934855 ± 13	55,934845 ± 8
	$C_3H_4O-Fe^{56}$	91,381 ± 15	55,934835 ± 15	
	$C_2H_2NO-Fe^{56}$	78,790 ± 24	55,934850 ± 24	
	$C_2O_2-Fe^{56}$	54,990 ± 9	55,934840 ± 9	
	$C_4H_9-Fe^{57}$	135,085 ± 11	56,935342 ± 11	
	$C_3H_7N-Fe^{57}$	122,500 ± 10	56,935351 ± 10	
	$C_3H_5O-Fe^{57}$	98,684 ± 8	56,935357 ± 8	
$_{26}Fe^{57}$	$C_2H_3NO-Fe^{57}$	86,104 ± 17	56,935361 ± 17	56,935351 ± 8
	$Fe^{56}H-Fe^{57}$	7,325 ± 7	56,935345 ± 11	
	$Fe^{57}-Fe^{56}$	1000,491 ± 39	56,935336 ± 40	
	$C_3H_8N-Fe^{58}$	132,382 ± 12	57,933294 ± 12	
	$C_3H_6O-Fe^{58}$	108,576 ± 13	57,933290 ± 13	
$_{26}Fe^{58}$	$C_2H_4NO-Fe^{58}$	95,999 ± 13	57,933291 ± 13	57,933291 ± 8
	Ni^{58}†$-Fe^{58}$	2,059 ± 32	57,933283 ± 33	

*Isotope mass values for copper are taken from [1].

† Ni^{58} mass value is taken from [1].

TABLE 3

NN	Isotope	Data from the present paper, at units	Mass-spectroscopic data, at units			Date from [1] at units
			paper [8]	paper [2]	paper [3]	
1	2	3	4	5	6	7
1	Ti^{46}	45,952602 ± 8	45,952335 ± 40	45,952632 ± 2		45,952632 ± 3
2	Ti^{47}	46,951691 ± 8	46,951733 ± 90	46,951759 ± 3		46,951768 ± 3
3	Ti^{48}	47,947900 ± 8	47,947817 ± 60	47,947947 ± 1		47,947950 ± 2
4	Ti^{49}	48,947864 ± 8	48,947914 ± 50	48,947866 ± 2		48,947870 ± 2
5	Ti^{50}	49,944864 ± 8	49,944781 ± 30	49,944789 ± 2		49,944786 ± 4
6	V^{50}	49,947167 ± 8	49,947421 ± 120*	49,947166 ± 2		49,947164 ± 4
7	V^{51}	50,943951 ± 8	50,944221 ± 50	50,943979 ± 2		50,943961 ± 3
8	Cr^{50}	49,946032 ± 8	49,946112 ± 60	49,946050 ± 2		49,946054 ± 4
9	Cr^{52}	51,940472 ± 8	51,940450 ± 90	51,940513 ± 2		51,940513 ± 3
10	Cr^{53}	52,940593 ± 8	52,940777 ± 70	52,940651 ± 2		52,940653 ± 3
11	Cr^{54}	53,938927 ± 8	53,939085 ± 200	53,938875 ± 2		53,938882 ± 4
12	Mn^{55}	54,937935 ± 8	54,938232 ± 110	54,938058 ± 2		54,938050 ± 4
13	Fe^{54}	53,939546 ± 8	53,939785 ± 50	–	53,939611 ± 14	53,939617 ± 5
14	Fe^{56}	55,934850 ± 8	55,934823 ± 100	–	55,934945 ± 4	55,934936 ± 4
15	Fe^{57}	56,935343 ± 8	56,935380 ± 90	–	56,935415 ± 7	56,935398 ± 5
16	Fe^{58}	57,933291 ± 8	57,933498 ± 400	–	57,933321 ± 4	57,933282 ± 5

*See [11].

Part 4 Coulomb Energies

Coulomb Energies and Nuclear Shapes

Joachim Janecke

The University of Michigan,
Ann Arbor, Michigan USA

One of the first estimates of nuclear radii was obtained from experimental Coulomb energies. Bethe (1), over 30 years ago, showed that the energy difference between the ground states of mirror nuclei is essentially due to a difference in electrostatic energy. Assuming spherical charge distributions of uniform density, charge radii were extracted. Many experimental and theoretical papers on Coulomb energies have appeared since, but muonic X-ray and electron scattering experiments (2,3) have contributed considerably more to the understanding of the size and shape of nuclear charge distributions. Only very recently Nolen and Schiffer (4) have reestablished the importance of Coulomb energy data for obtaining information about nuclear sizes. They showed that the Coulomb displacement energy between any nuclear state, a ground state for example, and its analogue state in the neighboring proton-rich isobar depends strongly on the radial distribution of the neutron excess in the ground state. Thus, experimental Coulomb displacement energies, now known even in heavy nuclei (5), present a powerful tool for determining distributions of neutrons.

Let us write the displacement energy ΔE between neighboring analogue states as a sum of several terms

$$\Delta E = \Delta E_{DIR}\,(1-\varepsilon) + \Delta E_{SO} + \Delta E_{CORR}\ . \qquad (1)$$

Here, ΔE_{DIR} represents the difference in the direct Coulomb energy, $\varepsilon = -\Delta E_{EXCH}/\Delta E_{DIR}$ is the so-called exchange factor which accounts for the reduction due to the anti-symmetrization of the many-particle wave function, and ΔE_{SO} is the electro-magnetic spin-orbit energy associated with the nuclear magnetic moments. The quantity ΔE_{CORR}, finally, is a correction term which results from the numerous small and not so small charge-dependent effects which have been recognized in recent years, such as Coulomb perturbations, rearrangement of the core, charge-dependent nuclear forces, nuclear correlations, isospin impurities of the core, the dynamic proton-neutron mass effect, finite size corrections, etc. It appears that not all corrections are fully understood, a fact that may account for the so-called Coulomb energy anomaly. Considerable progress towards understanding these contributions has been made in recent years (4, 6-12), particularly for the ^{41}Sc-^{41}Ca mirror pair, but quantitative agreement still appears to be lacking.

The purpose of the present investigation was to derive a Coulomb energy equation for medium-heavy and heavy nuclei based upon realistic density distributions for the proton core and the neutron excess. It was hoped that a least-squares analysis of the more than 100 experimental values with $Z \geq 28$ would provide insight into the properties of such quantities as the radius of the neutron excess, the neutron halo, deformation effects and the isotope, isotone and isobar shift of the nuclear charge radius. No attempts were made at a detailed treatment of the correction terms. It was hoped that by including them as a single adjustable quantity, insight, for example on its dependence on A, could be obtained. The results for spherical nuclei will be presented in a condensed version. Details are contained in a forthcoming publication (13). The results for deformed nuclei are preliminary. They have been obtained in cooperation with J. P. Draayer and will be published at a later time (14).

The calculation of the direct, exchange and spin-orbit terms will be based on the assumption that the configuration of the analogue state is obtained exactly by applying the isospin-lowering operator T_- to the wave function of the ground state. Thus, one has to calculate the electrostatic energy of the "excess proton" in the field of the core protons (equal to that of the ground state), whereby the spatial distribution of the "excess proton" is identical to that of the neutron excess in the ground state. The assumption that the isobaric analogue state is obtained exactly by applying the T_- operator is known to be not quite correct, and any correction to the displacement energy arising from this assumption will have to be included in the correction term ΔE_{CORR}.

Fermi distributions

$$\rho(\vec{r}) = \frac{\rho(0)}{1 + \exp\left\{\dfrac{r - \eta c_o(1 + \Sigma\, \alpha_{\lambda\mu} Y_{\lambda\mu})}{a(1 + \Sigma\, \alpha'_{\lambda\mu} Y_{\lambda\mu})}\right\}} \tag{2}$$

will be used throughout the present work to describe the matter density distribution of the protons and the neutron excess in the ground states of nuclei. The sums Σ have to be taken over λ and μ. Here, $c_o = r_o A^{1/3}$ with a constant value r_o and η is a quantity which ensures density conservation in the nuclear interior, often referred to as volume conservation (for nuclei with sharp surfaces). The quantity η depends on a, c_o and the various deformation parameters $\alpha_{\lambda\mu}$ and $\alpha'_{\lambda\mu}$. In the general case, the half density radius $c = \eta c_o(1 + \Sigma\, \alpha_{\lambda\mu} Y_{\lambda\mu})$ and the diffuseness $a(1 + \Sigma\, \alpha'_{\lambda\mu} Y_{\lambda\mu})$ are assumed to be angle dependent.

A special case will be considered first. For spherical distributions without an explicit dependence on the neutron excess we have

$$c = \eta c_o \tag{3}$$

$$R = \sqrt{\frac{5}{3}}\, \langle r^2 \rangle^{1/2} = c_o \left[\eta^5 + \frac{10\pi^2}{3}\, \eta^3 \left(\frac{a}{c_o}\right)^2 + \frac{7\pi^4}{3}\, \eta \left(\frac{a}{c_o}\right)^4\right]^{1/2} \tag{4}$$

with $$\eta = 1 - \frac{\pi^2}{3}\left(\frac{a}{c_o}\right)^2 + \frac{\pi^6}{81}\left(\frac{a}{c_o}\right)^6 . \tag{5}$$

The expressions for the half density radius c and the equivalent radius R are known (15). Elton (2), by using the experimental information from muonic X-ray and electron scattering experiments, derived $r_o \approx 1.135$ fm and $a \approx 0.513$ fm for the nuclear charge radii. Any Coulomb energy equation should reproduce the experimental displacement energies with R and c given by eqs. (3) and (4) and the above values for r_o and a.

The direct Coulomb energy and Coulomb displacement energy can be calculated from

$$E_{DIR} = (1 - \frac{1}{Z})\, e \int_0^\infty \rho(r) V(r) 4\pi r^2 dr = \frac{Z(Z-1)e^2}{c} \frac{(\frac{c}{a})}{2[F_2(\frac{c}{a},\infty)]^2} \int_0^\infty \frac{[F_2(\frac{c}{a},\xi)]^2}{\xi^2} d\xi \tag{6}$$

$$\Delta E_{DIR} = \frac{e}{N-Z} \int_0^\infty \rho_{EXC}(r) V_{CORE}(r) 4\pi r^2 dr = \frac{Ze^2}{c} \frac{(\frac{c}{a})}{F_2(\frac{c}{a},\infty) F_2(\frac{c'}{a},\infty)} \int_0^\infty \frac{F_2(\frac{c}{a},\xi) F_2(\frac{c'}{a},\xi)}{\xi^2} d\xi \; . \tag{7}$$

Here, generalized Fermi integrals defined by

$$F_n(k,x) = \int_0^x \frac{\xi^n}{1 + e^{\xi - k}} d\xi \tag{8}$$

have been introduced. Properties of these integrals are well known (16,17) in the limit $x \to \infty$. A study of the properties of the generalized integrals in conjunction with proper expansions leads to expressions for E_{DIR} and ΔE_{DIR} which are given in ref. (13). Certain characteristics of the expression for ΔE_{DIR} can be well recognized in the limits $a \to 0$ or $f \to 1$ where $f \equiv R'/R$ is the ratio between the equivalent radii of the neutron excess and the proton core. For $a \to 0$ one obtains

$$\Delta E_{DIR} \approx \frac{6}{5} \frac{Ze^2}{R} \frac{5f^2-1}{4f^3} \; . \tag{9}$$

Thus, ΔE_{DIR} depends on R <u>and</u> $R'=fR$. If R or R' is kept constant and the respective other radius is varied, one obtains

$$\frac{\delta \Delta E_{DIR}}{\Delta E_{DIR}} \approx - \frac{5f^2-3}{5f^2-1} \frac{\delta R'}{R'} \qquad \text{if } R = \text{const} \tag{10}$$

$$\frac{\delta \Delta E_{DIR}}{\Delta E_{DIR}} \approx - \frac{2}{5f^2-1} \frac{\delta R}{R} \qquad \text{if } R' = \text{const} \; . \tag{11}$$

Both equations combined give

$$\frac{\delta R}{R} + \frac{5f^2-3}{2} \frac{\delta R'}{R'} \approx 0 \qquad \text{if } \Delta E_{DIR} = \text{const.} \tag{12}$$

This equation shows that it is impossible to use experimental Coulomb displacement energies and search independently for the radii of the core <u>and</u> the neutron excess. Not unless information exists or assumptions are made about the radius of the neutron excess can charge radii be determined. Not unless the charge radius is known can the radius of the neutron excess or the neutron halo be determined. The present investigation uses the latter approach with the charge radii required to be described by eqs. (4) or (5).

In the limit $f \to 1$ one obtains

$$\Delta E_{DIR} \approx \frac{6}{5}\,\frac{Ze^2}{R}\,(1 + 18.030\,(\tfrac{a}{R})^3 + \cdots) \quad (13)$$

which shows that the diffuseness correction to the direct Coulomb displacement energy is quite small if the equivalent radius is used in the expression.

Using similar mathematical procedures, analytical expressions for the exchange factor ε and the electromagnetic spin-orbit energy ΔE_{SO} have been obtained. The equation for the exchange factor is based on an expression derived in the statistical Fermi gas model by Bethe and Bacher (1) many years ago. The interesting result is that the diffuseness correction to the otherwise well-known expression for ε is of first order in (a/R) and therefore appreciable. Nolen and Schiffer (4) compared exchange factors calculated for different nuclear model and found the results from the statistical model to disagree considerably from the predictions of the other models. This discrepancy is removed when the diffuseness of the nuclear surface is taken into consideration. The exchange factor ε decreases from about 4% at Z=20 to 2% at Z=80.

The calculation of the electromagnetic spin-orbit energy ΔE_{SO} requires an evaluation of the spin-orbit sum $S = \Sigma\,\vec{\ell}\,\vec{\sigma}$ (summed over the configurations of the excess neutrons). Note that $S = S_N - S_Z$, where the subscript indicates summation over all configurations of N neutrons or Z protons, respectively. Assuming a single particle shell model with some mixing between subshells, the line labelled 0% in Fig. 2 is obtained. A least-squares analysis was performed for the 42 energies (taken from ref. (18)) of nuclei which are presumably nearly spherical in shape (Z or N equal to 28, 50, 82, 126 plus or minus 0 or 1 unit) and a standard deviation of only 60 keV was obtained. Moreover, a strong correlation between the differences between the calculated and experimental energies and the term ΔE_{SO} was recognized. This correlation is shown in Fig. 1. A more realistic result is obtained when large amounts of excitations into higher shell model orbits are introduced for all nuclei. The occupation probability has to be changed from a simple step function to a more realistic function such as a Fermi function. Figure 2 shows $\overline{S}_N$ for p=10% and 30% excitations into higher orbits. Using p as variable parameter, a χ^2-fit showed a significant improvement for p>15% with a broad minimum near $p \approx 35\%$. The standard deviation is decreased to slightly over 20 keV. It may be remembered that theoretical estimates (19-23) for core excitations in the regions of ^{16}O, ^{40}Ca, ^{208}Pb often amount to over 50%.

The expression $\overline{S}_N$=0.6N (dotted line in Fig. 2) has been used in all the following analyses. The use of this expression yields ΔE_{SO}

$\approx$ - 20 keV for all nuclei which differs by less than ±10 keV from the values calculated with any p>15%. Figure 3 shows on the left hand side the difference between the experimental and calculated Coulomb displacement energies for several values of f and $\Delta E_{CORR}=0$. The individual points are shown only for f=1.12 and are otherwise represented by the straight lines. The lines deviate greatly from zero. Therefore, the graph must be re-interpreted: if $f \equiv R'/R$ is equal to 1.12, for example, there must exist a correction term ΔE_{CORR} which decreases linearly with A from about 550 keV in the light nuclei to about 350 keV in the heavy nuclei. Otherwise, the calculated and experimental values would not agree. The right hand side of Fig. 3 shows in a similar way the dependence on the parameters r_o and a.

The question arises as to which values for the ratio $f \equiv R'/R$ are most reasonable. If f is given fixed values and ΔE_{CORR} is allowed to vary linearly with A, the standard deviation obtained from a least-squares analysis exhibits a broad minimum near f=1.125. It thus appears that values for f between 1.10 and 1.15 are most reasonable. The rather limited knowledge about ΔE_{CORR} does not permit any more definite statement about the value of f. However, it will be shown that for f=1.125±0.025 one obtains very reasonable values for the neutron halo.

If the ratio f and the radius of the proton core (ground state charge radius) are assumed to be known, the neutron halo H can be calculated from

$$H \equiv \langle r_n^2 \rangle^{1/2} - \langle r_p \rangle^{1/2} = \left(\sqrt{\frac{Z+(N-Z)f^2}{N}} - 1 \right) \sqrt{\frac{3}{5}}\, R \qquad (14)$$

with

$$R = c_o\left\{1 + \frac{5\pi^2}{6}\left(\frac{a}{c_o}\right)^2 - \frac{7\pi^4}{24}\left(\frac{a}{c_o}\right)^4 + \frac{1003\pi^6}{1296}\left(\frac{a}{c_o}\right)^6\right\};\ c_o = r_o A^{1/3}. \qquad (15)$$

Figure 4 shows a plot of H as a function of A for the nuclei along the stability line. The three heavy lines have been obtained using r_o=1.135 fm, a=0.513 fm and f=1.100, 1.125 and 1.150, respectively. The thin line has been calculated by Myers and Swiatecki (24,25) on the basis of the droplet model. Excellent agreement exists over the entire range. A comparison for H obtained from a variety of procedures for a selected set of nuclei ranging from ^{48}Ca to ^{208}Pb is given in ref. (13). Again, excellent agreement exists between the results from this work and those from the droplet model. In addition, the results obtained from Hartree-Fock calculations are in excellent agreement. The values obtained by Nolen and Schiffer (4) based on wavefunctions generated in Woods-Saxon potential wells also agree quite well. The good agreement between the results for the different methods suggests that the calculated values for the neutron halo are indeed close to the true values.

Deformation effects will be discussed next. The results are preliminary. Figure 5 presents as a function of A the individual values for the difference between the experimental and calculated displacement energies using all data with Z≥28. Properly adjusted values for f and ΔE_{CORR} were used in the calculation. For the nuclei near magic lines which are presumably nearly spherical in shape (filled points), the difference is close to zero and the standard deviation is about 20 keV. For the other nuclei which are presumably statically or dynamically deformed (rotational and vibrational nuclei), the experimental energies are systematically lower by up to 200 keV. A quantitative description of these effects will be presented below.

The electrostatic potential of any spherical or non-spherical charge distribution $e\rho(\vec{r})$ is given by the well known expression

$$V(r) = \sum_{\lambda\mu} \frac{4\pi}{2\lambda+1} Y_{\lambda\mu}(\theta,\phi) \int \frac{r_<^\lambda}{r_>^{\lambda+1}} Y^*_{\lambda\mu}(\theta',\phi') e\rho(\vec{r}')d\tau' \ . \tag{16}$$

Starting with this equation and performing a sequence of partial integrations, the direct Coulomb displacement energy can be given in the symmetric form

$$\begin{aligned}\Delta E_{DIR} &= \frac{e}{N-Z} \int \rho_{EXC}(\vec{r})\, V_{CORE}(\vec{r})d\tau \\ &= 4\pi \sum_{\lambda\mu} \int_0^\infty \frac{\mathcal{M}_{EXC}(E\lambda\mu,r)\mathcal{M}^*_{CORE}(E\lambda\mu,r)}{r^{2\lambda+2}}\, dr \ .\end{aligned} \tag{17}$$

Generalized electric multipole operators $\mathcal{M}_{CORE}$ and $\mathcal{M}_{EXC}$ have been introduced. They are defined by

$$\mathcal{M}_{CORE}(E\lambda\mu,r) = e \int_{4\pi}\int_0^r r^\lambda Y_{\lambda\mu}(\theta,\phi)\rho_{CORE}(\vec{r})r^2 dr d\Omega \tag{18}$$

$$\mathcal{M}_{EXC}(E\lambda\mu,r) = \frac{e}{N-Z} \int_{4\pi}\int_0^r r^\lambda Y_{\lambda\mu}(\theta,\phi)\rho_{EXC}(\vec{r})r^2 dr d\Omega \ . \tag{19}$$

If the radial integration is carried out to infinity, the generalized operators reduce to the ordinary electric multipole operators. For spherical distributions only the monopole element is different from zero. If the spherical distribution is of the Fermi type, $\mathcal{M}(E00,r)$ is directly proportional to the generalized Fermi integral $F_2(c/a,r/a)$ which was introduced earlier, and eq. (17) is reduced to eq. (7). In the case of an angular dependent Fermi distribution as given by eq. (2), $\mathcal{M}(E\lambda\mu,r)$ can be related to generalized Fermi integrals and their derivatives. None of these general relationships will be presented here. Instead, the results for the simpler distribution

$$\rho(\vec{r}) = \frac{\rho(o)}{1 + \exp\left\{\dfrac{r - \eta c_o(1+\beta Y_{20})}{a}\right\}} \tag{20}$$

will be discussed including a least-squares analysis.

Equation (20) can be expanded in the form

$$\rho(\vec{r}) = \rho_o(r) + \rho_2(r)\, Y_{20}(\theta,\phi) + \cdots \ . \tag{21}$$

Using eq. (17) and keeping only the lowest order correction terms in β one obtains

$$\Delta E_{DIR} = \frac{6}{5}\frac{Ze^2}{\eta c_o}\eta^3\ \eta'^3\left\{B_1 + B_2\left(\frac{a}{c}\right)^2 + B_3\left(\frac{a}{c}\right)^3 + B_4\left(\frac{a}{c}\right)^4 + 4\frac{\beta^2}{4\pi}\right\} \quad (22)$$

with $c_o = r_o A^{1/3}$ and

$$\eta = 1 - \frac{\pi^2}{3}\left(\frac{a}{c_o}\right)^2 + \frac{\pi^6}{81}\left(\frac{a}{c_o}\right)^6 - \frac{\beta^2}{4\pi} \quad (23)$$

$$\eta' = 1 - \frac{\pi^2}{3}\left(\frac{a}{hc_o}\right)^2 + \frac{\pi^6}{81}\left(\frac{a}{hc_o}\right)^6 - \frac{\beta^2}{4\pi}\ . \quad (24)$$

The coefficients B_1, B_2, B_3 and B_4 (not given here) depend on c'/c and a/c. The ratio $h \equiv c'/c$ between the half-density radii of the neutron excess and the proton core can be expressed in terms of the ratio $f \equiv R'/R$ introduced earlier. For $\beta=0$, eq. (22) reduces to the expression for spherical nuclei which was used before. An inspection of eq. (22) shows that in lowest order the net effect of the deformation β is to multiply ΔE_{DIR} (spherical) by the factor $(1 - \beta^2/4\pi)$. The form of eq. (22) with eqs. (23) and (24), however, is more convenient to use when higher order corrections in β, the diffuseness correction to β as well as additional shape parameters are included.

The deformation correction to the exchange energy is zero in the present approximation. Thus, the exchange factor ε has to be multiplied essentially by $(1 - \beta^2/4\pi)^{-1}$.

In order to subject the over 100 data on Coulomb displacement energies for spherical and non-spherical nuclei with $Z \geq 28$ to a least-squares analysis, it is necessary to introduce an analytical expression for the dependence of β^2 on Z and N. It should be noted at this point that β^2 represents the square of the static deformation of rotational nuclei but also the expectation value $\langle\beta^2\rangle$ for the dynamic deformation of vibrational nuclei in their ground states (zero point vibration).

An approximate analytical expression for β^2 of deformed nuclei can be derived following theoretical considerations by Bertsch (26) on nuclear hexadecapole moments. Assuming aligned wave functions with strong spatial correlations between the particles (equatorial orbitals are filled last), one obtains an expression for $\beta^2(Z,N)$ which is cubic in Z and in N. The heavy line in Fig. 6 shows that the known deformations in the regions of nuclei from A=150-190 and A>220 are well described by this expression. The dynamic deformations $\langle\beta^2\rangle$ of vibrational nuclei in the regions of nuclei from A=70-80 and A=95-105 are better represented by a distribution which is more symmetric. Therefore, a simpler but completely empirical expression which is only quadratic in Z and in N has been introduced instead. Again, Fig. 6 shows the comparison with the known values of $\langle\beta^2\rangle$ for nuclei along the stability line.

The experimental Coulomb displacement energies for the nuclei with $Z \geq 28$ have been subjected to a least-squares analysis. The ratio $f \equiv R'/R$ and the correction term ΔE_{CORR} were fixed at their previous values, and several deformation strength parameters were varied to optimize the agreement between the experimental and calculated energies. (Practically no data exist for nuclei beyond ^{208}Pb.) Agreement with a standard deviation of 32 keV was obtained for the over 100 data. This result is quite satisfactory. Figure 6 shows the comparison between the resulting values for β^2 (thin line) and the experimental results (data points and thick line). The comparison indicates that the factor $(1-\beta^2/4\pi)$ underestimates the dependence of the direct Coulomb displacement energy on the deformation by about 20-50%. It is expected that the inclusion of higher order terms in β, the diffuseness correction to β^2 and additional shape parameters will rectify the situation.

The isotope and isotone shift of nuclear charge radii and related quantities will be discussed briefly as the last topics. There exist four reduced partial derivatives of the equivalent charge radius R which are defined by

$$\text{isotope shift} \qquad \gamma_N \equiv \frac{3A}{R}\frac{\partial R}{\partial N} \qquad (Z=\text{const}) \qquad (25)$$

$$\text{isotone shift} \qquad \gamma_Z \equiv \frac{3A}{R}\frac{\partial R}{\partial Z} \qquad (N=\text{const}) \qquad (26)$$

$$\gamma_A \equiv \frac{3A}{R}\frac{\partial R}{\partial A} = \gamma_N \frac{\partial N}{\partial A} + \gamma_Z \frac{\partial Z}{\partial A} \qquad (27)$$

$$\text{isobar shift} \qquad \gamma_{T_z} \equiv \frac{3A}{R}\frac{\partial R}{\partial T_z} = \gamma_N - \gamma_Z \qquad (A=\text{const}). \qquad (28)$$

Relationships between these quantities are indicated. The quantity γ_A is generally calculated along the line of stability, and the partial derivatives $\partial N/\partial A$ and $\partial Z/\partial A$ can simply be calculated using

$$T_z = N - \frac{1}{2}A = \frac{1}{2}A - Z = A^{5/3}/\,(266.9 + 2A^{2/3}). \qquad (29)$$

The definition of the four coefficients is such that for $R=r_oA^{1/3}$ with r_o=const we obtain $\gamma_N = \gamma_Z = \gamma_A = 1$ and $\gamma_{T_z} = 0$. Extensive experimental information exists on the isotope shift coefficient γ_N, while in contrast very little is known about the isobar shift coefficient γ_{T_z} which appears to be more important theoretically.

Two components contribute to the various shift coefficients γ. The first one is due to the isospin-dependent contribution to the nuclear potential (shell model and optical) (29,30). This added potential contains the factor $\vec{t}\,\vec{T}$. Therefore, the nuclear potential well for protons becomes deeper with increasing neutron excess. The protons become more tightly bound, the tail of the wave function is reduced and the radius becomes smaller. The exact opposite is true for neutrons. All other effects excluded, the isobar shift of the charge radius should therefore be negative. Similarly, the isobar shift of the neutron radius should be positive. The coefficients γ_{T_z} are also related to the nuclear compressibility (2).

The second contribution to the various shift coefficients γ results from the deformation of the nuclear surface. The requirement that the density in the nuclear interior is conserved results in a change of the equivalent radius with deformation. The partial derivatives γ are strongly affected by this change.

The dependence of Coulomb displacement energies on the deformation has already been discussed but not their explicit dependence on the neutron excess due to the first of the just mentioned effects. If we denote the direct Coulomb displacement energy along the line of stability by $\Delta E_{DIR}^{(o)}$, a Taylor expansion gives

$$\Delta E_{DIR} = \Delta E_{DIR}^{(o)} + \frac{\partial \Delta E_{DIR}}{\partial T_z} \Delta T_z \tag{30}$$

where $\Delta T_z = T_z - \tilde{T}_z$ with $\tilde{T}_z$ defined by eq. (29). Making use of the definition for the isobar shift and of eqs. (10) and (11) one obtains

$$\Delta E_{DIR} = \Delta E_{DIR}^{(o)} \left\{ 1 - \frac{2\gamma_{T_z}^{(o)} + (5f^2-3)\, \gamma_{T_z}^{(o)\prime}}{3A\,(5f^2-1)} \Delta T_z \right\} . \tag{31}$$

Here $\gamma_{T_z}^{(o)}$ and $\gamma_{T_z}^{(o)\prime}$ are the coefficients for the isobar shift of the proton core (charge radius) and the neutron excess (all deformation effects are excluded).

A least-squares analysis of all data with $Z \geq 28$ based on eq. (31) was performed. The results show that the coefficient of the correction term proportional to ΔT_z must be very small. While surprising at first, this result is easily understood. Coulomb displacement energies depend on the charge radius R and the radius R' of the neutron excess. A decrease in R with increasing neutron excess is accompanied by an increase in R' and the change in ΔE_{DIR} is small (see eq. 12). A very approximate numerical value for the coefficient $\gamma_{T_z}^{(o)}$ for the isobar shift of the charge radius can be obtained from the tables of Elton (2). Using a value $\gamma_{T_z}^{(o)} = -0.75$, a search on the isobar shift $\gamma_{T_z}^{(o)\prime}$ of the neutron excess was performed. The result is $\gamma_{T_z}^{(o)\prime}/\gamma_{T_z}^{(o)} \approx -0.5$ which is very reasonable since it requires the neutron radius to increase with increasing neutron excess as one expects from the isospin-dependent contributions to the nuclear potential.

The preceding considerations make it possible to estimate all four coefficients γ_N, γ_Z, γ_A and γ_{T_z} of the nuclear charge radius (deformation effects included) for any nucleus. For the equivalent charge radius we use

$$R = c_o \left[\eta^5 \left(1 + 10 \frac{\beta^2}{4\pi}\right) + \frac{10\pi^2}{3} \eta^3 \left(\frac{a}{c_o}\right)^2 + \frac{7\pi^4}{3} \eta \left(\frac{a}{c_o}\right)^4\right]^{1/2} \tag{32}$$

with $c_o = r_o A^{1/3}$ and

$$\eta = \left\{ 1 - \frac{\pi^2}{3} \left(\frac{a}{c_o}\right)^2 + \frac{\pi^6}{81} \left(\frac{a}{c_o}\right)^6 - \frac{\beta^2}{4\pi} \right\} \left\{ 1 + \frac{\gamma_{T_z}^{(o)}}{3A} \Delta T_z \right\}^{2/5} . \tag{33}$$

The coefficients γ can now be calculated by numerical differentiation.

Figure 7 shows the well known Brix-Kopfermann diagram. It presents experimental values for γ_N as a function of N. This particular diagram is taken from a review article by Wu (31). The heavy solid curves were obtained for nuclei along the stability line by using the above procedure. A value of $\gamma_{T_z}^{(o)} = -0.75$ was used in both cases, while the deformation strength parameters are those extracted from the tables (upper graph) or from Coulomb energies (lower graph). The upper curve gives better agreement, of course. Except for the region from N=85-90 with exceptionally high isotope shifts, the agreement is very good. The general shape is well reproduced including the small values around N=50 and N=82 with the subsequent sharp rise.

Figure 8 presents the predicted curves for γ_Z, γ_N, γ_A and γ_{T_z} for nuclei along the stability line as a function of A. The two curves for γ_N on the left and right hand side correspond to the upper and lower curve in Fig. 7. Experimental values for γ_Z, γ_A and γ_{T_z} are very scarce and a detailed comparison is not possible.

A phenomenological Coulomb energy equation has been derived. The equation includes the effects of the diffuse surface and the deformation of the shape. It has been used in a least-squares analysis of the experimental Coulomb displacement energies but can also be applied to the calculation of atomic masses. Agreement with a standard deviation of about 20 keV is obtained for the data on Coulomb displacement energies of nearly spherical nuclei with $Z \geq 28$; agreement with a standard deviation of about 30 keV is obtained for the data on all nuclei with $Z \geq 28$. Large amounts of excitations into higher shell model orbits for all nuclei are required to achieve this agreement. Values of $f \equiv R'/R$ (ratio of the equivalent radii of the neutron excess and the proton core) between 1.10 and 1.15 appear most reasonable. Neutron halos have been calculated and are found in very good agreement with the results from other calculations. The Coulomb displacement energies of statically and dynamically deformed nuclei (rotational and vibrational nuclei) are found to be smaller by up to 200 keV than corresponding spherical nuclei. Equations have been developed which describe the deformations as well as the isotope, isotone and isobar shifts of nuclear charge radii with reasonable accuracy. The isobar shifts of the neutron excess and the proton core are found to have opposite sign.

1. BETHE, H. A., and BACHER, R. F., Rev. Mod. Phys. 8, 82 (1936); BETHE, H. A., Phys. Rev. 54, 436 (1938).

2. ELTON, L. R. B., Nuclear Radii, Landolt-Börnstein, New Series Vol. I/4 (Springer, Berlin, Heidelberg, New York, 1967) p. 1.

3. HOFSTADTER, R., and COLLARD, H. R., ibid, p. 21.

4. NOLEN, J. A., and SCHIFFER, J. P., Ann. Rev. Nucl. Sci. 19, 471 (1969).

5. ANDERSON, J. D., WONG, C., and MC CLURE, J. W., Phys. Rev. 138, B615 (1965).

6. AUERBACH, N., HÜFNER, J., KERMAN, A. K., and SHAKIN. C. M., Phys. Rev. Letters 23, 484 (1969).

7. AUERBACH, E. H., KAHANA S., and WENESER, J., Phys. Res. Letters 23, 1253 (1969).

8. AUERBACH, E. H., KAHANA, S., SCOTT, C. K., and WENESER, J., Phys. Rev. 188, 1747 (1969).

9. WONG, C. W., Nucl. Phys. A151, 323 (1970).

10. DAMGAARD, J., SCOTT, C. K., and OSNES, E., Nucl. Phys. A154, 12 (1970).

11. NEGELE, J. W., Nucl. Phys. A165, 305 (1971).

12. NGUYEN VAN GIAI, VAUTHERIN, D., VENERONI, M., and BRINK, D. M., (to be published).

13. JÄNECKE, J., (to be published).

14. DRAAYER, J. P., and JÄNECKE, J., (to be published).

15. ELTON, L. R. B., Nucl. Phys. 5, 173 (1958) and 8, 396 (1958); MEYER-BERKHOUT, U., et al., Ann. Phys. (N.Y.) 8, 119 (1959).

16. RHODES, P., Proc. Roy. Soc. (London) A204, 396 (1950).

17. WILSON, A. H., The Theory of Metals (Cambridge, At the University Press, 1958) p. 330.

18. JÄNECKE, J., Chapter 8 in Isospin in Nuclear Physics, edited by D. H. Wilkinson, (North-Holland, Amsterdam 1969).

19. BROWN, G. E., and GREEN, A. M., Nucl. Phys. 75, 401 (1966).

20. ROST, E., Phys. Letters 21, 87 (1966).

21. GERACE, W. J., and GREEN, A. M., Nucl. Phys. A93, 110 (1967).

22. AGASSI, D., GILLET, V., and LUMBROSO, A., Nucl. Phys. A130, 129 (1969).

23. VERGADOS, J. V., Phys. Letters 34B, 458 (1971).

24. MYERS, W. D., Phys. Letters 30B, 451 (1969).

25. MYERS, W. D., and SWIATECKI, W. J., Ann. Phys. 55, 395 (1969).

26. BERTSCH, G. F., Phys. Letters 26B, 130 (1968).

27. STELSON, P. H., and GRODZINS, L., Nucl. Data Tables A1, 21 (1965).

28. LÖBNER, K. E. G., VETTER, M., and HÖNIG, V., Nucl. Data Tables A7, 495 (1970).

29. LANE, A. M., Nucl. Phys. 35, 676 (1962).

30. PEREY, F. G., and SCHIFFER, J. P., Phys. Rev. Letters 17, 324 (1966).

31. WU, C. S., International Nuclear Physics Conference, Gatlinburg (1966) p. 409.

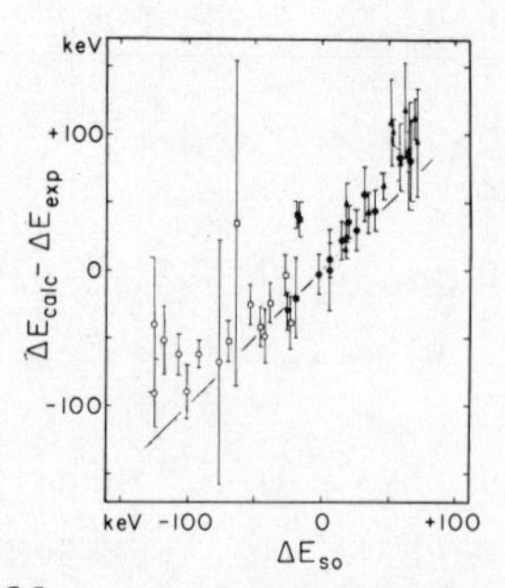

Fig. 1. Plot of the difference between the calculated and experimental Coulomb displacement energies as a function of the electromagnetic spin-orbit interaction correction term ΔE_{SO}. The results shown are for r_o=1.135 fm, a=0.513 fm, f=1.144, ΔE_{CORR}=550 keV. Only semi-magic nuclei and nearly semi-magic nuclei have been included. The filled and open triangles, circles and squares denote: ▲ Z=28,29; ○ N=49, 50,51; ● Z=49,50,51; □ N=81,82,83; ■ Z=81,82,83.

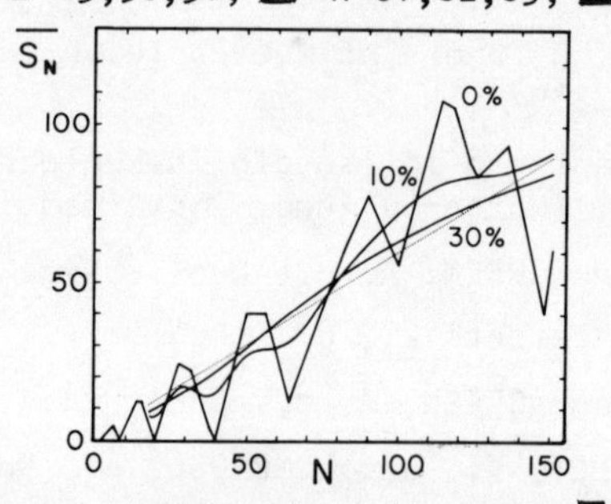

Fig. 2. Plot of the averaged spin-orbit sums $\overline{S_N}$ obtained for percentages of p=0%, 10% and 30% excitations into higher shell model orbits. The dotted straight line represents the approximation $\overline{S_N}$ = 0.6 N.

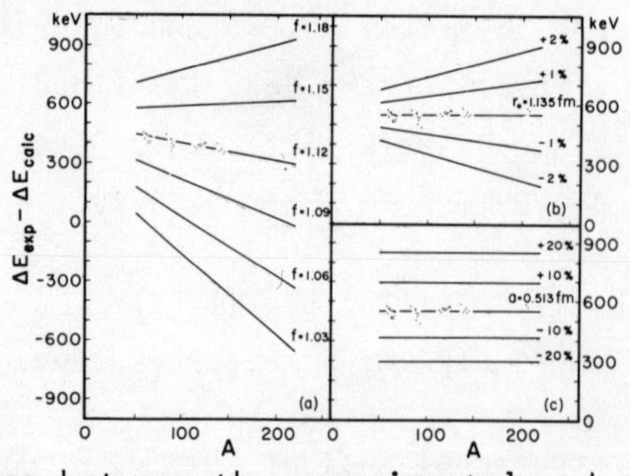

Fig. 3. Difference between the experimental and calculated Coulomb displacement energies as a function of A with (a) variable ratio f≡R'/R and r_o=1.135 fm, a=0.513 fm, ΔE_{CORR}=0 keV; (b) variable radius parameter r_o and f=1.144, a=0.513 fm, ΔE_{CORR}=0 keV; (c) variable diffuseness a and r_o=1.135 fm, f=1.144, ΔE_{CORR}=0 keV.

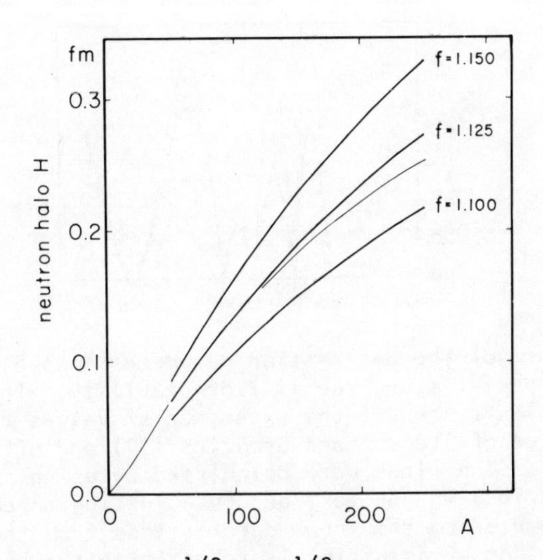

Fig. 4. Neutron halo $H \equiv \langle r_n^2 \rangle^{1/2} - \langle r_p^2 \rangle^{1/2}$ for nuclei along the stability line as a function of A. The three heavy lines are for f=1.10, 1.125 and 1.15. The thin line is derived from the droplet model of Myers and Swiatecki (24,25).

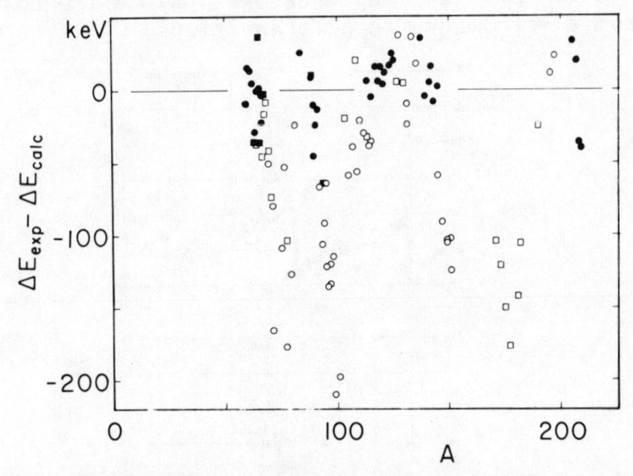

Fig. 5. Plot of the difference between experimental and calculated energies as a function of A for r_o=1.135 fm, a=0.513 fm, f=1.144 and ΔE_{CORR}=550 keV. Only the filled circles and squares for the nuclei near magic lines were used in the analysis. The energies for all other nuclei are indicated by open circles and squares. Circles and squares represent values with experimental uncertainties less than 30 keV and less than 100 keV, respectively. The experimental energies of rotational or vibrational nuclei (nuclei with a static or dynamic deformation in the ground state) are considerably smaller than the energies calculated on the basis of spherical charge distributions.

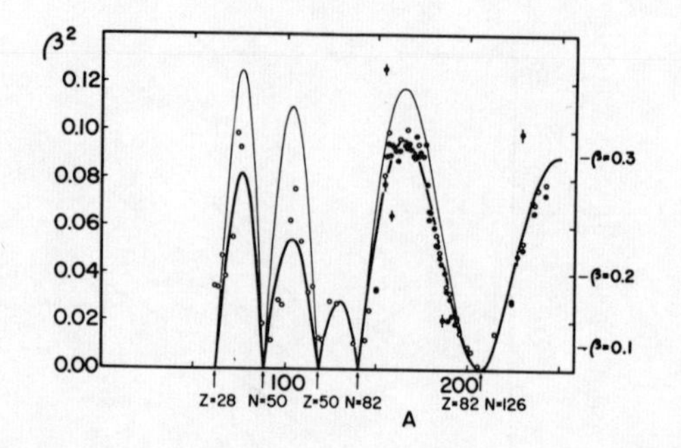

Fig. 6. Square of the deformation parameter β as a function of A for nuclei along the line of stability. The open and filled circles represent the experimental values taken from the Tables of Stelson and Grodzins (27) and of Löbner et al. (28). The lines were calculated by using the functions described in the text and by adjusting an empirical strength parameter to the known deformations from the tables (heavy line) or by adjusting it to optimize the agreement between the experimental and calculated Coulomb displacement energies (thin line). Note that β_q^* of ref. (28) and β_2 of ref. (27) are defined differently and are related according to $\beta_q^* = \beta_2\{1 - (9/56)\sqrt{(5/\pi)}\,\beta_2\}$. For vibrational nuclei the data of ref. (27) represent $<\beta^2>$ since the data are derived from electromagnetic E2 transitions.

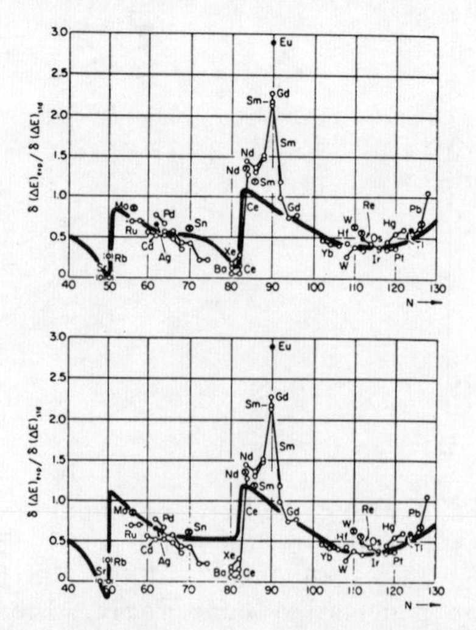

Fig. 7. Brix-Kopfermann diagram for the coefficient γ_N of the isotope shift of nuclear charge radii (taken from the review article by C. S. Wu (31)). The heavy lines represent estimates for nuclei along the line of stability. They were calculated using the functions described in the text with the strength parameters derived from the experimental deformations (upper diagram) or the experimental Coulomb displacement energies (lower diagram), respectively.

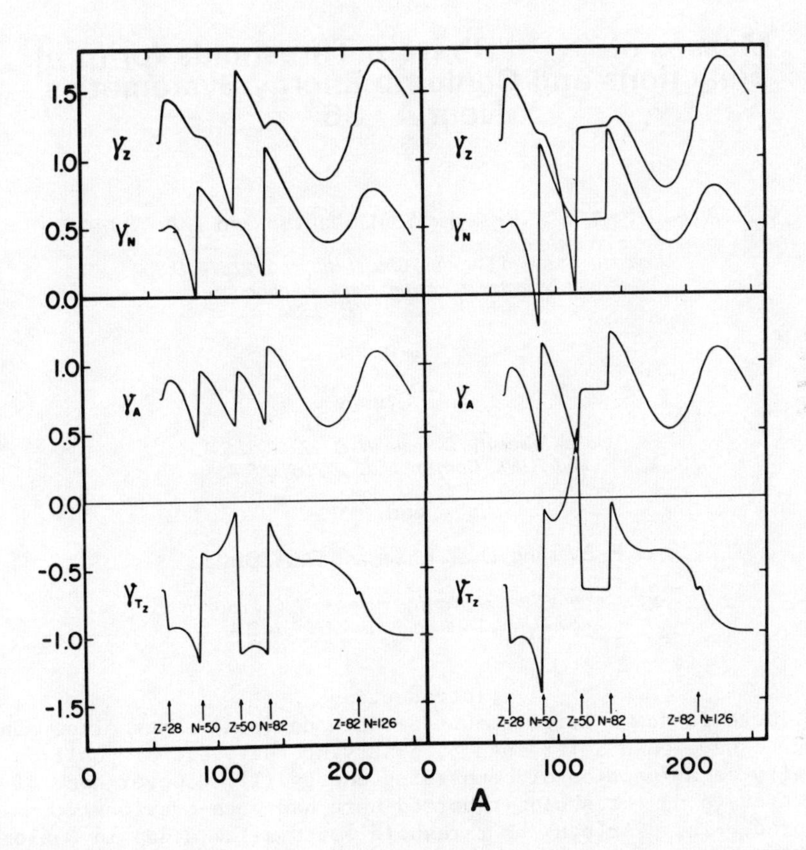

Fig. 8. Calculated coefficients for the isotone shift γ_Z, the isotope shift γ_N, the derived quantity γ_A and the isobar shift γ_{T_z} for nuclei along the stability line. As in Figs. 6 and 7, the lines were obtained by using the functions described in the text with the empirical strength parameters derived from the known deformations (left hand side) or the experimental Coulomb displacement energies (right hand side).

Masses of ^{36}K, $^{80,82}Rb$ from Thresholds for (p,n) Reactions and Coulomb Energy Systematics Near A=36*

A. A. Jaffe,† G. A. Bissinger, S. M. Shafroth, and T. A. White‡

University of North Carolina, Chapel Hill, N.C. 27514, U.S.A. and Triangle Universities Nuclear Laboratory Durham, N. C. 27706, U.S.A.

and

T. G. Dzubay

Duke University, Durham, N. C. 27706, U.S.A. and TUNL, Durham, N. C. 27706, U.S.A.

and

F. Everling, D. W. Miller and D. A. Outlaw

North Carolina State University, Raleigh, N. C. 27607, U.S.A. and TUNL, Durham, N. C. 27706, U.S.A.

1. Introduction

There is now a large amount of data on (p,n) thresholds, many of which have been determined by observing the yield of positron activity as a function of bombarding energy.(1) However none of the three (p,n) thresholds reported here had been previously measured. The $^{36}Ar(p,n)^{36}K$ threshold would allow a gap in Coulomb-displacement energy systematics to be filled and the $^{80,82}Kr(p,n)^{80,82}Rb$ thresholds would permit, in conjunction with other information, the prediction of disintegration energies for ^{78}Rb and ^{82}Y which are far off the stability line.

2. Experimental Equipment Common to All Three Threshold Measurements

The TUNL FN Tandem Van de Graaff supplied proton beams of typically 200 nA for the threshold measurements. A mechanical chopper placed near the focus of the 38° magnetic analyzer interrupted the proton beam at suitable intervals. A signal provided by a micro-switch and later a photodiode at the chopper started an on-line computer storing pulse height and time information during

*Work supported by the U. S. Atomic Energy Commission.

†Work done while on leave from The Hebrew University, Jerusalem, Israel.

‡Summer research participant from Furman University.

beam-off periods using program FCHOP. (2) Typically 32 sequential 128 channel beta spectra were stored. The pulse-height information arising from positron decay of the residual nucleus was provided by a 4 x 5 cm plastic scintillator (NE 201).

3. Mass 36 Systems

A prediction for the mass of ^{36}K had been made by Kelson and Garvey (3) and an indirect mass determination had been made by Berg et al. (4) It was based on Coulomb-displacement energy systematics and locating the lowest T = 1 state in ^{36}Ar at 6.6119 ± 0.0009 MeV. An unpublished (5) Q-value for the reaction ^{36}Ar(^{3}He,t)^{36}K combined with the Berg et al. results lead to a significant discrepancy in Coulomb displacement energy systematics. A new measurement (6) of the Q-value for the ^{36}Ar(^{3}He,t)^{36}K reaction is in much better agreement with Coulomb energy systematics and the present (p,n) threshold result. The present measurements were carried out with a 200 nA beam of protons which was interrupted with a repetition rate of about 0.3 Hz, and was used to bombard enriched ^{36}Ar gas of 99.9% isotopic purity which was contained in a 2.5 cm diameter cell whose axis was vertical. It had 6.3 μm thick tantalum entrance and exit windows and the beam was stopped 3 m behind the cell by a graphite stop surrounded with paraffin.

Measurements were made in beam energy steps of 30 keV between 13.94 and 14.12 MeV and in 500 keV steps up to 30 MeV. Above ~15 MeV the Cyclo-Graaff (7) consisting of a fixed energy 15 MeV H$^-$ ion AVF cyclotron injecting into the FN Tandem Van de Graaff was used. The gas cell was also filled with CS_2 vapor, and measurements were made in 10 keV steps between 13.88 and 13.94 MeV. Additional measurements were made using a high resolution magnetic analyzing system with proton beams which were varied between 14.0 and 15.5 MeV, and which were incident on a 25-μm thick Kapton cell filled with ^{36}Ar. The peak ^{36}Ar(p,n)^{36}K yield was found at E_p = 17.5 MeV. The decay of β^+ events which had energies greater than 5 MeV was observed for five half-lives for the measurements between E_p = 16.5 and 18.5 MeV. The average half-life for the ground state of ^{36}K was found to be 341 ± 6 ms. This is somewhat larger than a previous value of 265 ± 25 ms. (4)

The yield from the (p,n) reaction was determined by fitting the data, which were stored on magnetic tape, with a decay curve having a 341 ms half life. The results of the (p,n) threshold measurements for ^{32}S and ^{36}Ar are shown in Fig. 1. The ^{32}S(p,n)^{32}Cl yields were deduced using a presently determined half-life of 305 ms for ^{32}Cl. The difference between the two thresholds is 103 ± 20 keV, and when combined with a known value of 13.899 ± 0.014 MeV for the ^{32}S(p,n) threshold, (8) a value of 14.002 ± 0.024 MeV for the ^{36}Ar(p,n) threshold was deduced. Using the mass excess value for ^{36}Ar from the tables of Wapstra and Gove (9) a mass excess of -17.395 ± 0.023 MeV is deduced for ^{36}K. This is in acceptable agreement with the (^{3}He,t) measurement of Dzubay (6) et al. of -17.32 ± 0.04 MeV, and it is in good agreement with the -17.42 ± 0.07 MeV prediction of Kelson and Garvey (2). By assuming that the analog of the ground state of ^{36}K occurs at an excitation energy of 6612.7 ± 0.8 keV in ^{36}Ar (this determination is discussed below) a Coulomb-displacement energy of 7.02 ± 0.03 is deduced. This value fills a gap in the available data on Coulomb-displacement energies between $T_z = 0$ and $T_z = -1$ members of the T = 1 multiplets which are now known up to A = 42.

The present value of ΔE_c is in good agreement with the systematics which are plotted in Fig. 2. The trend of the lower curve in Fig. 2 is expected to be generally parallel to the upper curve if it is shifted to the left by two mass units. Two striking exceptions to this rule are evident near the shell closures at A = 16 and 40. This is presumably due to the changes in the spin-orbit contribution to the Coulomb-displacement energy.

There is considerable structure in the excitation curve for ^{36}Ar(p,n) between threshold and 15.5 MeV. This may be due to resonances in the compound nucleus ^{37}K, at excitation energies of about 15 MeV. One might speculate that the structure is due to the excitation of T = 5/2 states. According to Endt and Van der Leun(10) the first T = 5/2 state has been identified in ^{37}Cl at an excitation of 10.24 MeV. From this one can conclude that the lowest T = 5/2 state in ^{37}K could be seen at E_p = 13.9 ± 0.5 MeV on ^{36}Ar. It is not known if the isospin impurities in ^{37}K are large enough to allow the doubly isospin forbidden T = 5/2 resonances to be observed. It would be useful to investigate this phenomenon in other reaction channels leading to the same compound states.

A search was made for the ^{36}Ar(p,2n)^{35}K threshold using the experimental setup which is described above. According to the ^{35}K mass prediction of Kelson and Garvey (3) the (p,2n) threshold should occur at about E_p = 28.5 MeV. We scanned our data in this energy region, but no new decay component could be detected. Recent work with the (p,2n) reaction by Cerny et al. (11) indicates that the cross sections are too small near threshold to be seen by the present technique.

The decay of the J = 2^+, T = 1 ground state of ^{36}K to its analog state in ^{36}Ar is possible by both Fermi and Gammow-Teller transitions. Hence its log (ft) value should be no larger than 3.5. A previous measurement determined a 25 ± 5% branch in the beta decay spectrum to the T = 1 level in ^{36}Ar at 6.61 MeV. (4) This number can be combined with the present half life of 341 ± 6 ms and a decay energy of 12.82 MeV to yield a log (ft) = 3.8. This is too large, and it casts a doubt on the T = 1 assignment of the level at 6.61MeV.

We remeasured this branching ratio using an 80 cm^3 Ge(Li) detector, (12) which was a factor of 40 larger than that used in the earlier study. A 12 hour bombardment at a current of 20 nA with E_p = 18 MeV was made on a 12.5 x 1.4 cm diam. cylindrical cell with 25 μm thick Kapton windows which was filled with 1/2 atm of ^{36}Ar. A 4096-channel spectrum was taken of the 0 to 8 MeV gamma ray energy interval. The computer data array was also divided into four 1024-channel time blocks, and a search was made for gamma rays having the correct 341 ms half life. The resulting decay scheme for ^{36}K is shown in Fig. 3. The present gamma ray transition energies in ^{36}Ar are in excellent agreement with those of Berg et al., (4) and the present uncertainties are smaller, By using the present 43% branching ratio for the transition to the T = 1 state at 6.6129 MeV one now has an acceptable log (ft) = 3.49.

4. Mass and Half Life of ^{80}Rb

The mass of ^{80}Rb has been investigated by only one group [Hoff et al. (13)] whose measurement of the positron end-point energy had an uncertainty of 0.5 MeV, and yielded a determination of its mass having the same uncertainty. Thus a threshold measurement of the ^{80}Kr(p,n)^{80}Rb reaction in conjunction with the accurately known value for the ^{80}Kr mass should yield a value of the ^{80}Rb mass at least 10 times more accurate than was previously known.

A half an atmosphere of enriched Kr gas (obtained from the Mound Laboratory of the Monsanto Research Corporation) (17.6% ^{78}Kr, 51.3% ^{80}Kr, 28.8% ^{82}Kr, 2.0% ^{83}Kr and 0.2% ^{84}Kr) was introduced into the vertical gas cell with the tantalum entrance and exit windows. Tantalum was chosen because it gave rise to essentially no troublesome beta radiation. Bombardments were done for 30 seconds and then on receipt of a pulse from the chopper the beta spectrum was taken every five seconds for 32 equal time intervals. Since the beta end point energy of ^{82}Rb is 3.3 MeV and ^{80}Rb is 4.7 MeV, only those beta pulses above 3.3 MeV were taken for the ^{80}Rb threshold measurement. This procedure gave a practically background-free threshold result (Fig. 4). A least-squares analysis of the half-life of the positron activity above the ^{82}Rb beta endpoint energy as well as lower energy positrons including the 80 sec ^{82}Rb background yielded a half-life of 29 ± 2 sec. A least squares fit of the yield of the ^{80}Rb activity to the 2/3 power vs. E_p is also plotted in Fig. 4. In this way, after correcting for energy-loss in the Ta window, the threshold value E_p = 6566 ± 20 keV is obtained. This corresponds to a Q = -6484 ± 20 keV and using the most recent value (9) for the mass of ^{80}Kr a mass excess value for ^{80}Rb of -72,195 ± 21 keV is obtained.

5. Mass and Half-life of ^{82}Rb

A redetermination of the mass of ^{82}Rb was also of special interest. In this case more information was available from positron end-point energy measurements. With the exception of a paper by Beck (14) this data has been summarized by Artna (15) who adopted a value of 4170 ± 30 keV for the ^{82}Rb disintegration energy. Recently Raman and Pinajian (16) studying the γ decay of ^{82}Rb following β^+ decay concluded that the disintegration energy must be at least 200 keV larger than the adopted value. Thus a determination of the ^{82}Kr(p,n)^{82}Rb threshold energy would be of interest.

The same equipment and enriched Kr gas was used as for the ^{80}Kr(p,n)^{80}Rb threshold experiment. A run one MeV above threshold was used to determine the half-life of ^{82}Rb. The major source of background positrons was due to ^{81}Rb but these were long lived and of low energy. The run was done below the ^{80}Rb threshold so pulses of heights corresponding to betas of greater than 1.4 MeV were used to determine the half-life. A value of 80 ± 3 sec was obtained.

Fig. 5 shows the results of the ^{82}Kr(p,n)^{82}Rb threshold determination. In this case, the positrons whose energies were greater than 1.4 MeV were summed. These sums were then fitted to a 30-minute background and an 80-second component due to ^{82}Rb. Analysis followed as with ^{80}Rb; the normalized yield was fitted to the two-thirds power and led to 15 keV uncertainty. In this case, however, this power law only seemed to hold for the first 100 keV above threshold of NMR frequency 14,473 Hertz or a beam energy of 5507 ± 2 keV. After subtracting a window correction of 285 ± 13 keV, the threshold for the $^{82}Kr(p,n)^{82}Rb$ reaction is 5222 ± 20 keV leading to Q = -5159 ± 20 keV, and a mass excess for ^{82}Rb of -76,215 ± 21keV.

6. Disintegration Energy Systematics in the Mass 80 Region

In Fig. 6 a revised beta-disintegration energy graph is plotted essentially in the manner of Way and Wood except with mass number A instead of neutron number (17) as abscissa. It is based on the latest mass table by Wapstra and Gove (9) with the addition of new values of 6.2 ± 0.1 MeV for ^{74}Se - ^{74}Br (Ref. 18) and 4.88 ± 0.05 MeV for ^{76}Se - ^{76}Br (Ref. 19). By using the present mass measurements, it becomes possible now to predict by extrapolation the beta-disintegration energy of ^{78}Rb - ^{78}Kr to be Q^+ = 6.9 ± 0.4 MeV and the one of ^{82}Y - ^{82}Sr to be Q^+ = 7.9 ± 0.4 MeV.

7. Summary of Results

The following table summarizes the results of the present work.

	$^{80}Kr(p,n)^{80}Rb$	$^{82}Kr(p,n)^{82}Rb$	$^{36}Ar(p,n)^{36}K$
Natural abundance of target	2%	12%	0.3%
Enrichment used	51%	29%	99.9%
Half-life of product*	29 ± 2 sec	80 ± 3 sec	0.341 ± 0.006 sec
$E_{threshold}$*	6,566 ± 20 keV	5,222 ± 20 keV	14,002 ± 24 keV
Precision of target mass*	± 6 keV	± 5 keV	± 1.1 keV
Q value*	-6,484 ± 20 keV	-5,159 ± 20 keV	-13,618 ± 23 keV
Mass excess of product†**	-72,195 ± 21 keV	-76,215 ± 21 keV	-17,395 ± 23 keV
Disintegration energy*	5,702 ± 20 keV	4,377 ± 20 keV	12,836 ± 23 keV

* Present work

† Present work combined with 1971 mass tables (Ref. 9).

** Differences between these numbers and those in the abstract are due to use of the 1971 mass tables.

8. References

1. FREEMAN, J. M., JENKIN, J. G., MURRAY, G. and BURCHAM, W. E., Phys. Rev. Letters 16, 959 (1966).

2. TILLMAN, P. W. and JAFFE, A. A., unpublished, 1970.

3. KELSON, I. and GARVEY, G. T., Phys. Letter 23, 689 (1966).

4. BERG, R. E., SNELGROVE, J. L., and KASHY, E., Phys. Rev. 153, 1165 (1967).

5. MATLOCK, R. G., Ph.D. thesis, University of Colorado, 1965 (unpublished).

6. DZUBAY, T. G., et al., Phys. Letters 33B, 302 (1970).

7. NEWSON, H. W., PURSER, F. O., ROBERSON, N. R., BILPUCH, E. G., WALTER, R. L., and LUDWIG, E. J., Bull. Am. Phys. Soc. 14, 533 (1969).

8. OVERLEY, J. C., PARKER, P. D., and BROMLEY, D. A., Nucl. Instr. Methods 68, 61 (1969).

9. WAPSTRA, A. H. and GOVE, N. B., Nucl. Data Tables, 9, No. 4-5 (1971).

10. ENDT, P. M. and VAN DER LEUN, C., Nucl. Phys. A105, 1 (1967).

11. CERNY, J., Bull. Am. Phys. Soc. 16, 530 (1971).

12. BASKIN, A. B., JOYCE, J. M., SHAFROTH, S. M., and WHITE, T. A., Bull. Am. Phys. Soc. 15, 1346 (1970).

13. HOFF, R. W., HOLLANDER, J. M., and MICHEL, M. C., J. Inorg. Nucl. Chem. 18, 1 (1961).

14. BECK, E. Nucl. Instr. and Meth. 76, 77 (1969).

15. ARTNA, A., Nucl. Data 1b, No. 4 (1966).

16. RAMAN, S. and PINAJIAN, J. J., Nucl. Phys. A125, 129 (1969).

17. EVERLING, F., GOVE, N. B., and VAN LIESHOUT, R., Nucl. Data Sheets, National Academy of Sciences-National Research Council, Washington, D. C., pages NRC-61-3-142 through 149 (1961).

18. LADENBAUER-BELLIS, I. M. and BAKHUA, H., Phys. Rev. 180, 1015 (1969).

19. DZHELEPOV, B. S., et al., 21st Moscow Conf. Abstr. p. 51 (1971).

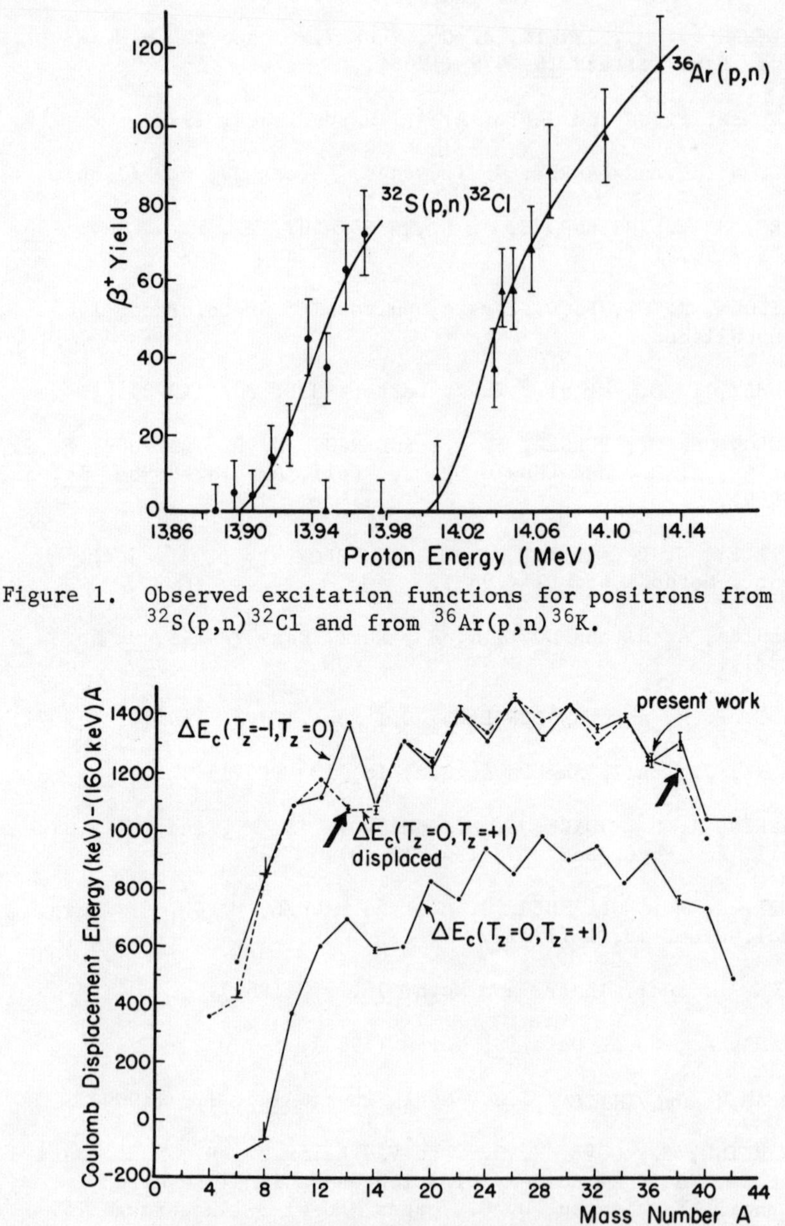

Figure 1. Observed excitation functions for positrons from $^{32}S(p,n)^{32}Cl$ and from $^{36}Ar(p,n)^{36}K$.

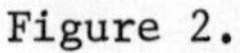
Figure 2. Coulomb-displacement energy between the two adjacent members of T = 1 triplets versus mass number. A linear sequence (160 keV)A is subtracted in order to eliminate the steep rise. The dashed trend represents the lower one after displacement by 2 mass units to the left and an upward shift to coincide with the upper trend.

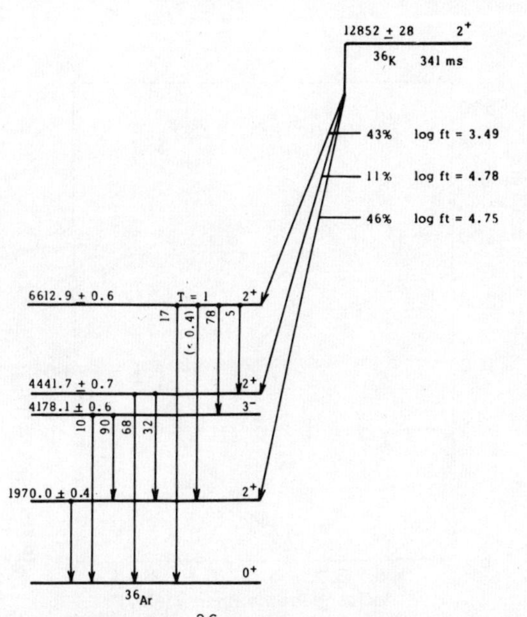

Figure 3. Decay scheme for ^{36}K showing gamma transitions in ^{36}Ar.

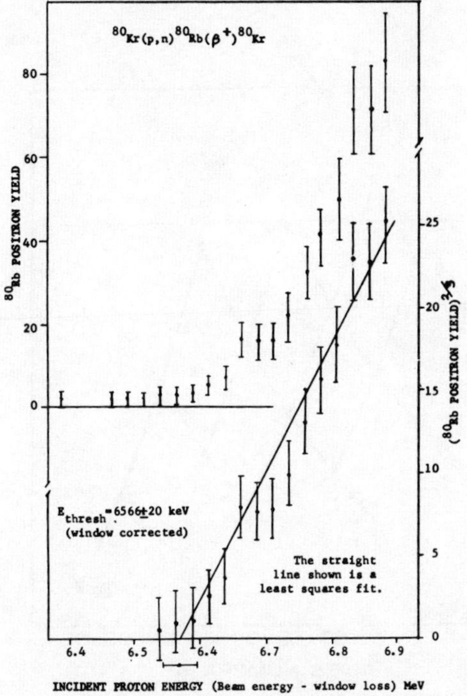

Figure 4. ^{80}Rb positron yield from the ^{80}Kr(p,n)^{80}Rb(β^+)^{80}Kr reaction.

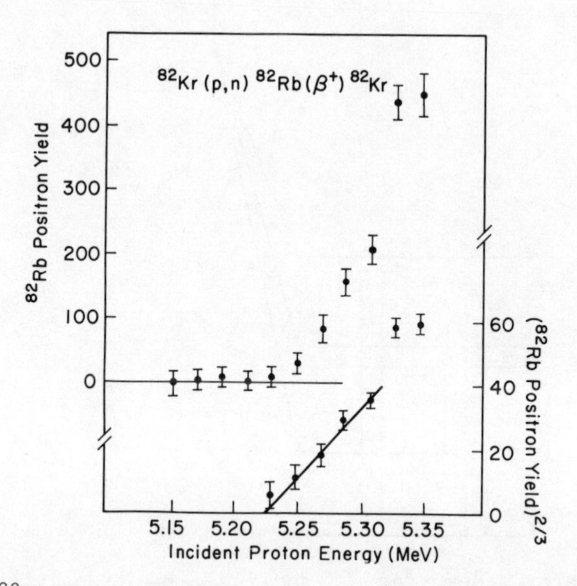

Figure 5. ^{82}Rb positron yield from the $^{82}Kr(p,n)^{82}Rb(\beta^+)^{82}Kr$ reaction.

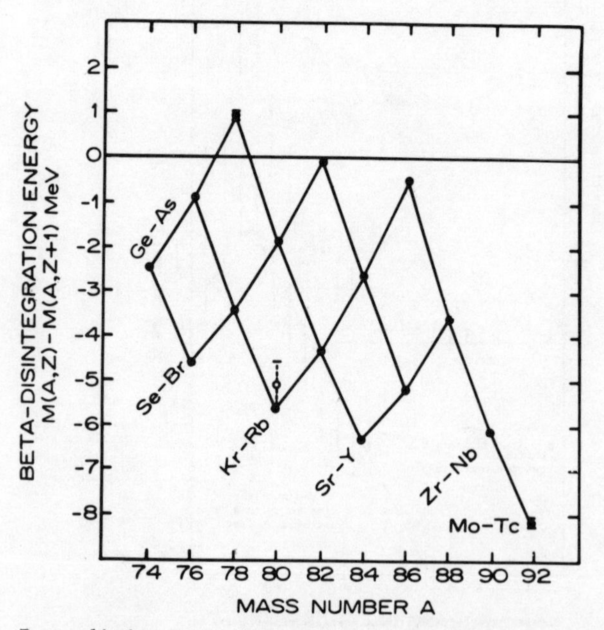

Figure 6. Beta-disintegration energy systematics. The open circle is the previous data for ^{80}Kr - ^{80}Rb. From such a diagram the disintegration energy for ^{78}Kr - ^{78}Rb and ^{82}Sr - ^{82}Y can be predicted.

Part 5 Mass Formulae and Mass Calculations

Shell Model Masses and Nuclear Structure

N. Zeldes

Racah Institute of Physics, Hebrew University, Jerusalem

1. The importance of nuclear structure considerations

Semi-empirical mass equations are at present our only means of obtaining without too much labour numerical values for large numbers of nuclear masses. This makes the subject important for studies of nuclei far from stability, as described by Drs. Cerny, Klapisch, Nielsen and Scott during this conference.

In the semi-empirical approach, a nuclear model furnishes the functional dependence of the nuclear mass on N and Z, whereas some coefficients entering the relation are determined by adjustment to the experimental data. However, the latter are concentrated in a narrow strip in the N-Z plane, and they are therefore usually insufficient for a significant determination of all the coefficients in a complicated mass equation, different terms of which might behave similarly along the stability valley. Thus, some terms may be altogether omitted, without worsening the goodness of fit of the equation to the known masses. However, two such different and equally-well-fitting equations will usually give very different results on extrapolation. Therefore, the problem of retaining important terms whose necessity is not indicated by the experimental masses becomes of major importance for the applicability of mass equations to extrapolation. Attempting to solve this, one is necessarily led to use information concerning nuclear structure which is derived from experimental sources other than mass systematics.

The present talk describes in some detail the experimental basis and connections of a shell model mass equation.

2. The nuclear potential

The basis of the shell model are the magic numbers. Nuclear separation energy systematics (as in fig. 6 of ref. (1)) is so similar to atomic ionization potential systematics (as in fig. 5 of ref. (2)) that one is immediately led to interpret the discontinuities at the magic numbers as shell effects. Fig. 2-12 of ref. (3) shows the reduced density of higher excited nuclear levels at the magic numbers, which also naturally indicates the existence of closed shells in a nuclear potential at these numbers.

In order to check this interpretation one has in the first place to find a potential reproducing the magic numbers. The short range of the nuclear forces leads one to try a potential resembling the nuclear mass distribution, like a square well or more realistically a Woods-Saxon potential, with a radius following the experimental $A^{1/3}$ dependence. Indeed, adding a spin-orbit term of the right magnitude and A-dependence to such a potential, one obtains shells filled at exactly the magic numbers. Such a potential and its energy levels as function of A along the stability line are shown in figs. 2-22 and 2-30, respectively, of ref. (3).

To further confirm the validity of this potential, one looks at low-lying excited levels of nuclei with one nucleon added to or missing from the proposed closed shells. There one should find single-particle and single-hole levels, corresponding to the last few subshells around the Fermi level. This is indeed the case, as shown in figs.3-2a to 3-2f in ref. (3).

Optical model calculations likewise confirm the existence of an average nuclear potential like the above. However, to account for nucleon scattering from nuclei one has to assume an energy dependence of the potential, as shown in fig. 2-29 of (3). Indeed, fig. 3-3 of (3) shows experimental single-nucleon levels around Pb^{208} and corresponding levels calculated from the realistic energy-independent potential of fig. 2-30 mentioned above, and it is clear, that whereas there is good agreement near the Fermi level, the deep lying levels and also the higher unoccupied ones do not agree. However, this might also indicate that one should not interpret these experimental levels as single-particle levels in a static potential, and higher-order residual interactions contribute to their energies in a significant way, such as will be explicitly considered in the sequel.

In addition, there is variation of the optical potential with N and Z. That such should be the case is also clearly indicated by the variation of nuclear radii with N and Z, revealed by electron scattering and by isotopic shifts in atomic spectra, and discussed by Dr. Janecke here.

3. The pairing scheme

To calculate the energy of the nucleus as an aggregate of nucleons in a potential well, one should specify the nuclear state in more detail. Here again nuclear spectra supply useful hints. Ground state spins of even-even and odd-A nuclei suggest that there is a strong pairing of identical nucleons in $J_{12} = 0$ states. This is Mayer's(4) single-particle approximation, given a better mathematical formulation using Racah's seniority concept (5).

Fig. 1 shows the low-lying levels of the even-even Zr^{90} nucleus, of its odd-A neighbours Zr^{91} and Nb^{91}, and of the odd-odd Nb^{92}. One notices the relatively high energy of about 2 MeV needed to break the tight coupling of the two protons in the Zr^{90} $p_{1/2}{}^2$ (0^+) ground state, to form the excited $g_{9/2}{}^2(2^+)$state. (Incidentally, the excitation energy of the 1st excited o^+ level is roughly twice the distance between the single-proton $p_{1/2}$ and $g_{9/2}$ levels, which is known from the spectrum of Y^{89}, and the excitation energy of the 5^- level is roughly compounded of the energy needed to break the $p_{1/2}{}^2(0+)$ ground state proton pair and the energy needed to transfer one of its members to the $g_{9/2}$ subshell). This situation is rather general in nuclei not too far from closed shells, and confirms the above assumption of strong pairing interaction.

We next consider Zr^{91} and Nb^{91}. Their 1st excited single-neutron and proton levels, respectively, are considerably lower than the Zr^{90} 2^+ level. This is again a general rule for nuclei not too far from closed shells: the low-lying levels of odd-A nuclei are usually single-nucleon levels, obtained from the ground state by lifting the odd nucleon from one subshell to another, while leaving intact its saturated-pairs even-even core. Thus they have lower excitation energies than the 1st 2^+ levels of even-even neighbours.

Finally, the low-lying levels in Nb^{92} consist of an even-parity multiplet with J=2,3,...,7 and an odd multiplet with J=2,3. Their excitation energies are considerably lower than that of the 1st excited $1/2^+$ level of Zr^{91}. Thus they could naturally arise by coupling the odd $d_{5/2}$ neutron of Zr^{91} with the ground $g_{9/2}$ proton and 1st-excited $p_{1/2}$ proton of Nb^{91}, to form the even and odd multiplets, respectively. The spins of the multiplets seem to confirm this interpretation. This is an illustration of the odd-group model for odd-odd nuclei not too far from closed shells: their low-lying levels are obtained by coupling the odd-neutron and odd-proton groups of their odd-A neighbours, without changing their internal state, to form a multiplet of close lying levels. The odd-neutron and odd-proton states are called the parents (6) of the odd-odd state. This interpretation is confirmed in many cases by the agreement of the odd-odd magnetic moment with the value calculated from the experimental moments of the odd-A neighbours according to the vector model.

A further confirmation can be obtained by using the center of mass theorem of nuclear spectroscopy (5), which says that when two multiplets are based on a well defined parent each, the energy distance between their centers of gravity defined by $\Sigma(2J+1)E_J/\Sigma(2J+1)$ should equal the energy distance between the parents. This is a generalization of Landés interval rule in atomic LS multiplets. It is illustrated in fig. 2 where the Ni^{62} ground and 1st excited states are considered parents of the Cu^{63} g. s. and excited multiplet, respectively.† Many such cases are encountered, which gave the impetus to the excited-core model of DeShalit and Braunstein (presently Gal). In the case of Nb^{92} the theorem is but roughly satisfied.

Summarizing, level schemes as in fig. 1 lead to the following pairing scheme: ground states of even-even nuclei consist of saturated $J_{12} = 0$ nucleon pairs filling shells in a potential well, odd-A nuclei consist additionally of the unpaired nucleon, whereas odd-odd nuclei consist of odd-neutron and odd-proton groups as described above, coupled together to form the total J.

The energy of such a nucleus, consisting of j_n^n neutrons and j_p^p protons outside closed shells of N_o neutrons and Z_o protons can be written as (2,5)

$$E_{pair}(N_o+n, Z_o+p) = E_o + nc + \frac{1}{2}n(n-1)d + [\tfrac{1}{2}n]\pi \qquad (1)$$

$$+ pc + \frac{1}{2}p(p-1)D + [\tfrac{1}{2}p]\Pi$$

$$+ npI^o + \frac{1}{4}(1-(-1)^n)(1-(-1)^p)I'$$

where the coefficients multiplying powers of n and p are given linear combinations of the two-body interactions matrix elements $(j_1 j_2 J|V_{12}|j_1 j_2 J)$. They are independent explicitly* of n, p and J, except I', which depends on

---

† See, however, remark in the caption of fig. 2.

* There is a N- and Z- dependence of all matrix elements due to the corresponding variation of the nuclear potential, mentioned in part 2.

them as

$$I'(n,p,J) = I''(J) + \frac{(2j_n+1-2n)}{2j_n - 1} \cdot \frac{(2j_p+1-2p)}{2j_p - 1} \cdot I'''(J) \quad (2)$$

where $I''(J)$ and $I'''(J)$ are the contribution of the odd and even multipoles, respectively, to the average neutron-proton interaction $(j_n j_p J|V_{np}|j_n j_p J)$.

Incidentally, this means that the excitation energies of levels of odd-odd multiplets with respect to their center of mass should vary bilinearly with n and p. An approximate behaviour of this kind is shown in fig. 3. A further check on the above interpretation of the odd-odd Nb multiplets is furnished by the Pandya transformation, relating in the present case Nb^{96} to Nb^{92} energies. The agreement with experiment is here very good (7).

What is the sign of I'? Since $\Sigma(2J+1)I'(J)$ vanishes according to the center of mass theorem, $I'(J)$ of the ground state should be negative (8). With positive signs of d and D and negative I^{0}, π, Π and I' (2), eq.(1) predicts four parabolic mass surfaces, curving upwards, the even-even one lowest and the odd-odd highest and somewhat nearer to the odd-A surfaces, as in fig. 1 of (1).

The I' term is responsible for the zig-zag behaviour of the Sn isotonic and S_p isotopic lines (9), as shown in full lines in fig. 4. However, as $\Sigma(2J+1)I'(J)$ vanishes, the zig-zag should disappear when the mass of the center of gravity of g.s. odd-odd multiplets is used in the calculation of the separation energies. The broken lines in fig. 4 confirm this to certain extent.

Eq. (1) can be further related to spectra by calculating the two-body matrix elements $(j_1 j_2 J|V_{12}|j_1 j_2 J)$ from experimental level schemes of one and two nucleon configurations, and comparing their appropriate linear combinations with the parameters of (1) as obtained directly from ground state masses. The agreement is usually good, confirming that the simple quadratic eq. (1) is merely one aspect of a more comprehensive physical scheme, describing the complex quantum-mechanical nuclear system.

4. Subshell mixing and deformation

Eq. (1) refers to a well defined configuration in a heavy nucleus with different neutron and proton filling shells. However, in heavier nuclei, where many neighbouring subshells fill simultaneously, configurations mix, which leads to departures from the spherical shell picture presented above.

What configurations should mix? the tight pairing suggests mixing of configurations consisting entirely of saturated pairs in neighbouring subshells. However, this cannot be the whole truth, as on such mixing the quadrupole moment will always have approximately a single-proton value in odd-Z nuclei, and Z/A^2 times less for odd-N. This contradicts experiment, as shown in fig. 6-37 of ref. (10). A second consequence of not breaking saturated pairs is independent deformation of neutrons and protons (1), contradicting experiment as well. Thus, pair-breaking should also be allowed for.

Configuration mixing affects eq. (1) in two ways. First, the diagonal matrix elements of the energy will contribute sums of terms like (1) for the various subshells, plus mutual subshells interaction. However, they can all be summed up to an equation like (1), provided corresponding parameters in different subshells are approximately equal (2). This seems to be the situation in most cases, as demonstrated by the nearly-equal slopes of $s_{1/2}$ and $d_{3/2}$ S_n systematics in fig. 5. Thus, one should be able to ignore the difference between subshells as a fine structure superimposed on the main trends of mass systematics. Indeed, no subshell effects are displayed by the masses at all(2).

As for the non-diagonal elements, their contribution to the energy may be summed up by using symmetries between particles and holes (11) in the form

$$E_{def}(N_o+n, Z_o+p) = \Sigma\alpha_{kl} n^k(\delta-n)^k p^l(\Delta-p)^l \tag{3}$$

$$+ \Sigma\beta_{kl} n^k(\delta-n)^k p^l(\Delta-p)^l (n-\delta/2)(p-\Delta/2)$$

where δ and Δ are the number of neutron and proton states in their respective filling major shells. Some discussion of these oscillating terms has been presented at Leysin (12). They are responsible for the bump in the S_{2n} lines around N=90, illustrated in fig. 14 of (12) and by Dr. Barber here. They can also reproduce the deviations from parabolic trends shown by Dr. Nielsen.

5. Light nuclei

For nuclei with neutrons and protons in the same shell, eq. (1) has to be modified to take account of isospin conservation. The resulting eq. (36.12) of ref (5) is quadratic in n+p = a, with a pairing term and T(T+1) "Wigner" symmetry term. This last one is responsible for the main qualitative difference between eq. (36.12) of (5) and our eq. (1), namely a cusp along the line N=Z. This is shown in fig. 6 for A=21.

Shell mixing, symmetric between particles and holes, adds extra terms, of the form (11,12)

$$E_{def}(A_o+a,T) = \sum_{kl} \alpha_{kl} \frac{T^{l-k}(D/4 - T)^l}{(D/2 - T)^k} a^k(D-a)^k \qquad (4)$$

where $A_o = N_o + Z_o$ and D is the total number of nucleon states in the major filling shell. These terms are indicated experimentally by oscillations superposed on the otherwise linear Q_α systematics, as in our fig. 7 and in fig. 20 of (12), which are the Q_α analogues of the S_{2n} fig.14 of (12) mentioned above.

1. Zeldes, N., Ark. Fys. 36, 361(1966).

2. Zeldes, N., Nuclear Physics 7, 27 (1958).

3. Bohr, A. and Mottelson, B.R., Nuclear Structure Vol. I (Benjamin, 1969).

4. Mayer, M. G. and Jensen, J.H.D., Elementary Theory of Nuclear Shell Structure (Wiley, 1955).

5. De-Shalit, A. and Talmi, I., Nuclear Shell Theory (Academic Press, 1963).

6. Condon, E.U. and Shortley, G.H., Theory of Atomic Spectra (Cambridge, 1935).

7. Comfort, J.R., Maher, J.V., Morrison, G.C. and Schiffer, J.P., Phys. Rev. Letters 25, 383 (1970).

8. De-Shalit, A., Phys. Rev. 105, 1528(1957).

9. Gove, N.B. and Yamada, M., Nuclear Data A4, 237(1968).

10. Segre, E., Nuclei and Particles (Benjamin, 1964).

11. Liran, S., Wagman, J. and Zeldes, N., in preparation.

12. Comay, E., Liran, S., Wagman, J. and Zeldes, N., CERN 70-30(1970 p. 165.

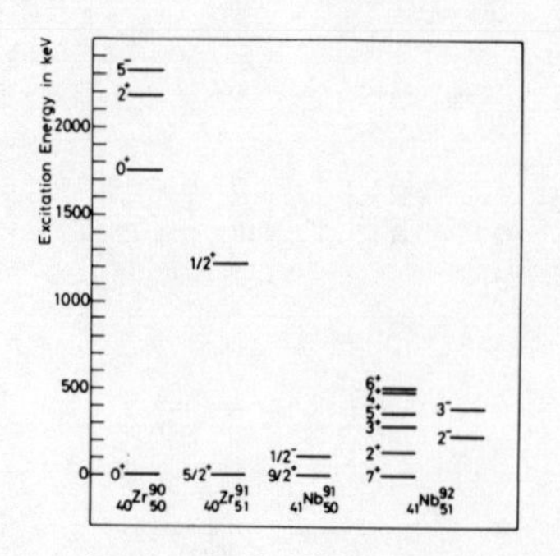

Fig. 1. Typical excitations of even-even, odd-A and odd-odd nuclei near magic numbers.

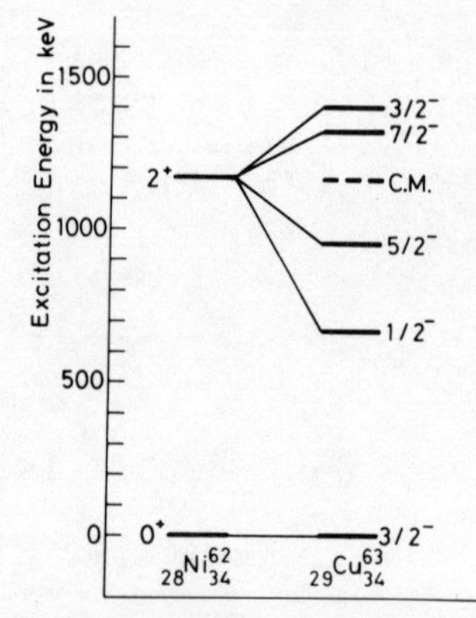

Fig. 2. Illustration of Lawson and Uretzky's center of mass theorem (recent measurements (Nuclear Data B2-3-31) determined the spin of the highest level as $5/2^-$ rather than $3/2^-$. Nevertheless, the theorem is very clearly illustrated by this older level scheme).

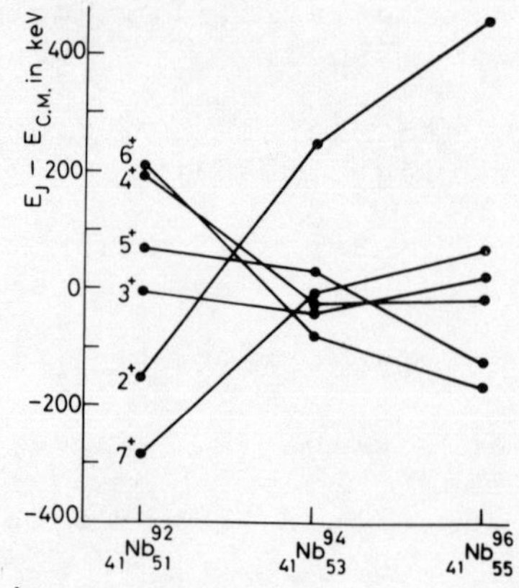

Fig. 3. Systematics of ground state multiplets in odd-odd Nb isotopes.

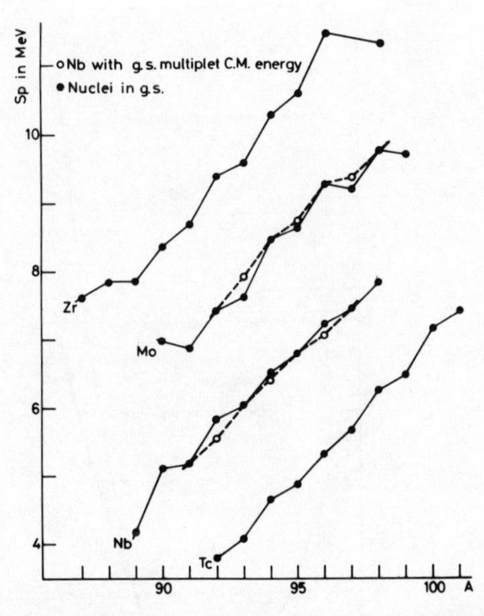

Fig. 4. Straightening of the Nb and Mo Sp isotopic lines by using the odd-odd Nb masses of the center of gravity of g.s. multiplets.

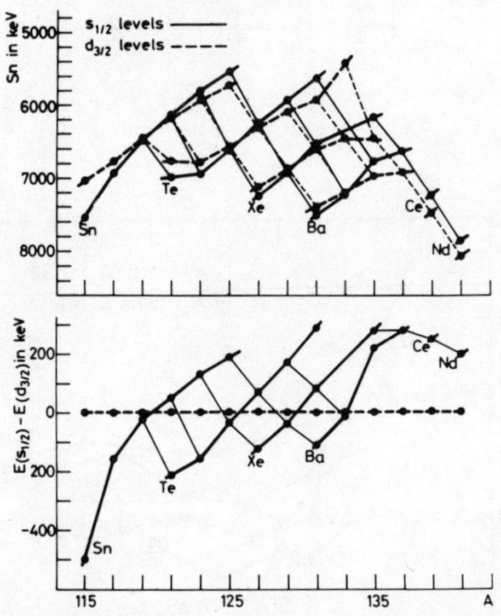

Fig. 5. Lower part: systematics of $s_{1/2}$-$d_{3/2}$ excitation energies in odd-N isotopes of Sn,Te, Xe, Ba, Ce and Nd. Upper part: fine structure of neutron separation energies of the above nuclides when distinguishing between different subshells. Thick (thin) lines connect isotopes (isotones).

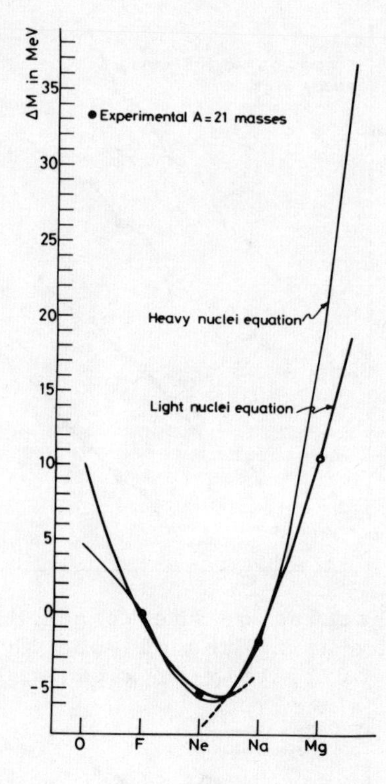

Fig. 6. Comparison of the shell model light and heavy nuclei mass equations for A=21 isobars.

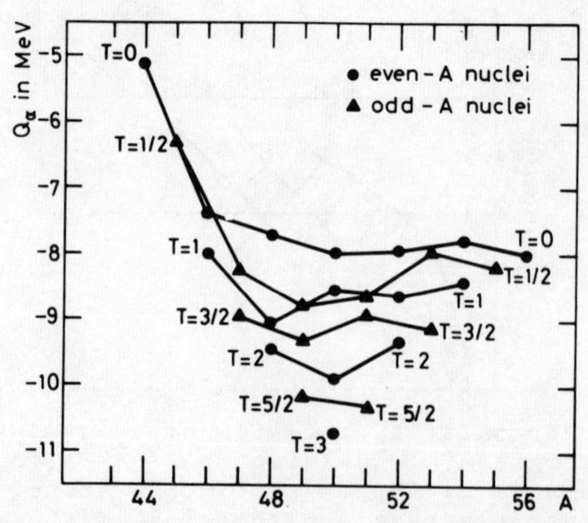

Fig. 7. Q_α values connecting $1f_{7/2}$ nuclei, as function of A. Line connects nuclei with the same T and T_Z values.

A Deformable Mass Formula and Fission Data*

P. A. Seeger

Los Alamos Scientific Laboratory, University of California, Los Alamos, New Mexico 87544

ABSTRACT

Development within the past decade of improved shell model and deformation corrections to the liquid-drop mass law has not only improved its accuracy but also allowed use in more aspects of nuclear physics, such as calculation of fission barriers. Additional experimental data are thus available as input for determination of mass-formula parameters. Preliminary results including fission barrier heights and fission Q-values as input data will be presented. Statistical tests have been performed on fits to the 1971 mass table and to the input reactions and doublets, indicating a possible discrepancy between the two forms of data.

1. The Mass Formula

At the Winnipeg conference in 1967 the author presented (1) a mass formula which used the Strutinsky normalization of a simply parametrized Nilsson-model level diagram to obtain shell and deformation corrections to the liquid-drop formula. This formula was further refined by including nuclear shapes with $P_4(\cos\theta)$ as well as $P_2(\cos\theta)$ terms, and by removing some self-imposed restrictions on the Nilsson levels. This latter work was reported fully at the Leysin conference (2) in 1970. The only changes made in the present work are conversion of the surface-symmetry term to the denominator form (3) and inclusion of a curvature term. The binding energy is represented by

$$B(Z,A) = \alpha A - \beta(N-Z)^2/A \cdot \frac{1}{1 + \zeta B_S(\varepsilon,\varepsilon_4)/A^{1/3}} - \gamma A^{2/3} B_S(\varepsilon,\varepsilon_4)$$

$$\pm \frac{\delta}{\sqrt{A}} - 0.864 \frac{Z^2}{R_0 A^{1/3}} \left\{ B_C(\varepsilon,\varepsilon_4) - \frac{0.76361}{Z^{2/3}} - \frac{2.453}{R_0{}^2 A^{2/3}} \right\}$$

$$+ \phi A^{1/3} B_L(\varepsilon,\varepsilon_4) + S(N,Z,\varepsilon,\varepsilon_4) + 7 \exp(-6|N-Z|/A)$$

$$+ 14.43 \times 10^{-6} Z^{2.39} \quad . \qquad (1)$$

*Work performed under the auspices of the United States Atomic Energy Commission.

Details, in particular of the shell and deformation function $S(N,Z,\varepsilon,\varepsilon_4)$, are given in Ref. (2). The quantity $B_L(\varepsilon,\varepsilon_4)$ is the integrated surface curvature, $\oint(\frac{1}{R_1} + \frac{1}{R_2})\,dS$, normalized to a sphere of the same volume.

Although the changes in the formula are slight, the changes in the fit are significant, as can be seen by comparing column 1 of Table I with Ref. (2). The standard deviation is reduced from 0.733 to 0.714 MeV. These fits were made to 1134 mass excesses for $N \geq 20$ and $Z \geq 20$ from the 1964 (4) and 1967 (5) mass tables.

TABLE I

Fits to the Mass Formula

	1	2	3	4	5	6
	1964 Table	1971 Table	1971 No δ>1	D+R	D	R
α	16.305	16.312	16.327	16.157	16.31	15.34
β	44.2	43.4	44.4	44.5	40.6	41.2
γ	23.57	23.58	23.73	22.76	23.23	19.69
ζ	5.81	5.62	5.84	6.00	4.87	5.63
δ	9.45	9.58	9.58	9.45	10.8	9.49
ϕ	5.07	5.06	5.40	3.95	3.93	14.6
R_n	1.232	1.227	1.238	1.252	1.262	1.238
R_o	1.1610	1.1590	1.1599	1.1738	1.153	1.230
n	1134	1553	1024	2077	553	1524
d.f.	1126	1545	1016	2069	545	1516
χ^2	5214	5648	5300	47683	14439	29648
σ	0.714	0.704	0.698	0.508	0.553	0.466

	7	8	9	10	11	12
	Near Stable	Not Stable	D+R Var. c^2	Table+ Fission	D+R+F	R+F
α	16.40	16.27	16.153	16.061	16.073	15.67
β	42.3	44.8	44.3	42.7	43.4	43.5
γ	24.00	23.40	22.79	21.91	22.17	20.15
ζ	5.2	5.99	5.96	5.55	5.75	6.01
δ	9.6	9.5	9.45	9.56	9.45	9.37
ϕ	5.5	4.9	4.04	2.13	2.88	3.4
R_n	1.238	1.236	1.253	1.229	1.251	1.236
R_o	1.152	1.165	1.1750	1.1728	1.177	1.240
c^2			0.9296			
n	494	530	2077	1598	2122	1569
d.f.	486	522	2068	1590	2114	1561
χ^2	2712	2541	47468	6876	48511	30871
σ	0.711	0.684	0.507	0.774	0.512	0.475

2. Fit to the 1971 Mass Table

The new mass adjustment (6) contains 1553 data with $N \geq 20$ and $Z \geq 20$; the results of a fit to these data are given in column 2 of Table I and the deviations plotted in Fig. 1. In the right-hand portion of the figure, the weighted frequency distribution of the errors is compared to a normal curve of the same variance. Even without seeing the plot vs A, one finds from the χ^2-test that there is infinitesimal probability that the distribution is normal. The effect of the correlations among the data is to increase the variance without extending the tails of the distribution. It is all too clear that systematic discrepancies persist; this indicates that the mass formula is incomplete and seriously affects the statistical arguments which follow.

The variance-ratio, or F-test (7,8), may be used to check the hypothesis that two independent estimates represent the same variance. If $s_1{}^2$ and $s_2{}^2$ are variance estimates corresponding to f_1 and f_2 degrees of freedom, then

$$s_1{}^2/s_2{}^2 \propto F(f_1,f_2) = \frac{\chi^2(f_1)/f_1}{\chi^2(f_2)/f_2} \tag{2}$$

where the symbol $\propto$ is used to mean "is distributed as." Percentage points of F are tabulated for many values of f_1 and f_2. For larger f_1 and/or f_2, the relations

$$F(f_1,\infty) = \chi^2(f_1)/f_1 \ , \tag{3}$$

expectation
$$E[F(f_1,f_2)] = \frac{f_2}{f_2 - 2} \tag{4}$$

variance
$$V[F(f_1,f_2)] = \left(\frac{f_2}{f_2 - 2}\right)^2 \frac{2}{f_1} \frac{(f_1 + f_2 - 2)}{(f_2 - 4)} \tag{5}$$

are useful; also, F approaches a normal distribution for large f's.

Applied to columns 1 and 2 of Table I,

$$(0.714)^2/(0.704)^2 = 1.024 \propto F(1126,1545) \approx N(1.0013,0.0555);$$

since randomness would produce estimates this different 69% of the time, we conclude that there is no significant improvement in the fit.

The F-distribution is also useful for comparing the sums of residuals (distributed as χ^2) obtained with different numbers of degrees of freedom (9). For instance, 529 of the mass table entries involve an extrapolated energy. If these are omitted, as in Table I column 3, χ^2 is reduced from 5684 to 5300. But

$$\frac{(5648 - 5300)/529}{5648/1545} = 0.18$$

K

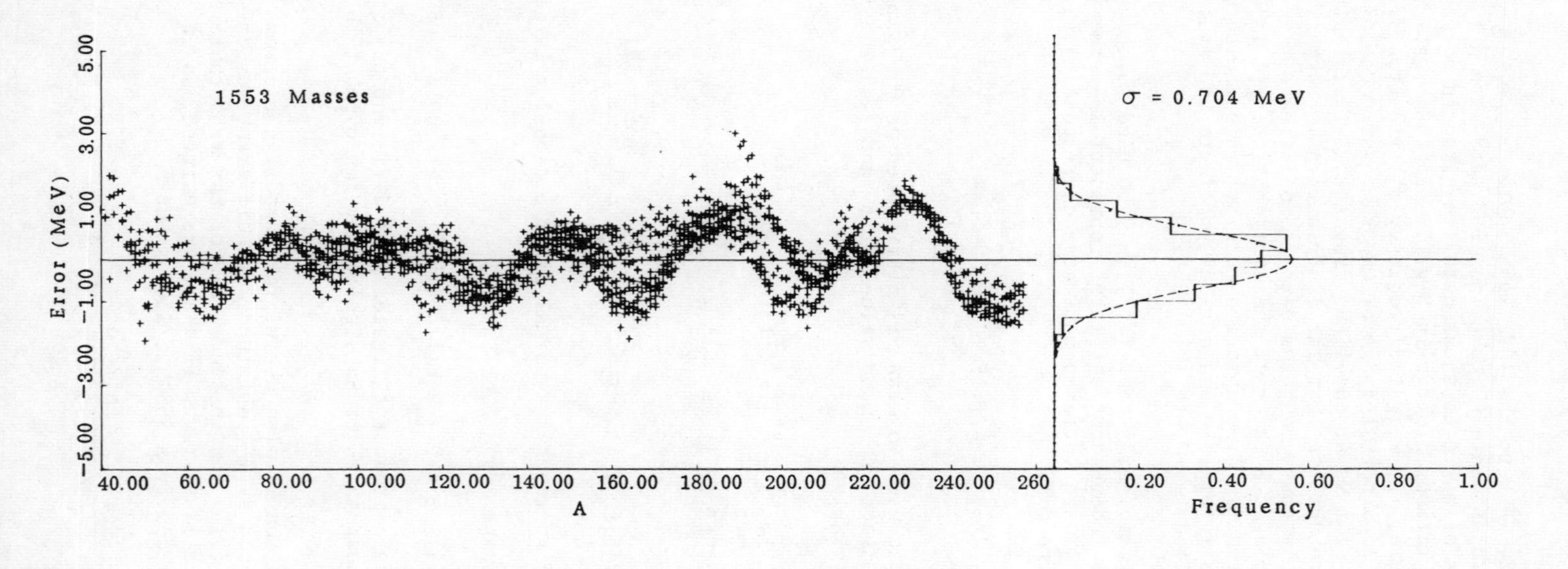

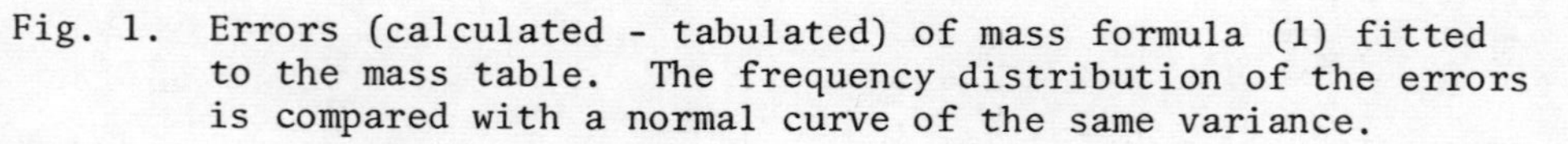

Fig. 1. Errors (calculated - tabulated) of mass formula (1) fitted to the mass table. The frequency distribution of the errors is compared with a normal curve of the same variance.

is an impossibly small figure in the distribution F(529,1545); the additional mass excesses add nothing to the fit.

3. Fit to Mass Table Input Data

The effects of correlations in the mass table are difficult to include in the calculations (9); they can be eliminated by using the mass table input data instead. In the region of interest, the data used as input to the 1971 evaluation include 1524 reactions and decay energies and 553 mass-spectroscopic doublets; the 1000 reference masses and the data indicated as being estimates from systematics, unnecessary, or false were omitted. The fit is shown in column 4 of Table I. Fits were also made separately to the doublets and to the reactions; see columns 5 and 6 of Table I and Figs. 2 and 3. The frequency distribution in Fig. 2 is consistent with the normal, but that in Fig. 3 is not, indicating a probable systematic difficulty in the fit to the reactions.

Four of the parameters fitted to reactions appear to be different from those fitted to doublets: α, γ, R_0, and ϕ. However, these are all strongly correlated, and none are well determined by the reactions because the largest changes in A and Z are 4 and 2, respectively. We can form an independent estimate of the variance of the doublets-plus-reactions fit by pooling the separate fits, giving $\sigma = 0.491$ MeV for 2061 degrees of freedom. By a two-sided F-test, the hypothesis that the two estimates are equal is acceptable at any significance level above 84%. We can also partition the χ^2 of column 4 by subtracting separate contributions due to the doublets and reactions, and perform another F-test:

$$\frac{(47683 - 14439 - 29648)/8}{47683/2069} = 19.5 \propto F(8,2069)$$

shows that there is a definite systematic difference in the fit of Eq. (1) to the doublets or reactions. The suggestion I wish to make is that the data themselves disagree.

The doublet data deal with stable nuclei, and the reactions with nuclei further from the stability line. To investigate whether the formula behaves differently, separate fits were made to portions of the mass table on and off the stability line; the results are given in columns 7 and 8 of Table I. Subtracting the two separate fits from the total, in the same manner as above:

$$\frac{(5300 - 2712 - 2541)/8}{5300/1016} = 1.13 \propto F(8,1016)$$

gives no indication of any difficulty or inconsistency in the formula itself.

The most obvious possible discrepancy between doublets and reactions is the value of c^2, assumed (6) to be $c^2 = 0.931504$ MeV/u. A new fit was made with c^2 as a free parameter; the resulting value was

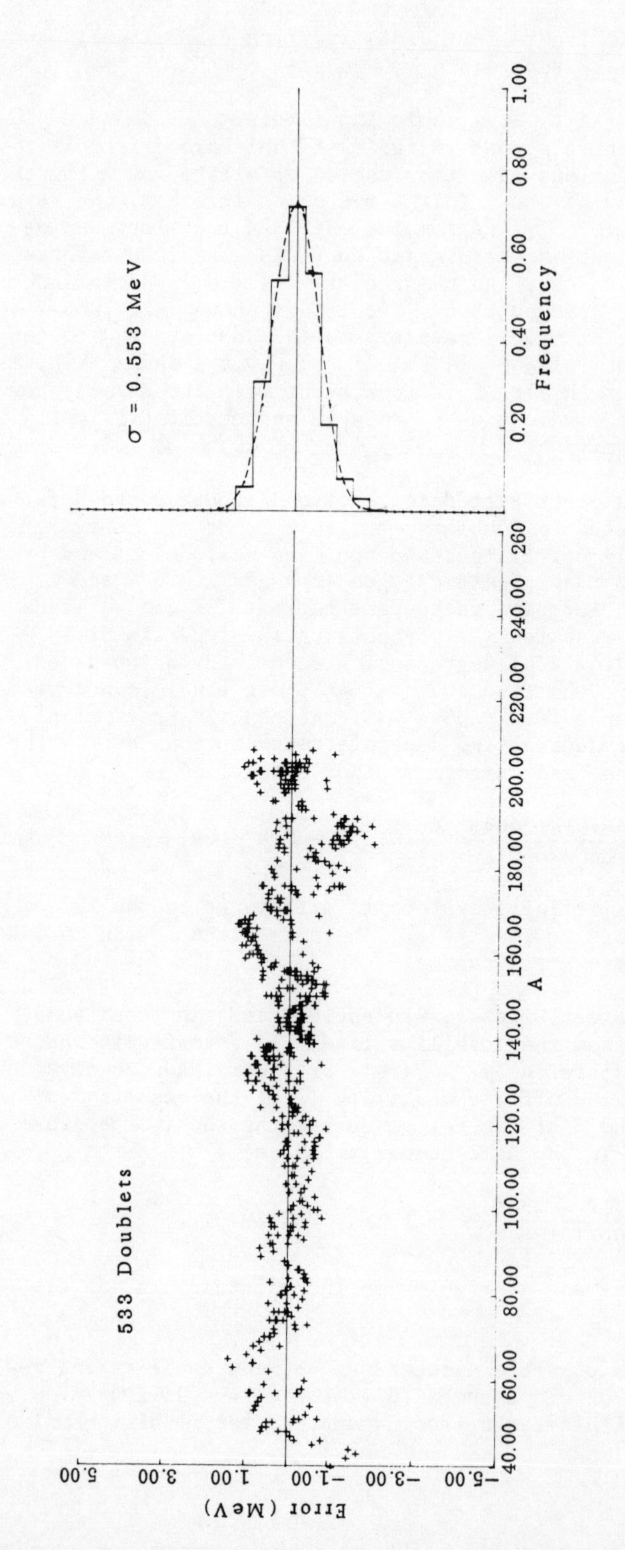

Fig. 2. Errors (calculated - tabulated) of mass formula (1) fitted to mass-spectroscopic doublets. The frequency distribution of the errors is compared with a normal curve of the same variance.

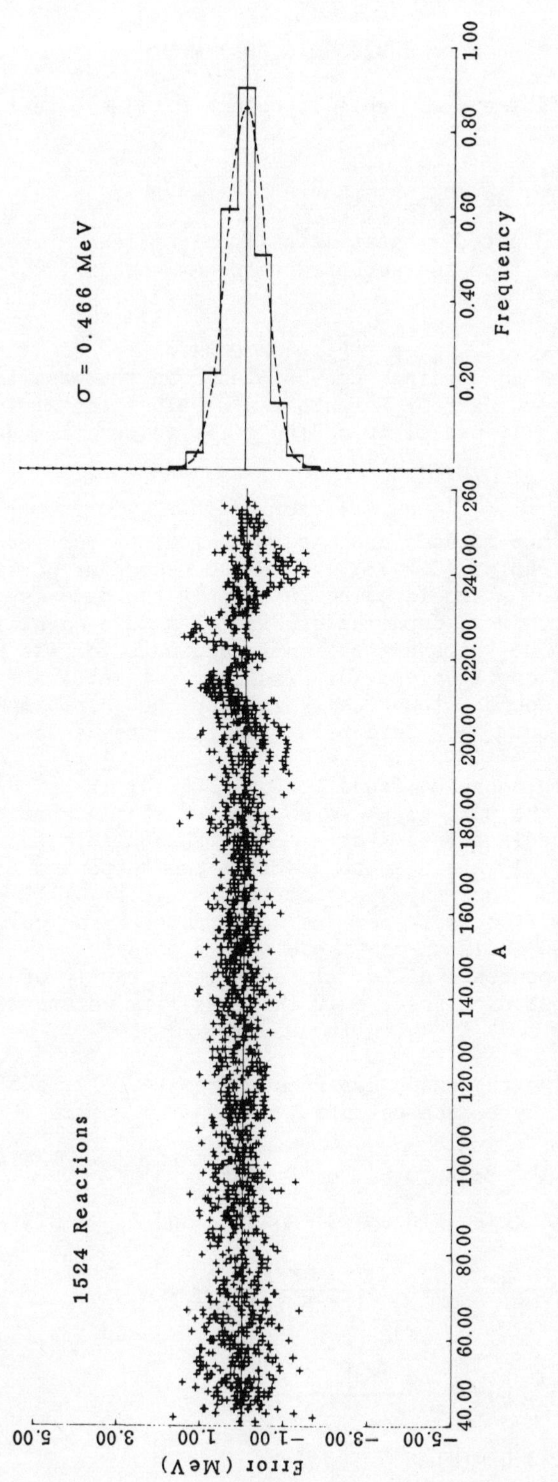

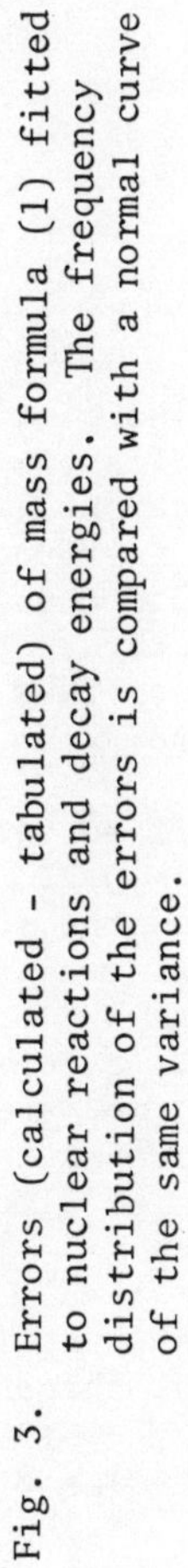

Fig. 3. Errors (calculated - tabulated) of mass formula (1) fitted to nuclear reactions and decay energies. The frequency distribution of the errors is compared with a normal curve of the same variance.

$$c^2 = 0.9296 \pm 0.0007 \text{ MeV/u} \quad , \tag{6}$$

with the coefficients of Table I, column 9. The F-test for this parameter,

$$\frac{(47683 - 47468)/1}{47683/2069} = 9.3 \propto F(1,2069) \approx \chi^2(1) \quad ,$$

shows it is undoubtedly "statistically significant"; the value obtained is lower than the accepted value at the 99.7% level of significance.

4. Fission Data

Additional constraints can be placed on the mass-formula fit by including fission barrier heights and Q-values in the input data. This is especially useful in making a fit to reaction data.

a. Barrier Heights

Two distinct methods are used to calculate barrier height. For heavy fissile nuclei ($Z > 88$), the shape-dependent parts of Eq. (1) are searched for a saddle point in exactly the same way the ground state is found. More detailed studies of saddle-point shapes (10) indicate the α_2-α_4 parametrization is adequate for the relatively small deformations involved for fissility parameter $x \gtrsim 0.75$. The barriers thus found automatically include the shape dependences B_S, B_C, and B_L in Eq. (1). Zero-point energies are, as usual, ignored.

If no saddle can be found for $\varepsilon < 0.5$, or if $Z \leq 88$ or fissility $x < 0.72$, the pure liquid-drop without shell corrections is used to find the saddle point. For $x > 0.5$, the calculations of Cohen and Swiatecki (11) are used to describe the shape and find B_S and B_C; for $x < 0.5$, the shape parametrization of Nix (12) is used. Effects of curvature were not included in the shape calculations, so B_L has not been included for these saddle points. Despite the more complicated dependence of Eq. (1) on B_S, the family of saddle-point shapes is almost unchanged; only the fissility parameter need be redefined.

Abstracting the shape dependence from Eq. (1), and representing general shapes by ε, the potential is

$$V(\varepsilon) = B_S(\varepsilon)\, E_S + B_C(\varepsilon)\, E_C + E_{sy}(1 + B_S(\varepsilon)\, \zeta/A^{1/3})^{-1} \tag{7}$$

where $E_S = \gamma A^{2/3}$, $E_C = 0.864\ Z^2/R_0A^{1/3}$, and $E_{sy} = \beta(N - Z)^2/A$.

$$\frac{dV}{d\varepsilon} = E_S \frac{dB_S}{d\varepsilon} + \frac{dB_C}{d\varepsilon} - \frac{\zeta/A^{1/3}}{(1 + B_S(\varepsilon)\zeta/A^{1/3})^2}\, E_{sy}$$

If we define

$$E_S^{\,o}(\varepsilon) = E_S - \frac{\zeta/A^{1/3}}{(1 + B_S(\varepsilon)\zeta/A^{1/3})^2}\, E_{sy} \tag{8}$$

and $x(\varepsilon) = E_C/2\, E_S^{\,o}(\varepsilon)$, (9)

we obtain the usual stationary-value equation

$$\frac{1}{E_s{}^o(\varepsilon)}\frac{dV(\varepsilon)}{d\varepsilon} = \frac{dB_s(\varepsilon)}{d\varepsilon} - 2\ x(\varepsilon)\ \frac{dB_c(\varepsilon)}{d\varepsilon} = 0 \tag{10}$$

for which the solutions at constant x, viz., $\hat{\varepsilon}(x)$, $\hat{B}_s(x)$, and $\hat{B}_c(x)$ are tabulated. Consider the variation in $x(\varepsilon)$ near $\hat{\varepsilon}$, for $x \sim 0.7$:

$$\frac{dx}{x} = -\frac{dE_s{}^o}{E_s{}^o} = \frac{E_{sy}}{E_s{}^o}\frac{(\zeta/A^{1/3})^2}{(1 + B_s(\hat{\varepsilon})\zeta/A^{1/3})^3}\,dB_s \sim 0.010\ dB_s$$

$$\frac{dx}{d\varepsilon} \sim 0.007\ \frac{dB_s}{d\varepsilon} \sim 0.005\ \frac{dB_c}{d\varepsilon}$$

which has been assumed small enough not to affect seriously the results for $\hat{B}_s(x)$ and $\hat{B}_c(x)$ found by iteration.

Application of the calculations to experimental fission thresholds is not always clear. The calculated lowest barrier may be of a spin and parity which is not accessible to the experiment. For instance, low-energy neutrons on ^{235}U $(7/2^-)$ cannot excite 0^+ states, which the "ground state" at the saddle surely is. Also, cases where the second hump of the barrier is higher must be avoided. Thus a careful selection of the data must be made; the 12 entries in Table II include photofission (13), (d,pf) and (t,pf) studies (14), and (α,f) thresholds (15,16)--two odd-Z targets and the photofissions are included despite spin uncertainties because they have been used historically, and the large experimental uncertainties should cover the calculational difficulties. Six additional thresholds not yet included in this calculation are found in Ref. (17).

TABLE II

Fission Threshold Data

Reaction	-Q (MeV)	Calculated		
		α_2	α_4	x
^{232}Th (γ) $^{232}Th^*$	5.90±0.10	0.23	0.03	
^{238}U (γ) $^{238}U^*$	5.60±0.10	0.25	0.00	
^{238}Pu (γ) $^{238}Pu^*$	6.10±0.10	0.22	0.02	
^{240}Pu (γ) $^{240}Pu^*$	6.00±0.10	0.25	-0.02	
^{239}Pu (d,p) $^{240}Pu^*$	0.645±0.070	0.25	-0.02	
^{234}U (t,p) $^{236}U^*$	1.91±0.10	0.22	0.02	
^{235}U (d,p) $^{236}U^*$	0.96±0.10	0.22	0.02	
^{233}U (d,p) $^{234}U^*$	0.74±0.10	0.23	0.02	
^{197}Au (α,γ) $^{201}Tl^*$	24.0 ±1.5			0.642
^{197}Au (α,γ) $^{201}Tl^*$	21.4 ±2.0			0.642
^{206}Pb (α,γ) $^{201}Po^*$	24.0 ±2.0			0.669
^{211}Bi (α,γ) $^{215}At^*$	25.2 ±2.0			0.679

b. Fission Q-Values

An enormous number of experiments have measured average properties of fission, but none has ever measured enough parameters of any single event to determine unequivocally the Q-value of a specific reaction of the form

$$\begin{aligned}(Z-z,A-a) + (z,a) + E_o &\rightarrow (Z,A)^* \\ &\rightarrow (Z_1,A_1')^* + (Z_2,A_2')^* + E_K' \\ &\rightarrow (Z_1,A_1) + (Z_2,A_2) + \nu(n + \overline{E}_n) + E_\gamma + E_K \qquad (11)\end{aligned}$$

where the second line represents the initial mass split, and the last line includes the post-neutron (pre-β-decay) fragments in their ground states, ν neutrons, and gamma and kinetic energy. All of these have been measured separately, and in principle could all be measured simultaneously. Workers at Oak Ridge (18,19) have measured A_1, A_2, ν_1, and ν_2, but not Z_1 and Z_2; estimating the latter from $Z_L = (A_L'/A)\ Z + 0.5$, all possible cases have been taken from their graphs for which all Z's and A's are nearly integers; see Table III. Measurements at Saclay (20) determined Z_1 and ν_1 but not A_1'; they used $A_L' = (Z_L/Z)\ A - 2$. Their interger cases are also included in Table III. Neither of these groups measured $\overline{E}_n$ or E_γ; we took $\overline{E}_n$ from Terrell (21) and E_γ was assumed distributed as the sum of two square distributions between 0 and the separation energy of the next neutron in each fragment, from the mass table (6). Similar correlations of measurements must exist in lab notebooks around the world, but we have not found any others in the literature. Much better measurements are needed before this type of data can be used in the mass table evaluation. Such data would be of value in the evaluation because they connect the upper reaches of the table with its heartland, and also bring in nuclides not presently included. As shown in the final column of Table III, there are distinct systematic discrepancies between each of these experiments and table. This may be due in part to inconsistencies between the reactions and couplets on which the table is based.

5. Fits Including Fission Data

The last three columns of Table I give the fits to the mass table + fission data, to reactions and doublets + fission data, and to reactions only + fission data. From F-tests, there is essentially zero probability that the fission data are consistent with the mass table, but the fission data are not necessarily inconsistent with the reactions and doublets. Comparing columns 6 and 12 of Table I, it will be seen that when fission is included, the terms which were poorly determined by reactions alone are closer to the values obtained by fit to the mass table. It is possible that more complete and better fission data would allow the mass formula to be derived from reaction data alone, without the intermediate step of the mass table.

Acknowledgments

The author is very grateful to A. H. Wapstra and N. B. Gove for providing him with a deck of the 1971 mass evaluation and a tape of the input data. I also wish to thank J. R. Nix and S. C. Burnett for helpful discussions on the theoretical and experimental aspects of fission.

TABLE III

Fission Q-Values

Reaction	Experimental Q (MeV)	Table Q (MeV)	ΔQ (MeV)
$^{233}U + p \rightarrow {}^{87}Kr + {}^{141}La + 6n$	170.3±2.8	159.46	
$\rightarrow {}^{88}Kr + {}^{140}La + 6n$	170.8±2.8	159.77	
$\rightarrow {}^{97}Zr + {}^{132}I + 5n$	180.6±2.8	172.50	8.3±2.7
$\rightarrow {}^{111}Rh + {}^{117}Cd + 6n$	170.6±2.8		
$\rightarrow {}^{112}Pd + {}^{116}Ag + 6n$	170.8±3.2	164.5	
$^{233}U + p \rightarrow {}^{83}Se + {}^{146}Pr + 5n$	174.4±2.6	156.13	
$\rightarrow {}^{84}Se + {}^{145}Pr + 5n$	177.3±3.4	159.39	
$\rightarrow {}^{89}Rb + {}^{140}Ba + 5n$	180.0±3.1	168.82	
$\rightarrow {}^{92}Sr + {}^{137}Cs + 5n$	182.4±3.4	173.35	12.3±4.3
$\rightarrow {}^{93}Sr + {}^{136}Cs + 5n$	180.9±2.8	170.17	
$\rightarrow {}^{102}Zr + {}^{128}Sb + 4n$	186.9±3.2	180.2	
$^{238}U + p \rightarrow {}^{85}Se + {}^{149}Pr + 5n$	161.7±3.0		
$\rightarrow {}^{99}Zr + {}^{136}I + 4n$	173.5±2.2	180.2	
$\rightarrow {}^{100}Zr + {}^{135}I + 4n$	176.6±3.3	183.2	-6.9±0.4
$\rightarrow {}^{101}Nb + {}^{134}Te + 4n$	176.9±3.3	184.4	
$^{209}Bi + \alpha \rightarrow {}^{87}Kr + {}^{118}In + 8n$	100.8±2.7	87.75	
$\rightarrow {}^{93}Y + {}^{112}Pd + 8n$	101.1±3.4	90.13	
$\rightarrow {}^{96}Zr + {}^{110}Rh + 7n$	107.5±3.1	96.03	
$\rightarrow {}^{98}Nb + {}^{108}Ru + 7n$	107.5±3.1	94.9	12.2±1.2
$\rightarrow {}^{99}Nb + {}^{107}Ru + 7n$	106.7±2.9	94.3	
$\rightarrow {}^{101}Mo + {}^{105}Tc + 7n$	106.6±3.0	93.70	
$^{252}Cf \rightarrow {}^{96}Y + {}^{152}Pr + 4n$	197.0±3.0		
$\rightarrow {}^{97}Y + {}^{152}Pr + 3n$	204.0±3.0		
$\rightarrow {}^{98}Zr + {}^{148}Ce + 6n$	189.0±3.0	179.6	
$\rightarrow {}^{99}Zr + {}^{148}Ce + 5n$	190.0±3.0	184.8	
$\rightarrow {}^{99}Zr + {}^{149}Ce + 4n$	192.0±3.0		
$\rightarrow {}^{100}Zr + {}^{150}Ce + 2n$	209.0±3.0		
$\rightarrow {}^{103}Mo + {}^{144}La + 5n$	194.0±3.0	190.3	6.2±2.4
$\rightarrow {}^{104}Mo + {}^{144}La + 4n$	207.0±3.0	198.1	
$\rightarrow {}^{105}Mo + {}^{145}La + 2n$	217.0±3.0		
$\rightarrow {}^{110}Rh + {}^{136}I + 6n$	194.0±3.0	189.99	
$\rightarrow {}^{111}Rh + {}^{136}I + 5n$	204.0±3.0		
$\rightarrow {}^{112}Rh + {}^{137}I + 3n$	213.0±3.0		

1. SEEGER, P. A., in Proceedings of the Third International Conference on Atomic Masses, R. C. Barber, ed. (University of Manitoba Press, Winnipeg, 1967), p. 85.

2. SEEGER, P. A., in Report CERN 70-30, p. 217 (1970).

3. MYERS, W. D. and SWIATECKI, W. J., Ann Phys. (NY) 55, 395 (1969).

4. MATTAUCH, J. H. E., THIELE, W., and WAPSTRA, A. H., Nucl. Phys. 67, 1 (1965).

5. WAPSTRA, A. H., KURZECK, C., AND ANISIMOFF, A., in Proceedings of the Third International Conference on Atomic Masses, R. C. Barber, ed. (University of Manitoba Press, Winnipeg, 1967), p. 153.

6. WAPSTRA, A. H. and GOVE, N. B., Nuclear Data Tables 9, 265 (1971).

7. BROWNLEE, K. A., Statistical Theory and Methodology in Science and Engineering, (Wiley, New York, 1960).

8. BEYER, W. H., ed., Handbook of Tables for Probability and Statistics, (Chemical Rubber, Cleveland, 1966), p. 19.

9. BREITENBERGER, E., in Nuclidic Masses, W. H. Johnson, Jr., ed. (Springer-Verlag, Vienna, 1964), p. 91.

10. NILSSON, S. G., TSANG, C. F., SOBICZEWSKI, A., SZYMAŃSKI, Z., WYCECH, S., GUSTAFSON, C., LAMM, I-L., MÖLLER, P., and NILSSON, B., Nucl. Phys. A131, 1 (1969).

11. COHEN, S. and SWIATECKI, W. J., Ann. Phys. (NY) 22, 406 (1963).

12. NIX, J. R., Nucl. Phys. A130, 241 (1969).

13. HYDE, E. K., The Nuclear Properties of the Heavy Elements, Part III, (Prentice-Hall, Englewood Cliffs, 1964), p. 23.

14. BRITT, H. C., RICKEY, F. A. JR., and HALL, W. S., Phys. Rev. 175, 1525 (1968).

15. HUIZENGA, J. R., CHAUDHRY, R., and VANDENBOSCH, R., Phys. Rev. 126, 210 (1962).

16. BURNETT, D. S., GATTI, R. C., PLASIL, F., PRICE, P. B., SWIATECKI, W. J., and THOMPSON, S. G., Phys. Rev. 134, B952 (1964).

17. CRAMER, J. D., Report LA-4198; EANDC(US)131"A," (1969).

18. BURNETT, S. C., FERGUSON, R. L., PLASIL, F., and SCHMITT, H. W., Phys. Rev. C 3, 3 (1971).

19. PLASIL, F., FERGUSON, R. L., and SCHMITT, H. W., in Second IAEA Symposium on The Physics and Chemistry of Fission, (IAEA, Vienna, 1969), p. 505.

20. NIFENECKER, H., FREHAUT, J., and SOLEILHAC, M., ibid., p. 491.

21. TERRELL, J., Phys. Rev. 113, 527 (1959); Phys. Rev. 127, 880 (1962).

Mass Formula for Low Charge Atomic Nuclei

S. Ludwig

Institut für Theoretische Physik, Würzburg, W. Germany

G. Süßmann

Sektion Physik, Theor. Physik, München, W. Germany

E. R. H. Hilf

Institut für Theoretische Physik, Würzburg, W. Germany

"Low charge atomic nuclei" or "extrem neutron rich" nuclei means nuclei far below the valley of β-stability in vacuum, where the number of neutrons is relatively large with respect to the number of protons.

The existing mass formulae, as for example the droplet model formula are derived starting with the assumption that the asymmetry $I=(N-Z)/A$ of the nucleus is small with respect to one. Therefore one cannot expect, that an extrapolation of the results for the binding energies into the range of low charge nuclei will be right.

The droplet model, which was supported by investigations of Thomas Fermi systems gives good results for nuclei along the β-valley, but in order to find a model, which could be reliably extrapolated, one has to start again with Thomas Fermi calculations without the assumption $I<<1$.

The Thomas Fermi method itself can also be improved concerning the description of density inhomogenities.

We started with a two nucleon central force containing 4 free unknown constants, one for each type of exchange force. In order to get saturation we added as usual a momentum dependent part, which contains once more 4 free constants.

$$w^{12} = \sum_{S=0}^{1} \sum_{T=0}^{1} \left[V_{ST}\,\delta_a(|\underline{r}|) + \frac{1}{2m}\underline{p}K_{ST}\,\delta_a(|\underline{r}|)\underline{p} \right] P_S Q_T$$

with $P_0 = \frac{1}{2}(1-p),\ P_1 = \frac{1}{2}(1+P),$

where $\underline{P} := \frac{1}{2}(1 + \underline{\sigma}^{(1)} \cdot \underline{\sigma}^{(2)})$ is the Bartlettoperator

and $Q_0 = \frac{1}{2}(1-Q).\ Q_1 = \frac{1}{2}(1+Q),$

where $Q := \frac{1}{2}(1 + \underline{\tau}^{(1)} \cdot \underline{\tau}^{(2)})$ is the Heisenbergoperator

$\underline{r}$ = relative distance; $\underline{p}$ = relative momentum.

To study the effect of improvements in the Thomas Fermi method we first took for the dependance on the relative distance of the two nucleons a contactpotential $\delta(|r|)$, because of simpler analytic treatment. But also a broadent contactpotential, which takes care of the finite range of the force and from which one gets one additional free constant, was taken. Other potentials are principally possible but cause a greater numerical effort.

Now one has to calculate the expectation value of this two particle operator, which yields to the expression

$$<H_2> = \frac{1}{2} \int dq_1 \int dq_2 \int dq_1' \int dq_2' <q_1'|\rho|q_1> <q_2'|\rho|q_2> \cdot \{<q_1q_2|w^{12}|q_1'q_2'> - <q_1q_2|w^{12}|q_2'q_1'>\}, \tag{2}$$

containing the mixed density

$$<q|\rho|q'> := \sum_{\lambda_{occupied}} u_\lambda^*(q')u_\lambda(q), \tag{3}$$

where u(q) are eigenfunctions.

With $|q> = |\underline{r}, m_s, m_t>$ and

$$<\underline{r}'|n_t|\underline{r}> := <\underline{r}'m_s t|\rho|\underline{r}m_s t> + <\underline{r}'-m_s t|\rho|\underline{r}-m_s t>$$

one gets

$$<H_2> = \frac{1}{8} \int d\underline{r}_1 \int d\underline{r}_2 \int d\underline{r}_1' \int d\underline{r}_2' \sum_{t_1t_2} \sum_{s_1s_2} <\underline{r}_1'|n_{t_1}|\underline{r}_1><\underline{r}_2'|n_{t_2}|\underline{r}_2> \cdot \{<\underline{r}_1s_1t_1\underline{r}_2s_2t_2|w^{12}|\underline{r}_1's_1t_1\underline{r}_2's_2t_2> - <\underline{r}_1s_1t_1\underline{r}_2s_2t_2|w^{12}|\underline{r}_2's_2t_2\underline{r}_1's_1t_1>\}. \tag{3'}$$

We have summed up over the spins but not over the isotopic spins in order to have the possibility for independent proton and neutron density distributions. So $<r|n_t|r>$ is the density of nucleons at position $\underline{r}$ with isotopic spin t, no matter if they have spin up or spin down.

One then has to express the mixed density

$$n_t(\underline{r}, \xi) := >\underline{r} - \frac{1}{2}\xi|n_t|\underline{r} + \frac{1}{2}\xi> = \sum_{\mu=1}^{N_t} \varphi_\mu\left(\underline{r} - \frac{1}{2}\xi\right)\varphi_\mu^*\left(\underline{r} + \frac{1}{2}\xi\right) \tag{4}$$

by the normal density $n_t(\underline{r}) := \sum_{\mu=1}^{N_t} \varphi_\mu(\underline{r})\varphi_\mu^*(\underline{r})$, which is done in the usual Thomas Fermi method by taking plane waves for the one nucleon eigenfunctions and replacing the sum over the occupied levels $\sum_{\mu=1}^{N_t}$ by the integral $2\cdot\Omega/(2\pi)^3 \int d^3k$ over the sphere in the wave number with radius $k_{Ft}(\underline{r}) = (3\pi^2 n_t(\underline{r}))^{1/3}$

This yields the result

$$n_t(\underline{r}, \xi) = n_t(\underline{r}) \quad (k_{Ft}(\underline{r}) \cdot |\xi|), \tag{5}$$

where $(x) := 3/x^2 \cdot \left(\frac{\sin x}{x} - \cos x\right).$

The assumption that the one-nucleon eigenfunctions can be replaced locally by plane waves is a very crude approach concerning the density inhomogenities. If one calculates for example in this approach the kinetic energy of the nucleon system, one gets the usual expression

$$E_{kin} = \frac{3}{5}\frac{\hbar^2}{Lm}\sum_t \int d\underline{r}\, n_t^{5/3}(\underline{r})\ (3\pi^2)^{2/3} \tag{6}$$

but not in addition the well known Weizsäcker correction-term

$$E_W = \frac{\hbar^2}{2m}\sum_t \int d\underline{r}\, |\nabla(\sqrt{n_t(\underline{r})})|^2. \tag{7}$$

As Macke[+)] has shown, it is possible to get the Weizsäcker correctionterm by a generalisation of the Thomas Fermi method, which takes better into account the density inhomogenities. He did not use plane waves as eigenfunctions, but started from the ansatz

$$\psi_\mu(\underline{r}) = \sqrt{\det\left(\frac{\partial y_i}{\partial r_k}\right)}\ \varphi_\mu(y(\underline{r})) \tag{8}$$

where $\underline{y}(\underline{r})$ is a transformation of the r-space into the $\underline{y}$-space with $\Delta y_k \cdot \Delta y_l = f_k \delta_{kl}$ and the $\varphi_\mu(y)$ are plane waves in the $\underline{y}$-space, so that the $\psi_\mu(r)$ are again orthogonal. This ansatz is a loss of generality with respect to the Hartree-Fock method, where one has to minimize the total energy with respect to all N one-nucleon wave functions, but it is richer than the usual Thomas Fermi method, which starts from the density-distribution and thus only 1 free function.

+) Ann. d. Physik 17, 1955)

Here we have three free functions $\underline{y}(\underline{r})$ and the density is simply the functional determinant

Our single nucleon eigenfunctions can be called "deformed plane waves".

Calculating the expectation value of the kinetic energy yields the Weizsäcker correctionterm as well. In the energy expression one has the functional determinant of the transformation, but no longer the three free functions itself. In the same way it is possible to calculate the mixed density. One may evaluate the functions $y_i(r + \frac{1}{2}\xi)$ in Taylor series and can cut off beyond second order in ξ, because we have only derivations of second order in ξ for ξ going to zero in the energy expression. The resulting expression for the mixed density is

$$n_t(\underline{r}, \underline{\xi}) := \left(n_t\left(\underline{r} + \frac{1}{2}\underline{\xi}\right) n_t\left(\underline{r} - \frac{1}{2}\underline{\xi}\right)\right)^{1/2} \quad (k_{Ft}(\underline{r})|\underline{\xi}|) \tag{9}$$

By using this expression, we now get in addition all gradient correctionterms in the potential part of the energy. In the last two equations the results are shown for radial symmetric density distributions. Underlined are the additional gradient correctionterms resulting from the more symmetric form of the mixed density. The first expression gives always the result for contact-potential, while the second expression shows the additional terms resulting from the finite range of the force which is indicated by the occurence of the additional constant a^2.

$$U_V^{(0)} = \frac{1}{4}\int d\underline{r}\{(3V_{10} + V_{01})n_t n_{-t} + V_{01}(n_t^2 + n_{-t}^2)\}$$

$$\begin{aligned} U_V^{(2)} = \frac{a^2}{48}\int d\underline{r}\Big\{ &-(V_{00} + 3V_{11} + 3V_{10} + V_{01})\,\partial n_t \cdot \partial n_{-t} \\ &+ (V_{00} + 3V_{11} - 3V_{10} - V_{01}) \\ &\cdot\Big(\frac{3}{5}n_t n_{-t}(n_t^{2/3} + n_{-t}^{2/3})(3\pi^2)^{2/3} \\ &+ \frac{1}{2}\partial n_t\,\partial n_{-t} + \frac{1}{4}n_t n_{-t}\left[\frac{(\partial n_t)^2}{n_t^2} + \frac{(\partial n_{-t})^2}{n_{-t}^2}\right]\Big)\Big\} \end{aligned} \tag{10}$$

with $\partial n_t := \dfrac{dn_t(r)}{dr}$; $a^2 : 4\pi\int dr\, r^4\, \delta_a(r)$

$$U_K^{(0)} = \frac{\hbar^2}{2m}\frac{1}{16}\int dr \Big\{ (3K_{10}+K_{01})\Big(-\partial n_t \partial n_{-t} + \frac{3}{5} n_t n_{-t} (n_t^{2/3} + n_{-t}^{2/3})(3\pi^2)^{2/3} + \frac{1}{2}\partial n_t \partial n_{-t} + \frac{1}{4} n_t n_{-t} \Big[\frac{(\partial n_t)^2}{n_t^2} + \frac{(\partial n_{-t})^2}{n_{-t}^2} \Big] \Big) + K_{01}\Big(-(\partial n_t)^2 - (\partial n_{-t})^2 + \frac{6}{5}(n_t^{5/3} + n_{-t}^{5/3})(3\pi^2)^{2/3} + (\partial n_t)^2 + (\partial n_{-t})^2 \Big) \Big\}$$

$$U_K^{(2)} = \frac{\hbar^2}{2m}\frac{a^2}{96}\int dr \Big\{ (3K_{10}+K_{01})\Big(\frac{1}{2}\partial n_t \partial n_{-t} - \frac{1}{2}\partial^2 n_t \partial^2 n_{-t} + \frac{3}{5}(n_t^{5/3}\Delta n_{-t} + n_{-t}^{5/3}\Delta n_t)(3\pi^2)^{2/3} - \frac{1}{2}\Delta n_t \Delta n_{-t} + \frac{1}{4}\Big[\frac{(\partial n_t)^2}{n_t}\Delta n_{-t} + \frac{(\partial n_{-t})^2}{n_{-t}}\Delta n_t \Big] \Big) + K_{01}\Big(\frac{1}{2}(\Delta n_t)^2 + \frac{1}{2}(\Delta n_{-t})^2 - \frac{1}{2}(\partial^2 n_t)^2 - \frac{1}{2}(\partial^2 n_{-t})^2 + \frac{6}{5}(3\pi^2)^{2/3}(n_t^{5/3}\cdot\Delta n_t + n_{-t}^{5/3}\cdot\Delta n_{-t}) - \frac{1}{2}(\Delta n_t)^2 - \frac{1}{2}(\Delta n_{-t})^2 + \frac{1}{2}\frac{(\partial n_t)^2}{n_t}\Delta n_t + \frac{1}{2}\frac{(\partial n_{-t})^2}{n_{-t}}\Delta n_{-t} \Big) \Big\} \tag{11}$$

with $\Delta n := \partial^2 n_t + \frac{2}{r}\partial n_t$.

One sees, that the 8 at the beginning unknown free constants appear only in 5 different combinations, so that we have to fit only these 5 ones to the known masses of stable nuclei. Once this fit is done one could use this formula for calculations of the binding energies of neutronrich nuclei with more confidence, because of the more exact treatment of the asymmetry $I:=(N-Z)/A$.

This project is in progress, making use of a similar work of Graef & Hilf without gradient corrections.

Is there a Relation between the Nuclear Binding Energy and the Parameters of the Variable Moment of the Inertia Model?*

G. Scharff-Goldhaber

Brookhaven National Laboratory, Upton, New York 11973, U.S.A.

Successful models have been obtained by comparing the atomic nucleus with 1) the atom (shell model (1947)), 2) the molecule (liquid drop) (collective models of strongly deformed nuclei (1951/2) and of "vibrational nuclei" (1955)), and 3) superconductivity (pairing plus quadrupole model (1958)). 4) A phase transition between closed shell nuclei and even-even nuclei containing one or several pairs of nucleons was found [1,2]. The latter are governed by the equation $\partial^3 - \partial_o \partial^2 = \frac{J(J+1)}{2C}$, where ∂ denotes the moment of inertia, J the angular momentum and ∂_o and C are parameters characteristic for a nucleus (N,Z). ∂_o is found to be highest at the centers of both neutron and proton shells, where $\partial_o \sim A^{5/3}$. When the closed shell is exceeded by only ~ 4 nucleons, ∂_o becomes negative and then gives a measure of the rigidity of the nucleus. The "stiffness parameter" C is highest for stable nuclei of each element. C decreases toward zero with increasing Coulomb (or symmetry) energy. It appears to be smaller in regions where the "gap" is relatively small. - It is tempting to compare ∂ with the density ρ, J with the pressure p, and the number of nucleons exceeding a closed shell (measured by ∂_o) with the temperature T of a real gas (Van der Waals equation). The region near the phase boundary appears to be characterized by large fluctuations of ∂ and ∂_o at constant J. Since the shell model received an important confirmation from the appreciably higher binding energies of closed shell nuclei, the next challenge is to relate ∂_o and C to binding energies.

* Work performed under the auspices of the U.S. Atomic Energy Commission.
(Abstract of invited paper not read because of illness).

1. MARISCOTTI, M.A.J., SCHARFF-GOLDHABER, G., BUCK, B., Phys. Rev. 178, 1864 (1969).

2. SCHARFF-GOLDHABER, G and GOLDHABER A.S., Phys. Rev. Lett. 24, 1349(C) (1970); see also Physics Today 23, November 1970, 17(1970).

Mass Extrapolation and Mass Formulae

K. Bos and A. H. Wapstra

Institute for Nuclear Physics Research (IKO),
Amsterdam, the Netherlands

1. Introduction

For several reasons, the interest of nuclear physicists is continuously shifting towards nuclei further and further removed from stability. In low mass nuclides one wants to study nuclei with high values of the isospin. For very heavy ones, the matter of the possible existence of an island of stability near at mass number 300 is a popular subject. And at intermediate mass number, the possible existence of new regions of deformation receives much attention.

For most experimental studies in these regions, it is desirable to have estimates of binding energies (or, what amounts to the same, atomic masses) of nuclides in their ground states. Essentially three approaches have been made. Probably the oldest[1)] (Bethe, 1936), is trying to devise a mass formula representing as well as possible the masses of known nuclides; one then hopes that by inserting values of Z and A for the unknown nuclei one gets good estimates of their masses. A second[2)] is to estimate atomic mass differences (alpha or beta decay energies, single or double proton or neutron binding energies) by extending systematics of known values. A third method, easier to computerize, is use of formulae giving relation between masses of adjacent (in a N vs. Z plot) nuclei for stepwise extrapolation[3)].

Each of these methods has its peculiar difficulties. As shown in fig. 1. systematic differences remain between any existing mass formula and experimental data, suggesting that mass tables derived from these formulae can still be improved. With the third method, computerized extension, unavoidable erroneous values among the experimentally determined masses cause progressively larger deviations in extrapolation, as can be amply demonstrated Judicious use of the second method, especially if several kinds of systematics are used simultaneously (Wapstra and Gove[4)]) can avoid this difficulty, but the process can not easily be computerized and far extrapolation becomes progressively less dependable.

Already some time ago, it occurred to us that a better solution might be found by combining the first and the last approach. One has then, first, to select a mass formula of sufficient quality, secondly to set up a systematic of the differences with the actual masses, and in the third place to extrapolate this systematics. Steps in the direction of such a solution are discussed below.

2. Requirements for selection of the mass formula

One of us (A.H. Wapstra) has noticed before that for practical use, a mass formula should not in the first place represent actual masses as well as possible, but rather mass differences between adjacent nuclei. He therefore suggested that, in order to derive best values for the constants in such a formula, one should not adjust it to the experimental mass values (which, moreover, are strongly correlated in a statistical sense) but to mass differences (mainly β and α decay energies but also many binding energies of one or two nucleons). One can then, rather easily, devise a procedure that makes the formula represent mass differences as well as possible. This requirement also allows more influence of data for nuclides far removed from stability (which is of course of primary importance if one wishes to use the formulae for extrapolation), since though few masses are known there, a considerable number of alpha and beta decay energies have been measured in that region.

It is rather remarkable that exactly the same requirement applies to the present purpose. For easy extrapolation, it is not primarily necessary that the differences between actual values and formula are small but that they do not vary rapidly.

Just because of this, the treatment of the odd-even staggering term (the δ-term) is rather important. In all proposed formulae, this term is a single additive term to the main part giving a relatively small correction that, however, varies very fast with $Z, N, A = Z+N$ or $I = N-Z$. The above observation that adjustments should be made primarily to mass differences rather than to masses themselves therefore applies especially to this term.

We even propose that the adjustment of the pairing-term should be kept separate from the determination of the constants in the rest of the mass formula. This would be most easy if a table could be devised giving best values of the pairing energies.

3. The pairing energy

For our purpose we will write the pairing energy:

	for N	Z
$\delta = o$	even,	even
$\delta = -\delta_N$	odd	even
$\delta = -\delta_Z$	even	odd
$\delta = -\delta_o$	odd	odd

The above definitions differ from the usual ones, as e.g. $S_Z = M(Z,N)-\frac{1}{2}M(Z-1,N)-\frac{1}{2}M(Z+1,N)$ in the (realistic) case that the mass surfaces are not nearly straight (see fig.2). Determining δ this way, this curvature yet introduces difficulties where it is very pronounced. We expect that use of a mass formula offers again help: instead of the masses themselves, we plot differences between actual masses and the part of a decent mass formula remaining after removal of its pairing **term.** Examples of such plots are given in fig. 3, using the mass formulae of Seeger[5]) and of Zeldes[6]). Somewhat to our surprise, the results depend on the mass formula chosen. We hope that further study will give an explanation of this effect and a possibility for its removal. We then plan to construct a table of these pairing energies. With its help, we will study further the extrapolation procedure proposed in the introduction.

1. H.A. BETHE, Rev. Mod. Phys., 8, 82 (1936).

2. J.H.E. MATTAUCH, W. THIELE, A.H. WAPSTRA, Nucl. Phys., 67 (1965).

3. G.T. GARVEY, W.J. GERACE, R.L. JAFFE, I. TALMI, I. KELSON, Rev. Mod. Phys., 41, 4 (1969).

4. A.H. WAPSTRA, N.B. GOVE, Nuclear Data 9 (1971) 357.

5. P.A. SEEGER, Proc. of the Conference on the properties of Nuclei far from the Region of Beta-Stability. Leysin, Switzerland, 1970.

6. N. ZELDES, A. GRILL, A. SIMIEVIC, Shell-model semi-empirical nuclear masses, Mat. Fys. Skr. Dan. Vid. Selsk, 3, no. 5 (1967).

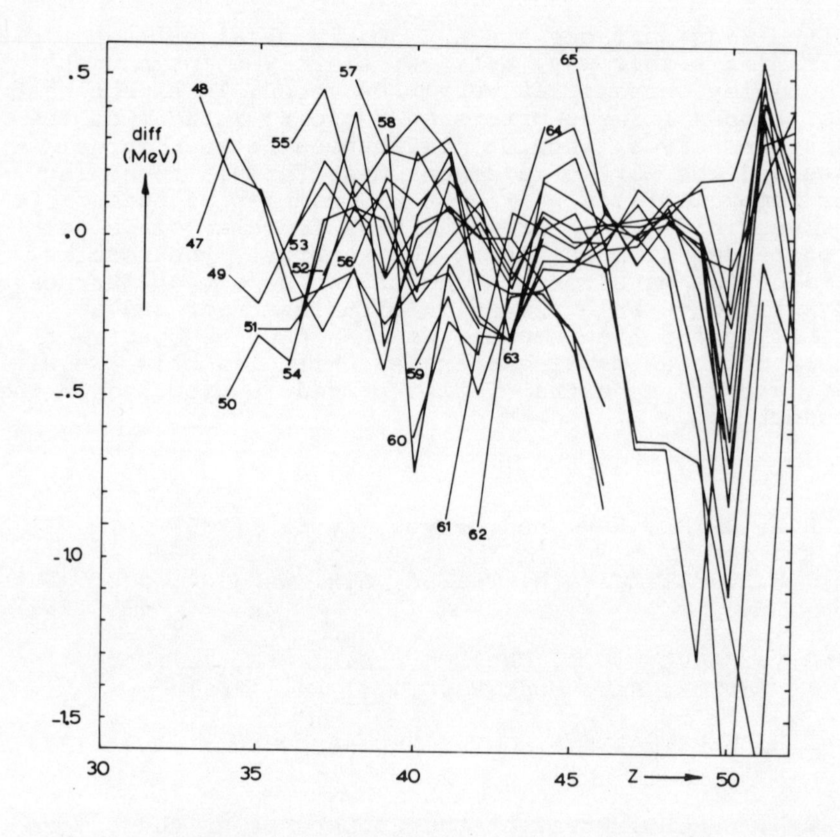

Fig. 1. Differences between atomic masses and the mass formula of Zeldes. The lines connect isotones; the neutron numbers are given.

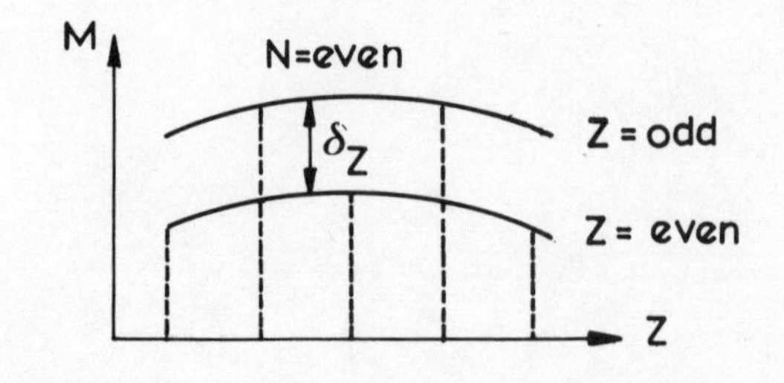

Fig. 2. Definition of δ_Z.

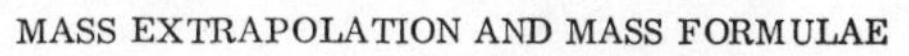

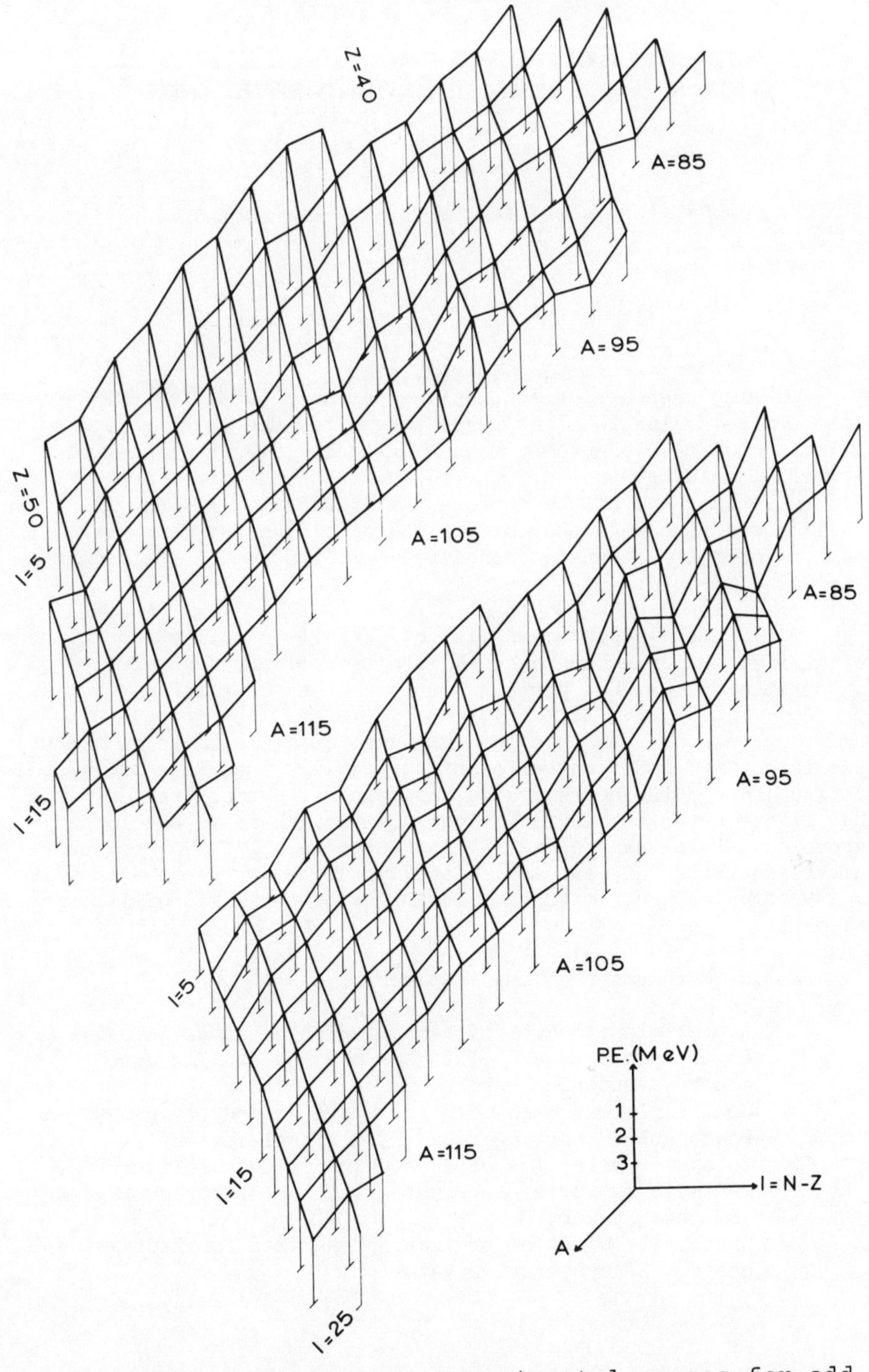

Fig. 3. Differences between experimental masses for odd mass nuclei and the pairing-term free part of a mass formula (upper part: Seeger, lower part: Zeldes). This kind of diagram is supposed to reveal differences between neutron and proton pairing energies (δ_N and δ_Z).

Problems Connected with Calculations of Atomic Masses from Experimental Data*

N. B. Gove

Oak Ridge National Laboratory, Oak Ridge, Tennessee 37830, U.S.A.

1. Introduction

In many respects, the techniques used in preparing the 1971 Atomic Mass Evaluation (1-4) were taken from the several papers relating to the 1964 Atomic Mass Adjustment (5-7) and the 1961 Atomic Mass Adjustment (8,9). Indeed, the connections are very close, in that one of the authors (A. H. Wapstra) of the '71 version was also an author of the '64 and '61 adjustments and both authors of the '71 version benefited from frequent assistance of the Mainz group.

The present paper, then, is not a review of all the problems of mass adjustments but rather a selection of some (hopefully) interesting items that pertain mainly to the '71 effort.

Most of the "new" problems are due to the increase in amount, precision, and interconnection of the data. In earlier times, the data could readily be separated into groups, by mass region, which had little or no connection. Now, however, there is one large group containing all the nuclei connected to all the other nuclei. After some study of this and related problems, we decided to write a new computer program to take advantage of the IBM 360/91 computer at ORNL.

We added a number of new features:

1. A preliminary check of each Q-value or doublet to see if it is "reasonable." This was most helpful in catching gross keypunching errors.
2. Semi-empirical mass formula, where needed, to produce a reasonable "starting value" for a nucleus.
3. Automatic selection of data which can be pre-averaged. The reasons for pre-averaging will be discussed in the following paper.
4. Graphical output of separation energies to display the choices of systematics values.

---

* Research supported in part by the U. S. Atomic Energy Commission under contract with the Union Carbide Corporation.

5. The chaining procedure for secondary masses. This is perhaps the most important new feature of the program and is discussed below. With this technique, the normal matrix and its inverse are reduced from > 1700 squared to about 600 squared.
6. An edit procedure for either temporary or permanent change of the data list.

2. Connections List

As the input data is processed, a list is built containing for each nucleus all the nuclei it is connected to. The program then assigns a degree of 9 (unconnected) to each nucleus except ^{12}C, ^{1}n, and ^{1}H which are assigned a degree of 1 (primary). Next all nuclei which are connected to ^{12}C via hydrocarbon doublets are assigned 1. Then the list is scanned for possible reassignments. If a nucleus is connected to a primary nucleus, it is assigned a degree of 2. If it is connected to two primary nuclei, then it becomes primary too. If any nucleus is connected to two or more nuclei of lower degree, then it comes primary.

This scan is repeated until no more changes occur in one scan. Then any nuclei which still have 9 are removed from the adjustment. These are unconnected and would spoil the matrix inversion process.

Then all non-primary nuclei are temporarily removed and put in a special list, containing for each nucleus, the nucleus of lower degree to which it is connected and the mass difference and uncertainty of that connection as determined from the relevant data card (or average of several data cards).

3. Inversion of the Normal Matrix

After the pre-averaging is done and all the secondary reactions are set aside, the program forms a coefficient matrix, K, which contains for each doublet or reaction a row showing the coefficient of each nucleus in that equation. Thus, $Kx = q$, where x is the unknown adjustment vector and q is the vector of Q-values minus the Q-values calculated from the starting masses. Naturally, most elements of K are 0. Following the standard least-squares procedure, the normal matrix $A = K^{+}WK$ is formed, where W is a diagonal weight matrix. The weights are the inverse squares of the uncertainties. The answer is given by $x = A^{-1}K^{+}Wq$, where A^{-1} can also be used to determine the uncertainties of the adjusted masses and linear combinations.

The inversion of the normal matrix A is a major part of the program. Including all 1700 or so nuclei in the matrix would make the inversion procedure prohibitively expensive. Keeping only the primary nuclei reduces the normal matrix to about 640 by 640 which is a size that can be inverted. However, we took advantage of the weakness of the links around mass 107 to prepare the adjustment in two overlapping parts, thus, inverting two small matrices (about 380 x 380 and 300 x 300) rather than one large one.

The inversion was done by the Choleski method which factors the matrix into triangular matrices. The normal matrix is symmetric so only slightly more than half of it need be computed.

Only the computed part is in core storage and the matrix is inverted in place. A 380 x 380 normal matrix can be inverted using a core storage space of 380 x 381/2, or 72390 words, plus the program. We use 64-bit words so this requires some 5 million bits of core memory.

The inversion takes about 10 minutes of computer time. The inversion accuracy is checked by matrix multiplication. Errors in masses caused by inversion inaccuracy are less than 10^{-14} u.

4. The Chaining Procedure

With the inverse matrix it is possible to find the right uncertainty for any linear combination. Therefore, the preparation of the various output tables is straightforward for cases where all nuclei involved are primary. For secondary nuclei a somewhat more devious method is employed. We make use of the fact that nearly all the desired output numbers (S_p, S_n, $Q\alpha$, etc. plus the input Q-values) can be adequately expressed as the mass difference between two nuclei plus a constant.

Each nucleus in a desired combination is checked to see if it is primary or secondary. If it is secondary, the program looks up the connected nucleus and the mass difference and uncertainty. The connected nucleus replaces the original nucleus in the desired combination while the mass difference and uncertainty are set aside to be added to the final Q-value (the uncertainty is added quadratically). This process is repeated until a primary nucleus is reached.

If there is another secondary nucleus, it is also chained until it either reaches a primary or intercepts the previous chain. If two different primaries are reached, the inverse matrix is used as before to find the relevant uncertainty.

There are a very few cases where this procedure returns an uncertainty that is slightly too large. For example, ^{24}Mg (^{3}He, ^{6}He) ^{21}Mg is measured with precision comparable to that of ^{6}He itself. The proper uncertainty can be obtained by treating ^{6}He and ^{21}Mg as primary before the adjustment program is run.

5. Mass-to-Energy Coefficient

We decided to convert all energy measurements to the mass scale because the "anchor" to the whole system is ^{12}C = 0 in the mass scale. (The mass numbers can be added just before printout but are not used elsewhere in the program. Thus it is convenient to set ^{12}C = 0 instead of the more familiar ^{12}C = 12.)

At present there does not seem to be any completely right way to treat the uncertainty in the mass-to-energy conversion coefficient, $X = 931504 \pm 10$, which is a provisional value pending the next adjustment of the fundamental constants. One could ignore the uncertainty, treating X as a constant, in which case it would not matter whether the adjustment was done in mass units or energy units. Or one could treat X as a variable and $X = 931504 \pm 10$ as one of the input equations. Or one could increase the effective uncertainty of each reaction by QU_x, where Q is the Q-value, U_x is the ± 10 in X. It turns out that the answers are almost the same for each approach but in future adjustments this problem may be more serious. What we actually did was to treat X as a constant when the equations are being set up but when the output values are printed, those that are to be printed in keV have their uncertainties increased:

$$V_{keV} = V\ X \pm (UX + VU_x)^{1/2}$$

where V_{keV} is the output value in keV, V is the internal value in mass units, U is the uncertainty in V in mass units, and U_x is the uncertainty in X, the mass-to-energy conversion coefficient.

This procedure increases the uncertainties of reactions in a slightly unfair manner and in a very few cases one can (just barely) see the effect. For example, $^{103}Rh(n,\gamma)\ ^{104}Rh$ has a measured Q-value of 6.9992 ± 0.3; the output of the program is 6.99920 ± 0.31.

6. Problems for Future Adjustments

The complexity of future adjustments will probably increase. The interconnection between atomic masses and the fundamental constants is increasing. Perhaps future versions of the program will need to process calibration information as well as the Q-values.

As the number of primary values increases, it may become desirable to think up new ways to reduce the matrix size. For example in the present adjustment, if a nucleus is connected to two primaries, it thereby becomes primary too. However, it may be possible to combine the two Q-values to form a single link and temporarily remove that nucleus from the primary list. Similarly, a chain of nuclei connected at both ends to primaries maybe collapsed to a single link and later expanded again after the normal matrix is inverted.

References

1. WAPSTRA, A. H. and GOVE, N. B., The 1971 Atomic Mass Evaluation, Part I, Atomic Mass Table, Nuclear Data 9, 267 (1971).
2. WAPSTRA, A. H. and GOVE, N. B., The 1971 Atomic Mass Evaluation, Part II, Nuclear Reaction and Separation Energies, Nuclear Data 9, 299 (1971).
3. WAPSTRA, A. H. and GOVE, N. B., Evaluation of Input Values; Adjustment Procedures, Nuclear Data 9, 355 (1971).

4. GOVE, N. B. and WAPSTRA, A. H., Systematics of Separation and Decay Energies, Nuclear Data 9, 455 (1971).
5. MATTAUCH, J. H. E., THIELE, W. and WAPSTRA, A. H., 1964 Atomic Mass Table, Nuclear Physics 67, 1 (1965).
6. MATTAUCH, J. H. E., THIELE, W. and WAPSTRA, A. H., Consistent Set of Q-Values, Nuclear Physics 67, 32 (1971).
7. MATTAUCH, J. H. E., THIELE, W. and WAPSTRA, A. H., Adjustment of Relative Atomic Masses, Nuclear Physics 67, 73 (1971).
8. KONIG, L. A., MATTAUCH, J. H. E. and WAPSTRA, A. H., New Relative Nuclidic Masses, Nuclear Physics 31, 1 (1962).
9. KONIG, L. A., MATTAUCH, J. H. E. and WAPSTRA, A. H., 1961 Nuclidic Mass Table, Nuclear Physics 31, 18 (1962).

The 1971 Atomic Mass Evaluation

.A. H. Wapstra

Instituut voor Kernphysisch Onderzoek, Amsterdam

1. Introduction

Together with dr. N.B. Gove of Oak Ridge National Laboratories, the present author very recently completed a set of papers (1-4) in which atomic masses and related quantities of interest in nuclear physics (such as binding energies of one or two nucleons, or alpha and beta decay energies) are calculated in a consistent way from experimental data. The present paper is a short commentary on the meaning of the given values and the given estimates of their precision.

2. The experimental data

The quantities to be calculated are masses of atoms in their atomic and nuclear ground states. Table 1. gives a summary on the character of the experimental data with a bearing on this calculation. We will first discuss the mass spectroscopic data which are, essentially, ratios of masses of atomic or molecular ions. The fact that they concern ions and not atoms does not require corrections within the present experimental precision; it might conceivably in future calculations.

At the present level of precision in the determination of atomic masses, the influence of the data in group 1.1 of table 1. outweights that of the other data. Therefore, atomic mass values can most accurately be expressed in terms of a unit defined as a convenient fraction of the mass of a certain kind of atoms. Since always $12\ n > m$, the error in the mass of the C_nH_m combination is lowest if ^{12}C is selected for this purpose. Thus, the choice of the atomic mass unit 1u as 1/12 of the mass of a ^{12}C atom is very fortunate. Also, in this scale all known atomic masses differ by less than 0.1u from a whole number, the atomic mass number.

It should be mentioned here that the mass spectroscopic data given in litterature are not the measured mass ratios. The precision with which the mass of at least one of the two fragments is known is such that the data can, without noticable loss of accuracy, be converted in a mass difference, and this is therefore done without exception.

TABLE 1.

Experimental Data for Atomic Mass Calculations.

1. Mass spectrometric measurements

   1.1 $^{A}Z - {}^{12}C_n\,{}^{1}H_m$ type [r)]

   1.2 $^{A}Z - {}^{A'}Z'$ type [r)]

2. Reaction energy measurements

   2.1 charged particle measurements [s)]

   2.2 gamma ray measurements [s)]

3. Fundamental Constant Measurements

[r)] Mass Doublets exist in which one or both of the molecular fragments contain atoms such as ^{2}D, ^{14}N, ^{16}O, ^{19}F, ^{35}Cl or ^{37}Cl. Since their masses are known with high precision, these doublets are grouped with 1.1 and 1.2 for the present discussion. The few cases that cannot thus be grouped are unimportant for this discussion.

[s)] Another classification can be made into nuclear reactions (induced by bombarding nuclei) and decays (spontaneous emission of particles or gamma rays).

Where reaction or decay energies involve nuclear excitation, it is usually possible to convert the measured data to relations between ground states. In several cases one has to keep in mind, however, that the error value attached to the result does not account for uncertainty in the assignment of the transition to the correct final state.

Nuclear reaction energies are expressed in energy units (eV) and thus have to be converted in mass units before they can be combined with mass spectroscopic results. The conversion factor is c^2/F in which c is the velocity of light and F the Faraday constant. In our work[3)], a value 931504 ± 10 keV/u was adopted. We have checked that the precision given is such that this factor can be treated as a constant in our calculation[5]. A considerable complication would arise if this would not be the case.

In the first place, the quantities primarily measured in nuclear reactions are not energies. Simplifying slightly, in particle measurements momenta are measured (unit Wb/m), whereas gamma rays as occurring in (p,γ) and (n,γ) are measured in a scale essentially (5) based on the adopted energy of annihilation radiation, and therefore on the mass of the electron. To convert these data either to energy values (eV) or to atomic mass differences (u), one requires values for the same fundamental constants that influence the conversion factor mentioned above. Thus, secondly, the measurements in group 3 in table 1. would become connected to our input data in a complicated way - as can also be understood directly from the fact that measurements usually called determinations of Avogadro's number are in reality mass measurements of a known number of atoms of a certain kind, though this number may be determined through deposition by a certain current in a known time.

As a last point, it may be mentioned that the sum of the mass numbers does not change in nuclear reactions. Nuclear reaction energies may, therefore, be computed from mass excesses (M-A) in exactly the same way as from the masses themselves. Mass excesses can be assigned much smaller errors than the masses themselves if both are given in the energy scale (the contribution of the error in A times the conversion factor given above then disappears). For this reason we[1] have tabulated M-A in keV too, though this quantity has no simple physical meaning.

3. Application of the least square methods

We will not discuss the assumption that the least squares method is a sufficient and effective way for obtaining best values and error estimates from measured values and (root mean square) errors of linear combinations of these quantities. We can then limit our discussion to the two problems of (a) how to check whether the reported data can indeed be considered best values and good estimates of RMS errors, and (b) what will be the consequence of accepting wrong values among the input data.

We then first note that, in the χ^2 test, the least squares method contains a check for determining the presence of wrong values or, what is nearly the same, underestimated errors. The value of χ^2 is defined as the sum of the squares of the differences between the input values and the adjusted ones divided by the relevant input errors:

$$\chi^2 = \Sigma\ (V/\sigma)^2$$

If this value is significantly larger than the number of

degrees of freedom N-n (N = number of measured combinations, n = number of calculated masses), then serious deviations are present. The system of data under consideration can be transformed into one having the correct value of χ^2 by multiplying all initial RMS-errors by

$$R_e/R_i = \sqrt{\chi^2/(N-n)} \text{ (the consistency factor)}$$

This, however, is a very rough correction method. Probably more often than not, but certainly in our system of about 3500 data on 1200 atomic masses, serious deviations occur far more probably in a limited number of measurements than in a uniform way. It is therefore imperative that a considerable effort is devoted to locating and eliminating such deviations; in fact, this has been the most time-consuming task in the whole evaluation.

We then first notice that not all data are equally important for our purpose: many can be omitted with only minor influence on the finally obtained values and errors. As an example, we consider the reactions

$$^{24}Mg(n,\gamma)^{25}Mg,\ ^{24}Mg(p,\gamma)^{25}Al \text{ and } ^{25}Mg(p,n)^{25}Al,$$

measured with reported RMS-errors of 1.5, 1.0 and 6 keV respectively. By combining the first two, a value for the last one is obtained with an error of 1.8 keV (according to the law of propagation of errors). Since weights in least squares calculations are reversely proportional to squares of errors, this result outweights the directly measured one by a factor 11; it can therefore be omitted with little loss. Because of the overriding need for good checks on all really important values, we have decided not to use in our calculation data outweighted by more than a factor 10 by relatively simple combinations of other data. (The exact criterion used is, in fact, a little less restrictive especially where mass spectroscopic results are involved, since we feel that the latter are especially important and somewhat less likely to show individual errors). Thus, the number of relevant input data is reduced by almost 800 items. Yet, though not used in the determination of the masses, these cases have been carefully summarized in our tables of input values and compared there with the values following from the calculation.

A rather evident observation has led to a further reduction applied here for the first time. Consider the group of data on the mass difference between ^{28}Si and ^{29}P: two values for the reaction

$^{28}Si(p,\gamma)^{29}P$, one for $^{28}Si(d,n)^{29}P$ and one for

$^{28}Si(\tau,d)^{29}P$ (as usual, τ stand for ^3He), all with errors far larger than those in the masses of the reaction particles involved. The values do not quite agree: $R_e/R_i = 1.4$, and therefore give a too large contribution to χ^2 for the total mass adjustment. Yet, this contribution does not tell anything about the quality of data not concerned with this specific mass difference. We therefore decided that such sets of parallel reactions were to be averaged, and only the average (with, where necessary, the error corrected by the local consistency factor) was to be used in the final calculation. Again, for mass spectroscopic results an exception was made: no preaveraging was applied to them.

A second necessity for application of a χ^2 test on subgroups arises for mass-spectroscopic data. Most authors quote errors in these values obtained from the widths of distribution of results from repeated measurements. Though this is recommendable in as far as it makes for well defined, relevant quantities, it means that no allowance is made for systematic errors. As already noticed earlier (6) separate least squares analysis of groups of data from the same laboratory yield χ^2 tests with a bearing on the ratio of systematic and stochastic errors: by multiplying the last one with the consistency factor resulting from the analysis, one gets a better approximation of the real errors. Of course, this procedure can correct only for part of the systematic error. A further test is obtained in combining these data with nuclear reactions. The result of this comparison for nuclides with mass number $A < 100$ will be discussed below; it does not indicate necessity of a further correction. This is not so, though, for a group of mass spectroscopic doublets measured in Russia and reported in some six papers (3). Here, the consistency factor was found to be about 2.5, but the "pseudo"-consistency factor (3) found in combination with a number of very accurate and, it is thought, dependable reaction energies is 4.3 (It should be mentioned that the same test applied to results given in two additional recent publications by the same group gives a factor of only 1.3).

It is realized that each procedure proposed to correct for the evidently underestimated errors is somewhat arbitrary. Our choice was to indicate all data in such groups by a ground symbol (e.g. R1-- R6 and R7,R8 for the Russian groups mentioned) and to write the adjustment program in such a way that a correction factor (in our case chosen to be 1.5, 2.5 or 5) is applied to errors of all data within each group.

In the present situation, no group checks are desired for reaction data, but spot checks are. They are carried out in two fases. Before adding new data to the data files, they are compared with existing data. And in each trial-adjustment, a list is printed out of values V/σ (see definition of χ^2). If V/σ is larger than 2, the original paper is checked carefully for hidden mistakes, which are then indeed found fairly often. In several cases, such mistakes can be corrected; then, the corrected value is used in stead of the reported one. The reason for the correction is properly indicated, e.g. (transition not to ground state but to a known excited level). If the occurance of a mistake is clear but no correction can be made, the value considered is not used in the calculation; it is, though, printed out (properly indicated) in the list of input data and compared there with the output values.

4. Application of systematics

The checks on the data in the preceding chapter are based only on considerations of a mathematical nature, concerning the compatibility of different data. Yet, there is a class of data that is not checked at all by such considerations. In order to show this, we present a graphical representation of a typical part of the collection of measured data (fig. 1.). We then notice that, e.g., ^{176}Tm occurs only in one connection in the beta decay energy of this nuclide. But also data occuring in a chain of connections linked only on one side to other data, e.g.

$$^{178}\mathrm{Re}(\beta^+)^{178}\mathrm{W}(\varepsilon)^{178}\mathrm{Ta}(\beta^+)^{178}\mathrm{Hf},$$

belong to this class of "secondary" data. Eliminating secondary data from the input files does not affect "primary" data on masses in any way. Thus, the quality of the calculation is determined primarily by the data on the primary nuclides. Table 2. shows the breakdown in primary and secundary data and nuclides, and also the number of data judged unimportant or not dependable.

Of course, it is very desirable to have a check on the validity of secondary data; and if it could be one that could also give an additional check on the primaries for which V/σ is large, this would be very wellcome. Such a test is provided by the systematics of nuclear masses and derived quantities. For discussing this, we remember that nuclear binding energies B or mass excesses if plotted as a function of the numbers Z and N of neutrons and protons (or of A = N+Z and I = N-Z) are found on four rather smooth surfaces, one for each combination of N or Z even or odd. The relatively small distances between these surfaces (the pairing energies) are even very smooth functions of Z and N. As a consequence, systematics of the separation energies of two protons B(Z,N)-B(Z-2,N) or of two neutrons B(Z,N)-B(Z,N-2), or the alpha decay energy which is equal to B(Z-2,N-2)-B(Z,N) but for an additional constant, also vary smoothly. As an example, fig. 2. shows the two-neutron separation energies in the same region as fig. 1. In the neighbourhood of shell closures (here at N=82), a rather fast change occurs, and at N=88 where it is known that the shape of the nucleus changes rather suddenly, the two-neutron separation energies show a characteristic behaviour; but these two effects too occur in a systematic way. The oldest systema-

TABLE 2.

Numbers of calculated masses and input data

	Masses	Doublets	Reactions	Total Data
Primary	640	680	1220	1900
Secondary	560	-	620 a)	620
B	-	-	40	40
U	-	200	590	790
F	-	30	140	170
Syst.	550	-	-	-
Total	1750	910	2610	3520

(For the meaning of "primary" and "secondary" see paragraph 3, end. The data marked, F, U and B have not been used in the calculations because (F) their correctness is in doubt, (U) they are of much less weight than other, dependable data, (B) for other reasons, see text. The data marked "Syst." are not experimentally determined but obtained from interpolation in systematics of binding- or decay-energy systematics.

a): the excess over the number of masses is caused by the fact that "parallel" reactions occur: see paragraph 3).

---

tics used for the purpose of checking mass data is that of beta decay energies (7). Pairing energy causes a slight complication in this systematics, but it is still very useful.

In cases where "secundary" mass data correspond to large deviations in these systematics (in deciding this, all four kinds mentioned were used), the same line of action was followed as for primary data causing discrepancies. In one respect, though, decision not to use such a value is more serious than for a primary nuclide: now, no mass value can any longer be calculated for the corresponding nuclide. On the other hand, noticing the deviation means essentially that a more probably mass value can be suggested from these systematics.

Replacement of secondary data by systematics values is primarily done again in cases where sufficient grounds are judged present for distrust; in judging this, relatively small deviations from systematics alone are not thought sufficient. Yet, there are a number of such cases where it is attractive to replace experimental values by systematic ones. One example is the decay energy of $^{124}In(\beta^+)$ which causes a 1000 keV deviation in all systematics. Though this deviation is not large compared with the reported precision (800 keV), it is felt to be rather certain that here the value derived from systematics is more nearly correct. Another case is the decay energy of $^{154}Tb(\beta^+)$ reported as 3560 $\pm$ 50 keV whereas systematics indicates a value nearer to 3400 keV. Here the difference is large in terms of the reported precision but small in absolute magnitude. Yet, it causes a very pronounced change in various systematics, among them in fig. 2. In cases like these, we have added the value derived from systematics to our file of input data and labeled the experimental value in such a way that our computer program has the option of using either the one or the other. In our present final tables, the systematics values have been used.

Fig. 1. shows that, at present, rather large collections exist of data giving accurate values for differences in nuclear masses which, themselves, are not determined by experimental data; an example is the decay chain $^{187}Hg(\alpha)^{183}Os(\alpha)^{179}W$. Application of systematics in such cases is especially sensitive since choice of one decay energy here determines systematics data for three nuclides. Because of this, we have extended application of systematics for this purpose to include all these cases.

In principle, further extension could be made. We have not done so for two reasons. In the first place, the procedure cannot easily be computerized, and it is very time consuming. But in the second place, we think that further extrapolations should if possible be guided by theoretical expectations on the course of the four binding energy surfaces. In a separate paper (8) we discuss some steps we consider to take in this direction in preparation of a future mass calculation.

5. Dependability of the 1971 mass values

Some indication about the dependability of the final values is obtained in the χ^2 test applied in the final least squares adjustment. For practical purposes, but also because no connections happen to be present for data below and above mass number 106, the adjustment was made in two parts; the corresponding values for the consistency factors are 1.17 and 0.95 and thus quite satisfactory. A check on the absence of a certain kind of systematic errors is obtained by making a separate adjustment in which the mass-energy conversion factor was used as a variable (3) (5). The output value agrees well with the results of measurements of fundamental constants.

Yet, it remains rather certain that a larger fraction of the over 2500 accepted items will differ by more than the finally adopted errors from the correct values than would follow from systematics alone. Personally, I feel that authors quoting large errors just for certainty sin more than those who try to estimate their errors as accurately as possible and in doing so underestimate unsuspected systematic errors, even though this causes the situation mentioned above.

Indeed, a number of indications are present that, at least in some regions, the real errors in the given masses of the 1971 mass table are rather much larger than the errors following from the least squares adjustment. Thus, recent doublet measurements on ^{57}Fe deviate by about 25 keV from older ones. The last form part of a series of measurements which agree well with nuclear reaction data. Thus, it has to be concluded that either the new value is wrong by the amount mentioned (almost seven times the error claimed for this measurement), or that a systematic error is present in the earlier results.

Another case is found at mass numbers around 90. The data below this number agree well with one another, and so do the data starting with A = 90. Yet, the reaction bridge ^{89}Y(n,γ)^{90}Y(β^-)^{90}Zr, containing two very precise and dependable reaction energies, differs by 45 keV from the value calculated from the above data. We are inclined to believe therefore that the errors in the masses around A = 90 are of the order of 20 keV, though the values following from the least squares adjustment are about five times less.

An indication that underestimates by similar factors occur even at the lowest mass numbers is obtained in the new mass spectroscopic measurements of dr. L.G. Smith (9). Atomic masses for about half a dozen light nuclides derived from these data (see table 3) deviate by up to about five errors from values in the 1971 table. Though the somewhat limited possibility of statistical checks on Smith's data makes it difficult to asses already now the degree of validity of the very low errors assigned to them, I do not really doubt that his values are more nearly correct than those in the 1971 tables; they will certainly play an important part in any future atomic mass evaluation. Yet, the major part of the mass excesses in the 1971 table will not be very significantly altered by these new results.

References

1. A.H. WAPSTRA and N.B. GOVE, "Atomic Mass Table", Nuclear Data 9, 265 (1971).

2. A.H. WAPSTRA and N.B. GOVE, "Nuclear Reaction and Separation Energies", Nuclear Data 9, 294 (1971).

3. A.H. WAPSTRA and N.B. GOVE, "Evaluation of input values; Adjustment procedures", Nuclear Data 9, 355 (1971).
4. N.B. GOVE and A.H. WAPSTRA, "Systematics of Separation and Decay Energies", Nuclear Data 9, 455 (1971).
5. A.H. WAPSTRA, Proc. Intern. Conf. Precision Measurements, Gaithersburg 1970.
6. F. EVERLING, S.A. KONIG, J.H.E. MATTAUCH and A.H. WAPSTRA, Nuclear Physics 25, 177 (1961).
7. K. WAY and M. WOOD, Phys. Rev. 94, 119 (1954).
8. K. BOS and A.H. WAPSTRA, Fourth International Conference on Atomic Masses and Fundamental Constants, Teddington 1971.
9. L.G. SMITH, Fourth International Conference on Atomic Masses and Fundamental Constants, Teddington 1971; Phys. Rev. 4C, 22 (1971).

Comparison of mass excesses of some light nuclei of the 1971 adjustment (A = 1 to 29) and a new adjustment with extra input of Smith doublets and the latest Japanese doublets.

	1971	1971 + SM + J.
^{1}N	8071.69 ± 0.10 keV	8071.48 ± 0.09 keV
	8665.22 ± 0.09 μu	8665.00 ± 0.04 μu
^{1}H	7289.22 ± 0.09 keV	7289.07 ± 0.08 keV
	7825.22 ± 0.04 μu	$7825.05_1 \pm 0.01_2$ μu
^{2}D	13136.27 ± 0.16 keV	13135.90 ± 0.14 keV
	14102.22 ± 0.09 μu	$14101.81_8 \pm 0.02_3$ μu
^{3}T	14950.39 ± 0.23 keV	14950.01 ± 0.17 keV
	16049.78 ± 0.17 μu	16049.33 ± 0.04 μu
^{3}He	14931.74 ± 0.22 keV	14931.40 ± 0.17 keV
	16029.71 ± 0.17 μu	16029.35 ± 0.04 μu
^{4}He	2424.74 ± 0.27 keV	2424.99 ± 0.06 keV
	2603.04 ± 0.29 μu	2603.31 ± 0.06 μu
^{8}Be	4941.4 ± 0.5 keV	4941.87 ± 0.13 keV
	5304.70 ± 0.6 μu	5305.25 ± 0.13 μu
^{14}N	2863.83 ± 0.14 keV	2863.46 ± 0.04 keV
	3074.42 ± 0.14 μu	$3074.02 \pm 0.02_9$ μu
	C.F. = 1.098	C.F. = 1.123
	$P.C.F._Q$ = 1.091	$P.C.F._Q$ = 1.097
	$P.C.F._M$ = 1.114	$P.C.F._M$ = 1.169

For the meaning of the pseudo-consistency factors for reactions-Q and mass doublets-M see ref. (3). Input consistency factor for new Smith doublets 1.5.

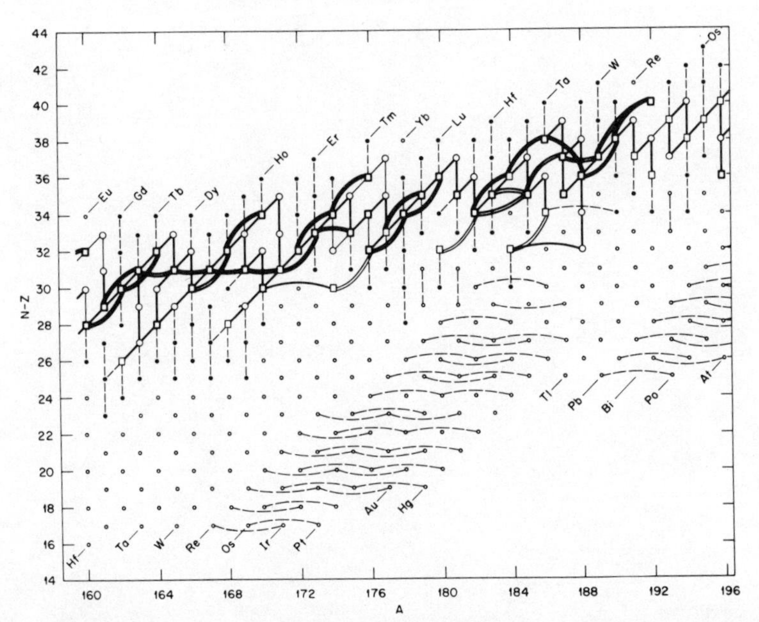

Fig. 1. Diagrammatic representation of part of the input data. Nuclei represented by squares are contained in doublets of type 1.1 of Table 1, whereas double lines represent type 2.1 doublets. All other lines represent reactions or decays. Thick (single or double) lines represent data of primary importance (most precise data) in this special region. Dotted lines are "secondary" data, as defined in part 3; large and small open circles and dots represent nuclides connected by "primary" data, not connected, and connected by "secondary" data alone.

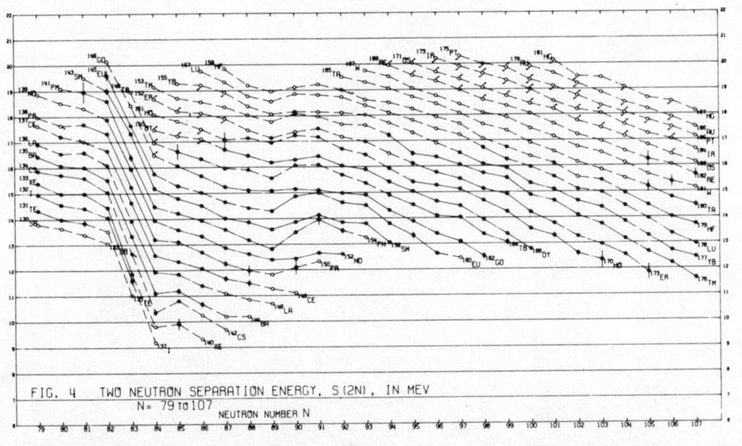

Fig. 2. Example of two-neutron separation energy systematics. Points represent precise values, larger errors are indicated by vertical bars, open circles are values estimated from this (and other) systematics; arrows near such circles indicate that two or more values had to be estimated simultaneously (see text). Lines join values belonging to isotopes of the same element.

Part 6 Velocity of Light

Progress with a Determination of the Speed of Light

C. C. Bradley, J. Edwards, D. J. E. Knight, W. R. C. Rowley, P. T. Woods

Introduction

A programme at the National Physical Laboratory is in progress for the purpose of making a determination of the speed of light (c) to an accuracy of one in 10^8 by measuring the frequency and wavelength of an infra-red laser emission. This accuracy is essential if the value for c is to be used to unify the standards of frequency and length, the former being at present realizable to one in 10^{12} and the latter to one in 10^8. The presently accepted value is 299 792.50 ± 0.10 km s^{-1} as determined by K.D. Froome at the National Physical Laboratory (Proc. Roy. Soc., A247, 109 (1958)).

In part the present experiment is an extension of Froome's method in which the frequency and wavelength of a 72 GHz klystron source were measured. A major limitation in accuracy in the microwave method was the precision in calculating a diffraction correction for the wavelength measurement which required a large air-path interferometer. In order to reduce the diffraction correction to the order of one in 10^8, for conveniently sized interferometer optics, it is necessary to increase the frequency to the 30 THz (10 μm) region. The CO_2 laser is an extremely convenient source for this frequency since it is powerful, well developed and can be stabilized to the order of one in 10^9. Thus the present method shown schematically in Figure 1 is concerned with making frequency measurements of the CO_2 laser emission and either measuring its wavelength directly or up-converting to the visible to produce side bands whose wavelength can be measured after the method of Z. Bay and G.G. Luther (Appl. Phys. Letters, 13, 303 (1968)). The CO_2 laser frequency is related to the caesium frequency standard at 9 GHz by a succession of harmonic mixing and beat frequency detection stages using a number of submillimetre wave lasers.

The following papers are a summary to date (September 1971) of the progress towards the speed of light measurement using this technique.

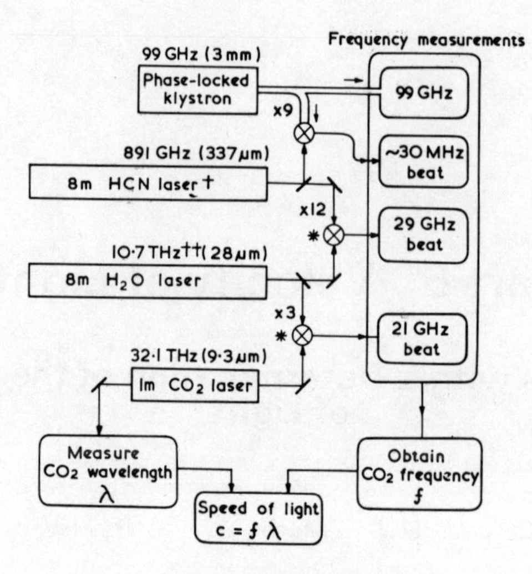

Fig. 1. The NPL speed of light programme.

Progress with a Determination of the Speed of Light

I. Far Infra-red Frequency Measurements and Two-laser Mixing

C. C. Bradley

Division of Electrical Science,
National Physical Laboratory, Teddington, England

and

G. J. Edwards and D. J. E. Knight

Division of Quantum Metrology,
National Physical Laboratory, Teddington, England

Until recently the direct measurement of frequencies in the infra-red spectral region has not been possible because of the lack of suitable detectors and mixers with response times of the order of 10^{-13} to 10^{-14} seconds. Through the development over the last three or four years of the metal-oxide-metal diode and Josephson junction point contact devices such measurements are now feasible.

Frequency measurements in the microwave region are made in a straightforward manner by reference to a standard frequency synthesized from the caesium time standard at 9 GHz using harmonic generation and mixing, usually in metal-semiconductor diodes. Submillimetre and far infra-red lasers provide suitable bridging between the microwave and infra-red regions. This report is concerned with absolute frequency measurements, spectral width and stability of the HCN laser oscillating at 805, 891 and 964 GHz and the H_2O laser, oscillating at 2.53 THz as a stage in the measurement of the frequency of a CO_2 laser at 32 THz which is part of the speed of light programme described above.

The first direct frequency measurement of a far infra-red laser emission was by Hocker *et al.* in 1967 (1). They determined the frequency of the HCN laser at 891 GHz by mixing with a high harmonic of a millimetre wave klystron in a commercial silicon-tungsten point contact diode. By tuning the klystron a suitable beat frequency is obtained and can be measured; thus,

$$891 - (12 \times 74.2)\ \text{GHz} \simeq 30\ \text{MHz beat signal.}$$

The frequency of the phase locked klystron is obtained by reference to an appropriate frequency standard. The measurement of the HCN laser was important because it provided the first link between microwave and essentially optical frequency sources. The success of the experiment depends largely on the optimization of the silicon-tungsten diode. Beyond about 1 THz the response of the latter falls off because of the combined effects of the spreading resistance r_s and shunt capacitance C_o

$$\text{i.e.} \qquad f_{\text{cut off}} \propto \frac{1}{r_s\, C_o} \,.$$

The former can be reduced by using a semi-conductor of higher mobility such as GaAs but in the case of this material this does not result in a higher cut-off frequency (2). The development of the metal-oxide-metal diode by Sokoloff et al.(3), Evenson et al. (4) and Green et al. (5) has extended response times to the 10^{-14} seconds (60 THz) region. In these a very finely pointed tungsten whisker makes contact with a metal such as nickel. The rectification process is by tunnelling through the oxide barrier. Figure 1 compares the DC characteristics of tungsten-silicon and tungsten-nickel diodes.

In the process of evaluating suitable materials for these diodes it has been found that the substitution of graphite in the form of pencil 'lead' for nickel gives the best results for the range 1 to 2.5 THz (6). The whole range of pencil lead hardness from 4H to 8B was used with little variation in the results. It has been found that there is no serious response fall off between 1 and 2.5 THz with these diodes. (At the present time there is little mixing data beyond 2.5 THz.) The main criterion for success using any of the diodes mentioned above is the sharpness of the tungsten whisker which is typically 2 to 3 mm long and about 5 μm in diameter. The point is etched down to a tip of 100 nm diameter. Normally a good whisker can be used for at least one day's experiments,but not usually for more than two or three days. The usual configuration for the metal-oxide-metal and metal-oxide-graphite diodes is a free-space arrangement. The nickel or graphite post is mounted on a micrometer differential screw movement and the tungsten whisker is welded to a copper wire (6).

It has been demonstrated by Matarrese and Evenson (7) that these diodes, when receiving radiation in free space, act as travelling wave antennas. Using conventional theories it may be shown that the optimum angle for coupling to the diode is given approximately by

$$\theta_{max} = \cos^{-1}\left(1 - \frac{0.371}{L/\lambda}\right)$$

where L is the effective length of the antenna, λ is the wavelength and θ is the angle between the antenna and the Poynting vector of the radiation. This is an important consideration when mixing two laser emissions of different frequencies on the diode (Figure 2).

The high frequency circuit parameters for the diode are complicated by the open structure and no attempt has been made to control them to any degree. The effect of biasing the diode has often been found to have a marked effect on the beat signals as will be described later. In all the experiments described below some microwave power is used to 'make up' appropriate frequency differences (see Table I). For this purpose it is found satisfactory to have an open-ended waveguide mounted vertically above the horizontal whisker and some few mm above it. (The position of the waveguide is usually not critical within reasonable limits.)

TABLE I

Frequency mixing using tungsten to graphite diodes

Laser	Frequency GHz	Mixing scheme[+] GHz	Beat signal/noise dB[++]
HCN	891	Δf = 891-(9×99)*	30
	805	Δf = 805-891+(3×28.7)*	45
	964	Δf = 964-891-(2×36.5)*	35-55
H_2O	2528	Δf = 2528-(3×891)-(4×36)*	35
		2Δf - 2nd harmonic of beat	8

*klystron source

[+]Δf in range 1-100 MHz

[++]best signal/noise in 10 kHz bandwidth

HCN laser measurements

The frequency of the HCN laser at 891 GHz was measured using the technique of Hocker *et al*. (1). The microwave source was a phase-locked 74 or 99 GHz klystron and the harmonic generation and mixing was in either a silicon-tungsten, nickel-tungsten or graphite-tungsten diode. 30 MHz beat frequencies were generated and measured. The details of the accurate measurement of the frequency and a discussion of the associated problems are given in the following paper II. In these measurements the laser was 3 m long with a Fresnel number of 1.4. The near-confocal cavity was mounted on invar spacers. The hollow cathode discharge was 0.7 A at 0.1 torr pressure in a CH_4/N_2 gas mixture and the output of about 15 mW was coupled out using a 45° inclined polythene beam divider. The presence of very stable striations in the discharge was necessary for obtaining a minimum frequency width of the radiation. The latter was about 7 to 10 kHz full width at the -3 dB point, although the beat signals were somewhat wider than this due to a contribution from the microwave radiation (see associated paper II). Signal to noise ratios of about 30 dB were obtained for the beat signal using a 2 dB noise-figure pre-amplifier and a 10 kHz band width (Table I).

As well as the 891 GHz radiation the HCN laser emits two other fairly strong lines at 805 and 964 GHz. These can be mixed with 891 GHz and suitable microwave radiations in a metal-oxide-graphite diode to produce strong beat frequency signals. The contribution to the laser emission frequencies width from broadening by cavity instabilities and discharge fluctuations is proportional to the frequency. Therefore, if the laser emits two frequencies simultaneously ν_1, ν_2 which are close to each other, to first order approximation the width of a beat signal between these will be proportional to $\frac{\nu_1-\nu_2}{\nu_1}$ times the width of the single emission. In this case e.g. 891 and 964, a reduction of about 10 times from this cause will be obtained. The observed width was about 3 kHz (full width at the -3 dB point). The laser used for these experiments was 8 m long with a near-confocal invar spaced cavity. The output coupling was the double mirror Michelson unit (8). Up to 100 mW at 891 GHz and some three or four times less than this at 805 and 964 GHz were obtained. The results are shown in Table I and Figure 3. The effect of biasing the diode was investigated for 964 and 891 GHz mixing and it was found that optimizing produced up to 5 dB gain in S/N for most of the harmonics seen and the 6-8th harmonics were only seen with bias on the diode. Harmonics of the beat signals from 805 and 891 GHz mixing were also observed but have not been listed here.

HCN laser stabilization

Since the HCN laser is an important step in extending the frequency scale to the infra-red a technique was developed to lock its output at 891 GHz to an absorption line in difluorethylene ($CH_2 = CF_2$) which is approximately 0.5 MHz below the laser line centre (9). The frequency of the laser is modulated at 400 Hz by mounting one of the cavity mirrors on a PZT crystal. The modulated output is divided and passed through two cells, one containing $CH_2 = CF_2$ at 0.04 torr and the other evacuated. The two transmitted signals are subtracted and the difference is amplified and finally detected by a phase sensitive detector. The output of the latter is used as an error signal and applied to a motor driven potentiometer which in turn supplies a correcting voltage to the PZT crystal after DC amplification. The locking stability was 2 in 10^8 r.m.s. A small correction factor is applied in measuring the absolute frequency of the absorption line centre because of the differential technique (Table II).

H_2O laser measurements

The water vapour laser emits a number of strong lines under DC excitation in the range 1.5 THz (220 μm) to 10.7 THz (28 μm). For the speed of light programme the 2.53 THz (118.6 μm) and the 10.7 THz lines are of most immediate interest. Each of these can be related to the HCN laser at 891 GHz by harmonic mixing, that is third and twelfth respectively. In the first case it will be seen from Table I that approximately 144 GHz of microwave frequency are required to produce suitable IF frequencies. This was obtained as

TABLE II

Precise frequency measurements

	Frequency MHz	r.m.s. error
HCN laser	890 760.2	± 0.2
Difluoroethylene (CF_2=CH_2) absorption	890 759.60	± 0.08
H_2O laser (set to Lamb dip)	2 527 952.5	± 1.5

the fourth harmonic of a 36 GHz klystron signal using an open ended waveguide arrangement as described previously. The H_2O laser was 6 m long with a Fresnel number of 0.75 with a near-confocal cavity. The discharge was 1.4 A at 0.2 torr pressure and the output was coupled out using a polypropylene beam divider in a double mirror Michelson arrangement. Approximately 20 mW of power at 2.53 THz were obtained (10). The radiations from this laser and from the 8 m long HCN laser were focussed onto the tungsten-graphite diode using an off-axis paraboloid mirror and a T.P.X. lens respectively (Figure 2). The coupling to the diode was optimized using the considerations described above. Typical DC signals of 200 mV at 891 GHz and 10 mV at 2.53 THz were observed. Beat signals were obtained relatively easily although their optimization depended greatly on the relative powers of the laser and 36 GHz radiations and the diode contact.

The details of the signal to noise ratios and of the frequency measurement at 2.53 THz are given in Table II. The latter was measured for the H_2O laser output centred on the Lamb dip. The error quoted mostly arises from the uncertainty in the 891 GHz frequency since this was not monitored directly during these experiments. The value quoted in Table II agrees well with that reported previously by Frenkel (2 527 952.8 ± 0.1 MHz) in a direct microwave laser mixing experiment (11). The width of the observed beat signal was about 75 kHz (full width at half height). This would indicate a width for the 2.53 THz emission of about 50 kHz or 2 in 10^8 after allowing for the klystron and 3 × 891 GHz contributions. The frequency pulling effect of the second mirror in the Michelson coupling-out unit was investigated and found to be approximately 100 kHz at 2.53 THz for tuning through a complete $\lambda/2$ on either laser (after allowing for the third harmonic in the case of the HCN laser). Length changes due to vibration and thermal effects on this mirror would be at least an order of magnitude smaller than this and hence the influence of the second mirror should be less than one in 10^9.

In the case of the 10.7 THz emission the laser was 8 m long with a folded confocal cavity. For an effective cavity diameter of 5 cm, that is Fresnel number equal to 1.3, a large number of TEM_{ooq} and TEM_{lmq} modes were obtained and beats between these were observed on a tungsten-graphite diode. In order to reduce the number of off-axis modes an iris was placed over one mirror with an effective aperture of 3.5 cm which reduced the Fresnel number to 0.68. In this case only two or three TEM_{ooq} modes are obtained. (The Doppler gain width of the laser at this frequency is about 38 MHz and since the cavity frequency c/2L is about 18.6 MHz it is possible to have up to three TEM_{ooq} modes oscillating simultaneously.) The axial mode beat frequency was measured accurately in order to check the mode-pulling theory of Bennett (12). Thus for two modes ν_{ooq} and ν_{ooq+1}

$$(\nu_{ooq} - \nu_{ooq+1})_{observed} \simeq \frac{c}{2L}\left(1 - \frac{\Delta\nu_{cav}}{\Delta\nu_D}\right)$$

where L is the cavity length, $\Delta\nu_{cav}$ the passive cavity width and $\Delta\nu_D$ the Doppler width and c is the speed of light. Estimates of $\Delta\nu_{cav}$ from the reflectivity of the mirrors and beam divider indicated a value for $\frac{\Delta\nu_{cav}}{\Delta\nu_D} \simeq 0.008$. The experimentally observed value for $(\nu_{ooq} - \nu_{ooq+1})$ was 18.84 MHz and L was measured to be 7.89 ± 0.01 m, thus giving good agreement within experimental error with the above equation.

A more important phenomenon in regard to absolute frequency measurement was the observation of a splitting of the 18.6 MHz beat signal into two components about 60 kHz apart at maximum. This arises from mode pulling when the emitted lines are not symmetrical about the Doppler gain centre (Figure 4).

The techniques described above will be used to determine the frequency of the H_2O laser emission at 10.7 THz and of the CO_2 laser at 9.3 μm (32 THz).

References

1. HOCKER, L.O., JAVAN, A., RAMACHANDRA RAO, D., FRENKEL, L. and SULLIVAN, T., Appl. Phys. Lett., 10, 5 (1967).

2. BAKER, J.G., (In) Spectroscopic techniques for far infra-red, submillimetre and millimetre waves, (ed. D.H. Martin), North Holland, Amsterdam, 208 (1967).

3. SOKOLOFF, D.R., SANCHEZ, A., OSGOOD, R.M. and JAVAN, A., Appl. Phys. Lett., 17, 257 (1970).

4. EVENSON, K.M., WELLS, J.S., MATARRESE, L.M. and ELWELL, L.B., Appl. Phys. Lett., 16, 159 (1970).

EVENSON, K.M., WELLS, J.S. and MATARRESE, L.M., Appl. Phys. Lett., 16, 251 (1970).

5. GREEN, S.I., COLEMAN, P.D. and BAIRD, J.R., Proceedings of the symposium on submillimetre waves. Polytechnic Institute of Brooklyn, 369 (1970).

6. BRADLEY, C.C., EDWARDS, G.J. and KNIGHT, D.J.E., Conference on infrared techniques held at University of Reading, September 1971, IERE Conference Proceedings No. 22, 377 (1971).

7. MATARRESE, L.M. and EVENSON, K.M., Appl. Phys. Lett., 17, 8 (1970).

8. WELLS, J.S., EVENSON, K.M., MATARRESE, L.M., JENNINGS, D.A. and WICHMAN, G.L., Nat. Bur. Stds. Technical Note 395 (1971).

9. BRADLEY, C.C., KNIGHT, D.J.E. and McGEE, C.R., Electronics Lett., 7, 381 (1971).

10. BRADLEY, C.C. To be published.

11. FRENKEL, L., SULLIVAN, T., POLLACK, M.A. and BRIDGES, T.J., Appl. Phys. Lett., 11, 344 (1967).

12. BENNETT, W.R. Jr., Phys. Rev., 126, 580 (1962).

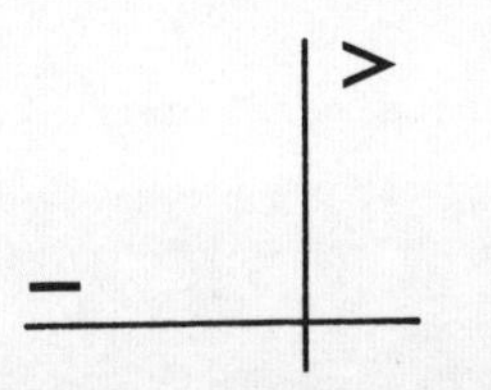

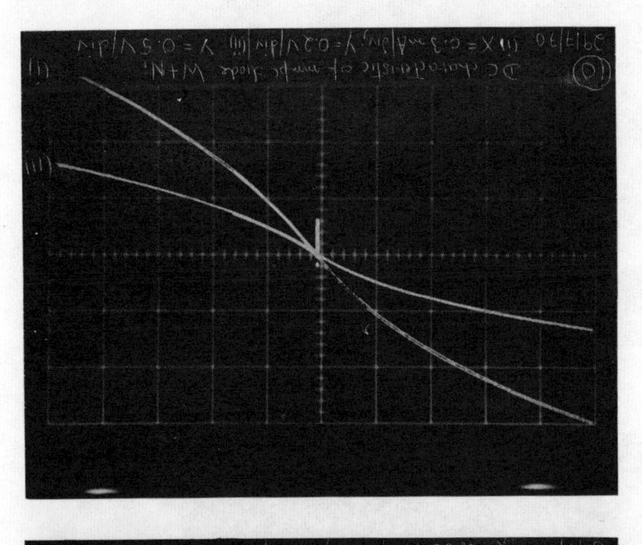

(b)

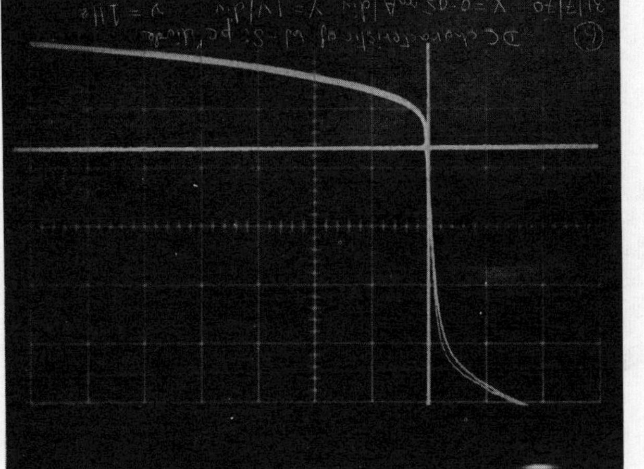

(a)

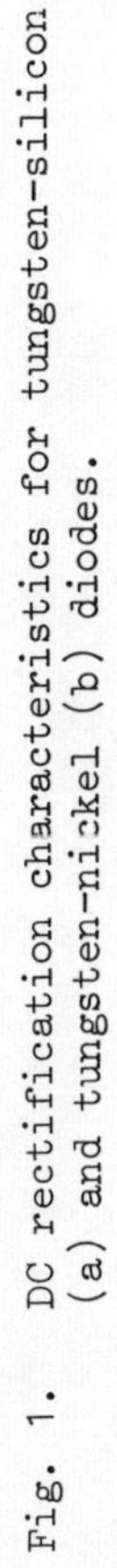

Fig. 1. DC rectification characteristics for tungsten-silicon (a) and tungsten-nickel (b) diodes.

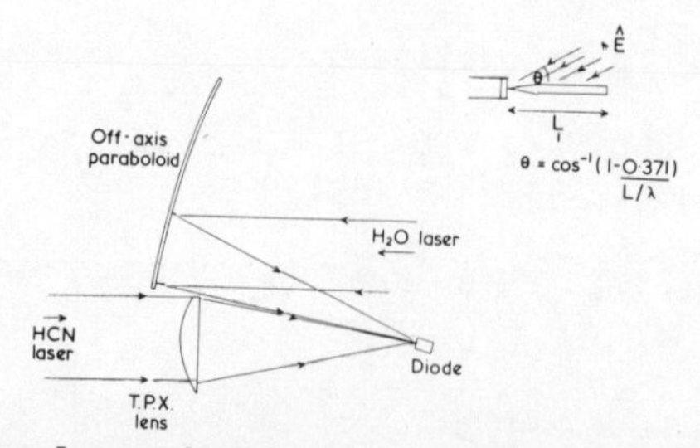

Fig. 2. Focussing laser beams onto a metal-oxide-graphite diode.

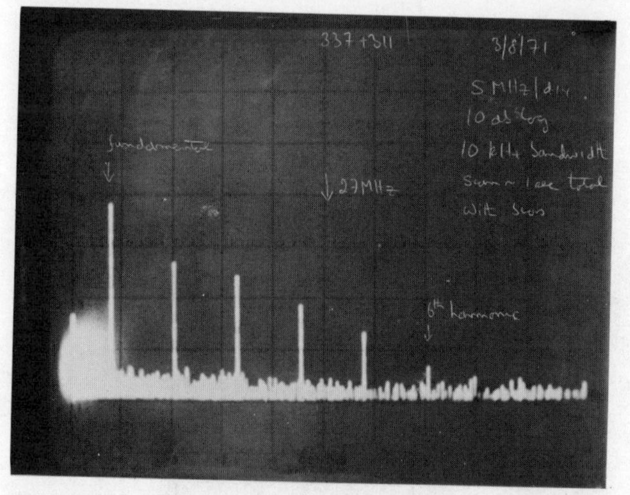

Fig. 3. Fundamental and harmonics of the beat frequency between 964 and 891 GHz HCN laser emissions (5 MHz/division).

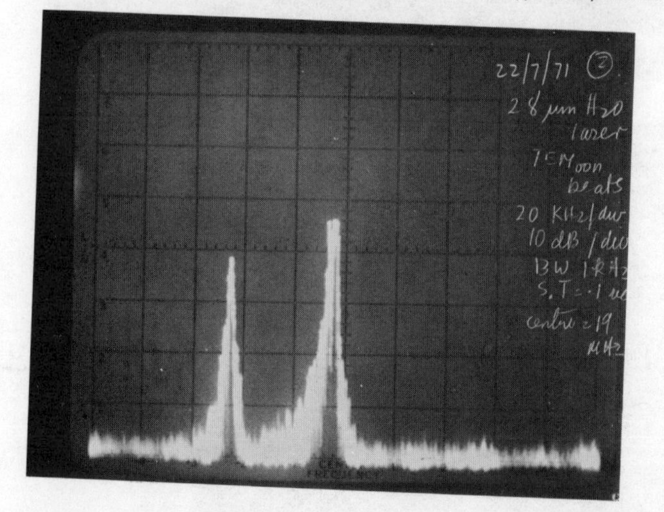

Fig. 4. Mode-splitting at 10.7 THz for axial mode beat signals, centre frequency 19 MHz (dispersion 20 kHz/division).

Progress with a Determination of the Speed of Light

II. Techniques for High-accuracy Frequency Measurement in the Far Infra-red

T. G. Blaney and C. C. Bradley

Division of Electrical Science,
National Physical Laboratory, Teddington, England

and

G. J. Edwards and D. J. E. Knight

Division of Quantum Metrology,
National Physical Laboratory, Teddington, England

1. Problems in achieving high accuracy

The scheme for measuring the CO_2 laser frequency (shown in the Introduction) consists of three separate frequency-mixing experiments using two lower-frequency lasers as transfer oscillators. If an accurate measurement on a stabilized CO_2 laser (1) is required, either all the experiments must be performed together, within the thermal drift limitations of the transfer oscillators, or the transfer oscillators must be stabilized too. Although having achieved about 1 in 10^7 stabilization of the HCN laser (2) we favour simultaneous frequency measurement using automated techniques, but achieving this is highly dependent on the fragile frequency-mixing diodes.

A possible simplification is suggested by recent harmonic-mixing experiments with Josephson junctions (3,4). High-order harmonic mixing with laser frequencies in the 3 to 10 THz region is conceivable (in spite of the superconducting energy gap being equivalent to a lower frequency) so that it might be possible to achieve mixing with just one transfer oscillator. (Owing to the 26.5 to 33 THz spread of closely-spaced CO_2 lines, and about 30 potential transfer oscillators in the 1 to 11 THz region, there are many alternative schemes for reaching CO_2 frequencies using 1, 2 or 3 transfer oscillators (5).)

There remains a problem in the measurement of the transfer-oscillator frequency because of the effect of phase noise when a crystal-oscillator reference frequency is multiplied to the region of 1 THz. The multiplied phase noise eventually swamps the coherent signal. We have overcome this difficulty by dividing the laser frequency (in effect) to the microwave region. This is achieved by phase locking a microwave oscillator to the laser and not to a quartz-crystal oscillator in making the frequency comparison (4). We have shown that this technique works for high-order mixing in a Josephson junction. In addition (since the frequency mixer is so vital in these measurements) we have devoted much effort to understanding and improving the metal-oxide-metal point contact for mixing two laser frequencies. Results of experiments are included in the associated paper I.

2. Phase noise on multiplied quartz-crystal oscillator frequencies

In measuring laser frequencies with microwave oscillators phase-locked to a quartz-crystal oscillator both we (6) and other workers (7) have observed noise sidebands on the beat spectrum which are attributable to multiplied phase noise from the quartz-crystal oscillator.

An illustration is shown in Figure 1. The spectrum of a klystron phase-locked to an HCN laser is displayed with the microwave local oscillator free running in (a) and phase-locked to a 1 MHz crystal in (b). Phase noise blurs the spectrum in (b) so that there is no increase in resolution.

The sidebands can be reduced by using the best quality quartz-crystal oscillators or by using high frequencies in the region of 120 MHz at the expense of medium-term stability. In both cases it is inconvenient that the oscillators are essentially fixed-frequency, single-application devices.

We comment here that if an HCN laser is to be used for frequency multiplication, phase locking to a microwave oscillator itself phase-locked to a quartz crystal oscillator (8) is fundamentally undesirable. However our method of phase-locking to a higher frequency being measured might be applied.

3. Phase locking klystrons to an HCN laser

We have demonstrated the principles of phase locking the klystron to the laser by a simple alteration of the scheme by which we measure the 891 GHz HCN frequency. A klystron at 99 GHz formerly phase-locked to a tunable quartz-crystal oscillator at 15 MHz was phase-locked instead to the laser by feeding the beat between the klystron harmonic and the laser into the feedback loop. The scheme of the apparatus is shown in Figure 2. We used both a commercial silicon point-contact mixer in a waveguide mount and an open-structure metal-oxide-metal point-contact mixer. The laser frequency was counted by using a second loop to phase-lock a 9.9 GHz oscillator to the 99 GHz klystron, and feeding the 9.9 GHz frequency to commercial equipment. We used a computing counter to measure the Allan variance of the HCN laser. The value for 1 second counts was 2×10^{-9}.

4. Use of a Josephson-junction mixer

The use of a Josephson junction allowed us to extend phase-locking to higher-order harmonic mixing. We took an existing type of point-contact device (9) and applied both 36 GHz and HCN laser radiation to it through a common light-pipe. The cut-off frequency of the light-pipe prevented our using as low a microwave frequency as McDonald et al. (3). We were able to obtain strong beats in the frequency range 1 to 100 MHz as a result of 25th and 27th harmonic mixing with the 891 GHz and 964 GHz laser lines.

A comparison of the frequency mixing and phase locking with the three mixers is given in Table I. The locked beat amplitude

TABLE I

Phase locking klystrons to HCN laser lines with different mixers

Mixer	F_{laser} GHz	$F_{klystron}$ GHz	n	IF s/n* dB	T_{lock} mins
Silicon	891	98.9	9	13-18	60
MOM	891	98.9	9	15-20	20
Josephson	891	35.6	25	30-40	5
	964	35.7	27		

*Typical obtainable signal/noise in a 300 kHz bandwidth when phase-locked at 60 MHz.

from the Josephson junction tended to decay with time owing to a critical dependence on microwave power and d.c. bias. A less rapid decay with the metal-oxide-metal (MOM) diode resulted from changes in the contact itself. The magnitude of these effects are indicated in Table I by the shortened times (T_{lock}) for which phase-lock could be held, compared with the duration (limited by occasional flashing of the laser discharge) in the case of the silicon diode.

The Josephson junction mixer required more laser power (50 mW against 10 mW) and less microwave power (1 to 10 mW against 50 mW) than the room-temperature mixers. In contrast to these it operated for several days without any mechanical adjustment of the contact, until the helium ran out.

We looked for possible spurious effects in the use of the Josephson junction. The spectrum of the beat between the klystron and the laser when phase-locked to a 60 MHz quartz crystal reference showed 50 Hz sidebands at -15 to -30 dB, depending on the loop filter setting. These amounted to a deviation of the klystron from the laser reference by about 1 in 10^{11} and were entirely attributable to a spurious voltage reaching the klystron reflector.

The Josephson junction generated a comb of harmonics of the fundamental beat frequency, but no subharmonics that we could see. The 6th harmonic was visible, and phase locking was possible on harmonics up to the 4th.

5. Summary

The problem of achieving adequate frequency resolution has been overcome as a result of inverting the phase-lock scheme to allow comparison with multiplied quartz-crystal oscillator frequencies in the microwave region. The question of accuracy depends mainly on laser frequency drift together with the ability to measure all the necessary beat frequencies in a given time. In turn the number of lasers necessary and the repeatable observation of beat frequencies depends strongly on the performance of the mixers. In this context the Josephson junction and metal-oxide-metal point contacts show promise but need more development.

References

1. FREED, C. and JAVAN, A., Appl. Phys. Lett., 17, 53-6 (1970).

2. BRADLEY, C.C., KNIGHT, D.J.E. and McGEE, C.R., Electronics Lett., 7, 381-2 (1971).

3. McDONALD, D.G., RISLEY, A.S., CUPP, J.D. and EVENSON, K.M., Appl. Phys. Lett., 18, 162-4 (1971).

4. BLANEY, T.G., BRADLEY, C.C., EDWARDS, G.J. and KNIGHT, D.J.E., Phys. Lett., 36A, 285 (1971).

5. KNIGHT, D.J.E., J. Opto-electronics, 1, 161-4 (1969).

6. KNIGHT, D.J.E., Electronics Lett., 7, 383-4 (1971).

7. FRENKEL, L. and SULLIVAN, T., Trans. IEEE, MTT-17, 281-2 (1969).

8. CUPP, R.E., CORCORAN, V.J. and GALLAGHER, J.J., IEEE J. Quant. Electron., QE-6, 241-3 (1970).

9. BLANEY, T.G., J. Phys. E., in the press.

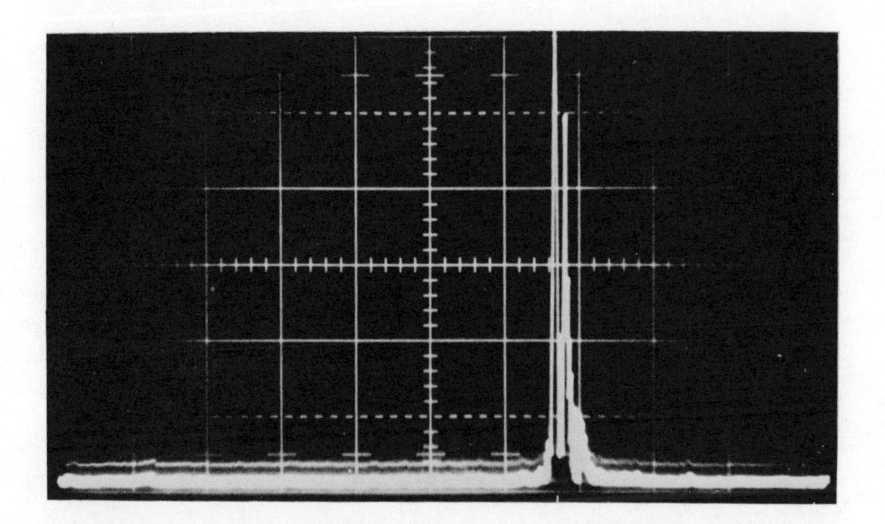

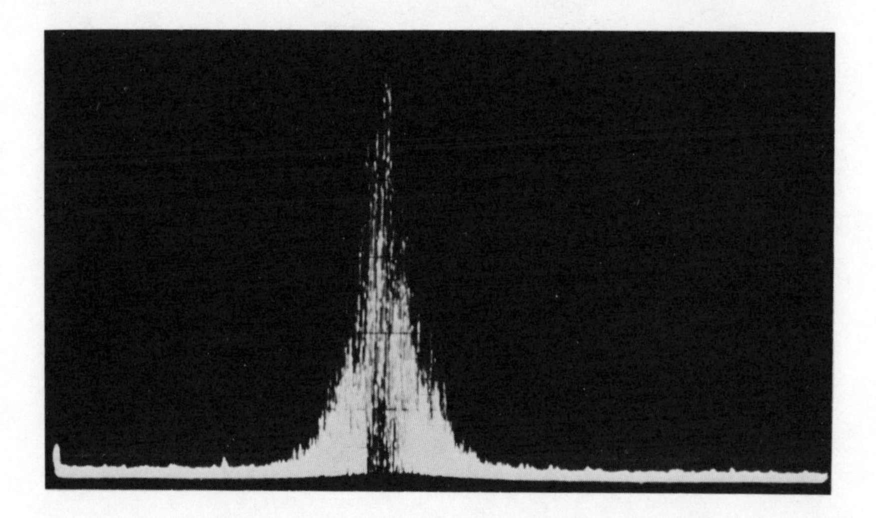

Fig. 1. An illustration of phase noise on a quartz-crystal oscillator frequency when multiplied to 99 GHz. A klystron phase-locked to an HCN laser provides a comparatively noise-free signal which is observed on a linear display on a spectrum analyser (with scan 20 kHz/division at 10 ms/division). The width of the signal results from the 99 GHz local oscillator signal in the spectrum analyser. (a) A single scan with the local oscillator free running, showing drift but not phase noise.
(b) A 1 second exposure with the local oscillator phase-locked to a multiplied 1 MHz quartz-crystal oscillator frequency, showing phase noise about a stable centre frequency.

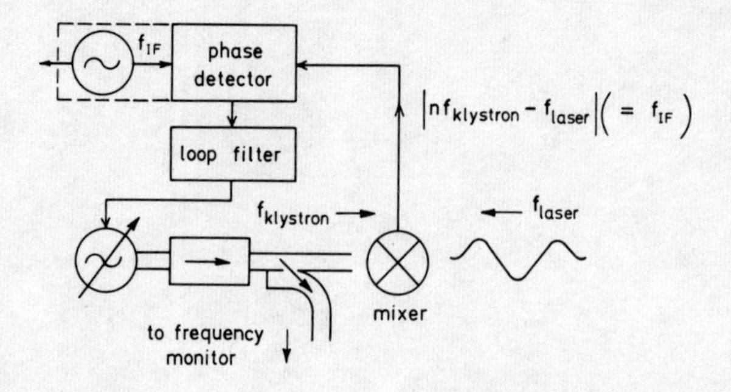

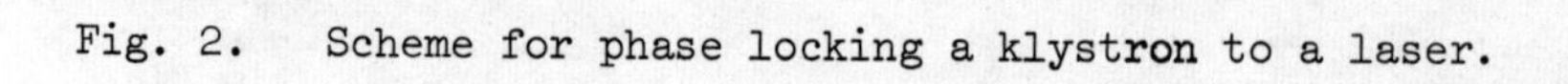
Fig. 2. Scheme for phase locking a klystron to a laser.

Progress with a Determination of the Speed of Light

III. Up-conversion of Carbon Dioxide Laser Radiation as a Means of Measuring Visible Laser Frequencies

W. R. C. Rowley and P. T. Woods

Division of Quantum Metrology,
National Physical Laboratory, Teddington, England

1. Wavelength measurement at 10 μm

Interferometric comparison of the carbon dioxide laser wavelength with the primary standard in the visible presents a number of special problems arising from the large wavelength difference. The first of these is the photodetector problem. The use of separate detectors means that particular attention must be paid to their uniformity of response for lateral and angular position variations. There are few materials with transmission properties in both spectral regions, and they are not readily available in the necessary quality and sizes. The use of separate components for the two radiations may thus be considered, such as variable path interferometers with the optics for the two radiations arranged coaxially or back to back. Such systems require stricter attention to stability and adjustment tolerances than do systems having common optical paths.

Direct wavelength comparison against the krypton-86 standard can only be carried out effectively at interferometric path differences less than 500 mm, because of the bandwidth of the standard. A measurement to 1 in 10^8 thus requires pointing to 0.0005 of the infra-red fringe spacing. This accuracy demands a good detection system, particularly if a two-beam interferometer is used, and very careful attention to systematic effects such as optical phase changes. The path difference limitation may be relieved by using a stabilized laser as source of reference wavelength in place of direct comparison with the primary standard.

The long wavelength of the carbon dioxide laser necessitates the use of correspondingly large optics in order to reduce diffraction effects to an acceptable level. In a conventional interferometer with the spatial coherence defined by a pinhole and lens system, an aperture of 50 mm gives rise to a diffraction correction of $10^{-8}\lambda$ for wavelengths in the 10 μm region. A correction of similar magnitude arises in an interferometer illuminated directly with coherent laser light of beam waist diameter w_o = 17 mm, which again demands optics of roughly 50 mm diameter.

Most interferometers reflect a significant amount of light coherently back into the source. Such light causes modulation of a laser which may adversely affect its frequency stabilization system. A visible He-Ne laser stabilized to the Lamb dip may be seriously affected by a reflection coefficient as small as 10^{-6}. Decoupling systems such as those employing circular polarization or Faraday rotation must usually be employed, or a rotating diffuser in those cases where the spatial coherence properties of the laser beam do not need to be preserved. Alternatively the decoupling problem may be avoided by using an interferometer which does not have a first order return beam, such as a circulating system or one using cube corner or cat's eye reflectors with the beams laterally displaced from the axis. Such interferometers have been used with great success in length measurement applications. A problem in precision wavelength measurement is that of achieving the necessary tolerance of alignment of the laser beams parallel to the axis of translation in interferometers which do not use a plane reflector to define the direction of the optical axis.

The above problems in precision wavelength measurement at 10 μm should not be over emphasized. They by no means rule out the possibility of carrying out such measurements with the required precision, they merely present a challenge to the experimenter. Nevertheless they increase the difficulty of precision interferometry sufficiently to stimulate the search for alternative approaches. One such approach is through up-conversion (1,2).

2. Up-conversion

The experimental arrangement for non-linear mixing of visible helium-neon and infra-red carbon dioxide laser radiations is shown in Fig. 1. A single frequency helium-neon laser, using a Fox-Smith cavity mode selector, emits about 10 mW at 0.633 μm which is passed through a narrow band interference filter and focussed onto a single crystal of proustite (3) cut and orientated for Type II phase matching conditions (4). The desired wavelength from the carbon dioxide laser is obtained with the help of an intra-cavity gas cell (using vinyl chloride for the 9.3 μm region). The radiation is passed through a variable attenuator gas cell containing chlorotrifluoroethylene and helium, and about 10 W is focussed onto the proustite crystal at a small angle (about 0.1 radian) to the visible radiation, avoiding the use of a beam splitter. The crystal is cooled with liquid nitrogen in order to reduce absorption of the helium-neon and carbon dioxide laser radiations. A fine angular adjustment allows optimization of the difference frequency intensity, approximately 0.1 μW having been observed after transmission through a Glan polarizer and filter which reduce the background helium-neon laser radiation to an insignificant level (less than 10^{-10} W). By more careful attention to orientation and focussing conditions within the crystal it is expected that approximately 1 μW of difference frequency power will be obtained and that sum frequency generation at 0.593 μm will also be observed.

3. Interferometry with up-converted light

The up-conversion process is essentially one of modulation by which sidebands are generated in the visible light, with separations corresponding to the frequency of the infra-red radiation. Let us consider the sideband of frequency $f_S = f_N - f_C$, where f_N is the frequency of the He-Ne laser, and f_C the frequency of the CO_2 laser which has been determined by a harmonic comparison experiment.

The wavelength ratio $R = \lambda_N/\lambda_S$ may be measured by interferometry, so that assuming the speed of light to be constant over the visible region we obtain

$$R = f_S/f_N = 1-f_C/f_N ,$$

giving

$$f_N = f_C(1-R)^{-1}.$$

Thus by measurement of the ratio of the wavelength of the visible laser light to its sideband, it is possible to determine the frequency of the visible laser radiation. A measurement of the wavelength of this visible laser light by direct comparison with the krypton-86 standard then leads to the value for the speed of visible light,

$$c = \lambda_N f_N.$$

The practical problem is to measure the ratio R with the necessary accuracy. For the value of $(1-R)$ to be determined to an accuracy of 1 in 10^8, the wavelength ratio R must be measured with a relative precision approximately 15 times better. This precision is greater than that normally used in wavelength comparisons with the krypton-86 primary standard, in which 1 in 10^8 uncertainty of the standard makes it fruitless to seek significantly higher precision. Nevertheless such comparisons are carried out at NPL with random errors corresponding to a standard deviation of 1.5 in 10^9. In the case of comparing up-converted light with that of a laser, however, the precision can be improved because the bandwidths of the radiations are very narrow, so that interference patterns of better contrast will be obtained and interferometers of much greater path difference may be used. Systematic errors of interpolation between interference fringe maxima, in the measurement of fractional parts of an order of interference, may be avoided by frequency adjustment of the sources so that their interference fringe maxima occur at the same path difference.

The outline of a wavelength intercomparison system is shown in Fig. 2. A Fabry-Perot interferometer of length t having photoelectric detection, receives the up-converted light and light of wavelength λ_1 from an iodine stabilized laser through a diffusing screen. This screen serves to decouple the optical systems and to destroy the spatial coherence of the beams so that they illuminate the interferometer similarly. The frequency difference Δf between the visible lasers is measured by beat frequency techniques.

In operation, the interferometer is maintained in vacuum with piezoelectric angle and length adjustments. Its length is controlled for maximum transmission of the stabilized laser wavelength λ_1, $t = m_1\lambda_1/2$. The high power visible laser is then tuned so that its sideband also has maximum transmission through the interferometer, $\lambda_S = \frac{2t}{m_S}$ so that $\lambda_1/\lambda_S = m_S/m_1$. If c is the speed of visible light, $f_N = \frac{c}{\lambda_N}$, $f_S = \frac{c}{\lambda_S}$, $f_1 = \frac{c}{\lambda_1}$.

The beat frequency observation gives $\Delta f = f_N - f_1$, so that

$\frac{\Delta f}{c} = \frac{1}{\lambda_N} - \frac{1}{\lambda_1}$, which gives $\frac{\lambda_1}{\lambda_N} = 1 + \frac{\lambda_1 \Delta f}{c}$.

From the up-conversion, $f_C = f_N - f_S$, thus $\frac{f_C}{f_1} = \frac{\lambda_1}{\lambda_N} - \frac{\lambda_1}{\lambda_S}$.

Combining this result with the interferometer and beat frequency observations,

$$f_1 = f_C \left[1 - m_S/m_1 + \lambda_1 \Delta f/c\right]^{-1}.$$

The speed of light is then determined by a measurement of the stabilized laser wavelength against the krypton-86 standard,

$$c = \lambda_1 f_1.$$

The term $\lambda_1 \Delta f/c$ due to the frequency difference between the visible lasers may be kept small by appropriate choice of interferometer length and does not require very accurate values for λ_1 or c. An uncertainty of 1 in 10^8 in the value of f_1 would result from an uncertainty of 300 kHz in the measurement of Δf.

References

1. GIORDMAINE, J., Phys. Rev. Lett., 8, 19 (1962).

2. PFITZER, E.K., RICCUS, H.D. and SIEMSEN, K.J., Optics Communications, 3, 277-8 (1971).

3. WARNER, J., Appl. Phys. Lett., 12, 222 (1968).

4. MIDWINTER, J.E. and WARNER, J., Brit. J. Appl. Phys., 16, 1135-42 (1965).

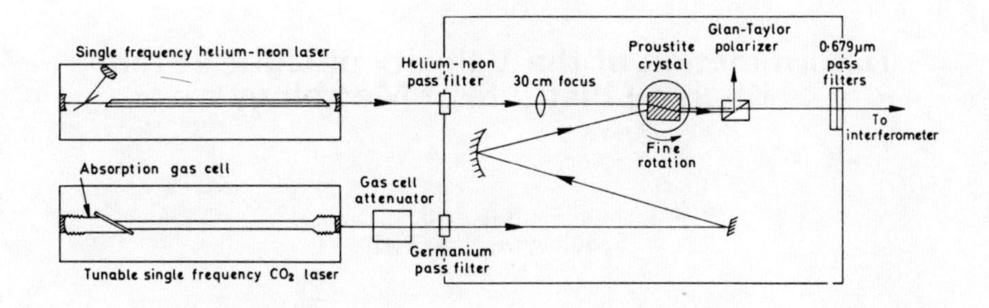

Fig. 1. Experimental arrangement to demonstrate non-linear mixing in proustite.

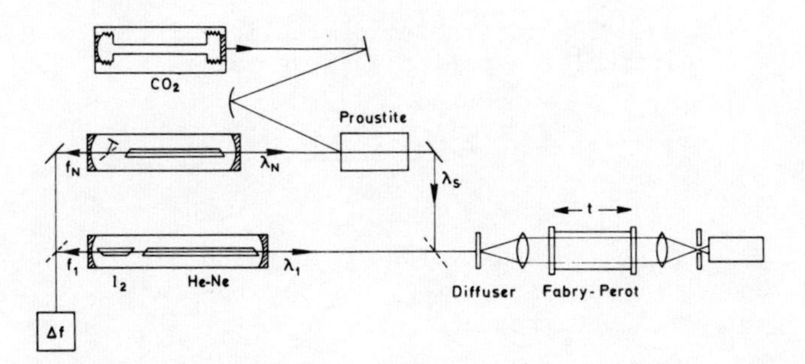

Fig. 2. Outline of system for interferometric measurement of up-converted light.

Determination of the Velocity of Light at the Kharkov Institute of Metrology

A. J. Leikin, I. V. Lukin, S. V. Sikora,
V. S. Solovyov, N. S. Fertik

Measurements of the velocity of light in vacuo were carried out at the Kharkov Institute of Metrology under the guidance of G.S. Simkin in 1958-1966 using the interference of millimeter waves. A two-beam interferometer with a closed end operating in the distant field region of wave propagation was used for this purpose (1,2).

The basic equation for the determination of the velocity of light in this particular experiment is as follows:

$$C_{vac} = 2\frac{(L - \delta)nf}{N} \tag{1}$$

where:

- C_{vac} - light velocity in vacuo
- f - frequency
- n - refractive index
- L - length of the baseline
- δ - diffraction correction
- N - number of wavelengths in the length of the baseline

A reflex klystron with $f = 36805$ MHz ($\lambda = 8$mm) was used as the microwave source for the interferometer. The klystron was stabilized parametrically. Subsequently a system of phase automatic frequency control of the klystron using the harmonic of a reference quartz generator was used.

The signal from the klystron was transmitted through a horn antenna with a matching lens onto a plane mirror mounted on a movable carriage. The reference signal was directed via the waveguide to a fixed path containing a phase shifter and an interference attenuator.

The UHF signals reflected in the measuring and reference channels were mixed by means of a double tee and then sent onto a crystal detector together with a UHF local oscillator signal. The resulting signal was shown on an indicator recording the minima of wave interference.

To increase the accuracy of measurements a baseline of 4447.8mm containing 1100 half wavelengths was used. The measurement of the base L was performed by the well-known method of optical multiplication (Vaisala's method), using 100mm standard quartz bars previously measured against the orange Kr^{86} line by means of a Kosters interferometer.

The refractive index was measured with a resonant refractometer of special design. The determination of n was made at only one place in the room so that the value obtained was not representative of the region where the propagation of the UHF waves took place. The final value of n was obtained by taking into account the difference between the mean value of temperature of the air along the path of wave propagation and that in the resonator.

The calculation of the diffraction correction was based on Huygen's principle. Integral equations determining the diffraction correction were obtained for several field distributions at the horn. The calculation of the correction was performed with a computer.

Measurements of the base L were made at different distances from the horn. The values obtained formed the following system of equations:

$$L = L_o + k\delta_1$$

$$L = L_o + k\delta_2$$

$$\cdots\cdots$$

$$L = L_n + k\delta_n$$

where L_1; L_2; .... L_n - are the measured values of the base length, L_o - value of the base length without diffraction, δ_1; δ_2; .... δ_n - calculated values of corrections, and k is a constant. Using the least squares method values of L_o and k were obtained. As the calculated values of the correction are not quite correct k is not equal to unity.

By varying the parameter a (the effective aperture of both the horn and the mirror) in the initial integral equation the values of δ_n were found at which k was nearest to unity. The values of the diffraction correction obtained did not differ from the computed ones more than by 12%. Such calculations were done for various cases of field distribution at the horn. The value nearest to 1 (1.0008) was obtained in case of $Hx = H_o$;

$$Ey = E_o \cos \frac{\pi y}{a}$$

RESULTS OF THE MEASUREMENTS OF THE VELOCITY OF LIGHT (1966)

The measurements were performed in a period of 8 months (December 1965 - August 1966). The total number of measurements was 70, with 6 different positions of the mirror relative to the horn.

The results of measurements are summarized in Table I.

The weighted mean appeared to be:

$$C = 299792.56 \pm 0.09 \text{ km.sec}^{-1}$$

The error given above (0.09 km.sec^{-1}) includes the varying part of the total error only. To determine the total error an analysis of possible errors was carried out, the results of which are given in Table II.

Hence it was found:

$$\sigma_{\Sigma} = 0.11 \text{ km.sec}^{-1}$$

$$\sigma_{rel} = 3.6 \times 10^{-7}$$

Thus, the results of measurements appeared to be:

$$299792.56 \pm 0.11 \text{ km.sec}^{-1}$$

Further work on the measurement of the velocity of light is directed towards achieving an accuracy of the order of 10^{-8}. To achieve such an accuracy we have considered (a) using shorter wavelengths to reduce errors due to diffraction, (b) conducting measurements in vacuo, (c) using laser interferometry for wavelength measurements.

At present preparations are in progress at the Kharkov Institute of Metrology for light velocity measurements in the infrared with laser sources. For this purpose the use of HCN lasers (λ = 337 μm), H_2O lasers (λ = 118.78 and 28 μm) and CO_2 lasers (λ = 10.6 μm), is suggested.

The main questions to be solved in these experiments are as follows: (a) measurement of the frequency of the laser radiation with an error of $\sim 10^{-8}$, (b) measurement of wavelengths (using the diffraction correction) with an error of $\sim 10^{-8}$.

At present an apparatus is being developed by means of which measurements of the frequency of the laser (λ = 337 μm) can be performed (3). Investigations showed that the errors introduced by the measuring apparatus did not exceed 10^{-10}. Harmonics of a 4 mm klystron were studied up to 1 THz. The results of frequency measurements of laser emission gave 890754 $\pm$ 0.22 MHz (3 x 10^{-7}). The error obtained is caused by the instability of the laser.

Table I

Distance between the mirror and the horn in mm	Mean value of the velocity of light in vac. (km.sec^{-1})	Number of measurements in a series	Standard deviation km.sec^{-1}	Statistical weight of a measurement	Diffraction correction km.sec^{-1}	Velocity of light in vacuo with diffraction correction km.sec^{-1}
9970-5492	299808.868	10	0.044	0.4	16.397	299792.471
10498-6020	6.908	10	0.028	1.0	14.310	2.598
11473-6995	3.870	10	0.052	0.3	11.400	2.470
11976-7498	2.817	12	0.052	0.3	10.229	2.598
12988-8510	0.903	10	0.044	0.4	8.376	2.537
13453-8975	0.251	12	0.031	0.8	7.680	2.571

Table II

	Source of errors	Absolute error $km.sec^{-1}$	Relative error, parts in 10^7
1	Variation of the results of measurements	0.092	3.1
2	Measurements of the standard bars	0.033	1.0
3	Measurements of the standard bars	0.033	1.0
4	Measurements of the coefficient of linear expansion of the bar material	0.009	0.3
5	Measurements of temperature of the bars	0.006	0.2
6	Frequency measurements	0.003	0.1
7	Measurements of the coefficient of linear expansion of the material of the refractometer resonator	0.009	0.3
8	Determination of the diffraction correction	0.015	0.5
9	Scale calibration	0.020	0.7
10	Temperature measurement during the determination of refractive index	0.003	0.1

At present work is in progress on phase automatic frequency control using as a reference the frequency of a secondary standard. Frequency measurements of lasers with $\lambda = 118$ μm, 28 μm and 10 μm are being continued, silicon diodes being used as mixers.

Wavelength measurements of lasers are being conducted with an interferometer with a maximum path difference of 3.5 m. by comparing IR-waves with He-Ne laser emission ($\lambda = 0.63$ μm). This line of the laser in its turn is found in terms of the orange Kr^{86} - line by means of a 100 mm Fabry-Perot interferometer. The measurements are carried out by the determination of the fringe number over the path of 3.5 m (Doppler interferometer) and by measuring this base by means of a reversible counter.

Two principles are used in the optical interferometer: a normal Michelson interferometer and a laser interferometer with optical feedback. The assessment of the diffraction corrections is made with a computer. The interferometric part of the experimental setup is placed in a vacuum chamber.

The suggested experiments are to be carried out in two stages:

1. Preliminary measurements to check the apparatus and to obtain an accuracy of 10^{-7} (in air).

2. Measurements in vacuo to achieve accuracies of the order of 10^{-8}.

The basic diagram of the experimental setup for the measurement of the velocity of light with a CO_2 laser is given in Figure 1. Further improvement suggests making measurements with lasers as intermediate components ($\lambda = 337, 84, 27$ μm). At present preparations for measurements of the velocity of light at $\lambda =$ 337 μm are being made.

Parallel experiments are planned with a two-mode He-Ne laser (Fig. 2). One of the main achievements in this field is the development of a system of automatic frequency control of inter-mode beats providing a stability not worse than 10^{-9}. The main problem in this case is that of measuring a base of considerable length (several tens of meters). To diminish this length it is suggested that a multipath delay line of variable length could be used in the measuring path. Calculations show that in this way it is possible to diminish the length of the base to several meters and to measure the base by means of an interferometer with a He-Ne laser.

A considerable advantage of the latter method is the absence of the error of a multiplicative nature due to diffraction. The expected accuracy of the velocity of light measurement is better than 5.10^{-8}.

References

1. SIMKIN, G.S., LUKIN, I.V., SIKORA, S.V. and STRELENSKY, V.E., Measurement Techniques No. 8, (1967).

2. SIMKIN, G.S., LUKIN, I.V. and SIKORA, S.V., Proceedings of the Khar'kov State Scientific Research Institute of Metrology, No. 2, Standards Publishing House, Moscow.

3. BONDAREV, V.P., LEIKIN, A.J., SOLOVYEV, V.S. and FERTIK, N.S., Proceedings of the 1st All-Union Conference on Metrology, September 1968.

4. SIKORA, S.V. and LUKIN, I.V., Proceedings of the 1st All-Union Conference on Metrology, September 1968.

5. LEIKIN, A.J., ZHABOTISKY, M.E., VALITOV, R.A., SOLOVYOV, V.S., BONDAREV, V.P., FERTIK, N.S. and TELEGIN, B.V., Measurement Techniques No. 11, (1970).

6. LEIKIN, A.J., SOLOVYOV, V.S., PAVLOV, V.G. and ZIMOCOSSOV, G.A. Measurement Techniques No. 6, (1970).

7. LEIKIN, A.J., SOLOVYOV, V.S. and PAVLOV, V.G., Problems in Radio-Electronics (Radio-measurement series), No. 5, (1970).

The Constancy of the Velocity of Light and Prospects for a Unified Standardization of Time, Frequency and Length

Z. Bay

National Bureau of Standards, Washington, D. C. 20234 U.S.A.

1. Introduction

The use of fundamental constants of nature for the definition of the units of physical measurements was suggested by M. Planck in 1906 (1). By assigning the value of unity to the four constants, c, h, k, and G the "natural units" of time, length, mass, and temperature can be defined. Albeit its great theoretical merit, that proposal could not be practical in 1906 since the large uncertainties in the known values of the constants would have propagated into the determination of the units.

After the advent of lasers C. H. Townes (2) directed attention to the future possibility of measuring optical frequencies and then connecting the units of length and time through the velocity of light. Practical proposals for a unified standardization of time and length via the velocity of light, c, were outlined by Z. Bay (3,4) and by Z. Bay and G. G. Luther (5,6). It was pointed out that the practical requirements for the realization of such a program are: 1) c should be known with the precision of the present length standard (a few parts in 10^9) (7); the precision of the time standard is more than 3 orders of magnitudes better (8); 2) it should be possible to measure optical frequencies to an accuracy greater than or equal to that of length measurements.

It was outlined in Refs. 3-6 that via microwave modulation of laser light (applicable to any laser lines) both requirements can be satisfied. Experiments to measure optical frequencies and the speed of light along those lines are underway and precisions obtained so far are promising.

In another line of development, optical frequencies are measured by directly mixing a high harmonic of a microwave frequency with the optical frequency on a point contact diode (9,10), or using a Josephson junction (11). Such direct measurements (with the use of intermediary heterodyne stages) are extended now into the infrared with the reasonable hope that, by multiplication in non-linear media and use of parametric amplification techniques, they can be further extended into the near infrared and into the visible. It is clear that such direct frequency measurements, if realized, can ultimately achieve the precision and accuracy of the time standard; thus they can a fortiori fulfill requirement 2.

The ability to measure optical frequencies in either way leads then to the possibility of the unified standardization of time interval, frequency and length such that the second is defined by the frequency standard and the meter is defined by an agreed upon value of c in meter/second.

The competing way of improving the definition of the meter is to stabilize laser frequencies and choose the most appropriate of them to define a new wavelength standard. There are good candidates now available (12,13) based on saturated molecular absorption.

The use of a fundamental constant of nature to reduce the number of independently defined units of measurements certainly has an aesthetic appeal (1-6,10,14). On the other hand, since metrology and spectroscopy are branches of experimental science, the acceptability of any such proposed scheme has to be judged on the basis of operational considerations.

The relative merits of the above two measuring systems have been discussed in Refs. 4,5, and 6. Their results will be extended and summarized below. But before doing that we turn to the question, which is of basic interest for physics and of primary importance for metrology: How good a constant is the speed of light?

On the basis of observational evidences we treat first the possible dependence of the vacuum velocity of light on frequency, then the constancy of c in moving frames.

2. Observational Limits For a Possible Dispersion of Light in Vacuum.

This topic is discussed in another paper (15), thus we give only a summary here.

Earlier astronomical observations of binaries and flare stars are now bypassed in accuracy by timing the arrival of pulses in different regions of the spectrum from pulsars. In particular, the time of flight of red, green, and blue light from the Crab nebula pulsar was found to be the same to within 5 parts in 10^{17} (16) and the time of flight of visible and x-ray (2-10 keV) pulses were found to be equal within 2 parts in 10^{15} (17).

To treat the dispersion, or lack thereof, in different parts of the spectrum we use a modified Cauchy expression of the form

$$n^2 = 1 + \frac{A}{\nu^2} + B\nu^2 \tag{1}$$

or

$$n^2 = 1 + A'\lambda^2 + \frac{B'}{\lambda^2} \tag{2}$$

where n is defined by $c_{phase} = c_o/n$ and c_o is the velocity of light in the absence of dispersion. The reason for that choice of $n(\nu)$ is first, that it describes the frequency dependence in any dispersive medium far from all resonances; and second, that both the A and B terms have separately been proposed on different theoretical grounds.

The special theory of relativity allows a non-zero A term, but requires B=0. This can be seen by applying the relativistic addition theorem of velocities to c_{phase} which leads to the transformation rule for n

$$n' = \frac{n + \beta}{1 + n\beta} \tag{3}$$

where $\beta = v/c$ and (for simplicity) the relative velocity, v, between the two systems is along the direction of light propagation. Simultaneously the frequency, ν, transforms by the relativistic Doppler formula

$$\nu' = \frac{1 + n\beta}{\sqrt{1 - \beta^2}} \tag{4}$$

Elimination of β from Eqs. (s) and (4) leads to

$$[n'(\nu')^2 - 1]\nu'^2 = [n(\nu)^2 - 1]\nu^2 \tag{5}$$

Now, if $n(\nu)$ is a dispersion function characteristic for vacuum (space), it should transform into itself, thus $n'(\nu') = n(\nu')$. From this it follows that $(n^2 - 1)\nu^2 = A$, where A is a relativistic invariant and therefore the only possible dispersion formula for vacuum is

$$n^2 = 1 + \frac{A}{\nu^2} \tag{6}$$

if the special theory of relativity is correct. (Note that no supposition concerning the nature of the waves or quantization of the waves was involved in the above derivation. Thus the dispersion formula of Eq. (6) is more general than was recognized before; it should be valid for all waves (e.g. gravity) if relativity theory holds.)

In particular, the linear generalization of Maxwell's equations for electromagnetic waves leads via Klein-Gordon or Proca equations to the dispersion formula of Eq. (6). The same dispersion formula is true for quantized matter waves of de Broglie, for which $A = -(m_o c^2/h)^2$, where m_o is the rest mass of the particle. Consequently a dispersion formula of the above type can also be attributed to a finite rest mass of the photon.

Although the principle of relativity rigorously rules out $B \neq 0$ in vacuum, we nevertheless retain the B term in Eq. (1) in interpreting data on the frequency dependence of the speed of light to allow for suggested breakdowns of Einstein's theory of relativity at very short wavelengths (18,19).

The results of several recent pulsar measurements interpreted according to Eq. (1) are summarized in Table I, together with other evidence, namely, measurements of the speed of very high energy (6 GeV) gamma-rays (23) and upper bounds determined for the restmass of the photon (24,25,26). From the calculated upper limits for A

TABLE I

Type of Measurement	Frequency Range (Hz)	$\Delta c/c_{visible}$	$A(Hz^2)$	$B(Hz^{-2})$	Ref.
Pulsar emissions:					
Radio wave	$1\text{-}4 \times 10^8$	$< 10^{-10}$	$< 10^6$		a
Visible	$5\text{-}8 \times 10^{14}$	$< 10^{-16}$	$< 10^{14}$	$< 10^{-45}$	b
X-ray	$4\text{-}24 \times 10^{17}$	$< 10^{-14}$		$< 10^{-50}$	c
γ-ray velocity	10^{24}	$< 10^{-3}$		$< 10^{-51}$	d
Photon mass limit	static field		< 1		e

Measurements providing upper bounds on the dispersion of the speed of light in different regions of the electromagnetic spectrum, and values of A and B to be used in interpolating to other regions of the spectrum via Eq. (1). References: a) 20,21,22; b) 16; c) 17; d) 23; e) 26.

and B the following conclusions can be drawn: 1) The speed of light in vacuum is constant to within one part in 10^{20} throughout the near infrared, the visible, and ultraviolet regions of the spectrum. 2) The speed of light in vacuum is constant to within one part in 10^{20} for all frequencies above 10^{10} Hz, if the special theory of relativity holds.

Now, if someone finds our above analysis too optimistic or that it ignores some valid theoretical arguments, he may rest assured on the observational evidence that c is constant throughout the visible spectrum to within 5 parts in 10^{17} (16). This accuracy exceeds that of any metrological measurement likely to be performed in the foreseeable future.

3. The Constancy of the Speed of Light in Moving Frames.

Any dependence of the speed of light on the variable motion of the frame of reference connected to the earth would have implications of basic character for metrology, since the unit of length is defined in terms of wavelengths of electromagnetic radiation. The very concept of wavelength standards tacitly assumes that there is no such dependence. It should be emphatically noted that any such dependence would affect length measurements in the same way in a scheme based on a chosen wavelength standard, as in the unified standardization system based on a chosen value of c.

Besides the principle of relativity, Einstein's special theory of relativity is based on the constancy of c, removing thereby any problems of the above kind if that theory proves to be correct.

H. P. Robertson (27) investigated the observational evidence of the special theory of relativity in 1949. He starts with a "rest-system", Σ, in which light is propagated rectilinearly and isotropically with constant speed c, which is independent of the motion of the source (this last assumption was verified (23) recently experimentally with high accuracy for source velocities close to c). He uses the usual assumption that the transformation into another system, S, moving with the uniform velocity, v, with respect to Σ, is linear. Next, he accepts Einstein's procedure for setting clocks in S, based on the constancy of c in Σ. Then he shows that if the metric in Σ is defined as

$$d\sigma^2 = d\tau^2 - \frac{1}{c^2}(d\xi^2 + d\eta^2 + d\zeta^2) \qquad (7)$$

then the metric in S will be

$$d\sigma^2 = g_o^2\, dt^2 - \frac{1}{c^2}[g_1^2\, dx^2 + g_2^2(dy^2 + dz^2)] \qquad (8)$$

where

$$g_o = (1 - v^2/c^2)^{\frac{1}{2}}\, a_o^o$$
$$g_1 = (1 - v^2/c^2)^{\frac{1}{2}}\, a_1^1 \qquad (9)$$
$$g_2 = a_2^2$$

and a_o^o represents the transformation coefficient between dt and dτ a_1^1 that between dx and dξ, and a_2^2 that between dy and dη and dz and dζ.

It should be noted that in Robertson's above derivation the principle of relativity was not used, and even though Einstein's procedure for setting clocks was adopted, his second postulate of the general constancy of c was not used either. Relativity (reciprocity between Σ and S) would make the transformation equations form a group, and together with the second postulate it would result in $g_o = g_1 = g_2 = 1$ as in special relativity.

For the experimental evaluation of the g coefficients Robertson lists three 'great optical experiments'. 1) The Michelson-Morley (MM) type experiments show that the velocity of light in the moving system is independent of the direction. This isotropy of light propagation in S makes $g_2 = g_1$, or $a_2^2/a_1^1 = (1 - v^2/c^2)^{\frac{1}{2}}$. This expresses the Lorentz-Fitzgerald contraction, but only in a ratio, thus it allows an additional isotropic change in the length dimensions of objects, dependent on a_2^2 which can be a function of v. Therefore, the MM experiments do not prove that the speed of light in moving systems is c (as often interpreted), they prove only that an isotropy in the light propagation is restored by Lorentz-contraction. 2) The experiments of Kennedy and Thorndike (28) (KT) (Michelson interferometer with unequal arms) prove that the velocity of light, as measured in the moving system, is independent of v, thus it is equal to c. The result of this (together with the MM experiments) is that $g_2 = g_1 = g_o$, and $a_1^1 = a_o^o = g_o/(1 - v^2 c^2)^{\frac{1}{2}}$, $a_2^2 = g_o$. Thus, there still could be an isotropic change in the length dimensions, but the speed of light would be c, due to a corresponding proper change in the rate of clocks. 3) The experiments of Ives and Stilwell (IS) (29) by the measurement of the transverse Doppler-shift, prove that $g_o = 1$. Thus, $a_o^o = a_1^1 = (1 - v^2/c^2)^{-\frac{1}{2}}$, and $a_2^2 = 1$.

Thus, the three experiments establish the Lorentz-transformation of special relativity, thereby depriving the 'rest-system', Σ, of any uniqueness. Robertson does not list the experimental uncertainties for the three experiments.

Up to now, the highest accuracies over the widest range of v have been achieved in the measurements of g_o, or "time dilatation". Some of the characteristic results are shown in Table II. The experiments listed are of three different types. 1) Transverse Doppler shift, observed with fast moving H atoms by Ives and Stilwell (29), and with Mossbauer gamma ray source on a centrifuge rotor by W. Kündig (30). 2) Azimuthal independence of the transverse Doppler shift. If the relative velocity between source and observer, u, forms an angle γ with v (velocity of the earth and of the laboratory) then aberrational effects introduce a Doppler shift proportional to $(vu/c^2) \cos \gamma$. The lack of such an azimuthal dependence in experiments was interpreted, based on discussions by Møller (31), to set upper limits for a classical 'ether drift" past the earth. Cedarholm and Townes (32) pointed out that the aberrational effects in such experiments are just compensated for by changes in the $1/2\ (v/c)^2$ term of the relativistic time dilatation formula and explain, thereby, the negative outcome of the experiments. Such experiments have been performed by Townes et al (33) with the use of ammonia masers, and Turner and Hill (34) and Champaney et al (35) using Mössbauer sources on fast centrifuge rotors. Thus these experiments, like the MM experiments, do not prove that c is a constant. We listed them in Table II, since we consider them to be the most precise checks of the time dilatation formula. 3) Prolongation of decay times of fast moving particles. In Table II we have listed the experiments of Farley et al (36) who measured the lifetime of high energy μ^- mesons, prolonged by a factor of about 12 as compared to the half life at rest. They found the relativistic formula to be valid to an accuracy of about 2%.

Table II

Type of Measurement		Precision %	Reference
Transferse Doppler	$\frac{1}{2}(v/c)^2$		
Ives and Stilwell	10^{-5}	3	29
Kundig	10^{-12}	1	30
Azimuthal Independence	uv/c^2		
Townes et al	2×10^{-10}	0.1	33
Turner and Hill	10^{-10}	0.04	34
Champeney et al	10^{-10}	0.02	35
Decay Time of μ^- meson			
Farley et al	$\tau/\tau_o \sim 12$	2	36

Measurements of the relativistic time dilatation.

The MM experiments of Joos (37) set an upper limit for $(a_2^2 - a_1^1)/a_2^2 = 1/2\ (v/c)^2$ of $\sim 1/400$ of the classically expected value which is $1/2 \times 10^{-8}$ for the orbital velocity of the earth. The experiments of Javan, Townes et al (38), using lasers, lower that upper limit to one part in 10^3.

It follows from the above discussions that these experiments (instead of setting limits to "ether drifts") show that the isotropy of light propagation in terrestrial laboratories is at least as good as a part in 10^{11} (37), or 5 parts in 10^{12} (38).

Kennedy and Thorndike (28) planned their experiments (in 1932, before the IS experiment in 1938) to investigate experimentally the "relativity of time". They take $a_2^2 = 1$ and show that Einstein's time dilatation formula is the only one compatible with their results, within the accuracy of 1/3 of the 1/2 $(v/c)^2$ term. This means according to the above interpretation that c, as measured in terrestrial laboratories is constant throughout the year to within 2 parts in 10^9.

Of course, metrologists could not be content if that uncertainty remains. There certainly are modern means to repeat the KT experiment. One method is to use a frequency stabilized laser and see that a cavity with stable parameters remains tuned to it throughout the year. The experiment would require much shorter times of observation and less stringent requirements for the constancy of the parameters if it is done in satellites, with their velocity vectors changing rapidly in space. It would be still faster to use a cavity on a centrifuge rotor and look for a possible azimuthal dependence. This experiment would not require (and would not be limited by) long term stability of frequency and temperature.

But before doing such experiments, or without doing them at all, we can invoke dynamic experiments, already performed, to diminish the uncertainty in a_2^2. We refer to the measurements of V. Meyer et al (39) who checked the relativistic mass formula for the electron in the region of energies such that, $m/m_o \sim 5$ and found Einstein's formula to be valid to within 5 parts in 10^4. The same authors rechecked the Sommerfeld fine structure splitting corrected for the Lamb-shift, and found the relativistic mass formula to be good to one part in 10^4. Thus we can trust that the mass formula is accurate to within a few parts in 10^4 of the $(v/c)^2$ term. The same was found to be true for the time dilatation formula, as seen in Table II.

Now, the change in mass, and time dilatation, are closely connected, as seen from the simple example of a rigid rotator. While it is transferred from one moving system into another (acceleration along the axis of rotation), angular momentum, $mr^2\omega$, is conserved. Since on the basis of the experimental mass and time formulas $m\omega$ is invariant in such a transformation, so should be r^2. For the orbital velocity of the earth the 1/2 $(v/c)^2$ term is 1/2 x 10^{-8}, then use of the above precisions leads to the conclusion that in a year's orbit a_2^2 cannot differ from 1 by more than 10^{-12}.

Thus our conclusion from present observational evidences is that light propagation is isotropic and its speed is constant to within the accuracy of a few parts in 10^{12} in terrestrial frames of references.

4. Accuracy Limits in the Measurement of Frequencies.

In any system of inertia, under steady (time independent) conditions of light propagation, a steady frequency emitted in point A will be observed as the same frequency in B, even if the velocity of wave propagation changes from point to point.

If the emission in A has a spectral distribution then propagational effects (dispersion, absorption, interference) may influence the spectral distribution seen in B. But in the case of very narrow (quasi monochromatic) spectra, emitted by stabilized lasers, it is relatively easy to assure in propagational effects and in the equipment at B, a spectral response much broader than that of the emission. In other words, corrections to be applied can be made negligible.

In "direct" frequency measurements, the comparison of two frequencies is made by generating a beat note of relatively small frequency on a nonlinear detector. If the difference between the two frequencies is very large, then a high order harmonic of one of them possibly followed by a heterodyne chain is used in order to make the frequency of the beat note small.

Thus, one comes to the conclusion that whenever those direct frequency measurements for the use of metrology and spectroscopy will be extended into the near infrared and into the visible, their accuracy can readily be made equal to their precision. The ultimate limit for precision and accuracy will be given by the stability of the optical frequency to be measured or that of the frequency standard, whichever of the two is poorer.

5. Accuracy Limits in the Measurement of Wavelengths.

No similar conclusions can be drawn for the accuracy of measurements based on optical wavelength. While the periodicity in time remains conserved in wave propagation, the periodicity in space changes in general from point to point.

The optical wavelength for a light oscillation is defined as

$$\lambda = c/\nu \tag{10}$$

where c is the vacuum velocity of light and ν is the frequency. The wavelength observed as a periodicity in space is $\lambda' = c'/\nu$, where c' is the phase velocity of the light wave propagated through a portion of three-dimensional space in the vicinity of a point. It is well known that even in vacuum (no dispersion) c' differs from c because of diffraction effects, thus corrections have to be applied to relate the measurements to the ideal λ, defined by Eq. (10).

In interferometry (Fabry-Perot cavities) the correction appears in the form:

$$L = (N + \varphi)\lambda/2 \tag{11}$$

showing that L, the length of the interferometer (in the case of full transparancy) is not an exact integral multiple of $\lambda/2$. The departure from the integer N (order number), φ, is mainly caused by diffraction phase shifts, phase shifts at reflection on the mirrors, and by small irregularities of mirror surfaces. The order of magnitude of φ is unity, thus in tabletop experiments ($L \sim 1$m, $N \sim$ a few times 10^6) the order of the corrections is 10^{-6}. Modern cavity theory (40) helps to treat diffraction effects. The reflection phase shifts are usually eliminated by making the measurements at two different lengths, L_1 and L_2, and relating them to a "fictitious" length $L_2 - L_1$. The uncertainties in the corrections, in particular those caused by mirror irregularities, have been treated in Ref. 4. It was found that, even with present day "superpolished" mirrors, it is hard to reach the one part in 10^{10} accuracy limit for L of a few meters. This holds, of course, for both length measurements and for interferometric comparisons of λ's.

Thus, in contrast to direct frequency measurements, the accuracy of measurements based on wavelengths is limited by wave propagational effects, however well defined wavelengths may be.

(This is the case at present with lasers stabilized to saturated molecular absorption (12,13). It is clear that such lasers are stabilized to a frequency, namely to the central frequency of the molecular absorption line. The parameter λ appears in no way in their operational characteristics. (Actually, in their absorption cell, λ' differs from λ by the order of a part in 10^6.) Long term frequency stabilities of as good or better than 1 part in 10^{11} have been reported (13) for the methane stabilized 3.39μ laser. This excellent stability and reproducibility is easily shown by observation of beat notes between two of them, not by the comparison of their λ's.)

Thus, there exists a clear distinction (we may call it "asymmetry") between frequency and wavelength measurements in favor of the first. This is based on the fact that while time runs conformally in a system of inertia (clocks can be synchronized and they run with the same rate), the physical wavelength (periodicity in space) is the result of three dimensional wave propagation effects.

6. Optical Frequency Measurements via Microwave Modulation Techniques. The Measurement of the Speed of Light.

It is shown in Refs. 4-6 that if by microwave modulation techniques (frequency ω), applied to a laser (optical frequency ν), the sidebands $\nu \pm \omega$ are generated and a passive Fabry-Perot cavity simultaneously is tuned to both sideband frequencies (order numbers N_+ and N_- resp.), then the optical frequency is related to the microwave frequency by the simple equation

$$\nu = \frac{N_+ + N_-}{N_+ - N_-}\,\omega \tag{12}$$

The length of the interferometer, L, and the vacuum speed of light, c, drop out of Eq. (12). The feature of this method is that it relates the optical to the microwave frequency in one step.

It is obvious that if simultaneously the wavelength λ, belonging to ν, is determined by reference to the present wavelength standard, ^{86}Kr, then $c = \nu\lambda$ will be known in the present m/sec system. Experiments are underway.

Another possibility, of using such modulation techniques for length measurements, was shown in Ref. 4. The length of the interferometer is an integral multiple of $\Lambda/2$, where $\Lambda = c/2\omega$ is the beat wavelength belonging to the microwave frequency 2ω. There are three interesting features of such length measurements.

1) The length is expressed by easily known small integral numbers of the beat wavelength.

2) As known from cavity theory (40) the diffraction phase shifts for the two sideband frequencies are equal for light propagation inside the cavity, thus they drop out, and the phase velocity of the

beat wave (group velocity of the microwave signal) is c. In this way a linear length scale is defined by the vacuum speed of light.

3) The length measurements are directly related via the velocity of light to frequency measurements and avoid any reference to an optical wavelength.

The questions of precision and accuracy are treated in Ref. 4. It turns out that in both optical frequency and length measurements based on microwave modulation techniques the limiting accuracies are the same as in length measurements and in λ comparisons, based on wavelength standards, however well defined.

(This result, without going into detail, appears to be obvious by remembering that these measurements, besides making use of the highly accurate frequency scale through the introduction of ω, are based on the wavelength comparison of two closely related wavelengths. Future improvements in interferometer techniques will also help in the same way in these measurements as they will in measurements based on wavelengths.)

7. Summary and Concluding Remarks.

1) Light propagation in vacuum is experimentally proven to be dispersionless to within 5 parts in 10^{17} in the visible spectrum and it can be taken with good confidence to be dispersionless to within 1 part in 10^{20} in the microwave, infrared and visible spectrum.

2) The velocity of light in frames of reference connected to the moving earth can be taken to be isotropic and constant to within a few parts in 10^{12}.

In both 1) and 2), the accuracy limits cited are those related to experimental uncertainties which currently bound any possible departure from the theory of relativity. The consequence of 1) is that the speed of light can be used in metrology without specifying the optical frequency; 2) implies that the "length of a rigid rod" can be expressed in terms of optical wavelengths (or time of flight of light) irrespective of the motion of the frame of reference. Both 1) and 2) together show that c is a constant, excellent for the use of standardization.

3) It can be assumed that in the not too distant future optical frequencies can be determined either by a) direct frequency measurements with ultimate accuracies given by that of the frequency standard, or by b) microwave modulation techniques the limiting accuracies of which are equal to those of interferometric length measurements or wavelength comparisons.

4) It can be assumed (on the basis of 3) that in the not too distant future the speed of light will be known in the present m/sec system to the accuracy of the present meter with respect to its definition.

5) There can be no doubt that if 3) and 4) materialize, the unified standardization of time, frequency, and length via an agreed upon value of c will be preferable to a system based on a frequency standard and on a wavelength standard. Indeed, under those conditions a standard wavelength, defined independently of the frequency standard, would not represent an additional advantage for the time-length measuring system. Instead, any other appropriate $\lambda = c/\nu$ could be used, where ν is measured with respect to (or locked to) the frequency standard.

6) Besides its theoretical simplicity and its versatility, a further advantage of the unified standardization system is that in case of future improvements of the standards only one unit of measurement has to be redefined. According to what was said in Sections 4 and 5, it should be the unit of time, the second, and not the meter. The meter will be automatically refined via c for the possible use of improvements in length measuring techniques.

7) If one of the frequency stabilized lasers turns out to be better in precision than the present frequency standard or others in prospect, and if its frequency can directly be connected to microwave frequencies, that laser could well be used as a new frequency standard. But the meter should not be tied to its wavelength. Instead, an agreed upon value of c should be adopted to be taken over as an invariable figure in future standardizations.

8. References

1. PLANCK, M., Theorie der Warmestrahlung, p.163, J. A. Barth, Leipzig, 1906.

2. TOWNES, C. H., Advances in Quantum Electronics, p.3. J. R. Singer, New York, Columbia University Press, 1961.

3. BAY, Z., Internal NBS Report, 1965, (unpublished).

4. BAY, Z., Proceedings of the International Conference on Precision Measurement and Fundamental Constants, Gaithersburg, Maryland, August 1970.

5. BAY, Z. and LUTHER, G. G., Appl. Phys. Letters, 13, No. 9, 303, (1968).

6. BAY, Z. and LUTHER, G. G., Proceedings of the International Conference on Precision Measurement and Fundamental Constants, Gaithersburg, Maryland, August 1970.

7. BAIRD, K. M. and SMITH, D. S., J. Opt. Soc. Am., 52, 508 (1962).

8. BECHLER, R. E., MOCKLER, R. G. and RICHARDSON, J. M., Metrologia 1, 114 (1965).

9. HOCKER, L. O., JAVAN, A., RAMACHANDRA RAO, D., FRENKEL, L. and SULLIVAN, T., Appl. Phys. Letters 10, 147 (1967).

10. EVENSON, K. M., WELLS, J. S. and MATARRESE, L. M., Appl. Phys. Letters, 16, No. 6, 251 (1970). See also EVENSON, K. M., WELLS, J. S. and MATARRESE, L. M., Proceedings of the International Conference on Precision Measurement and Fundamental Constants, Gaithersburg, Maryland, August 1970.

11. McDONALD, D. G., RISLEY, A. A., CUPP, J. D. and EVENSON, K. M., Appl. Phys. Letters, 18, No. 4, 162 (1971).

12. HANES, G. R. and BAIRD, K. M., Metrologia 5, 32 (1969).

13. BARGER, R. L. and HALL, J. L., Phys. Rev. Letters, 22, 4 (1969). See also BARGER, R. L., this conference.

14. O. COSTA DE BEAUREGARD, Metrologia, 4, No. 3, 144 (1968).

15. BAY, Z. and WHITE, J. A., to be published.

16. WARNER, B. and NATHER, E. R., Nature 222, 157 (1969).

17. RAPPAPORT, S., BRADT, H., and MAYER, W., Nature Physical Science, 229, 40 (1971).

18. BLOCKINTSEV, D. I., Physics Letters, 12, 272 (1964).

19. PAVLOPOULOS, T. G., Phys. Rev., 159, 1106 (1967).

20. Pulsating Stars, with introduction by SMITH, F. G. and HEWISH, A., N. Y. Plenum Press, 1968 (a Nature reprint).

21. CONKLIN, E. K., HOWARD, H. T., MILLER, J. S. and WAMPLER, E. J., Nature, 222, 552 (1969).

22. FEINBERG, G., Science, 166, 879 (1969).

23. ALVAGER, T., BAILEY, J. M., FARLEY, F. J. M., KJELLMAN, J. and WALLIN, I., Arkiv For Fysik, 31, 145 (1966).

24. GINSBURG, M. A., Soviet Astronomy, AJ7, 536 (1964) (Astronomickeskii Zhurual 40, 703 (1963).

25. SCHRODINGER, E., Proc. Roy. Irish Acad. A49, 135 (1943).

26. GOLDHABER, A. S. and NIETO, M. M., Phys. Rev. Letters, 21, 567 (1968).

27. ROBERTSON, H. P., Rev. Mod. Phys. 21, No. 3, 378 (1949).

28. KENNEDY, R. J. and THORNDIKE, E. M., Phys. Rev., 42, 400 (1932).

29. IVES, H. E. and STILWELL, G. R., J. Opt. Soc. Am., 28, 215 (1938); J. Opt. Soc. Am., 31, 369 (1941).

30. KÜNDIG, W., Phys. Rev. Ser. II, 129, No. 6, 2371 (1963).

31. MØLLER, C., Proc. Roy. Soc. (London) A270, 306 (1962).

32. CEDARHOLM, J. P. and TOWNES, C. H., Nature 184, 1350 (1959).

33. CEDARHOLM, J. P., BLAND, G. F., HAVENS, W. W., Jr., and TOWNES, C. H., Phys. Rev. Letters 1, 526 (1958).

34. TURNER, K. C. and HILL, H. A., Phys. Rev., 134, No. 1B, 252 (1964).

35. CHAMPENEY, D. C., ISAAK, G. R. and KHAN, A. M., Phys. Letters 7, No. 4, 241 (1963).

36. FARLEY, F. J. M., BAILY, J., BROWN, R. C. A., GIESCH, M., JOSTLEIN, H., van der MEER, S., PICASSO, E. and TANNENBAUM, M., Nuovo Cimento, XLV A, No. 1, 281 (1966). See also FARLEY, F. J. M., BAILY, J. and PICASSO, E., Nature, 217. 17 (1968).

37. JOOS, G., Ann. Phys., 7, 385 (1930).

38. JASEJA, T. S., JAVAN, A., MURRAY, J. and TOWNES, C. H., Phys. Rev., 133, No. 5, A1221, (1964).

39. MEYER, V., REICHART, W., STAUB, H. H., WINKLER, H., ZAMBONI, F. and ZYCH, W., Helv. Phys. Acta, 36, 981 (1963).

40. KOGELNIK, H. and LI, T., Proc. IEEE, 54, No. 10, 1312 (1966).

Part 7 Wavelength Comparisons

Intercomparison of Micrometer, Nanometer and Picometer Wavelengths

R. D. Deslattes and W. C. Sauder

National Bureau of Standards, Washington, D. C. 20234 U.S.A.

1. Scheme of measurements

We have undertaken to arrange a group of wavelength measurements in such a way as to assure that each is well tied to a standard wavelength common to all. [This wavelength is, in fact, the 606 nm emission from ^{86}Kr at the N_2 triple point which is the defined wavelength, although such a pedigree is unnecessary in this story.] By coordination with some aspects of the work of Earnest Kessler on the Rydberg (reported at this meeting) and of Horace Bowman and Randall Schoonover on density we plan to tie together the Rydberg (R_∞), Avogadro's constant (N_0), the electron's Compton wavelength ($\lambda_c = h/mc$), the scale of γ-ray wavelengths (eg, the 411 keV emission from Au) and the scale of x-ray wavelengths (as might be characterized by a value for Λ, the x-ray-metric conversion factor).

This hasn't been finished yet so that all that we can report at present is an outline of the structure of the measurement chain and partial and preliminary results in some cases. We begin with an overall systems-type description and then consider several components in detail.

In a gross overview we are confronted with a certain array of emission wavelengths (some generic and some ascribable only to given realizations of a particular source), a few lattice parameters (none of which need be taken to have generic significance in our engineering of the measurements) and a group of comparators or measuring engines. These comparators are variously interferometers, goniometers, spectrometers and balances.

We visualize the pattern of measurements as shown in Fig. 1. The simplest case, Fig. 1a, yields a fundamental constant ◇ by comparison of a specimen □ with a primary standard ◎ on a comparator D. Parts b and c suggest binary and ternary chains involving at least temporary secondary standards denoted by ○. A particular emphasis in our program is suggested by the dashed lines which indicate use of a common input standard. To the extent that we can realize this, many limitations associated with the standards drop out of our measurements.

In terms of the format just indicated, Kessler's Rydberg measurement (1) would appear as in Fig. 2. His report at this meeting details the pressure-scanned flat-plate Fabry-Perot system (PSFPFP) in which certain features of the HeII spectrum are compared with a ^{198}Hg reference. The other comparator indicated, namely the spacing-scanned flat-plate Fabry-Perot (SSFPFP) is also important for several of the following measurements and has not been reported at this meeting (2).

In its present incarnation the SSFPFP is arranged as in Fig. 3. It is a low finesse flat plate Fabry-Perot with 10 cm spacing between plates. At present interpolation is via a capacitance bridge as suggested in the figure and there is no provision for servo control of plate parallelism. We plan to replace the interpolation scheme by a laser frequency scale and to servo the plate parallelism as is done in the system reported by Richard Barger at this meeting (3). Even with present limitations, comparisons of coherent and incoherent visible sources are consistent to one or two parts in 10^9 and thus this is not a limiting factor in the applications considered here.

With this rather elementary exercise on a linear measuring engine, we have already taken care of comparing fractional micron wavelengths with one another. This supports the metrological truism that nearly equal magnitudes compare rather neatly. In contrast, consider what is required in our program to compare wavelengths six orders of magnitude apart.

Figure 4 outlines what is needed to tie the annihilation wavelength from positronium at 2 picometers to the same reference against which the Rydberg is determined, at 0.6 micrometers. There is a considerable story about trying to get a sharp annihilation line which we have reported elsewhere (4). Here we will emphasize the wavelength ratio problem, assuming that conditions have been arranged to make it a worthwhile exercise.

As can be seen, the first step (logically though not necessarily chronologically) is to tie a laser wavelength to ^{86}Kr in the SSFPFP comparator which was just described. The next step, though it involves in principle just another linear measuring engine, is more formidable. Namely, a crystal acting as an x-ray Moire grating is coupled to an optical Fabry-Perot. They are scanned to and fro over a common baseline by means of a suitably disposed linear engine. The output is a temporary secondary standard crystal spacing of about 1/5 nanometer which can then enter the next stage of the chain. (Since the x-ray interferometer is achromatic, there is no dependence on x-ray spectral purity.)

Other choices are available, but we have chosen to use a large angle two crystal instrument (LATXI) to transfer this crystal calibration to other crystals and elsewhere to x-ray wavelengths. This is an angular measuring engine which (as shown in section 2) is capable of self calibration by closure.

Finally, crystals calibrated in this way are supplied to a small angle two crystal instrument (SATXI) as is required for measurement of γ-ray wavelengths. It is this device which bridges the last two or three orders between fractional nanometers and picometers where our efforts presently terminate. In the following paragraphs, characteristics of each of these three measuring engines are described.

2. Measuring engines

Figure 5 illustrates the third generation linear engine for x-ray and optical interferometry. It is rather similar to the first generation device (5) but contains some refinements of construction, assembly and use. Because of an unanticipated degree of difficulty in cutting the very tough steel from which the device is fabricated, the holes defining the main flexure joints had to be severely lapped in the final stages of fabrication.

The total available range of 0.1 mm is not usable at present because of a slight overcorrection of the motion which occurred during lapping. Within a much more limited range of perhaps .01 mm measurements seem to have a noise level of about 1 picometer and to be dominated by laser frequency drifts and instability in the mode-matching optics. The open loop behavior of this system is suggested in Fig. 6 where the derivative of the optical signal (obtained by phase conscious rectification against the frequency modulated laser light) is shown along with the x-ray fringes. In practice the optical loop is closed yielding a much higher precision realized in determining the x-ray phase at the optical transparency maximum.

The crystal output from this x-ray/optical interferometer (or rather, for the moment its icon, an adjacent slab) is then fed to the large angle two crystal instrument (LATXI) for distribution of the results. This instrument (6) is now equipped with a two beam angle interferometer whose quarter phase count decodes at about 1/30 arc sec per count. The photograph in Fig. 7 shows the 1° incremental generator, x-ray detector and crystal holder. The angle interferometer is "below deck" and not visible. A recently completed calibration by Albert Henins yielded a precision to within 1 count (1/30 arc sec). The goniometric procedures thus will not limit the accuracy of our measurement chain at the sub one ppm level currently of interest.

The final comparator in this chain is the small angle double crystal instrument (SATXI) for γ-ray wavelength measurements. It has been described in some detail elsewhere (4) so only essential features will be indicated here. Its rather massive structure is suggested by the photograph in Fig. 8. The twin axes have a maximum driven excursion of $\pm 2\ 1/2^o$ and a refinement of motion approaching 10^{-5} arc sec. The dihedral angular excursions are reckoned by four Fabry-Perot corner cube cavities with finesse values in excess of 20. The arm lengths are such as to give an inter-order spacing of 0.25 arc sec. A tracing of a single cavity fringe system is indicated in Fig. 9 where the transparency curve widths of about 0.01 sec (FWHM) suggest the capability to go easily to 0.1 milli arc sec.

We plan to employ frequency shifting of the laser source to effect intra-order interpolation of the angle scale. There are several calibration algorithms under consideration of which at least two will be attempted. It is particularly attractive to consider temporarily expanding the driven range of the crystal tables to reach the inter-facial angle of (for example) a 30 sided polygon. The polygon and interferometer then calibrate each other by closure with a consequent overdetermination of the interferometer constant reducing autocollimation errors by a factor of $\sqrt{30}$. In addition, we can choose a nuclear γ transition which can be handled in high order of diffraction on the LATXI and reached in first order on the SATXI. This is obviously sufficient to establish an angular calibration provided either that the same crystal be used or that the crystals used have a determinate connection with one another.

3. Other schemes

Figure 10 suggests a direct application of the LATXI and a crystal related to the optical/x-ray interferometer to evaluate the x-ray wavelength scale. If the wavelength used has a "conventional" value on the x-ray scale (eg Mo Kα_1: 707.831 xu or W Kα: .2090100 Å*), then the suggested measurement yields Λ, the inter-scale conversion factor. This is a wholly trivial exercise with a future limited in usefulness by the large widths which x-ray lines seem unable to avoid.

Finally, Fig. 11 indicates the structure of density measurements under way by H. A. Bowman and R. L. Schoonover (7). The first phase, a diameter exercise suggested in the figure as interferometry, is actually much more complex. In addition to interferometry, mechanical gaging against end standards has been employed. There are as yet unresolved systematics in this phase which amount to about 3 ppm in density. The air weighing suggested by 'AIR BAL' is not difficult at this level. The outputs are 6 cm dia. balls of steel and quartz which are density standards. In principle, these objects may be compared with crystals of determined repeat distance in the indicated submarine balance (SUB BAL). Actually intermediate silicon density artifacts are employed as are a group of different density fluorocarbon fluids (to provide for overdetermination). Only on the basis of a gratuitous assumption about isotopic abundance can this operation be said to implicate Avogadro's constant, but that is in fact our objective.

4. Results - tentative conclusions

The effort just described has strongly interacting components. It would be best if all results were at hand and a discussion could be mounted on this structure. Alas there is little to prevent nature from non-invariant perversity or our own efforts from being somewhat uneven. The following discussion is based on tentative results from x-ray/optical interferometry, density measurements, $\lambda/2d$ values from the large angle two crystal instrument and some assumptions about isotopic abundance.

Results at hand are of a preliminary and incomplete character. Their import appears, however, to be already appreciable. The numerical results are the following:

(a) The 220 repeat distance of a specimen of modified float zone produced silicon is such that there are 1648.218 ±.002 repeat distances per half wavelength of ^{20}Ne laser light in helium at 1 atm and 25 °C. The implied metric value for this repeat distance is 1.920 170Å (2 ppm).

(b) When a sample taken adjacent to the above specimen is used to measure diffraction angles in various orders of (111) for Mo Kα_1 radiation, the implied value for λ(Mo Kα_1) = .709 320Å.

(c) Two samples taken adjacent to and on opposite sides of the above were compared indirectly (as noted above) to macroscopic artifact density standards. Their densities were 2.328 991 and 2.328 995 (when corrected to 25 °C). This difference of 2 ppm is significant in terms of the precision of measurement; there is also an unresolved systematic question which will probably require a larger estimate of error than might be suggested by this.

We can combine the results (a) and (b) and assume (as appears justified at present) that the "sampling" procedure is adequate. If λ(Mo Kα_1) is assigned a wavelength 707.844 xu (based on Cu Kα_1 = 1537.400 xu) then $\Lambda = \lambda g/\lambda s$ = 1.002 084(2). If this same emission is assigned the value suggested by Bearden (8) in Å* (.709 300Å*) then the implied value of Λ^* is 1.000 028(2).

We can also combine the results (a) and (c) and assume that the geochemical average atomic weight for Si (9) applies to the samples at hand (which is probably not justified). In this case N_o is fixed at 6.02217(8).

The atomic weight assumption is really terrible. We hope to do or encourage others to do some serious isotope enrichment which will take a long time. Meanwhile, we are pleading for some isotopic analysis on the samples which we have at hand.

In the context of adjusted values for constants over the past decade the results are as summarized in Table I (10). We feel that indicated estimates of error for the present results can be lowered with further observation and the elimination of certain rather easily fixed troubles. This refinement is not expected to alter the trends of Table I.

We wish to emphasize the preliminary and incomplete character of these results. They are especially deficient in regard to statistical control and error analysis. Efforts are underway to remedy these problems which will hopefully lead to firmer results over the next few months.

5. Acknowledgments

The density measurements are being carried on by Horace A. Bowman, Randall M. Schoonover and Leon Carroll. X-ray wavelength measurements on the LATXI and its interferometric calibration are due to Dr. Albert Henins. Our indebtedness to their outstanding work is evident in the discussion above.

TABLE I

Trends in constants

	1963[(1)]	1969[(2)]	1971[†]
N_0	6.02252(9)	6.02217(4)	6.02217(8)[††]
Λ (Cu xu)		1.002076(5)	1.002084(2)
Λ (Mo xu)	1.002080(6)		1.002103(2)
Λ^* (W Kα)		1.000020(6)	1.000028(2)

† This work, preliminary results and tentative error estimates.

†† Based on conventional isotope abundances, see Ref. 9.

(1) See Ref. 11.

(2) See Ref. 10.

6. References

1. KESSLER, E., NBS Special Publication. Proceedings of the International Conference on Precision Measurement and Fundamental Constants, Gaithersburg, Md., August 1970, and this conference.

2. MIELENZ, K. D., NBS Special Publication. Proceedings of the International Conference on Precision Measurement and Fundamental Constants, Gaithersburg, Md., August 1970.

3. BARGER, R., this conference

4. SAUDER, W. C., NBS Special Publication. Proceedings of the International Conference on Precision Measurement and Fundamental Constants, Gaithersburg, Md., Autust 1970; SAUDER, W. C., J. Res. NBS 72A, 91 (1968); SAUDER, W. C. and DESLATTES, R. D., ibid. 71A, 347 (1967).

5. DESLATTES, R. D., Appl. Phys. Letters, 15, No. 11, 386 (1969).

6. DESLATTES, R. D., Rev. Sci. Instr., 38, 815 (1967).

7. BOWMAN, H. A. and SCHOONOVER, R. M., J. Res. NBS 71C, 179 (1967).

8. BEARDEN, J. A., Rev. Mod. Phys., 39, 78 (1967).

9. HENINS, I. and BEARDEN, J. A., Phys. Rev., A135, 890 (1964).

10. TAYLOR, B. N., PARKER, W. H. and LANGENBERG, D. N., Rev. Mod. Phys., 41, 375 (1969).

11. COHEN, E. R. and DUMOND, J. W. M., Rev. Mod. Phys., 37, 537 (1965).

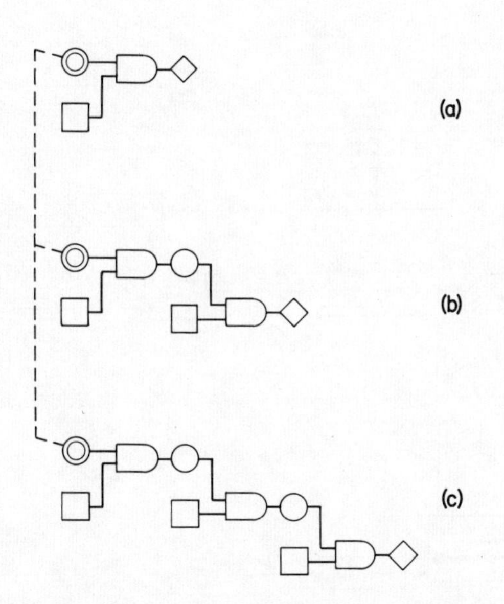

Fig. 1. Logical structures of constants measurements to be discussed. The symbol ◎ encloses such primary standards as the ^{86}Kr source and the kG; ○ will be used for secondary standards with or without generic significance as for example ^{20}Ne, ^{198}Hg, round balls and crystals; □ designates raw materials input to the measurements as might be optical, x-ray and γ-ray transitions and crystals. Comparators, indicated by D are interferometers, spectrometers and balances. The output values, ◇ are the expected results for R_∞, h/mc, N_o and Λ.

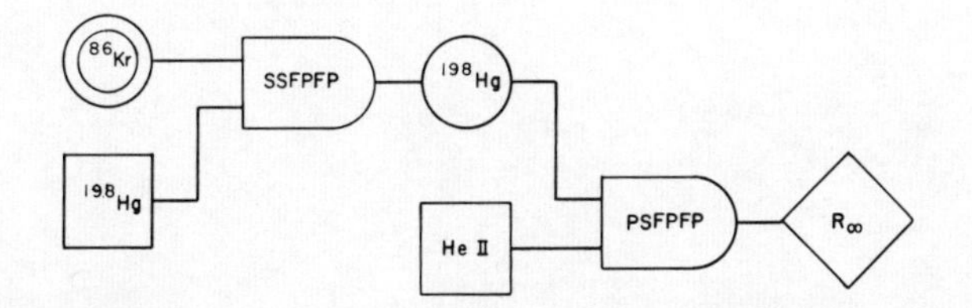

Fig. 2. Rydberg measurement chain. First comparator is spacing scanned flat plate Fabry-Perot (SSFPFP) while second is (actually 2) a pressure scanned flat plate Fabry-Perot (PSFPFP). Circles and squares are lamps of the indicated types.

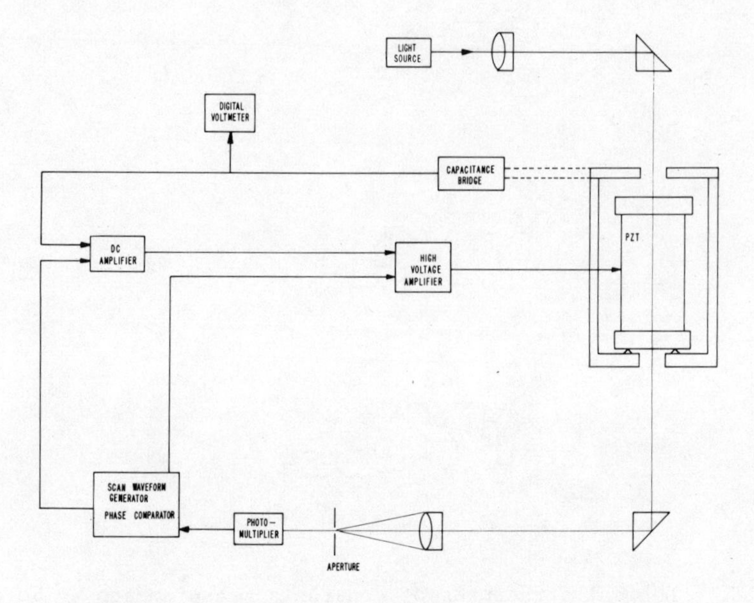

Fig. 3. Schematic of spacing scanned flat plate Fabry-Perot (SSFPFP).

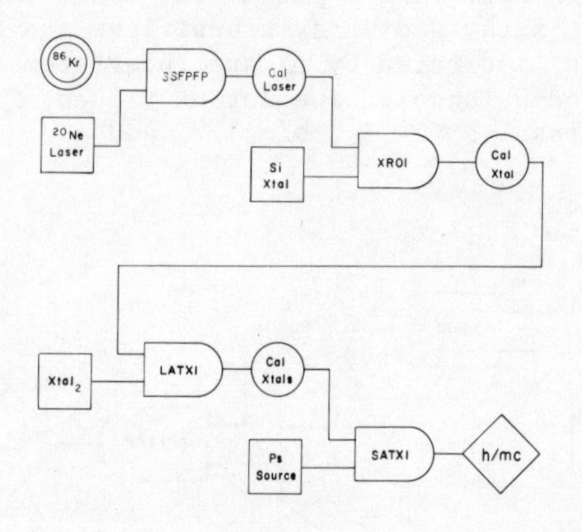

Fig. 4. Outline of measurement chain for the electron's Compton wavelength. Main comparators are an x-ray/optical interferometer (XROI), large angle two crystal instrument (LATXI) and a small angle two crystal instrument (SATXI).

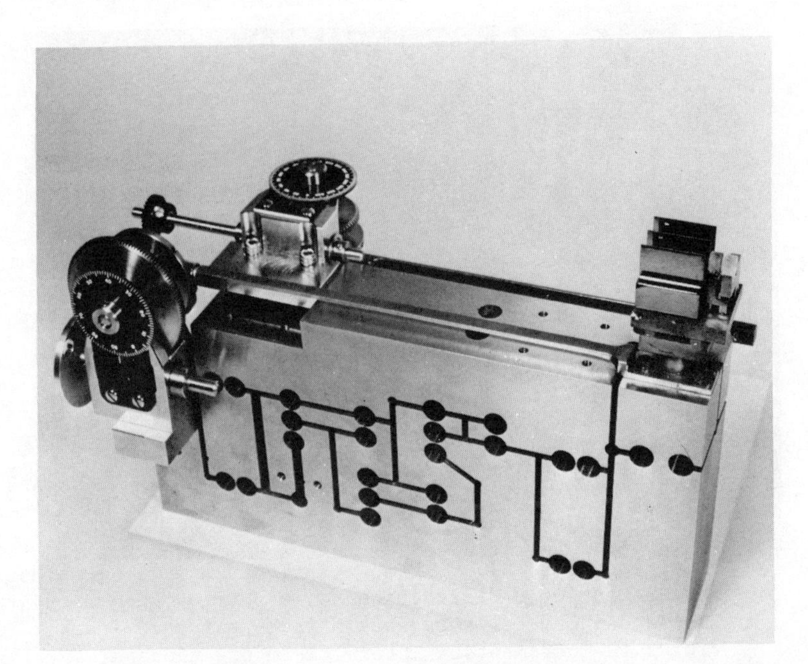

Fig. 5. Third generation x-ray/optical interferometer (XROI). Linear motion stage is a four flexure parallelogram driven by the indicated internal speed reducer. Required micro angular adjustments are effected by differential springs shown.

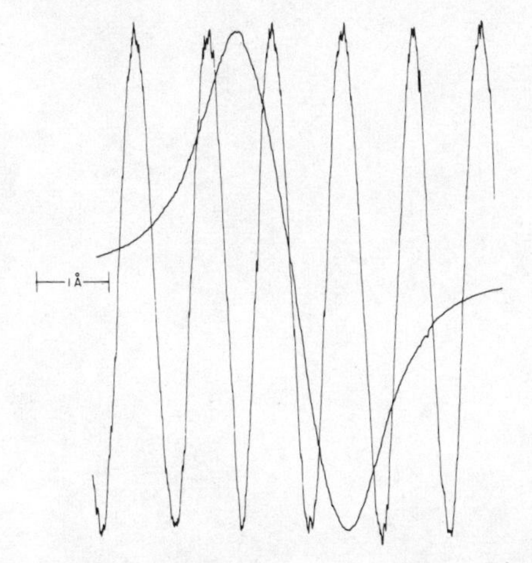

Fig. 6. Open loop characteristics of XROI. Dispersion shaped curve is the derivative of the optical 00ℓ optical maximum while the cosine curve is the two beam x-ray interferogram. Integration times are 2 sec for the x-ray channel and 1 sec for the optical channel. In practice, the optical channel is servoed to the attendant reduction of the noise indicate by about 1 order of magnitude.

Fig. 7. Photograph of LATXI. First crystal at left is fixed during measurement of a diffraction angle pair by second crystal. The second crystal surmounts a face gear pair indexer providing 1° increments. Intermediate positions are attained by sine arm and micrometer indicated. These positions are 'read' by a two beam angle interferometer below the level shown to the nearest 1/30 arc sec.

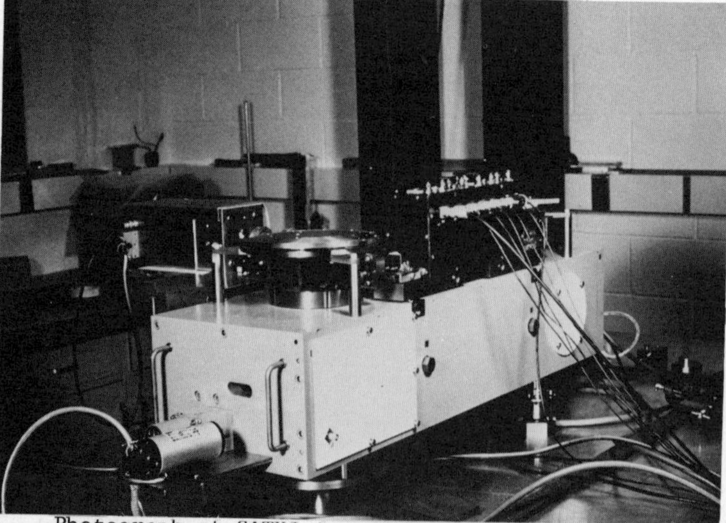

Fig. 8. Photograph of SATXI for measurement of γ-ray wavelengths. The two axes are symmetric and driven by micrometer actuated sine arms with piezoelectric verniers. These are contained within the housing visible. The optical components partly visible form four corner cube to flat Fabry-Perot cavities, the collection of positions of which specify the crystal dihedral angles.

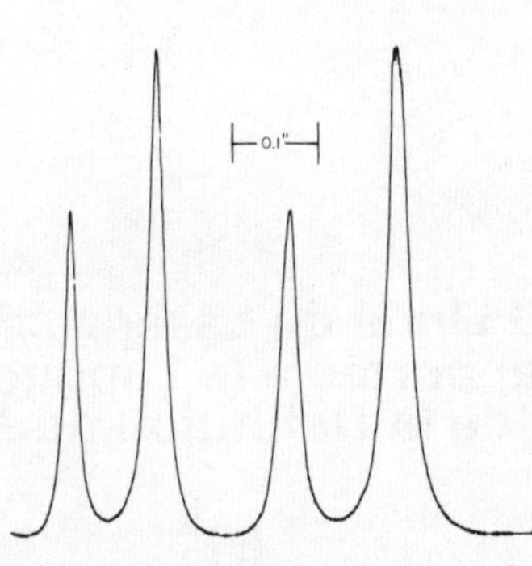

Fig. 9. Open loop display of optical signal from one of the four cavities determining intercrystal angles on SATXI. In use, the system is servoed to one of these peaks and interpolation achieved by frequency shifting of the laser source.

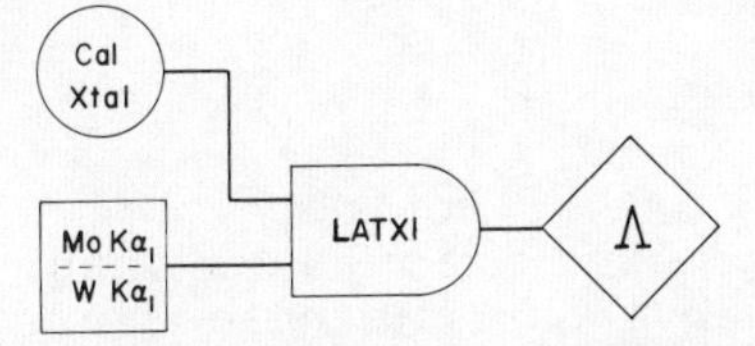

Fig. 10. Structure of measurement determining the metric to x-ray conversion factor. There are three x-ray scales which need be considered at present: on one, Mo $K\alpha_1$ has a wavelength of 707.831 xu; on the second, Cu $K\alpha_1$ has a wavelength of 1537.400 xu; on the third, W $K\alpha_1$ has a wavelength of 0.2090 100Å*. As is evident, questions regarding these scales do not impringe on values for the other constants in the present approach.

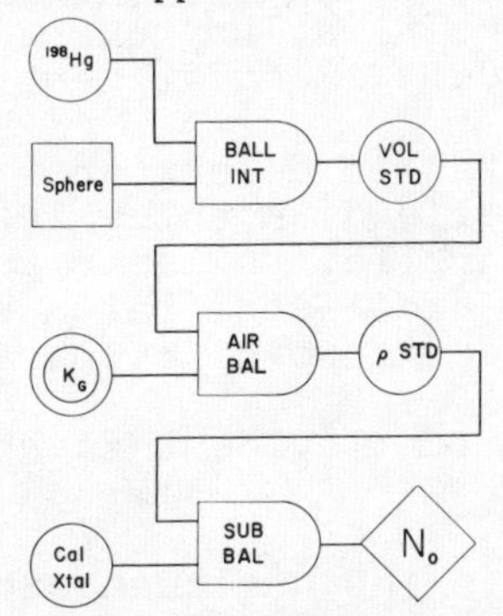

Fig. 11. Structure of measurements for density which implicate N_o. Ball interferometer (BALLINT) can use a calibrated laser source also, and can be replaced by a mechanical comparator. Quartz and steel ball type density standards have been compared with silicon artifacts in the submarine balance (SUBBAL) and these are used as secondary but more stable standards. The submarine balance employs a group of fluorocarbon fluids ranging from $\rho = 0.9$ to $\rho = 1.7$.

Longueur d'Onde du Laser Asservi sur le Methane Rapportée à la Longueur d'Onde Etalon de la Definition du Mètre

P. Giacomo

Bureau International des Poids et Mesures, Sèvres, France

1. Intérêt

Les qualités métrologiques du laser à hélium-néon (λ = 3,39 µm) asservi sur une raie d'absorption saturée du méthane ont été démontrées par R.L. Barger et J.L. Hall (1). Leurs expériences ont mis en évidence la stabilité de fréquence de cet oscillateur. La mesure de la longueur d'onde dans le vide de la radiation émise est importante à divers égards :

1 - On peut espérer que sa fréquence sera mesurée dans un proche avenir avec une incertitude inférieure à 1×10^{-6}. Si l'on connaît la longueur d'onde, la vitesse de la lumière s'en déduit immédiatement.

2 - Il est plus facile de comparer la longueur d'onde d'un laser à celle d'un autre laser qu'à celle d'une source traditionnelle. Le laser asservi sur le méthane fournirait un excellent étalon secondaire de longueur d'onde pour ce genre de comparaisons.

3 - C'est un candidat évident au rôle d'étalon primaire de longueur.

Ce dernier point exige quelques expériences préliminaires. Il faut en particulier démontrer que la longueur d'onde en question se prête bien aux mesures de longueur les plus précises. Or la longueur que l'on sait actuellement reproduire avec la plus grande précision, c'est la longueur d'onde de la radiation orangée du krypton, étalon primaire de longueur. Mesurer la longueur d'onde du laser par rapport à celle du krypton, c'est donc aussi mettre le laser à l'épreuve en tant qu'étalon de longueur.

2. Appareils et méthode (fig. 1)

Le laser stabilisé a été apporté en août 1970 au Bureau International par R.L. Barger, avec qui j'ai eu le plaisir de faire les premières expériences ; il a pu nous le laisser en prêt grâce à la générosité du J.I.L.A.

Nous disposons pour les mesures d'un autre instrument apporté en 1892 au Bureau International (2) par un illustre citoyen des Etats-Unis d'Amérique : l'interféromètre de Michelson, quelque peu modifié, mais toujours reconnaissable. Il est maintenu sous vide dans un caisson étanche et bien isolé thermiquement.

La lampe à krypton est du type Engelhard ; elle est utilisée dans les conditions recommandées par le C.I.P.M. (3) pour l'émission de la radiation étalon. Un monochromateur permet d'isoler les différentes radiations émises.

1 - Principe des mesures - On effectue une translation Δe du miroir mobile (Mm) ; on mesure les variations d'ordre d'interférence correspondantes, p_1 et p_2, pour les deux radiations de longueur d'onde λ_1 et λ_2. Dans le vide, on a

$$2\,\Delta e = p_1\lambda_1 = p_2\lambda_2 \qquad \text{donc} \quad \lambda_2 = \frac{p_1}{p_2}\,\lambda_1$$

Si les deux faisceaux tombent sur les miroirs avec le même angle d'incidence i, on a encore

$$2\,\Delta e \cos i = p_1\lambda_1 = p_2\lambda_2 \quad \text{et} \quad \lambda_2 = \frac{p_1}{p_2}\,\lambda_1$$

Les ordres d'interférence p_1 et p_2 sont déterminés par la méthode habituelle : mesure des parties fractionnaires pour plusieurs radiations et recherche des coïncidences. Les parties fractionnaires sont déterminées par la méthode dite "des quatre pointés" (4)(5)(6) : en mesurant le flux émergent pour quatre positions de la lame compensatrice (C) correspondant à des ordres d'interférence $p_0 \pm \delta p$ et $p_0 \pm 3\,\delta p$, on déduit par un calcul simple la valeur de p_0 (à un entier près). Cette méthode repose sur une propriété essentielle de l'interféromètre de Michelson : le flux émergent est une fonction pratiquement sinusoïdale de la différence de marche, avec toute la précision souhaitable, pourvu que la source soit honnêtement monochromatique, et cela même en présence des défauts inévitables (planéïté des miroirs, diaphragme isolateur, etc.). On obtient ainsi aisément la partie fractionnaire de p_0 à un dix-millième près pour la radiation du laser et à un millième près pour la radiation du krypton jusqu'à plus de 0,5 m de différence de marche. Pour cette dernière radiation, on peut montrer (7) que la précision sur p_0 est optimale au voisinage d'une différence de marche de ± 375 mm : on a utilisé systématiquement cette valeur pour les mesures les plus précises.

2 - Avantages de la méthode - Le fait que le flux soit fonction sinusoïdale de la différence de marche permet un étalonnage très précis de la commande de la lame compensatrice et de la linéarité des récepteurs utilisés.

La translation Δ e élimine les défauts de planéïté des miroirs ; l'utilisation des deux positions symétriques (différence de marche de + 375 mm et - 375 mm) élimine en outre tous les termes correctifs qui seraient fonction impaire de e.

3 - Problèmes propres à la mesure de λ = 3,39 µm - La lame séparatrice ainsi que le miroir dichroïque (D) utilisé pour superposer les deux faisceaux sont revêtus de couches diélectriques préparées spécialement pour cette expérience.

La commande (S) de la rotation de la lame compensatrice doit être spécialement soignée puisqu'elle doit assurer l'exploration d'un peu plus d'un ordre, pour λ = 3,39 µm, à un dix-millième d'ordre près. On a tenu compte du défaut résiduel de linéarité de cette commande dans les calculs.

Les fenêtres du caisson à vide, les lames séparatrice et compensatrice doivent être transparentes pour le visible et pour λ = 3,39 µm ; heureusement, des lames en silice fondue conviennent.

L'optique de sortie, qui définit l'angle d'incidence moyen des faisceaux, doit être parfaitement achromatique. On a utilisé une optique à miroirs (T) comportant essentiellement un diaphragme isolateur (Is) au foyer d'un miroir sphérique.

Les réflexions parasites sur les faces des lames séparatrice et compensatrice pourraient perturber notablement les interférences. Les faisceaux parasites, se combinant en amplitude avec les faisceaux principaux, avec des déphasages différents pour les deux radiations, pourraient introduire des erreurs systématiques appréciables. On élimine ces réflexions parasites en travaillant sous l'incidence de Brewster, avec une polarisation rectiligne convenablement orientée (polariseurs P_1 et P_2).

Par voie de conséquence, on ne peut utiliser, pour le découplage optique du laser, le polariseur circulaire habituel. Il a été remplacé par une cellule à effet Faraday (F) constituée par un grenat d'yttrium-iron fournissant, à la saturation magnétique, une rotation du plan de polarisation de 45°.

3. Précision, erreurs systématiques

1) La répétabilité des mesures est essentiellement limitée par la stabilité de l'interféromètre. Chaque mesure comporte une dizaine de déterminations de l'ordre d'interférence, pour chacune des deux radiations, à chaque différence de marche, pour l'interpolation des dérives. On en déduit un écart type, pour chaque mesure : il peut varier de 2×10^{-9} à 10×10^{-9} suivant les séries.

2) Les défauts de linéarité (commande de la compensatrice, réponse du récepteur infra-rouge) font l'objet d'une correction. L'erreur résiduelle ne doit pas dépasser 1×10^{-9}. Elle est aléatoire, comme la partie fractionnaire de l'ordre d'interférence mesuré.

3) L'indice de réfraction de l'air, pour une pression résiduelle de 10^{-5} mm de mercure, ne doit pas introduire d'erreur supérieure à 1×10^{-11}.

4) L'inclinaison moyenne des faisceaux serait la même pour les deux longueurs d'onde si le diaphragme isolateur (Is) était uniformément éclairé (en l'absence d'interférences) par les deux sources. C'est bien le cas, en première approximation, pour la lampe à krypton, mais non pour le laser. Avec un diaphragme isolateur de 0,2 mm de diamètre et une distance focale de 60 cm, on a, pour la radiation du krypton, $\cos i \simeq 1 - 7 \times 10^{-9}$; pour le laser, si le faisceau est parfaitement aligné et si les aberrations et la diffraction sont négligeables, $\cos i = 1$. On risque donc de commettre une erreur systématique de 7×10^{-9} sur la longueur d'onde mesurée pour ce dernier. Il s'agit en fait d'une limite que l'on n'atteint heureusement pas : On a pu vérifier, en utilisant un diaphragme de 0,1 mm de diamètre, que la variation apparente de λ_2 n'excède pas $+ 1 \times 10^{-9}$. On peut donc estimer que l'erreur systématique correspondante ne dépasse pas 1×10^{-9}.

5) Les deux sources éclairent les miroirs de l'interféromètre de façon très différente : la lampe à krypton fournit une densité superficielle de flux à peu près uniforme, tandis que le laser fournit un faisceau approximativement gaussien (limité à 27 mm de diamètre par le diaphragme d'entrée). L'effet de cette différence serait nul si l'interféromètre était parfait ; on peut craindre que, conjugué avec les défauts de planéïté, de

réglage du parallélisme des miroirs, de focalisation de l'optique de sortie, etc., il n'entraîne des erreurs systématiques. On a vérifié que la différence de distribution du flux sur les deux faisceaux n'entraîne pas d'erreur appréciable. En obturant partiellement le faisceau laser, sans modifier le faisceau de la lampe à krypton, on a obtenu :

pour des diaphragmes circulaires de 18 et 13,5 mm de diamètre, une variation apparente de 0 et - 6 x 10^{-9}, et pour une obturation centrale sur un diamètre de 15 mm, un effet nul ;

pour des diaphragmes obturant la moitié du faisceau, un effet moyen nul, avec des différences aléatoires liées au réglage du parallélisme des miroirs (écart type d'une mesure : 2 x 10^{-9}).

4. Résultats - Exactitude

On obtient sur 16 séries de mesures (fig. 2) une valeur moyenne pour la longueur d'onde dans le vide du laser asservi :

$$\lambda = 3\ 392,231\ 376 \text{ nm}$$

avec un écart type de la moyenne σ_m = 4 fm (soit 1 x 10^{-9}). On peut estimer que les erreurs systématiques n'excèdent pas 2 x 10^{-9}.

L'exactitude peut cependant être limitée par la radiation utilisée comme étalon.

En utilisant une lampe à krypton dans les conditions recommandées, on sait en effet qu'on reproduit la longueur d'onde de la définition du mètre à 1 x 10^{-8} près. Mais on sait, avec une même lampe à krypton, répéter une même mesure à 1 ou 2 x 10^{-9} près. On sait également que, pour l'interféromètre de Michelson, la longueur d'onde apparente de la radiation étalon varie, en fonction de la différence de marche, de plusieurs 10^{-9} en raison d'un léger défaut de symétrie de la raie spectrale du krypton (5) (8). L'extrapolation à pression et courant nuls conduit également à des valeurs de la longueur d'onde dans les conditions d'emploi normales qui peuvent différer d'une lampe à l'autre de plusieurs 10^{-9} (9)(10)(11).

Si l'on a effectué cette extrapolation avec la lampe utilisée, et si l'on fait choix, par convention, de la différence de marche de 375 mm dans un interféromètre à deux ondes, on peut estimer que la lampe à krypton fournit une longueur d'onde reproductible à quelques 10^{-9} près.

C'est pourquoi nous estimons que le résultat ci-dessus devrait être exact à 5 x 10^{-9} près.

Bibliographie

1. BARGER, R.L. et HALL, J.L., Phys. Rev. Letters, 22, 4 (1969).

2. MICHELSON, A.A., Travaux et Mémoires du B.I.P.M., 11, 3 (1895).

3. Comité International des Poids et Mesures, Procès-Verbaux, 49e session, 28, 71 (1960).

4. TERRIEN, J., Optica Acta (Paris), 6, 301 (1959).

5. ROWLEY, W.R.C. et HAMON, J., Rev. Opt., 42, 519 (1963).

6. CARRE, P., Metrologia, 2, 13 (1966).

7. TERRIEN, J., J. Phys., 19, 390 (1959).

8. ROWLEY, W.R.C., Comité International des Poids et Mesures, Comité Consultatif pour la Définition du Mètre, 3e session, 21 (1962).

9. TERRIEN, J. et HAMON, J., ibid., 76 (1962).

10. ROWLEY, W.R.C., ibid., 104 (1962).

11. BRUCE, C.F. et HILL, R.M., ibid., 118 ((1962).

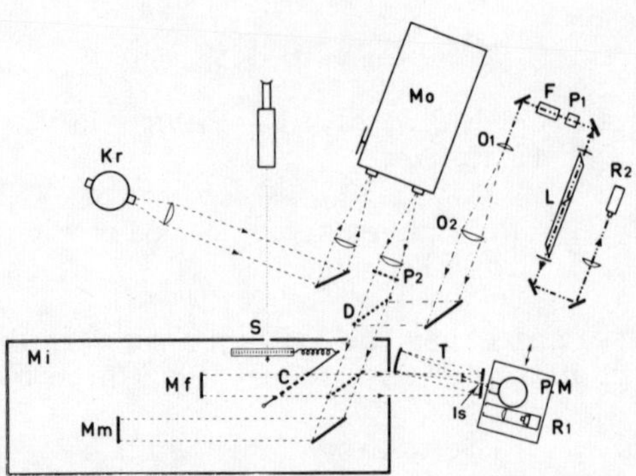

Fig. 1. Schéma optique de l'installation

C	lame compensatrice
D	miroir dichroïque
F	rotateur de Faraday
Kr	lampe à krypton 86
L	laser à $\lambda = 3{,}39$ µm
Mf	miroir fixe
Mi	interféromètre de Michelson sous vide ; diamètre des diaphragmes d'entrée et de sortie : 27 mm
Mm	miroir mobile
Mo	monochromateur à réseau
O1, O2	système optique transformateur d'étendue
P1	polariseur (glazebrook)
P2	polariseur (polaroïd)
P.M	photomultiplicateur pour la lumière visible (mesure de phase)
R1	récepteur pour $\lambda = 3{,}39$ µm (mesure de phase)
R2	récepteur pour l'asservissement du laser
S	système de commande et de lecture des petites rotations de la compensatrice
T	optique à miroirs de sortie (obturation centrale sur un diamètre de 5 mm ; diaphragme isolateur au foyer de diamètre 0,2 mm, longueur focale 0,6 m)
Is	diaphragme isolateur

λ
3392,231450 nm
400
1×10^{-8} 376
350
300
moyenne pondérée $\pm\sigma_m$

Fig. 2. Valeurs obtenues pour la longueur d'onde. Entre chaque groupe de mesures, diverses parties de l'appareil ont été modifiées (fenêtres, diaphragme isolateur, récepteur, ...). Les barres d'erreur correspondent à $\pm\ \sigma$.

Part 8 Fine Structure Constant

Review of Spectroscopic Data for Determining the Fine Structure Constant*

K. R. Lea

Department of Physics, Yale University, New Haven, Conn., U.S.A.

1. Introduction

The Sommerfeld fine structure constant α has played an important role in atomic spectroscopy ever since its introduction by Arnold Sommerfeld in 1916 (1). It originated in Sommerfeld's theory of the fine structure of the optical spectrum of atomic hydrogen. In a relativistic treatment of the Kepler problem in the Bohr theory, the quantity v_1/c (where v_1 is the velocity of the electron in the first Bohr orbit, and c is the velocity of light) was denoted by α. But its utility goes far beyond this particular case. We note the following features regarding α:

1. Dimensionless fundamental constant.
2. $\alpha = e^2/\hbar c \approx 1/137$, or, in SI units, $\alpha = (\mu_o c^2/4\pi)(e^2/\hbar c)$.
3. Basic coefficient in the electromagnetic interaction.
4. Expansion parameter in quantum electrodynamics (QED).
5. Cosmological significance? Group-theoretical interpretation?

To set the stage for a review of experimental measurements of α, Fig. 1 reminds us of the way experiment and theory are related. This figure shows the comparison between an experimentally measured quantity and the theoretically calculated value for that quantity. The experimental result depends on various standards, against which the measuring equipment is calibrated. The theoretical expression is of course dependent on the underlying principles of physics, such as the electromagnetic interaction and quantum electrodynamics, and also on numerical values of the fundamental constants, chief among which is α for the present discussion. The result of the comparison may lead to one of two conclusions: (i) "agreement," which is interpreted as experimental confirmation of theory and of the input value of α; or (ii) "disagreement", which implies a failure in one or more boxes. In this predicament, one course of action is to remove the disagreement simply by changing the value of α, i.e. deriving a new value for α. Indeed, an analysis of a considerable array of experimental data, <u>via the implied value of the fine structure constant</u>, was undertaken by Taylor, Parker, and Langenberg (2), in Tables 28 and 29 of their comprehensive review article. Today, however, I shall <u>adopt</u> that value for α which they obtained from data independent of quantum electrodynamic theory known as α_{WQED}, where $\alpha^{-1}_{WQED} = 137.03608(26)$. Experimental results

* Work supported in part by the U.S. Air Force Office of Scientific Research under AFOSR Contract No. F44620-71-C-0042.

will then be compared with theoretical numbers based on this "standard" value for α. Where discrepancies exist, it now appears likely that their origin lies not in the value for α, but rather in failures in other areas. In fact, the record of the past decade in this field shows discrepancies appearing and disappearing, with attempts being made to remove them at the various levels of boxes 2, 3, 4, and 5.

2. Methods of Measuring α

When we focus on experiments aimed at measuring α, we find that the methods of optical spectroscopy have long been dominant, giving way to microwave spectroscopy, as those techniques were developed. In the last few years, still other methods have come to the fore. Some of these different approaches are tabulated in the first column of Fig. 2. This list makes no claim to be exhaustive. The first three items, Lamb shift, fine structure, and hyperfine structure, comprise the subject I shall discuss shortly. Only a very brief mention will be made here regarding the remaining four entries in Fig. 2 which are really the province of other speakers at this conference.

The electron spin magnetic moment differs from the value of 1 Bohr magneton by a small amount (about 1 part in 10^3). This so-called "anomaly" is a quantum-electrodynamic effect, and can be calculated as a power series in α(3). Experimental measurements of the anomaly have been made on the charged leptons to as high a precision as 3 ppm (parts per million) in the case of the electron(4). These experiments _may_ be interpreted as a measurement of α, in which case excellent agreement is found with the "standard" α.

The ratio 2e/h is obtained from measurements of the current-voltage characteristic of a superconducting junction irradiated by microwave radiation. Using this a.c. Josephson effect, measurements of 2e/h have been made to a precision of 0.46 ppm (5). By a suitable combination of this result for 2e/h with other fundamental constants, a value for α has been derived good to 1.6 ppm. Here the precision is mainly limited by that of the other constants which enter the functional relationship between α and 2e/h, in particular the value of γ_p, the gyromagnetic ratio of the proton.

The use of a Josephson device in a superconducting ring which is physically rotated in a magnetic field is the basis of a method for measuring h/m (6). A precision of 6 ppm anticipated in the measurement of h/m will permit a determination of α to about 3 ppm, via the relation given in Fig. 2.

In an experiment under development at the NBS (7), the Compton wavelength of the electron is to be obtained by measuring the wavelength of the two-photon annihilation γ-rays of positronium. If the annihilation line can be measured to 1 ppm, then α can be derived to 0.5 ppm by means of the relation indicated in Fig. 2.

Returning to the earlier items on the list, namely Lamb shift, fine structure and hyperfine structure, we shall now discuss those spectroscopic experiments which have been important in determining the fine structure constant.

3. The Lamb Shift $\mathcal{S}$

The energy level diagram for the hydrogen atom given in Fig. 3 exhibits the Lamb shift interval $\mathcal{S}$, the fine structure interval ΔE, and the hyperfine structure interval $\Delta\nu$. According to the Dirac theory, the $2P_{1/2}$ and $2S_{1/2}$ levels should be degenerate. However, quantum electrodynamic theory leads to displacements of the levels, thereby raising the degeneracy. The most important contribution to the separation of $2P_{1/2}$ and $2S_{1/2}$ comes from the self energy correction to the bound electron. There are several other smaller contributions which have also been calculated. We note that the leading term contains α^3, c and R_∞.

The experimental observations of the Lamb shift date from the 1930's, when optical spectroscopic measurements, pushed to the limits of resolution set by Doppler broadening, furnished evidence that the level positions were not quite in accord with the Dirac scheme. In the late 1940's, Lamb and Retherford (8) performed a famous atomic beam experiment, in which the interval was definitely established, and was measured by radio-frequency resonance methods. Briefly, a beam of ground state hydrogen atoms was excited to the metastable 2S state, then traversed an rf field region, and subsequently impinged on a detector, sensitive to the metastable atoms. When the rf field was tuned to resonance for the electric-dipole transition $2S_{1/2} - 2P_{1/2}$, metastable atoms were quenched, returning to the ground state by spontaneous optical decay of the short lived $2P_{1/2}$ level. The corresponding decrease in metastable signal at the detector provided the means for tracing out complete resonance curves. Actually, the radio frequency was held constant, and the Zeeman effect produced by an applied, variable, magnetic field was used to scan the resonance curves. In translating the raw experimental data into the final quoted value for $\mathcal{S}$, a very remarkable and detailed analysis was undertaken. As a result, it proved possible for Triebwasser, Dayhoff and Lamb (9) to quote the value $\mathcal{S}(\mathrm{H}, n = 2) = 1057.77 \pm 0.10$ MHz which appears in Table I. The quoted "limit of error" of $\pm$ 0.1 MHz corresponds to 1 part in 1000 of the natural line width of 100 MHz. The observed width was actually greater, due to incompletely resolved hyperfine structure, and rf saturation effects. For comparison, in Table I we list the two most recent theoretical values, the earlier one of Appelquist and Brodsky (10) with a quoted "limit of error", and the new result of Erickson (11) with a one standard deviation error estimate. The difference between experimental and theoretical values is given in the last column, where we have chosen the more precise theoretical result for this comparison. Historically, the measured values of Lamb shifts have been regarded as tests of QED, and not as measurements of α. However, a knowledge of the Lamb shift is important in interpreting other experiments, to be reviewed shortly, which measure the fine structure interval $\Delta E - \mathcal{S}$. In addition, if one accepts the validity of QED theory, the new, small theoretical

TABLE I. Comparison of values of Lamb Shift interval $\mathcal{S} = n^2S_{1/2} - n^2P_{1/2}$

Atom	n	Experimental value (MHz)	Quoted error†	Ref.	Theoretical value (MHz)‡	Quoted error†	Ref.	Experiment − theory* (MHz)
H	2	1057.77±0.10	LE	9	1057.91±0.16	LE	10	−0.141±0.051
	2	1057.90±0.10	2σ	12	1057.911±0.011	1σ	11	−0.011±0.051
D	2	1059.00±0.10	LE	9	1059.17±0.22	LE	10	−0.271±0.056
	2	1059.28±0.06	1σ	13	1059.271±0.025	1σ	11	+0.009±0.065
H	3	313.6±5.7	LE	14				−1.3±2.9
	3	314.81±0.05	1σ	15	314.90±0.05	LE	10	−0.09±0.05
D	3	315.3±0.80	LE	16	315.27±0.06	LE	10	+0.03±0.27
H	4	133.18±0.59	1σ	17				+0.09±0.59
	4	$132.58^{+0.36}_{-0.45}$	1σ	18	133.09±0.02	LE	10	$-0.51^{+0.36}_{-0.45}$
D	4	133±10	LE	16	133.24±0.03	LE	10	−0.2±5
H	5	64.6±5.0	1σ	17	68.14±0.01	LE	10	−3.5±5.0
$^4He^+$	2	14 040.2±4.5	LE	19	14 044.5±5.2	LE	10	−4.6±2.3
	2	14 046.2±1.2	1σ	20	14 044.76±0.61	1σ	11	+1.4±1.3

TABLE I. (cont'd)

Atom	n	Experimental value (MHz)	Quoted error†	Ref.	Theoretical value (MHz)‡	Quoted error†	Ref.	Experiment − theory* (MHz)
$^{4}He^{+}$	3	4183.17±0.54	1σ	21	4184.3±1.6	LE	10	−1.25±0.57
					4184.42±0.18	1σ	11	
$^{4}He^{+}$	4	1751±25	LE	22				−18±13
	4	1766.0±7.5	1σ	23	1769.0±0.6	LE	10	−3.1±7.5
	4	1768.0±5.0	1σ	24	1769.087±0.076	1σ	11	−1.1±5.0
$^{6}Li^{++}$	2	63 031±327	LE	25	62 771±51	LE	10	+268±164
					62 763.4±9.1	1σ	11	
$^{12}C^{5+}$	2	$(780±8)x10^{3}$	1σ	26	$783x10^{3}$		26	$-(3±8)x10^{3}$

† LE stands for "limit of error". 1σ means one standard deviation.

‡ α^{-1} = 137.03608(26)

* For uniformity in assigning uncertainties in this column, all experimental and theoretical errors have been converted to one standard deviation (1σ); for this purpose, a limit of error was taken to be two standard deviations. Where two theoretical values are available, the one with the smaller error was used.

uncertainty of 0.011 MHz or 10 ppm in $\mathcal{S}$ would cause an uncertainty of only about 3.5 ppm in the value of α derived from it. The available experimental results given in Table I (refs. 9, 12-26) include (i) atomic beam experiments, some employing radio-frequency techniques, and others, level crossing; (ii) optical-microwave resonance experiments performed in "bottles" and in fast atomic beams; (iii) non-resonant electric field quenching methods for fast ionic beams. The last mentioned method seems at present the only feasible one for hydrogenic ions with $Z \geq 3$. The two very practical reasons for this are, firstly, the difficulty of producing, and confining a sufficient number of low-energy, one-electron ions of high Z, and secondly, the problem of generating sufficiently strong microwave electric fields at the very high frequencies called for. Nevertheless, some attempts are being made in the case of Li^{++} to observe a microwave resonance transition, but no successful results have yet been announced.

An examination of Table I shows a very satisfactory level of agreement between experiment and theory. The poorest agreement is found for the hydrogen and deuterium results of Triebwasser, Dayhoff and Lamb (9) published in 1953. For all the more recent Lamb shift data, one is led to conclude that QED has indeed been validated, a point of view also advanced by Brodsky and Drell in their recent review article (27) on the status of quantum electrodynamics.

4. The Fine Structure Intervals ΔE and ΔE-$\mathcal{S}$

Turning next to the fine structure interval ΔE, there are singularly few direct experimental measurements of this quantity. The method of optical level-crossing has been employed in two experiments on hydrogen (28,29). Changes in the resonance fluorescence pattern of atomic hydrogen are detected as a function of an external magnetic field, which is varied to bring two of the excited state Zeeman levels into coincidence. Since Professor Baird will shortly describe (29) the experiment carried out at Brown University, I need not go into further detail here. The results obtained for the fine structure interval ΔE are given in the first two lines in Table II. The comparison with the theoretical value (30) shows encouraging agreement to within one standard deviation.

The remaining entries in Table II relate to the interval ΔE-$\mathcal{S}$. The experimental results (16,17,21,24,31-34) have all been obtained using rf resonance, either in a beam or a bottle. The theoretical values (35,36) were obtained by subtracting the theoretical value for $\mathcal{S}$ from the theoretical value for ΔE, using the α_{WQED} value, as before. The three measurements claiming the highest precision are those for ΔE-$\mathcal{S}$ in hydrogen, n = 2. The first of these, by Kaufman et al. (31), came from a so-called "bottle" experiment, whereas the other two, by Shyn et al. (32), and by Cosens and Vorburger (33), were atomic beam experiments. Both methods have their strong points. In the bottle experiment, large signals can be obtained by electron bombardment of molecular hydrogen to create metastable atoms. The rf radiation is applied directly to the region where the metastables are produced. The resonant transitions are detected by the change

TABLE II. Comparisons of values of $\Delta E = n^2P_{3/2} - n^2P_{1/2}$ and of $\Delta E-\mathcal{S} = n^2P_{3/2} - n^2S_{1/2}$

Atom	n	Interval	Experimental value (MHz)	Quoted error†	Ref.	Theoretical value (MHz)‡	Quoted error†	Ref.	Experiment - theory* (MHz)
H	2	ΔE	10 969.6±0.7	1σ	28	10 969.026±0.042	1σ	30	+0.6±0.7
	2	ΔE	10 969.13±0.10	1σ	29				+0.10±0.11
H	2	$\Delta E-\mathcal{S}$	9 911.377±0.026	1σ	31	9 911.12±0.07	1σ	35	+0.262±0.050
	2	$\Delta E-\mathcal{S}$	9 911.250±0.063	1σ	32	9 911.115±0.043	1σ	36	+0.135±0.076
	2	$\Delta E-\mathcal{S}$	9 911.173±0.042	1σ	33				+0.058±0.060
D	2	$\Delta E-\mathcal{S}$	9 912.59±0.10	LE	34	9 912.85±0.08	1σ	35	-0.16±0.07
						9 912.749±0.049	1σ	36	
H	3	$\Delta E-\mathcal{S}$	2 933.5±1.2	1σ	17	2 935.19±0.02	1σ	35	-1.7±1.2
D	3	$\Delta E-\mathcal{S}$	2 935.2±0.8	LE	16	2 935.70±0.03	1σ	35	-0.5±0.4
H	4	$\Delta E-\mathcal{S}$	1 235.9±1.3	1σ	17	1 238.04±0.01	1σ	35	-2.1±1.3
	4	$\Delta E-\mathcal{S}$	$1\,237.79^{+0.23}_{-0.27}$	1σ	18				$-0.25^{+0.23}_{-0.27}$
H	5	$\Delta E-\mathcal{S}$	622.4±10.1	1σ	17	633.85±0.01	1σ	35	-11.5±10.1
$^4He^+$	3	$\Delta E-\mathcal{S}$	47 844.05±0.48	1σ	21	47 843.46±0.56	1σ	35	+0.69±0.56
						47 843.36±0.28	1σ	36	
$^4He^+$	4	$\Delta E-\mathcal{S}$	20 179.7±1.2	1σ	24	20 180.08±0.24	1σ	35	-0.3±1.2
						20 180.03±0.12	1σ	36	

†‡* See corresponding footnotes to TABLE I.

in the flux of Lyman - α radiation from this interaction region. The environment in which the metastable atoms are produced includes electrons and ions, and constitutes a relatively "dirty" setting in which to do precision measurements, as compared with the "clean" conditions prevailing in an atomic beam. Shifts of the line center are found in the bottle experiment, which depend on the electron beam current and on the gas pressure. Much of the strong signal has to be foregone in taking data at low pressures and small currents, so that an empirical extrapolation to zero values of these sources of perturbation can be made with confidence. This is by no means the only correction that has to be made to the data from the bottle experiment, but a careful examination of the possible sources of error has failed to come up with an explanation for what appears to be a serious systematic error in the measurements. In the two beam experiments, signals are small because of the relatively low density of metastable atoms in the beam. However the rf spectroscopy takes place in an environment free of electrons and ions, so no extrapolation of this sort is required. Unlike conventional atomic beam magnetic resonance machines, in which very small deflections of the beam have to be detected, the Lamb-Retherford style of beam machine selectively detects the number of metastable atoms remaining in the beam. Consequently, beams of relatively large cross section can be, and are, employed. However, the beam experiments have suffered from kinematic effects, in particular the presence of atoms with off-axis velocity components, and also the problem of the actual longitudinal velocity distribution in the beam. Only recently have these rather subtle sources of systematic error been discovered and corrected for (12). One finds (Table II) the agreement of these two beam experiments with one another, and with theory, to be good to fair. Interestingly enough, the measurement of $\Delta E - \mathcal{S}$ in the n = 2 state of deuterium by Dayhoff, Triebwasser, and Lamb (34) has not been repeated. This result was combined with the measurement of $\mathcal{S}$ by the same authors (9) to produce a value for ΔE, on which Cohen and DuMond relied rather heavily in their 1963 adjustment of the constants (37). The value for α which they obtained (α^{-1} = 137.0388(6)) has since been discarded, in view of the work by Taylor, Parker, and Langenberg (2), but in the interim, the puzzling discrepancy with the value of α derived from the hyperfine structure of hydrogen received a lot of attention. Now, instead, the results obtained by Triebwasser, Dayhoff and Lamb (9,34) have been called into question. Taylor _et al_. (30) have taken issue on the matter of discarding data. Robiscoe and Shyn (12) point out that their kinematic correction to the velocity distribution of the metastable atom beam is of less certain applicability to the apparatus of Triebwasser _et al_., although it could lead to a modest correction to the latter's results in the right direction. Clearly a case can be made for a modern remeasurement of the $\Delta E - \mathcal{S}$ interval in deuterium, in order to set the matter finally to rest.

5. The Hyperfine Structure Interval $\Delta\nu$

The next area to be discussed in spectroscopic measurements of the fine structure constant is the hyperfine structure of hydrogen. Table III shows some of the experimental results reported in the past 23 years (38-45). Unlike the situation in the Lamb shift and

TABLE III. Hyperfine Structure Interval, $\Delta\nu$, in Hydrogen

Experimental value (MHz) and quoted error	Publication date and authors	Ref.	Method
1420.410 (6)	1948, Nafe and Nelson	38	Atomic beam magnetic resonance
1420.405 73 (5)	1955, Kusch	39	Atomic beam magnetic resonance
1420.405 80 (6)	1956, Wittke and Dicke	40	Microwave absorption
1420.405 738 3 (60)	1962, Pipkin and Lambert	41	Optical pumping
1420.405 751 800 (28)	1963, Crampton *et al*.	42	Hydrogen maser
1420.405 751 785 (16)	1965, Peters and Kartaschoff	43	Hydrogen maser
1420.405 751 786 4 (17)	1966, Vessot *et al*.	44	Hydrogen maser
1420.405 751 768 (2)	1970, Hellwig *et al*.	45	Hydrogen maser

Theoretical value (reference 46)

1420.402 (7) using α^{-1} = 137.036 08 (26) (1.9 ppm)
and δ_N = −(34.4±3.1) ppm which incorporates δ_N (pol) = 0 ±3.0 ppm.

Solving for α α^{-1} = 137.035 91 (22) (1.6 ppm)

fine structure measurements, here the experimental precision has improved from 4.2 ppm in 1948 to 1.4 parts in 10^{12} in 1970. The experimental methods have also changed, as listed in the fourth column of Table III. Certainly, the hydrogen maser, developed in Professor Ramsey's laboratory at Harvard University, holds the world's record for the most precisely measured physical quantity. Among the desirable features possessed by the maser are (i) the long observation times possible for which the atoms remain in the storage bottle; (ii) the relative freedom of the atoms from perturbations inside the bottle, except, of course, for wall collisions; (iii) the fact that, since the average atomic velocity is close to zero, the first order Doppler shift is greatly reduced; and (iv) the very low noise character of maser amplification contributes greatly to the frequency stability. Unfortunately, the extraordinarily high precision obtained for $\Delta\nu$ cannot be carried over into the determination of a correspondingly precise value for α. In fact, the theoretical value (46) is only known to about the same precision as that of the 1948 experimental measurement of Nafe and Nelson (38). The theoretical precision is limited by uncertainty in the contribution of proton structure and proton polarizability to the hyperfine structure interval. These contributions are embodied in the quantity δ_N which we have taken to have the value $-(34.4 \pm 3.1)$ ppm. This incorporates a proton polarizability correction of 0 ± 3.0 ppm. This particular correction has received considerable attention in the last few years (46,47), and though no firm conclusions can be arrived at, it appears that the figure of ± 3 ppm should encompass its likely value. Accepting this, and adopting the standard value for α, we obtain $\Delta\nu$ (theoretical) = 1420.402(7) MHz. This result is seen to be in good agreement with the experimental value. As an alternative, we may take the experimental result, adopt the theoretical value for δ_N, and solve for α, with the result α^{-1} = 137.035 91 (22), which, of course, overlaps the "standard" α.

Table IV shows the high-precision measurements of hyperfine structure in the ground states and metastable states of H, D, and T, and $^3He^+$ (45,48-53). The H, D, and T ground states were measured by the hydrogen maser method; the $^3He^+$ ground state was measured using a quadrupole rf ion trap to hold the ions in place; the n = 2 metastable state hyperfine structures of H, D, and $^3He^+$ were all measured by atomic beam methods. The high precision of all these resonance experiments is achieved by virtue of the narrow line widths obtained, and the control of sources of perturbation that might shift the line centers. The theoretical values given in Table IV are written to display explicitly the nuclear correction factor, $1 + \delta$. The next-to-last column in this table lists the observed values for δ, obtained by equating the listed experimental and theoretical values, and solving for δ. Aside from the case of hydrogen, efforts to calculate δ have been none too successful, and subject to considerable variation (54,55). I have therefore not attempted to complete the last column in Table IV. Acknowledging this incompleteness in the theory, the remaining hyperfine structure measurements are not useful in determining α. However, by taking ratios, for example between $\Delta\nu$ values for n = 1 and 2 of a given isotope, the

TABLE IV. Comparison of values of Hyperfine Structure Intervals in Hydrogenic Atoms

Atom	n	Experimental value (MHz) and quoted error	Ref.	Theoretical value (MHz)†	δ(observed) (ppm)	δ(theory)* (ppm)
H	1	1420.405 751 768 (2)	45	$1420.4511(1+\delta_H)\pm0.0055$	-31.9 ± 3.9	-34.4 ± 3.1
H	2	177.556 86 (5)	48	$177.5625(1+\delta_H)\pm0.0007$	-31.8 ± 3.9	-34.4 ± 3.1
D	1	327.384 352 522 2 (17)	49	$327.3390(1+\delta_D)\pm0.0013$	$+138.5\pm3.9$	*
D	2	40.924 439 (20)	50	$40.918\ 79(1+\delta_D)\pm0.00016$	$+138.1\pm3.9$	*
T	1	1516.701 470 791 9 (71)	51	$1516.759(1+\delta_T)\pm0.006$	-37.9 ± 3.9	*
$^3He^+$	1	8665.649 867 (10)	52	$8667.50(1+\delta_{He})\pm0.03$	-213.5 ± 3.9	*
$^3He^+$	2	1083.354 99 (20)	53	$1083.586(1+\delta_{He})\pm0.004$	-213.2 ± 3.9	*

† $\alpha^{-1} = 137.03608(26)$. See ref. 2, especially sections IV B 1 and IV D, for $\Delta\nu$ in hydrogen.

* For recent discussions of theoretical values of δ, see references 47, 54, and 55. Except for hydrogen, calculations of δ are incomplete, and vary by up to 40 ppm. For $^3He^+$, see discussion by Williams, W.L., and Hughes, V.W., Phys. Rev. 185, 1251 (1969).

nuclear structure corrections cancel, and one can check the n-dependence of the radiative corrections in the theoretical expression. Alternatively, by taking the $\Delta\nu$ ratios of two different isotopes for the same n value, one gets off into the field of hyperfine structure anomalies.

It is apparent that the presence of the term in δ is a handicap to progress in determining α, but in the next atomic system to be discussed, the uncertain polarizability correction is absent. This is the situation for the muonium atom, in which the nucleus is a positive muon, which behaves like a Dirac particle, with spin 1/2. For this atomic system, the hyperfine structure computations have recently been improved by the inclusion of a higher order QED contribution (56), with the results shown in Table V. Two theoretical values are listed, which are based on different experimental values for the ratio of muon to proton magnetic moments (57,58). Despite the short lifetime of muonium of 2.2 μsec., high precision measurements of the ground state hyperfine structure have been made by two groups, at Chicago (57), and at Yale (59). The principle of these experiments involves inducing magnetic dipole transitions between pairs of levels, and detecting the occurrence of the transition through the change in angular distribution of the decay positrons. Since the muonium is formed by stopping a muon beam in an argon gas target at high pressures, the measurements have to be corrected for the pressure shift arising from this buffer gas. Recently, a quadratic dependence of this shift on pressure has been established (59). The hyperfine pressure shift for muonium conforms well with that independently measured for hydrogen and other isotopes (60). The first of the two experimental results in Table V was obtained by the Chicago group, employing a double resonance method in a magnetic field of 11.3 kG. The result from the Yale group was obtained by use of a single microwave frequency at a magnetic field of 0.01 G. The experimental results overlap, and are in good agreement with theory. One may regard this as verifying QED, or determining α. Moreover, by taking ratios with the corresponding values for $\Delta\nu$ in hydrogen, one can set an upper limit (56) to the elusive proton polarizability correction, which turns out to be consistent with the value zero adopted earlier.

6. Other Measurements and Future Outlook

Another short-lived atom that has been studied is positronium, e^+e^-. This is formed when positrons from a radioactive source are slowed down in argon gas. The large ground state fine structure interval (also referred to as the hyperfine structure interval) has not been measured directly, but has been obtained (61) by measuring a Zeeman transition in a magnetic field of about 8 kG. The transition is detected by the change in character of the annihilation γ-rays, as the static magnetic field is varied through resonance. A small extrapolation to allow for pressure shift has to be made here too. The latest experimental (61) and theoretical (62) results are shown in Table V. These agree to within one standard deviation of the experimental result (about 60 ppm), and thus confirm the underlying QED theory and Bethe-Salpeter formalism. Carlson (63) at Yale has

TABLE V. Comparison of Intervals in Other Atomic Systems

System	Interval	Experimental value (MHz)	Quoted error†	Ref.	Theoretical value (MHz)‡	Quoted error†	Ref.	Experiment - theory* (MHz)
Muonium	Hyperfine Structure	4463.302±0.009	1σ	57	4463.288±0.023	1σ	56	-0.017±0.024
		4463.311±0.012	1σ	59	4463.319±0.022	1σ	56	-0.008±0.025
Positronium	Fine Structure	203 403.±12	1σ	61	203 415±7	1σ	62	-12±14
^{4}He	$2^3P_0-2^3P_1$	29 616.864±0.036	1σ	64	29 623.047±?		68	-6.183
^{4}He	$2^3P_1-2^3P_2$	2291.72±0.36	LE	65				+3.11
		2291.200±0.022	1σ	66	2288.612±?		68	+2.588
		2291.196±0,005	1σ	67				+2.584
^{4}He	$2^3P_0-2^3P_2$	31 907.96±0.40	1σ	66	31 911.659±?		68	-3.70

†‡* See corresponding footnotes to TABLE I.

completed data taking with a new experimental apparatus with which he hopes to improve the precision to about 20 ppm. Even this still falls short of the precision of other experiments discussed earlier, so that its importance is predominantly as another test of QED, and not as a potential determination of α.

The remaining entries in Table V relate to the fine structure of the helium atom (64-67). Once again, the 2 or 3 ppm precision of the optical-microwave atomic beam measurements (64,67) cannot be exploited to yield α to $\sim$ 1 ppm because the theory (68) is insufficiently well developed. Not shown in Table V are further fine structure measurements in higher excited states of neutral helium (65,69) which are, likewise, not useful in the quest for α.

In the last few years, great strides have been made in spectroscopic measurements on muonic atoms (70). Because of its 207-fold greater mass than that of the electron, the (single) negative muon moves in a "Bohr orbit" of 207 times smaller radius. The screening effect of the remaining atomic electrons is very small, so that the muonic atom can be described as a hydrogen-like atom. Already, observations have been reported of the two-photon decay rate of the 2S metastable state of muonic helium (71), and the theory of the 2S-2P energy difference of muonic hydrogen has been worked out (72). It seems only a matter of time, and increased muon fluxes, before further significant experimental progress is made on muonic atoms. Although the outcome cannot be foreseen, the results will probably serve as tests of QED and muon-electron universality, rather than determinations of α.

In the relatively new field of beam-foil spectroscopy, observations of modulated exponential decay of excited atomic states have been made, and interpreted as quantum beats at frequencies corresponding to fine structure separations (73). These methods hold promise of measuring many fine structure intervals in H, He^+, _etc._, though as yet the precision is insufficient to challenge other methods of determining α.

In the search for greater precision in fine structure spectroscopy, ingenious ways to diminish line widths below natural width have been proposed, and in some cases demonstrated (74,15). Such line narrowing provides improved resolution and potentially improved line center location, assuming that the theory of the line shape is adequately known.

The prospect of harnessing a laser, and exploiting its unique properties as a tool in precision spectroscopic investigations is an attractive idea. However, I am not aware of any concrete proposals along these lines that pertain to fine structure measurements and the fine structure constant.

To conclude this review, we wish to draw attention to yet another way of deriving α. Wyler proposes that $\alpha = (9/8\pi^4)(\pi^5/2^4 5!)^{1/4}$, a formula obtained as the quotient of the volume elements of two rotation groups (75). The numerical value is $\alpha^{-1} = 137.036082...$ The agreement with the standard value of α^{-1} is uncanny, and prompts us to withhold any hasty, ill-informed criticism of the merits of Wyler's derivation. Instead, let posterity, as always, be the final judge!

Acknowledgements

It is a pleasure to acknowledge valuable consultations with several workers in the field, whose timely contributions have helped make this report as up to date as possible. I am also indebted to the National Science Foundation for the provision of an International Travel Grant towards transportation expenses incurred in attending this conference.

References

1. SOMMERFELD, A., Annalen der Physik (Leipzig) 51, 1 (1916). See also SOMMERFELD, A., Atombau und Spektrallinien (Friedr. Vieweg & Sohn, Braunschweig 1919) pp. 244, 331.
2. TAYLOR, B.N., PARKER, W.H., and LANGENBERG, D.N., Rev. Mod. Phys. 41, 375 (1969).
3. LEVINE, M.J., and WRIGHT, J., Phys. Rev. Lett. 26, 1351 (1971).
4. RICH, A., paper presented at this conference. WESLEY, J., and RICH, A., Phys. Rev. Lett. 24, 1320 (1970).
5. FINNEGAN, T.F., DENENSTEIN, A., and LANGENBERG, D.N., Phys. Rev. Lett. 24, 738 (1970). See also papers presented at this conference.
6. PARKER, W.H., and SIMMONDS, M., Proceedings of the International Conference on Precision Measurement and Fundamental Constants, edited by D.N. Langenberg and B.N. Taylor (National Bureau of Standards special publication 343, U.S. Government Printing Office, Washington, D.C., 1971).
7. SAUDER, W.C., Proceedings cited in ref. 6.
8. LAMB, W.E., Jr., and RETHERFORD, R.C., Phys. Rev. 79, 549 (1950).
9. TRIEBWASSER, S., DAYHOFF, E.S., and LAMB, W.E., Jr., Phys. Rev. 89, 98 (1953).
10. APPELQUIST, T., and BRODSKY, S.J., Phys. Rev. Lett. 24, 562 (1970).
11. ERICKSON, G.W., to be published and private communication.
12. ROBISCOE, R.T., and SHYN, T.W., Phys. Rev. Lett. 24, 559 (1970). ROBISCOE, R.T., Phys. Rev. 168, 4 (1968). ROBISCOE, R.T., and COSENS, B.L., Phys. Rev. Lett. 17, 69 (1966).
13. COSENS, B.L., Phys. Rev. 173, 49 (1968). VORBURGER, T.V., and COSENS, B.L., paper in Proceedings cited in ref. 6.
14. KLEINPOPPEN, H., Z. Physik 164, 174 (1961).
15. FABJAN, C.W., and PIPKIN, F.M., Phys. Letters 36A, 69 (1971).
16. WILCOX, L.R., and LAMB, W.E., Jr., Phys. Rev. 119, 1915 (1960).
17. FABJAN, C.W., PIPKIN, F.M., and SILVERMAN, M., Phys. Rev. Lett. 26, 347 (1971).
18. BROWN, R.A., and PIPKIN, F.M., BUll. Am. Phys. Soc. Ser. II, 16, 85 (1971). See also Proceedings cited in ref. 6.
19. LIPWORTH, E., and NOVICK, R., Phys. Rev. 108, 1434 (1957).
20. NARASIMHAM, M.A., and STROMBOTNE, R.L., Phys. Rev. A4, 14 (1971).
21. MADER, D.L., LEVENTHAL, M., and LAMB, W.E., Jr., Phys. Rev. A3, 1832 (1971).
22. BEYER, H.J., and KLEINPOPPEN, H., Z. Physik 206, 177 (1967).

23. HATFIELD, L.L., and HUGHES, R.H., Phys. Rev. 156, 102 (1967).
24. JACOBS, R.R., LEA, K.R., and LAMB, W.E., Jr., Phys. Rev. A3, 884 (1971).
25. FAN, C.Y., GARCIA-MUNOZ, M., and SELLIN, I.A., Phys. Rev. 161 6 (1967).
26. LEVENTHAL, M., and MURNICK, D.E., Phys. Rev. Lett. 25, 1237 (1970). LEVENTHAL, M., private communication.
27. BRODSKY, S.J., and DRELL, S.D., Annual Rev. Nucl. Sci. 20, 147 (1970).
28. WING, W.H., Ph.D. Thesis, Univ. of Michigan (1968), unpublished.
29. BAIRD, J.C., BRANDENBERGER, J.R., GONDAIRA, K-I., and METCALF, H., Phys. Rev. (to be published). See also BAIRD, J.C., paper presented at this conference.
30. Ref. 2, section IV.C.2.
31. KAUFMAN, S.L., LAMB, W.E., Jr., LEA, K.R., and LEVENTHAL, M., Phys. Rev. Lett. 22, 507 (1969).
32. SHYN, T.W., REBANE, T., ROBISCOE, R.T., and WILLIAMS, W.L., Phys. Rev. A3, 116 (1971).
33. COSENS, B.L., and VORBURGER, T.V., Phys. Rev. A2, 16 (1970).
34. DAYHOFF, E.S., TRIEBWASSER, S., and LAMB, W.E., Jr., Phys. Rev. 89, 106 (1953).
35. The theoretical value for $\Delta E-\mathcal{S}$ was obtained by subtracting $\mathcal{S}$ (given in ref. 10) from ΔE (given in ref. 30).
36. The theoretical value for $\Delta E-\mathcal{S}$ was obtained by subtracting $\mathcal{S}$ (given in ref. 11) from ΔE (given in ref. 30).
37. COHEN, E.R., and DUMOND, J.W.M., Rev. Mod. Phys. 37, 537 (1965).
38. NAFE, J.E., and NELSON, E.B., Phys. Rev. 73, 718 (1948).
39. KUSCH, P., Phys. Rev. 100, 1188 (1955).
40. WITTKE, J.P., and DICKE, R.H., Phys. Rev. 103, 620 (1956).
41. PIPKIN, F.M., and LAMBERT, R.H., Phys. Rev. 127, 787 (1962).
42. CRAMPTON, S.B., KLEPPNER, D., and RAMSEY, N.F., Phys. Rev. Lett. 11, 338 (1963).
43. PETERS, H.E., and KARTASCHOFF, P., Appl. Phys. Lett. 6, 35 (1965).
44. VESSOT, R.H., et al., IEEE Trans. Instr. Meas. IM-15, 165 (1966).
45. HELLWIG, H., et al., IEEE Trans. Instr. Meas. IM-19, 200 (1970).
46. Ref. 2, secs. IV.B.1 and IV.D. However, we take $\delta_N{}^2 = 0 \pm 3$ ppm.
47. DRELL, S.D., and SULLIVAN, J.D., Phys. Rev. 154, 1477 (1967).
48. HEBERLE, J.W., REICH, H.A., and KUSCH, P., Phys. Rev. 101, 612 (1956).
49. WINELAND, D.J., Ph.D. Thesis, Harvard University (1970), unpublished. See also CRAMPTON, S.B., ROBINSON, H.G., KLEPPNER, D., and RAMSEY, N.F., Phys. Rev. 141, 55 (1966).
50. REICH, H.A., HEBERLE, J.W., and KUSCH, P., Phys. Rev. 104, 1585 (1956).
51. MATHUR, B.S., CRAMPTON, S.B., KLEPPNER, D., and RAMSEY, N.F., Phys. Rev. 158, 14 (1967).
52. SCHUESSLER, H.A., FORTSON, E.N., and DEHMELT, H.G., Phys. Rev. 187, 5 (1969).
53. NOVICK, R., and COMMINS, E.D., Phys. Rev. 111, 822 (1958).
54. FOLEY, H.M., Atomic Physics, edited by B. Bederson, V.W. Cohen, and F.M.J. Pichanick (Plenum Press, New York, 1969), p. 509.
55. IDDINGS, C., Physics of the One-and Two-Electron Atoms, edited by F. Bopp and H. Kleinpoppen (North-Holland Publishing Co., Amsterdam, 1969), p. 203.

56. FULTON, T., OWEN, D.A., and REPKO, W.W., Phys. Rev. Lett. 26, 61 (1971).
57. DEVOE, R. et al., Phys. Rev. Lett. 25, 1779 (1970).
58. HAGUE, J.F. et al., Phys. Rev. Lett. 25, 628 (1970).
59. CRANE, T. et al., Phys. Rev. Lett. (to be published). See also Proceedings cited in ref. 6.
60. ENSBERG, E.S., and MORGAN, C.L., Proceedings cited in ref. 6.
61. THERIOT, E.D., BEERS, R.H., and HUGHES, V.W., Phys. Rev. Lett. 18, 767 (1967).
62. FULTON, T., OWEN, D.A., and REPKO, W.W., Phys. Rev. Lett. 24, 1035 (1970), and 25, 782 (1970).
63. CARLSON, E.R., private communication, and paper included in Proceedings cited in ref. 6.
64. KPONOU, A., HUGHES, V.W., JOHNSON, C.E., LEWIS, S.A., and PICHANICK, F.M.J., Phys. Rev. Lett. 26, 1613 (1971).
65. WIEDER, I., and LAMB, W.E., Jr., Phys. Rev. 107, 125 (1957).
66. LIFSITZ, J.R., Ph.D. Thesis, University of Michigan (1965), unpublished.
67. PICHANICK, F.M.J., SWIFT, R.D., JOHNSON, C.E., and HUGHES, V.W., Phys. Rev. 169, 55 (1968). LEWIS, S.A., PICHANICK, F.M.J., and HUGHES, V.W., Phys. Rev. A2, 86 (1970).
68. See Table I of ref. 64.
69. DESCOUBES, J.P., text cited in ref. 55, p. 341.
70. WU, C.S., text cited in ref. 55, p. 429. WU, C.S., and WILETS, L., Annual Rev. Nucl. Sci. 19, 527 (1969).
71. PLACCI, A., et al., Il Nuovo Cimento 1A, 445 (1971).
72. DI GIACOMO, A., Nucl. Phys. B11, 411 (1969); B23, 641 (1970).
73. ANDRA, H.J., Phys. Rev. Lett. 25, 325 (1970). BURNS, D.J., and HANCOCK, W.H., Phys. Rev. Lett. 27, 370 (1971).
74. ZU PUTLITZ, G., Comments At. Mol. Phys. 1, 74 (1969).
75. WYLER, A., Acad. Sci. Paris, Comptes Rendus 269A, 743 (1969). See also Physics Today, Aug. 1971, p. 17.

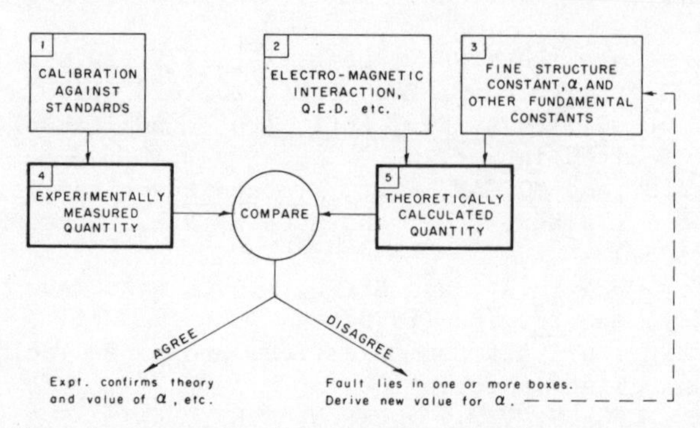

Fig. 1. Flow diagram showing inter-relation of experiment and theory in determining α.

REVIEW of METHODS for DETERMINING α

NAME	SYMBOL	SYSTEM	RELATION
Lamb Shift	S	H, D, He^+, Li^{++}	$S = \frac{8}{3\pi} \frac{\alpha^3 Z^4 c R_\infty}{n^3} \left(\ell n \frac{1}{(Z\alpha)^2} + \cdots \right)$
Fine Structure	ΔE	H, D, He^+, He	$\Delta E = \frac{(\alpha Z)^2 Z^2 c R_\infty}{2n^3} (1 + \cdots)$
Hyperfine Structure	$\Delta\nu$	$H, D, T, He^+, \mu^+e^-, e^+e^-$	$\Delta\nu(H) = \frac{16}{3} \frac{\alpha^2 c R_\infty}{n^3} \frac{\mu_p}{\mu_0} (1 + \cdots)$
"Anomalous" Magnetic Moment	$a = \frac{g-2}{2}$	e^-, e^+, μ^-, μ^+	$a = \frac{\alpha}{2\pi} - 0.328 \left(\frac{\alpha}{\pi}\right)^2 + \cdots$
ac Josephson Effect	$\frac{2e}{h}$	Various Superconducting Junctions	$\alpha^{-1} = \left[\frac{1}{4R_\infty} \frac{c\Omega_{abs}}{\Omega_{NBS}} \frac{\mu_p'}{\mu_0} \frac{1}{\gamma_p'} \frac{2e}{h} \right]^{1/2}$
Rotating Superconducting Junction	$\frac{h}{m}$	Sn or Al	$\alpha = \left[\frac{2R_\infty}{c} \left(\frac{h}{m}\right) \right]^{1/2}$
Compton Wavelength of the Electron	$\lambda_c = \frac{h}{mc}$	e^+e^-	$\alpha = \left[2R_\infty \left(\frac{h}{mc}\right) \right]^{1/2}$

Fig. 2. Summary of several different methods and atomic systems from which values of the fine structure constant may be derived.

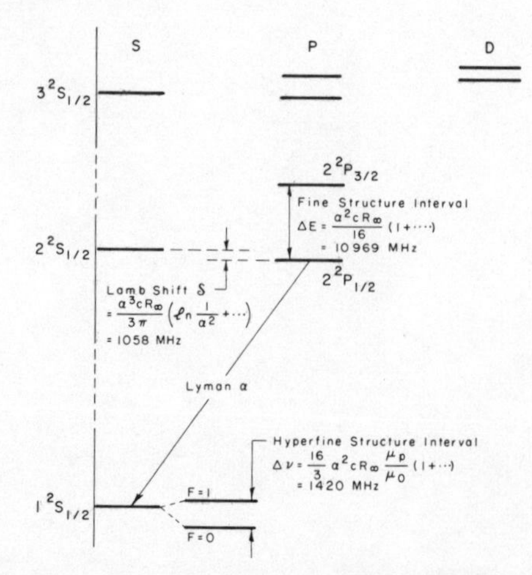

Fig. 3. Schematic energy level diagram showing some pertinent levels for the hydrogen atom. (Not to scale).

Muonium HFS Splitting as a source of Alpha*

D. Favart, P. M. McIntyre, D. Y. Stowell, R. A. Swanson,
V. L. Telegdi and R. DeVoe

University of Chicago, Chicago, Illinois

The purpose of this report is twofold, viz. (1) to review the information concerning α which one can derive, both in principle and in practice, from precision measurements of the hyperfine splitting, $\Delta\nu$, of muonium; (2) to describe a novel technique for measuring $\Delta\nu$ to high accuracy which we have recently developed at Chicago.

(1) <u>The hfs splitting of muonium</u>

We may write

$$\Delta\nu \sim \alpha^2 \text{ Ryc } g'_\mu \quad \{\text{QED}\} \qquad (1)$$

where only the relevant quantities have been exhibited; $g'_\mu \equiv g_\mu\,(m_e/m_\mu)$, and $\{\text{QED}\}$ are relativistic and quantum electrodynamics corrections to the simple Fermi formula. For the purposes of this discussion we shall assume (Ryc) and $\{\text{QED}\}$ to be known perfectly; the theoretical uncertainties in the latter quantity are presumably below the 1 ppm level.

The first difficulty in deriving α from $\Delta\nu$ stems from uncertainties in g'_μ. Whereas $g_\mu = 2(1 + a_\mu)$ is known to 0.3 ppm from the measurement of a_μ at CERN, direct measurements of m_e/m_μ are far too inaccurate ($\sim$ 100 ppm) to be of use. The information on g'_μ comes from measurements of the muon magnetic moment, μ_μ. These measurements determine either (a) f_μ/f_p, the ratio of muon and proton precession frequencies, or (b) $g_j(M)/g'_\mu(M)$, the ratio of $g_j(M)$ - the g-factor of the bound electron in muonium (M)-to g'_μ, also in muonium.

Measurements of f_μ/f_p were until recently not only somewhat inaccurate ($\sim$ 9 ppm), but also afflicted with a serious systematic uncertainty. The quantity f_μ/f_p equals μ_μ/μ_p only if the diamagnetic shielding corrections in the substance wherein the precession frequencies are compared (say, H_2O) are equal for protons and muons. Ruderman[1] had in fact suggested that these corrections were <u>different</u>; his considerations were motivated by the desire to bring the value of α derived from the early Yale determinations[2] of $\Delta\nu(M)$ in line with the value from $\Delta\nu$(hydrogen). Recent work of the Seattle-LRL group[3] has not only greatly increased the accuracy of f_μ/f_p, but has shown experimentally that the "chemical" shifts are much smaller than that suggested by Ruderman. The result of Ref. [3] is

$$f_\mu/f_p = 3.183\ 347\ (9) \qquad (3\text{ ppm}) \qquad (2)$$

*Research supported by NSF GP 29575 (Research).

A determination of $g_j(M)/g'_\mu(M)$ in Kr was performed by the Chicago group[4], through a measurement of two Zeeman frequencies in M. It turns out that $g_j(M)$ is affected by the collisions between M and the host gas atoms ("g_j-shift"), a phenomenon already studied for other atoms by optical pumping. Inasmuch as no direct determination of this shift is available, we adopted a semiempirical calculation[5] of the requisite correction (- 11 ppm), with the result

$$g_e/g'_\mu = 206.76282\ (8) \qquad (4\ \text{ppm}) \tag{3}$$

corresponding to

$$f_\mu/f_p = 3.183337\ (13) \tag{3'}$$

in excellent agreement with (2). No allowance for the uncertainties in the correction just mentioned is however made in (3); a generous allowance would be to assume that the correction lies between -5 and - 11 ppm.

By extending the novel technique to be described below to high field transitions, it will soon be possible to determine the g_j-shift directly and eliminate this uncertainty entirely.

A second source of error stems from the fact that while the vacuum value, $\Delta\nu(0)$, is of direct interest, experiments measure actually always $\Delta\nu(p)$, i.e. the hfs splitting at some finite pressure p. These quantities can be related by

$$\Delta\nu(p) = \Delta\nu(0)\ (1 + a\ p + bp^2 + \ldots) \tag{4}$$

where a is the customary fractional pressure shift coefficient (FPS), already well known for many atoms in various host gases from optical pumping studies. In these studies the pressure is generally so low (< 100 Torr) that the quadratic term b can well be neglected, and $\Delta\nu(0)$ obtained by linear extrapolation to zero density. Such an extrapolation procedure was applied by the Yale group to their muonium experiments, although these were carried out at high pressures, and lead to differing values for $\Delta\nu(0)$ from determinations in Ar and Kr[6] as well as to a-values which were different from those observed for hydrogen in Ar and Kr, although both theoretically and experimentally (from comparisons with D and T) no isotope shift of a is expected.

The Chicago group has performed a series of experiments[4,7,8] at sufficiently low pressures to keep the quadratic shift at a negligible level. Low pressures imply a small number of muon stops in the gas target, and the special techniques were developed to counterbalance the loss in count-rate by greater statistical power. By combining the low pressure data from Chicago and the (predominantly) high pressure data from Yale the coefficient b can actually be determined for both Ar and Kr, and it becomes evident that a linear extrapolation of the Chicago data does not affect $\Delta\nu(0)$ appreciably.

Very recently; the Yale group has obtained low-pressure data in Ar and accepted the existence of a quadratic shifts,[9] in contrast to their earlier point of view[10]. Thus the pressure-shift controversy is resolved; Fig. 1 summarizes the Yale and Chicago results obtained to date.

The novel technique which we shall briefly describe below yields (from a linear extrapolation of data at 2742 and 6949 Torr)

$$\Delta\nu(0,\mathrm{Kr}) = 4463.30133(395)\ \mathrm{MHz} \qquad (0.88\ \mathrm{ppm}) \qquad (5)$$

corresponding, with the moment as given by (2), to

$$\alpha^{-1} - 1 = 0.03638(19) \qquad (1.7\ \mathrm{ppm}) \qquad (6)$$

We note that (5) is in good agreement with the earlier Chicago result[4] [4463.3022(89) MHz] and with the most recent Yale value [4463.311(12)MHz]. It is hence legitimate to continue to fit the various data jointly.

<u>Chicago Kr data, linear fit</u>

$$\Delta\nu(0,\mathrm{Kr}) = 4463.3012(23)\ \mathrm{MHz} \qquad (0.5\ \mathrm{ppm}) \qquad (7)$$

$$a(\mathrm{Kr}) = -10.37(07)\ 10^{-9}\ \mathrm{Torr}^{-1} \qquad (8)$$

<u>Chicago and Yale Kr data, quadratic fit</u>

$$\Delta\nu(0,\mathrm{Kr}) = 4463.3028(25)\ \mathrm{MHz} \qquad (9)$$

$$a(\mathrm{Kr}) = -\ 10.56(9) \times 10^{-9}\ \mathrm{Torr}^{-1} \qquad (10)$$

$$b(\mathrm{Kr}) = 7.8\ (1.8) \times 10^{-15}\ \mathrm{Torr}^{-2} \qquad (11)$$

<u>Chicago and Yale Ar data, quadratic fit</u>

$$\Delta\nu(0,\mathrm{Ar}) = 4463.3075(57) \qquad (12)$$

$$a(\mathrm{Ar}) = -4.941(17) \times 10^{-9}\ \mathrm{Torr}^{-1} \qquad (13)$$

$$b(\mathrm{Ar}) = 7.8\ (2.1) \times 10^{-15}\ \mathrm{Torr}^{-2} \qquad (14)$$

<u>Chicago and Yale data, quadratic fit to Ar and Kr,</u>

<u>constraining $\Delta\nu(0)$ to give same value</u>

→

$$\Delta\nu(0) = 4463.3044(23)\ \mathrm{MHz} \qquad (0.5\ \mathrm{ppm}) \qquad (15)$$

$$a(\mathrm{Ar}) = -4.89(9) \times 10^{-9}\ \mathrm{Torr}^{-1} \qquad (16)$$

$$a(\mathrm{Kr}) = -10.61(8) \times 10^{-9}\ \mathrm{Torr}^{-1} \qquad (17)$$

$$b(\mathrm{Ar}) = 6.9(1.2) \times 10^{-15}\ \mathrm{Torr}^{-2} \qquad (18)$$

$$b(\mathrm{Kr}) = 8.6(1.7) \times 10^{-15}\ \mathrm{Torr}^{-2} \qquad (19)$$

It is seen that the "world average" (15) is no more accurate, and in substantial agreement with, the latest Chicago result, (7).

Fig. 2 shows all relevant data points and their extrapolation (9), (10) to zero density.

(2) Novel technique for measuring $\Delta\nu$

This technique, fully discussed in Ref. [8], is a modification of the zero-field field resonance technique of the Yale group. Because of its 3-level nature, that resonance has an unusually large width Γ for a reasonable saturation ($\Gamma \simeq 500$ kHz as compared to the "natural" line width of $\Gamma_0 \simeq 145$ kHz). The new method bears the same relation to the straightforward (Yale) resonance as the Rabi method in atomic beams does to its well-known modification by Ramsey. The muonium atoms are subjected to two coherent microwave pulses (lengths τ) separated by an interval T, and the muon polarization $\vec{P}$ is observed (through the positron decay asymmetry) at $t > T + 2\tau$. The relative phase of the two pulses is shifted (following Ramsey and Silsbee) by $\pm \pi/2$ for each successive muon stop, and the change in $\vec{P}$ is compared. Fig. 3 shows a typical curve obtained[8] by this approach. The improvement in line width is evident; the net gain in statistical power, as compared to the straightforward approach, is about a factor 25. It should be possible to push the accuracy of the result much farther in the near future.

References

1. RUDERMAN, M. A., Phys. Rev. Lett. 17, 794 (1966).

2. CLELAND, W. E., et al., Phys. Rev. Lett. 13, 202 (1964).

3. HAGUE, J. F., et al., Phys. Rev. Lett. 25, 628 (1970).

4. DEVOE, R., et al., Phys. Rev. Lett. 25, 1779b (1970).

5. HERMAN, R. M., private communications.

6. THOMPSON, P. A., et al., Phys. Rev. Lett. 22, 163 (1969).

7. EHRLICH, R. D., et al., Phys. Rev. Lett. 23, 513 (1969).

8. FAVART, D., et al., Phys. Rev. Lett. (in press)

9. CRANE, T., et al., Phys. Rev. Lett. 27, 474 (1971).

10. CRANE, P., et al., in High Energy Physics and Nuclear Structure, edited by S. Devons (Plenum Press, New York, 1970) p.677.

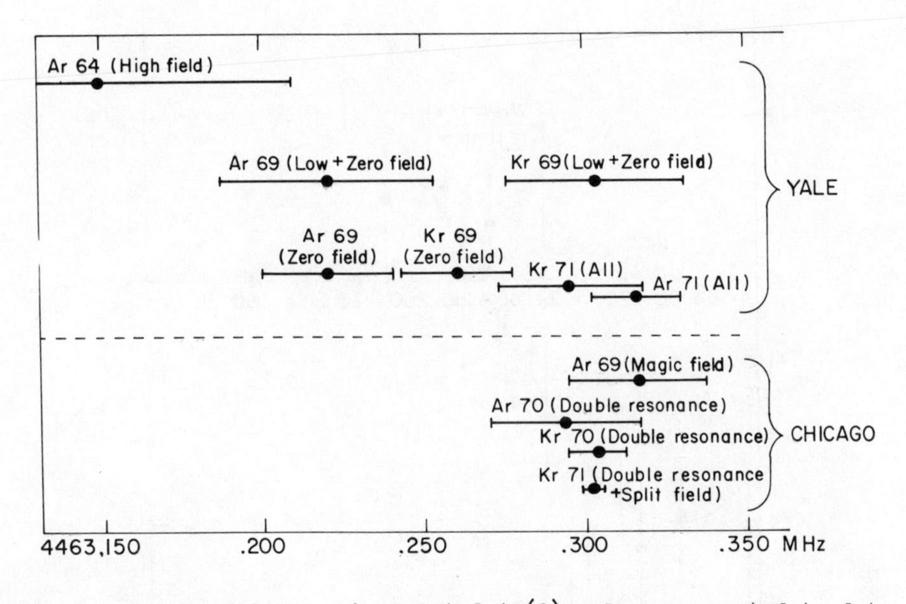

Fig. 1. Summary of all extrapolated $\Delta\nu(0)$ values presented to date.

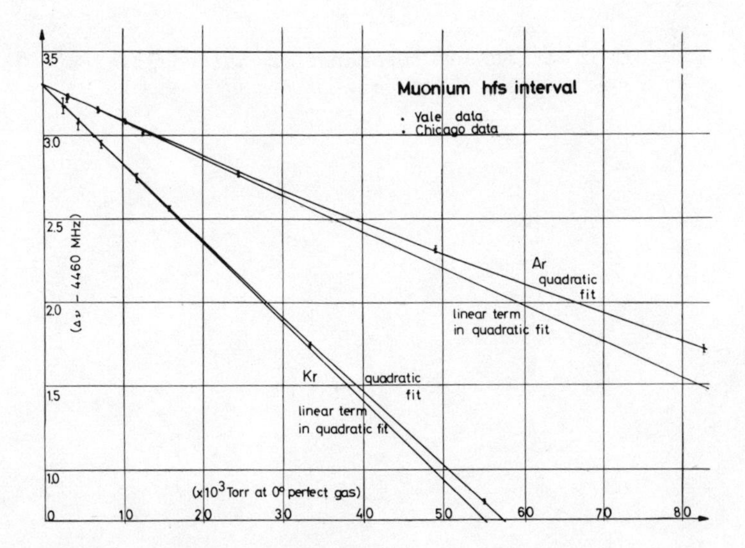

Fig. 2. Quadratic fits to all $\Delta\nu(p)$ data in Ar and Kr. Straight lines corresponding to the linear terms alone shown for comparison purposes.

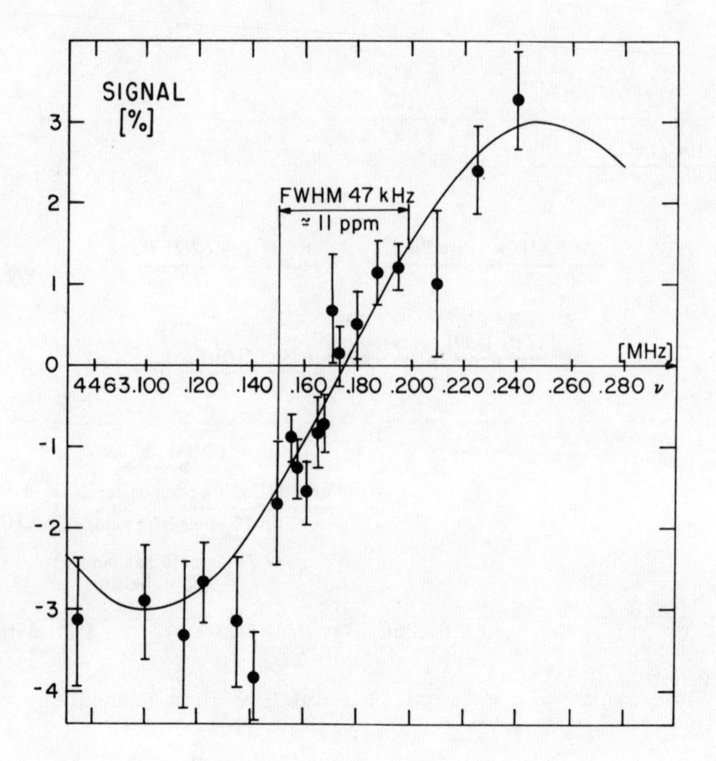

Fig. 3. Result of a "Ramsey resonance" obtained in Kr at 2742 Torr by Favart et al. (Ref. 8).

Electron Spectroscopy and the Fine Structure Constant

J. C. E. Jennings

Birkbeck College, London

1. Introduction

In the absence of ferromagnetic materials and of stray magnetic fields, the magnetic induction in a system of coils carrying a current I is $\underline{B} = I\,\underline{F}\,(x,y,z)$. An electron moves through a magnetostatic field with a constant scalar momentum p, and its trajectory in the field of a coil depends only on the initial conditions and on I/P, where $P = (B\rho) = p/e$ and e is the charge. In the absence of stray fields, therefore, any kind of iron-free spectrometer serves as a relative spectrometer, i.e. it enables P to be determined in terms of I, the required focusing current, and of an instrumental constant.

There are two well-known methods for finding this instrumental constant. In Method 1 the current is found which is necessary to focus electrons of known momentum, e.g. an internal conversion line whose momentum has already been determined in an absolute spectrometer. In Method 2, due to K. Siegbahn (1), the currents are determined which are required to focus electrons arising from a given γ-transition converted in either of 2 different levels of the same atom, e.g. the 239 keV γ-ray converted in the K and L1 levels of Th C(^{212}Bi). Suppose that I is the current for focusing the K-conversion electrons and P their $(B\rho)$ momentum, while kI is the current for the L1 conversion electrons, so that kP is their $(B\rho)$ momentum. Let ν be the wavenumber difference (known from the X-ray emission spectrum) between the K and L1 levels; then

$$h\nu/mc = \sqrt{1 + (keP/mc)^2} - \sqrt{1 + (eP/mc)^2}. \qquad (1)$$

Taking the left-hand side and also e/mc as known, equation (1) is an equation for P in terms of k; this is the basis of Method 2.

The purpose of the present paper is to draw attention to the possibility of using this equation to contribute to knowledge of $\lambda_c = h/mc$, the Compton wavelength of the electron (and hence of the fine-structure constant), provided that not only k is determined experimentally (relative spectrometry) but also P (absolute spectrometry). In fact the wavenumber difference is known in terms not of milliångströms but of x-units, so that, if ν' is its value in x.u.$^{-1}$ (times the appropriate power of 10, according as we work in MKSU or CGS EMU) then the working equation is:

$$\lambda_c'\,\nu' = (\lambda_c/\Lambda)\,\nu' = \sqrt{1 + (keP/mc)^2} - \sqrt{1 + (eP/mc)^2} \qquad (2)$$

where Λ is the length of the x.u. in terms of the mÅ, so that $\lambda_c' = \lambda_c/\Lambda$ is the Compton wavelength of the electron, measured in x.u.

2. Siday's Absolute Spectrometer

Absolute electron spectrometry has hitherto been based on the use of a uniform magnetic field, either normal to the plane of the principal ray (2) or substantially along the principal ray (3). A uniform field may be measured accurately in terms of the gyromagnetic ratio of the proton, and absolute spectrometry has been carried out in this way in a transverse field by Lindström (4) and in a longitudinal field by Jungerman et al.(5).

However, in these instruments a deviation from uniformity of the field produces a first-order effect upon the resulting determination of P. The late R.E. Siday (6) proposed a magnetic spectrometer whose accuracy as an absolute instrument does not depend on a detailed knowledge of the field distribution; it is based on the image rotation produced by a magnetic lens. Craig and Dietrich used this instrument both relatively (7,8) (to an accuracy of about 50 ppm) and absolutely (8,9) (to an accuracy of about 1 in 700). Since Siday's death in 1957, the work has been continued by Das-Gupta (10), R.J. O'Connor and the present author (unpublished).

The spectrometer (Fig.1) is a magnetic lens used at unit magnification, with an aperture in the mid-plane so small that the transverse spherical aberration is slightly less than the object width. The axis of the spectrometer is horizontal and lies in the magnetic meridian. The vertical component of the Earth's field is compensated by means of auxiliary coils. The source (usually the active deposit from emanating radiothorium) is carried on a wire, about 3 cm long and 25 µm in diameter, which intersects the axis of the lens perpendicularly. There are two solenoids, the inner having 29 213 turns and the outer 17 081; they are used either singly or in series, so that three independent determinations may be made. With any one lens, the current I_1 is found that is needed to focus onto the photographic plate the image due to a particular conversion line of (Bρ)/momentum P. The current is then reversed and requires to be set at a slightly different value, I_2; the difference between I_1 and I_2 arises chiefly from the effect of the horizontal component of the Earth's field, but partly also from the axial astigmatism of the lens, which is wound with imperfect rotational symmetry (10,12). To the paraxial approximation, the angle θ between the image lines is given by

$$\theta P = \int_{z_0}^{z_i} B\,dz \qquad (3)$$

where z_o and z_i are the axial coordinates of object and image. For an air-cored coil of n turns

$$\int_{z_0}^{z_i} B\,dz = \tfrac{1}{2}\,(I_1 + I_2)\,\mu_o\,(4\pi/\alpha)\,n\,(1 - \delta) \qquad (4)$$

where $\alpha = 4\pi$ in rationalized units, $\alpha = 1$ in unrationalized units.

An important feature of this absolute spectrometer is that object and image are substantially outside the field, so that $\delta \ll 1$, viz. for the outer coil $\delta \simeq 0.01$, for the inner coil $\delta \simeq 0.006$. In consequence the accuracy required in δ is some 100 times less than that desired for P. As δ is fairly insensitive to the turns distribution, it was calculated by Das-Gupta (10) from the geometry of the coil. O'Connor (11) has subsequently carried out the necessary search coil measurements to determine by numerical integration the value of $\int_{-\infty}^{z_0}(B/I)\,dz + \int_{z_i}^{\infty}(B/I)\,dz$, and hence to determine δ. As the remote field of the lens is weak, a search coil of very large area is needed. This was made by Das-Gupta from a design of the late H.R. Nettleton (13) (Fig.2), which is such that the flux linkage is independent of all derivatives of the field at its centre of order lower than B^{iv}; for the given problem, this and higher terms are known to be negligible.

The chief source of random error in P lies in the measurement of θ (eqn.3), which is about 160°, or 10^4 minutes of arc. Das-Gupta (10) used a goniometric microscope designed by C.F.Dietrich(14) the standard errors of determinations of θ on various plates (for the K-conversion line) lay between 6 and 8 ppm and the range of values of P for a total of 4 pairs of plates using the three different lenses was 3.6 ppm, indicating that very little systematic error was associated either with a change of lens or with a small change of focusing current.

To summarize, the special features of this spectrometer are

(1) the instrumental constant is a definite integral of the magnetic induction and is very weakly dependent on the form of the field;

(2) the Earth's field, if temporally constant during an exposure, has no effect on the result and

(3) although the magnitudes of the currents used for focusing must be accurately known, no first-order effect arises from their not being precisely the correct values for focusing, i.e. as I varies, I/θ stays constant.

3. Sources of Error

The lens current (70 to 200 mA) passes through a standard resistance (1 or 10 ohms), the voltage across which is compared potentiometrically with the mean e.m.f. of a group of Weston cells. We found (15), when the potentiometer current was stabilised, some anomalies in the relative e.m.f's. of the cells; it turned out that the stabiliser produced a.c. ripple which was rectified to different extents by different cells - a source of systematic error of which there is too little awareness among workers to whom absolute electrical accuracy is essential.

Evidently, for absolute values of P, I (known initially in NPL amperes) must be expressed in absolute units. However, eqn.2 for λ_c' contains P only in the combination eP/mc. Knowledge of e/m comes, at present, chiefly from the weak-field value of γ_p and it is worth noting that if this is expressed in terms of the same 'local' ampere (i.e. of the same place and vintage!) as that used for P, the ratio of this unit to the absolute ampere is eliminated.

Sources of systematic error that have been or are being investigated include the following:

(1) Electron-optical aberrations characteristic of a geometrically perfect coil have been fully described by Dietrich (9). We consider especially

(a) anisotropic distortion, accurately calculable and allowed for;

(b) anisotropic coma, also calculable for each annulus of the aperture; but the simultaneous presence of spherical aberration reduces its effect. This is being studied both experimentally (by varying the aperture) and computationally;

(c) the combined effect of spherical aberration and natural line width. Dijkstra and de Vries (16), for example, found the K-conversion line of the 239 keV transition in ^{212}Bi to have an energy width of about 60 eV, compared with 10 eV for the corresponding L1 line. Due to the spherical aberration of our lens spectrometer, the more energetic electrons of a broad line will produce a more intense image than the less energetic ones (Fig. 3); hence the angle θ of eqn.3 will be characteristic of an energy slightly higher than the mean. Since the K-line is so much broader than the L1 line, the ratio k between their measured momenta will be too small. We have devised the following method for obtaining a value of k which is not biased by the difference in the natural widths: the narrower line is artifically broadened by varying the focusing current appropriately during its exposure, so that it matches the broader line. For this purpose, these widths need to be known from high-resolution studies. The technique appears to be applicable in any iron-free spectrometer.

(2) Electron-optical aberrations arising from imperfect winding of the lens coil have been studied by Amboss and Jennings (12), especially

(a) anamorphism, i.e. a variation in magnification with orientation of the source. This effect is linearly proportional to the quadrupole field of the lens. It can be eliminated by taking the average of the rotations produced for two perpendicular settings of the source.

(b) Misrotation. If some of the turns of the coil are tilted, there will be a component of field perpendicular to the axis of the solenoid. Eqn.3 is then in error by an amount proportional to the square of this small effect. The amount has been estimated (12) at 2 or 3 ppm. O'Connor (11) has Fourier-analysed Hall-probe measurements of the transverse field and is determining the misrotation by numerical ray tracing. A search-coil technique is being developed (17) to the same end.

(3) A potential source of systematic error is the effect on δ from poles induced on ferromagnetic material by the lens field. This has been eliminated by the construction of a special laboratory in which steel reinforcement has been replaced by non-magnetic stainless steel, and all other iron and steel has been excluded.

(4) The material of the aperture of the inner surface of the spectrometer and of the source-bearing wire have all been varied in a series of tests, as were the period of activation of the source (cf. Graham et al.(18)) and its intensity; unfortunately these and other tests were all invalidated by photographic emulsion shrinkage, which had not been experienced by Das-Gupta (10). Further work will therefore be necessary before a definitive result can be quoted and residual systematic errors assessed.

4. Preliminary Results

Subject to these reservations, the following results of Das-Gupta (10) indicate the potential precision. They are based on eqns. 3 and 4 and are for the two lens coils in series

K-conversion line (Ellis' F-line)

From 2 pairs of plates mean $P = 1388.562 \pm 0.005$ NPL gauss cm.

L1-conversion line (Ellis' I-line)

From one pair of plates mean $kP = 1754.001 \pm 0.022$ NPL gauss cm.

These values are to be raised by +55 ppm for anisotropic distortion, by +18 ppm to convert the 1959 NPL ampere to absolute units and by +15 ppm (in the case of P), by +9 ppm (in the case of kP) for other electrical corrections. Hence $P = 1388.684$, $kP = 1754.145$ gauss cm; $k = 1.263\,171$, cf. Graham et al.(18), $k = 1.263\,204$ ($\pm$ 6).

According to Thomsen (19), for Bi $\nu^{-1} = 167.226$ mÅ* (hence $\nu'^{-1} = 166.8828$ x.u., since 209.010 mÅ* = 208.5811 x.u.). From Taylor et al. (20) Table XXXII, $e/m = 1.758\,8028 \times 10^{7}$ e.m.u. gm^{-1} and $c = 2.997\,9250 \times 10^{10}$ cm sec^{-1}. Hence $\lambda'_c = 24.2\,120$ x.u. (cf. Taylor et al.(20) Table XXXV : $\lambda'_c = 24.2\,128$ x.u.).

If, on account of the greater width of the F-line, we distrust P, but accept kP, and take k from Graham et al.(18) so that $P = kP/k = 1388.647$ gauss cm, then $\lambda'_c = 24.2\,143$ x.u.

The dependence of the error in λ'_c on those in k and P is given (for the F and I lines) by $d\lambda'_c/\lambda'_c = 5.1\, dk/k$ and $d\lambda'_c/\lambda'_c = 1.5\, dP/P$.

5. Acknowledgements

The experimental work has been carried out by Dr. D.K.Gas-Gupta (now at U.C.N.W.,Bangor) and Mr.R.J. O'Connor (now at Oxford). Thanks are due to the Governors of Birkbeck College for provision of the non-magnetic laboratory, to the former DSIR for a research grant, continued by the College, to Professors J.D. Bernal, W. Ehrenberg and J.B. Hasted for support, to Professor W. Ehrenberg, Dr.D.K.Butt and Dr.C.F. Dietrich for advice and to Messrs.F.H. Roberts and K.S. Jacobs for construction of the spectrometer.

References

1. SIEGBAHN, K., Ark. Mat. Astr. Fys. 30A, No. 20 (1944).
2. RUTHERFORD, E. and ROBINSON, H., Phil. Mag., 26, 717 (1913).
3. TRICKER, R.A.R., Proc. Camb. Phil. Soc., 22, 454 (1924).
4. LINDSTRÖM, G., Ark. Fys., 4, No. 1 (1951).
5. JUNGERMAN, J.A., GARDNER, M.E., PATTEN, G.G. and PEEK, N.F., Nucl. Instrum. Methods, 15, 1 (1962).
6. SIDAY, R.E., Proc. Phys. Soc., 54, 266 (1942).
7. CRAIG, H., Phys. Rev., 85, 688 (1952); CRAIG, H., Ph.D. Thesis, London (1952).
8. CRAIG, H. and DIETRICH, C.F., Proc. Phys. Soc. B 66, 201 (1953).
9. DIETRICH, C.F., Ph.D. Thesis, London (1951).
10. DAS-GUPTA, D.K., Ph.D. Thesis, London (1961).
11. O'CONNOR, R.J. (unpublished).
12. AMBOSS, K. and JENNINGS, J.C.E., J. Appl. Phys., 41, 1608 (1970): AMBOSS, K., Ph.D. Thesis, London (1959).
13. NETTLETON, H.R., (of Birkbeck College; ob. 1967), unpublished notes: cf. GRAY, A., Absolute measurements in Electricity and Magnetism (Macmillan, London, 2nd. Edition, 1921).
14. see: SIDAY, R.E., Physica, 18, 1063 (1952).
15. JENNINGS, J.C.E. and O'CONNOR, R.J., J. Sci. Instrum., 2, 353 (1969).
16. DIJKSTRA, J.H. and VRIES, C.de, Nucl. Phys. 23, 524 (1961).
17. JENNINGS, J.C.E. and PEPPITT, J.W. (unpublished).
18. GRAHAM, R.L., MURRAY, G. and GEIGER, J.S., Can. J. Phys. 43, 179 (1965).
19. THOMSEN, J.S., private communication (1965).
20. TAYLOR, B.N., PARKER, W.H. and LANGENBERG, D.N., Rev. Mod. Phys. 41, 375 (1969).

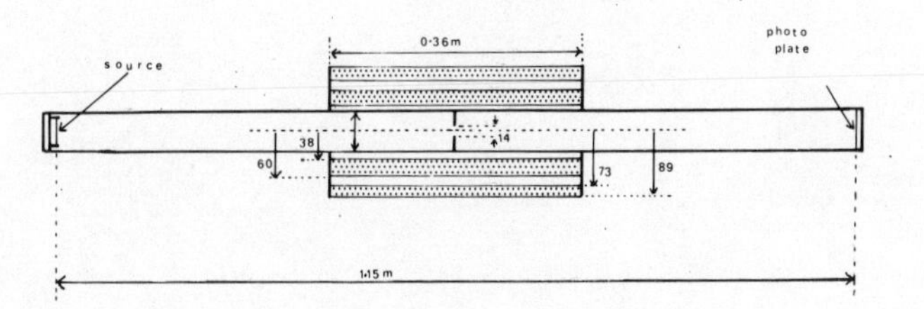

Fig. 1 Spectrometer. Dimensions in mm except where otherwise indicated.

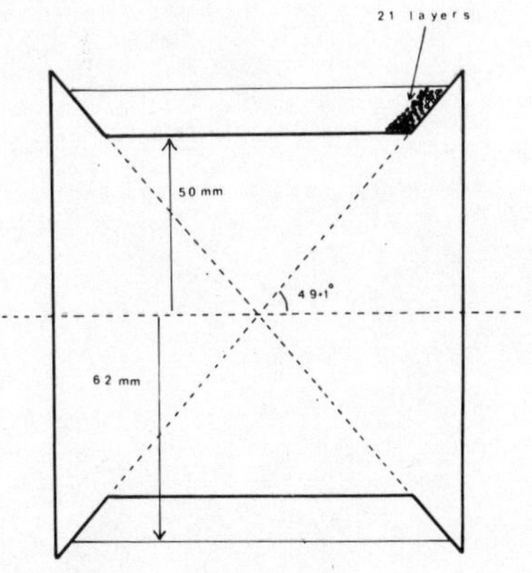

Fig. 2. Search coil. 3794 turns.
Total area 38 m^2. Tan 49.1° = $2/\sqrt{3}$.

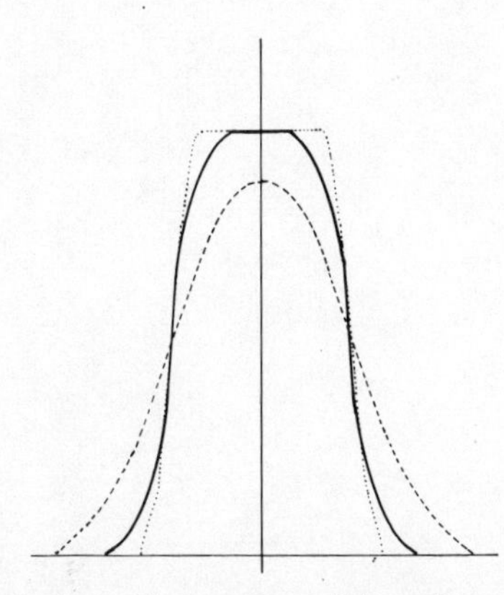

Fig. 3. Variation of intensity across line image.
Continuous curve is at the paraxial focus.
Dotted curve is at the marginal focus (Δz = LSA), i.e. the electrons are under-focused (more energetic).
Dashed curve is at Δz = -LSA, i.e. the electrons are over-focused (less energetic).

Part 9 2e/h

A Comparison of 2e/h Determinations at NSL Using Point-contact and Thin-film Josephson Junctions

I. K. Harvey, J. C. Macfarlane and R. B. Frenkel

CSIRO National Standards Laboratory, Sydney, Australia

(Presented by J. C. Macfarlane)

1. Introduction

Determinations of 2e/h by means of the a.c. Josephson frequency-voltage relationship have been reported by several groups of workers (1-7). Some of these determinations were carried out using thin-film Josephson junctions (1, 2, 6, 7) while others used point-contact junctions (4, 5, 7). Although no theoretical argument has been advanced to suggest that values of 2e/h obtained using different types of junction might differ, it is of interest to determine by experimental means whether any discrepancy exists at the present level of experimental precision. No such comparison has been reported since Parker et al. (7) indicated that there was no significant discrepancy at a precision of 0.9 p.p.m.

International comparisons of voltage standards in terms of Josephson frequency-voltage ratios obviously require that the 2e/h values quoted should not be dependent on the type of junction. This paper represents an interim report on comparative measurements of 2e/h which have been carried out at NSL using point-contact and thin-film junctions in otherwise identical experimental arrangements.

2. Junction Preparation and Properties

The point-contact junctions were formed by pressing a sharpened niobium wire against a flat niobium anvil, while immersed in liquid helium at 4.2 K. The point was resharpened and etched before each day's experiment. Critical currents of the junctions ranged between 15 and 40 mA. Further details of the point-contact junctions were given in an earlier publication (4).

Thin-film junctions of Pb-oxide-Pb were prepared on glass slides by evaporating 99.999% lead from a tantalum heater, in a vacuum near 1.3 mPa (10^{-5} torr). The vacuum chamber was pumped by a liquid nitrogen baffled diffusion pump with a liquid air trap in the backing line. Oxidation of the first lead film was carried out by heating the slide to 70°C in the coating chamber, in an atmosphere of pure dry oxygen at a pressure of 21 kPa (160 torr) for typically 2 hours. The thickness of the lead films was controlled to be in the range 70-120 nm by visually monitoring the transmission of light through the films. A protective coating of photoresist was applied to the slide after removal from the vacuum chamber.

The reported results were obtained using crossed-film junctions which provided well defined voltage steps at the 2 mV level required by the d.c. measuring apparatus. Junction dimensions were 1.0 × 0.3 mm, providing resonant frequencies in the range 8.5 to 9.5 GHz. Adequate range of voltage adjustment was achieved at 2 mV by varying the irradiating microwave frequency within the bandwidth of the junction resonance. The current ranges of the constant voltage steps were optimized by an applied d.c. magnetic field, and were of the order of 50 μA. Critical currents ranged from 5 to 14 mA.

Thin-film junctions presented measurement problems due to electrical interference, despite careful shielding of the apparatus against r.f. and magnetic fields. A screened room was not used. The point-contact junctions were much less affected by the interference, and regularly provided stable constant-voltage current steps exceeding 200 μA in amplitude.

The point contacts were operated at 4.2 K, and the thin films at approximately 2 K.

3. Electrical Measurements

The techniques employed in these measurements are essentially as described in (4), and full experimental details will be given in a forthcoming publication. For present purposes, the principles of the technique are outlined in Fig. 1.

The junction voltage V_J is adjusted to be close to either 1 mV or 2 mV, by proper choice of step number and microwave frequency. The 4 terminal resistors R_1 and R_2 connected in series act as a voltage divider. R_1 has a nominal value of 1 Ω. R_2 is a 1000 Ω resistor when the junction is to be operated at 1 mV. For 2 mV operation, R_2 consists of two 1000 Ω resistors in parallel. Divider current is supplied by a temperature stabilized mercury cell, and is servo-controlled so that the potential across R_2 is held within 0.05 p.p.m. of the e.m.f. of a reference standard cell in the same oil bath as the cells used to maintain the NSL standard volt. This servo-control system incorporates galvanometer G_2 with associated secondary amplifier S_2 driving an optical d.c. coupler. Servo error is continuously monitored and may be minimized when necessary by a manual adjustment.

The potential across R_1 is compared with the junction voltage, using galvanometer G_1. Junction voltage is adjusted by variation of the microwave frequency irradiating the junction to achieve a balance within several parts in 10^7. Final readings of balance are taken from recordings of the galvanometer G_1 deflections (see below).

The net thermal e.m.f. in the low-level circuit is compensated by a bucking-off technique. Fluctuations in the residual thermal e.m.f. place a limit on the useful resolution of the galvanometer reading. This source of random error in an experimental run is responsible for the major part of the uncertainty expressed as the standard deviation of the mean in Table I. (See Section 4)

The deflections of the galvanometers G_1 and G_2 are continuously recorded on a strip chart. The recorded deflection of G_1 is calibrated by making known variations in the frequency of the microwave irradiation of the junction, and observing the resultant change in deflection while the junction is biassed on a constant-voltage step. The effect of residual uncompensated thermal e.m.f.s is eliminated during a measurement by making periodic reversals of standard cell polarity, junction polarity, and divider current. Each experimental point in the day's measurement is obtained by recording the galvanometer G_1 deflection for three successive positions of the reversing switches (e.g. forward-reverse-forward). The number of such points obtained in each day's run is usually between 3 and 8, and the mean value of all the points obtained in the day's run is used to calculate 2e/h. We rejected all readings which were subject to thermal drifts greater than about 0.1 nV/min, and most of our results were obtained when the drift rate was about one-fifth of this figure.

Spurious e.m.f.s due to lead resistance were avoided by connecting the current, potential, and monitoring leads to the junction as shown in Fig. 1. The superconducting junction leads are represented by heavy lines. Comparisons of the divider components with the NSL series-parallel set are regularly carried out. From these measurements lines of best fit to the data pertaining to each resistor are derived. Systematic components of error due to errors in the series-parallel set and in the resistance comparisons will not significantly affect the comparison of relative values of 2e/h reported here.

The standard cell is calibrated against the NSL voltage standard before and after each day's run. Again, systematic errors in the definition of the NSL volt will not affect the present comparison of thin-film and point-contact results.

4. Results

The results of three runs using three different thin-film junctions, and nine runs using point-contact junctions, are summarized in Table I, in chronological order. The overall errors are estimated by compounding the standard deviation of the mean of the data for each day's run with the root-sum-of-squares of the quantities shown in Table II. In two cases (Day 305 and Day 326) the number of points was small, and the error was taken to be that of a single observation as estimated from other more extensive runs.

Fig. 2 shows a graph of the results.

The relative mean values of 2e/h as measured with point-contact and thin-film Josephson junctions are:

Point Contacts 483.593 74 ± .000 03 $MHz/\mu V_{NSL}$
Thin Films 483.593 74 ± .000 02 $MHz/\mu V_{NSL}$

TABLE I

Measured Values of 2e/h

Day No.	Junction Type (F = film) (P = point contact)	Step Number	2e/h $MHz/\mu V_{NSL}$ 483.593--	Std devn of mean of day's results p.p.m.	Overall error p.p.m.
293	P	104	82	.02	.09
305	P	104	68	.04	.09
307	P	104	73	.03	.09
313	P	104	77	.02	.09
322	F	112	74	.02	.09
326	F	113	71	.04	.09
334	F	116	75	.03	.09
340	P	52	67	.03	.09
343	P	52	61	.07	.11
349	P	52	80	.04	.09
357	P	52	83	.05	.09
362	P	52	78	.06	.10

Microwave frequency for all runs was in the range 8.5-9.5 GHz

TABLE II

Random Errors Estimated as One Standard Deviation

Source	p.p.m.
Determination of line of best fit to divider resistance ratio	< 0.01
Variation of temperature in divider oil bath	< 0.01
Variation of temperature gradients in divider oil bath	< 0.01
Frequency measurement	< 0.01
Error in standard cell to divider input potential control system	< 0.05
Uncertainty of cell voltage with respect to NSL voltage standard during experiment	< 0.05
Measurement uncertainty of standard cell with respect to voltage standard	< 0.02
Thermal voltages in lead to standard cell	< 0.02
Leakage in oil bath	< 0.01
Paralleling error for 500 Ω resistor	< 0.01
Scale calibration error for G_1	< 0.01
Root sum of squares	< 0.08

As pointed out above, the assigned errors do not include possible systematic errors in the national volt or in the divider ratio, as these are expected to enter into the experiments with both types of junction in the same sense, and consequently to have no net effect on the present comparison.

5. Conclusion

The standard deviations of the respective means are of the order of 0.05 p.p.m. We conclude that the means are identical to within 0.07 p.p.m., and that there is no significant difference at the present level of precision between point-contact and thin-film determinations of 2e/h. The achievement of greater precision must await better junction fabrication techniques, more effective electromagnetic screening, and improvements in stability of the NSL standard of e.m.f. Alternatively, a useful increase in precision might be achieved by operating thin-film and point-contact junctions simultaneously in the same cryostat and comparing their voltage outputs directly as has been done for example by Clarke (8) for comparing junctions between different superconductors.

References

1. FINNEGAN, T.F., DENENSTEIN, A., LANGENBERG, D.N., Phys.Rev.Lett., 24, 738 (1970).

2. FINNEGAN, T.F., DENENSTEIN, A., LANGENBERG, D.N., Phys.Rev.B, to be published.

3. PETLEY, B.W., MORRIS, K., Metrologia, 6, 46 (1970).

4. HARVEY, I.K., MACFARLANE, J.C., FRENKEL, R.B., Phys.Rev.Lett., 25, 853 (1970).

5. KOSE, V., MEICHERT, F., FACK, H., SCHRADER, H.J., PTB-Mitteilungen, 81, 8 (1971).

6. HARRIS, F.K., FOWLER, H.A., OLSEN, P.T., Metrologia, 6, 134 (1970).

7. PARKER, W.H., LANGENBERG, D.N., DENENSTEIN, A., TAYLOR, B.N., Phys.Rev., 177, 639 (1969).

8. CLARKE, J., Phys.Rev.Lett., 21, 1566 (1968).

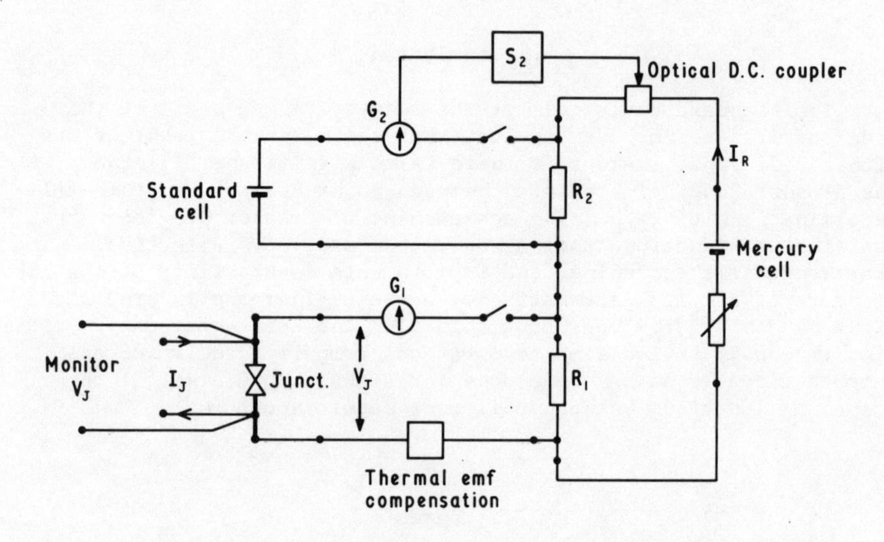

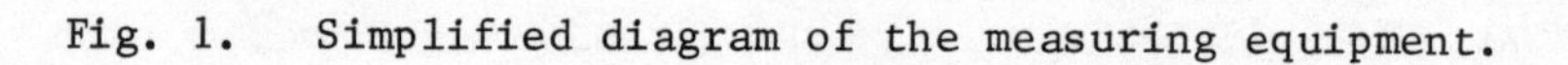
Fig. 1. Simplified diagram of the measuring equipment.

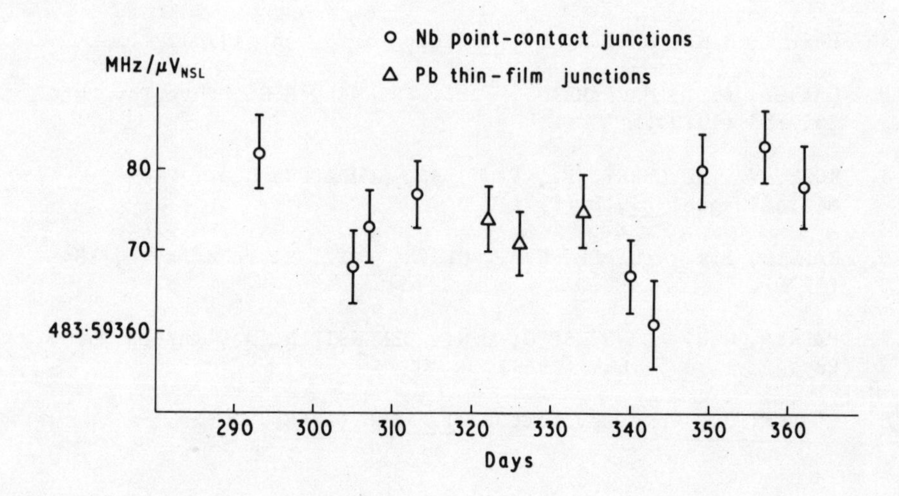

Fig. 2. Comparison of 2e/h values obtained using point-contact and thin-film junctions.

Recent Results of the *e/h*-measurement at PTB

V. Kose, F. Melchert, H. Fack, P. Gutmann

Electricity Division, Physikalisch-Technische Bundesanstalt in Braunschweig

It has been shown by several experimentalists (1,...,4) that the steplike current-voltage characteristic based on the Josephson voltage-frequency relation can be used to determine the fundamental physical constant e/h with high accuracy. This frequency-voltage relation is as far as we know from various experiments, independent of influences such as temperature, type of junction, superconducting material, magnetic field, step number, microwavepower, and so on. So it seems that the Josephson effect provides the most reliable method of maintaining the volt.

Last summer the National Bureau of Standards has suggested an intercomparison of the units of voltage maintained by BIPM, NBS, NRC, NSL, and PTB, in order to improve the comparison of the various e/h values which were measured with different apparatus and by using different techniques and different units of voltages in the different countries. This comparison is of great importance as it is the first time that e/h-values with an uncertainty of only a few parts in 10^7 were measured at different places and can now be compared with a desirably high accuracy.

Last month the PTB has participated in this intercomparison and at the same time we started a series of e/h-measurements.

Now we are going to describe how we measured e/h. Throughout we used point contact junctions which were formed by pressing a chemically etched niobium wire against a flat block of niobium. The contact of the junction was adjustable and the resistance was typical between 0.1Ω and 0.5Ω. The junction was radiated by microwaves of approximately 70.366 GHz. The 7th step was taken in order to achieve a dc voltage which was close to 1 mV. This chosen frequency is determined by the step number, the voltage of the standard cell and the divider ratio. At 1 mV level we observed step heights from 300 μA up to maximal 600 μA. Within the resolution of the galvanometer which is 3 parts in 10^8 we found no change in the voltage as the current was varied over 80% of the height of the step. Stable step amplitudes between 4 and 5 mV showed a current range

O*

of 20 μA. At lower frequencies the 4 and 5 mV steps were considerably smaller. So it seems that our theoretical result (5) is confirmed concerning the frequency dependence of step amplitudes.

Due to the thermal emf drift of about 0.1 nV/min, we had in our first e/h measurement an uncertainty of 1 part in 10^7. In order to minimize that, the potential leads were put into a sealed copper tube coated outside with styrofoam except at the lower end, where the tube had a good thermal contact to the helium bath. By this means the uncertainty was significantly reduced to 1 part in 10^8.

To phaselock the frequency of the 70 GHz Klystron to a quartz oscillator we extended the 10 GHz microwave system which was used earlier. Figure 1 shows a block diagram of our stabilization system. The frequency synthesizer provides a frequency selection of 1 Hz in the vicinity of 1 GHz. This enables us to adjust the Josephson dc voltage in digits of 1 part in 10^9 via a digital frequency variation of the 70 GHz Klystron. In the first mixer the 10th harmonic of 1 GHz was generated whereas in the second mixer the 7th from the X-band Klystron was taken. The intermediate frequencies were 10 MHz and 30 MHz respectively. As a reference we used a quartz oscillator which frequency was measured with respect to the PTB ceasiumbeam frequency standard to 1 part in 10^9. Although the whole microwave set up is in a screened cage it was useful to check the frequency spectrum of the stabilized Klystron outputs. This was done with a high resolution spectrum analyzer, the smallest band-width of which was 100 Hz. At 70 GHz all discrete sidebands were smaller than -40 dB whereas at 10 GHz they were less than -60 dB. The full linewidth of the 70 GHz signal at half power was 350 Hz or 5 parts in 10^9.

Figure 2 shows how the junction dc voltage U_J of 1 mV was compared with the standard cell voltage E_N of 1 V level. This was achieved by establishing a fixed resistance ratio of $R_1 : R_2 = 1:1000$. The resistive divider is composed of Hamon resistors which are connected in series with ten four terminal resistors of 10 nominal value. All divider resistors were immersed into an oilbath which was temperature controlled to 10^{-3}K. Our measurements showed that the Hamon resistors in the parallel arrangement (as shown in figure 2) were stable to better than 1 part in 10^7 during the time of two weeks. However the comparison of the 10Ω resistors with the series arrangement was checked before and after each day's run. The overall accuracy of the divider is at present 1 part in 10^7.

For the null detector system in the low voltage circuit we used two cascade-connected photocell-galvanometer amplifiers with a sensitivity of 10 mm/nV.

At the beginning of each measuring cycle we cancelled the thermal emf by bucking it with a voltage across an added resistor of $10^{-3}\Omega$. The zero reference of the Josephson zero-voltage characteristic was checked with respect to the dc current and the microwave power. This was done by varying the dc current from zero to about 90% of the critical current of the junction. The microwave power dependence was checked by biasing the dc current to approximately half of the critical current and by adjusting the power to a level, so that the zero step height always exceeded half of the step amplitude. In well prepared junctions we didn't find any change of the zero reference within the resolution of the null detector.

Figure 3 indicates the random distribution of 2e/h values from 50 runs, measured on two days by comparing the Josephson dc voltage with the same standard cell. On the horizontal axis the deviation of the individual 2e/h-values from the mean is plotted. n is the number of the runs. The day's result showed a one standard deviation of 1 to 2 parts in 10^7. In order to reduce this uncertainty to some extend we started some weeks ago to improve our temperature control and temperature measuring system of the standard cell box. The temperature accuracy will then be 10^{-3}K by using a calibrated platinum resistance thermometer. The remaining uncertainty due to this temperature effect should then become at a nominal temperature of 20 °C 4 parts in 10^8.

Finally it can be stated that with these improvements of our measuring system the overall accuracy of our measurements is between 1.4 to 1.8 parts in 10^7 depending on the final random error of the 2e/h values. The residual contribution to the overall uncertainty will primarily be associated with the stability of the standard cell and the divider in about equal parts. Compared with our first results (4), the overall uncertainty is now reduced by at least a factor of 2.

We would like to thank H.-J. Schrader, W. Hetzel for many helpful discussions and E. Staben and W. Leppelt for technical assistance.

References

1. FINNEGAN, T. F., DENENSTEIN, A. and LANGENBERG, D. N., Phys. Rev. Letters 24, 738 (1970).

2. PETLEY, B. W. and GALLOP, J. C., in Proceedings of the International Conference on Precision Measurement and Fundamental Constants, edited by D. N. Langenberg and B. N. Taylor (National Bureau of Standards Special Publication 343, U.S. Government Printing Office, Washington, DC, 1971) p. 147.

3. HARVEY, I. K., MACFARLANE, J. C. and FRENKEL, R. B., Phys. Rev. Letters 25, 853 (1970).

4. KOSE, V., MELCHERT, F., FACK, H. and SCHRADER, H.-J., PTB-Mitt. 81, 8 (1971).

5. FACK, H., KOSE, V. and SCHRADER, H.-J., Meßtechnik 79, 31 (1971).

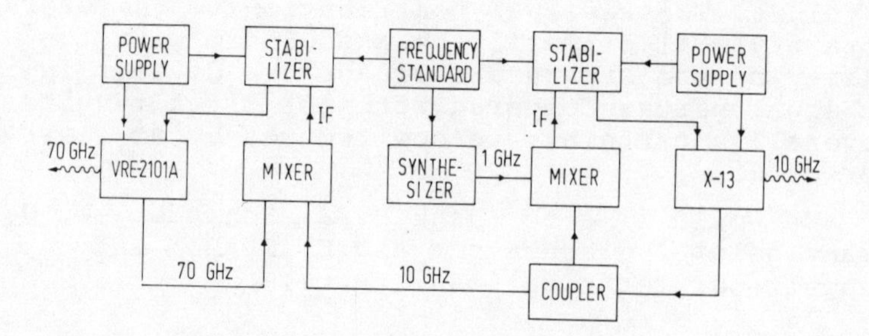

Figure 1. Block diagram of the microwave frequency stabilization system.

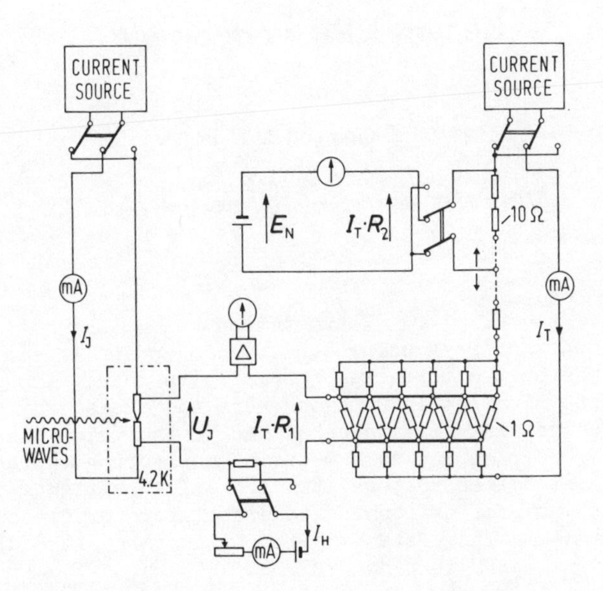

Figure 2. Block diagram of the dc measuring set up.

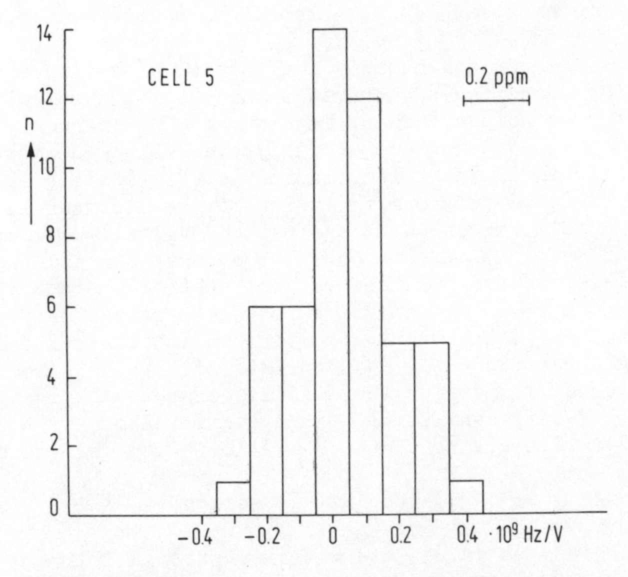

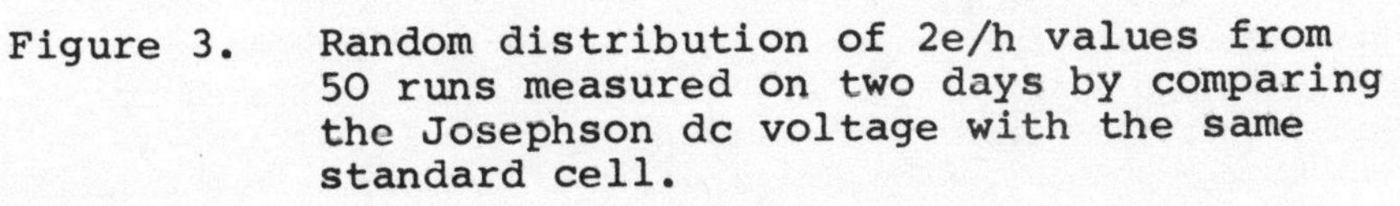
Figure 3. Random distribution of 2e/h values from 50 runs measured on two days by comparing the Josephson dc voltage with the same standard cell.

Recent NPL work on *e/h*

J. C. Gallop and B. W. Petley

Division of Quantum Metrology,
National Physical Laboratory, Teddington, England

1. Introduction

Measurements of 2e/h using the a.c. Josephson effect have been made at the NPL over a three year period with increasing precision. The latest determination is reported here and it is estimated that the overall uncertainty has been reduced to the point where the experiment may be regarded as a means of monitoring variations in the national maintained voltage standard. The existence of similar experiments at several national standards laboratories should enable the international voltage standards to be compared in addition to the international BIPM volt comparison.

2. The measurement technique

Two previous NPL determinations have already been described (1,2) and the present work will be reported more fully elsewhere. In this section the experimental procedure will be outlined together with the improvements which have led to an increase in precision.

The measurement essentially consists of measuring a microwave frequency and comparing as accurately as possible two voltages which differ by two or more orders of magnitude. The frequency may be measured adequately using commercially available equipment. A schematic diagram of the voltage comparison circuit is shown in Fig. 1. A standard cell e.m.f., which may be measured in terms of the national volt, is balanced against the voltage developed across the resistor R_1 and simultaneously the Josephson junction, biased on a microwave induced constant-voltage step, is balanced against the resistor R_2, using sensitive photocell galvanometer amplifiers as null detectors. The voltage comparison is thus reduced to determining the ratio R_1. R_1 consists of three 300 Ω coils which may be switched from series to parallel connection, giving an accurate 9 : 1 ratio whilst R_2 is a similar Hamon type resistor with ten 10 Ω coils giving a 100 : 1 ratio. With R_1 in the parallel mode $\left(\frac{300}{3}\,\Omega\right)$ and R_2 in series (10 × 10 Ω) their resulting resistance values are within 1 part in 10^4 of one another. Their values are compared to 1 part in 10^8 using a difference potentiometer which only needs to be accurate to 1 in 10^4 and a transfer voltage stable to 1 in 10^8. The latter is achieved by using a temperature controlled standard resistor and a constant current supply stabilized by reference to a standard cell. Thus, relying on the Hamon principle the ratio of R_1 in the series mode to R_2 in the parallel mode (very nearly 900 : 1) has now been determined with a precision approaching 1 in 10^8.

The technique of simultaneously balancing the junction and cell voltages against fixed Hamon resistors has the advantage of eliminating a wide voltage range potentiometer from the measurement. However it does require a reliable Josephson junction which has constant voltage steps to voltages above 1 mV. This is not attained by the average solder-drop type of junction used in the previous NPL determinations. A simple and reliable new type of junction has been developed for this purpose. It consists of two interlocking loops of Pb-Sn solder coated phosphor-bronze wire (0.46 mm diameter) which can be pulled into contact by a simple screw drive operated from outside the helium dewar. Tension is provided by forming the leads to the loops into small helical springs. Rotation of the screw allows the critical current for the junction between the super-conducting loops to be adjusted, and the microwave induced step structure to be optimized. Such junctions, which have a weak-link character, when mounted in a Q-band wave guide have shown steps 10 μA wide up to about 3 mV. This simple type of junction is easy to use for 2e/h work and is reasonably stable against mechanical vibrations. In this respect it is better than other adjustable point contact devices we have tried, perhaps due to the isolation provided by the small springs.

We have preferred to set up the measuring equipment in two adjacent screened rooms. One is the d.c. room and contains the helium dewar with Josephson junction, the Hamon resistors and difference potentiometer in a temperature controlled oil bath, null detectors, constant current controllers and a temperature controlled enclosure containing twelve standard cells. All power supplies in this room are derived from lead-acid batteries, apart from those for galvanometer lamps which are taken from well smoothed d.c. supplies. The other screened room contains the 36 GHz microwave source and frequency stabilizing and measuring equipment as well as <u>all</u> mains powered units. Fig. 2, showing the effect of opening the d.c. screened room door on the junction step shape, illustrates the necessity of using a screened room for the measurements at the present site of the experiment - without one the constant voltage steps do not have a zero slope region.

The standard cells are very frequently compared with one another <u>in situ</u> to an accuracy of 10 nV and by transporting the cell enclosure to the site of the national standard regular comparisons are made with the NPL maintained volt.

An important factor in determining the experimental precision is that the rate of change of the thermal e.m.fs which exist in the millivolt circuit must be minimized. Two improvements over previous techniques have been made. First, the d.c. screened room is temperature controlled to ±0.1 °C and secondly it is no longer necessary to break and remake a switch in the millivolt circuit whenever the current is reversed.

The implementation of the improvements outlined here has enabled the standard deviation of a single set of balances to be reduced to about 0.25 ppm.

3. Results

Using the equipment described above measurements have so far been made on three separate occasions spaced over a period of four months. One day's measurements consist of making a number of separate balances of the junction voltage against the e.m.f. of a standard cell selected from one of the twelve in the enclosure. All the standard cells are measured against one another immediately before and after the 2e/h measurement and against the national voltage standard within a few hours.

Fig. 3 shows the results of recent measurements and gives the e.m.f. of a particular standard cell derived from the Josephson effect experiment having assumed a value of $2e/h = 483\ 594.15\ GHz/V_{NPL}$. In addition it shows the values for the same cell obtained by direct comparison with the NPL voltage standard. Figure 4 shows the collected results of 2e/h measurements made at the NPL over a two and a half year period. This demonstrates the stability of the national maintained volt within the assigned uncertainties over that period.

1. PETLEY, B.W. and MORRIS, K., Metrologia, 6, (2), 46 (1970).

2. PETLEY, B.W. and GALLOP, J.C., Proceedings of the International Conference on Precision Measurement and Fundamental Constants, Nat. Bur. Stds. Special Publication 343 (1971).

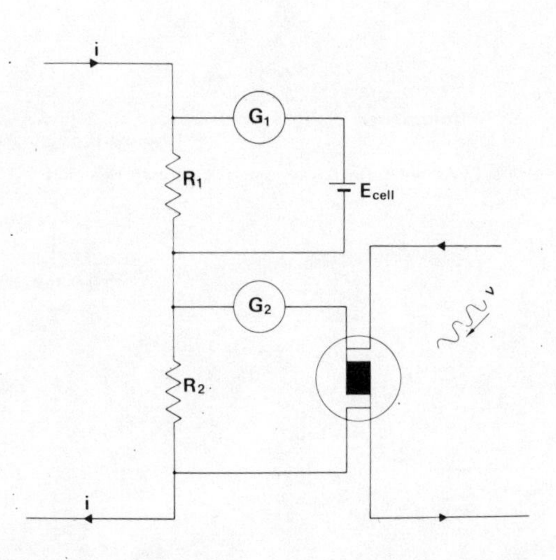

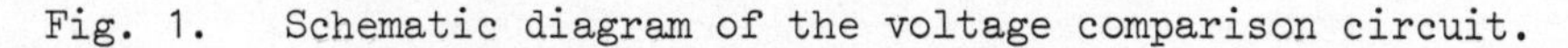

Fig. 1. Schematic diagram of the voltage comparison circuit.

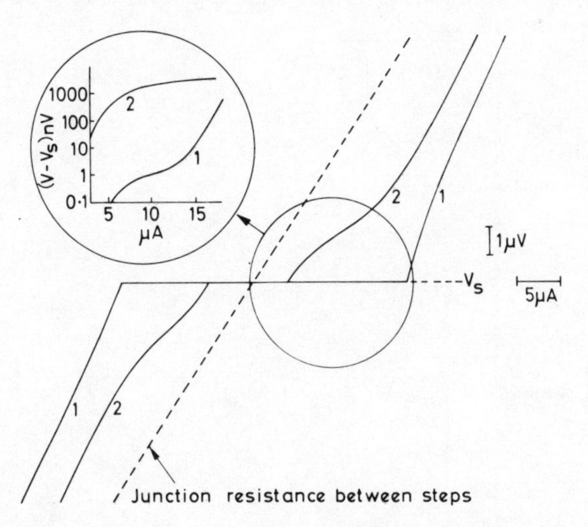

Fig. 2. The effect on the shape of step 15 (1.13 mV) of opening the door of the screened room containing the junction and d.c. equipment. Curve 1 door closed, curve 2 door open. The inset shows the shape of the step itself with a logarithmic voltage scale. Steps used in 2e/h measurements had a region 30-80 μA wide with no detectable slope, rather than the 10 μA shown here.

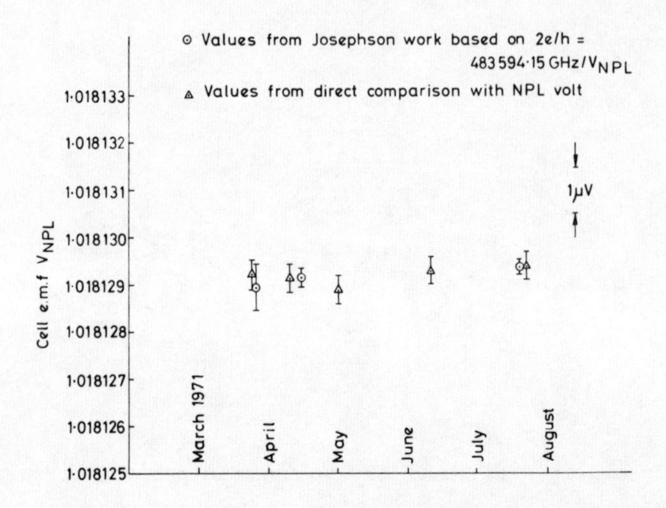

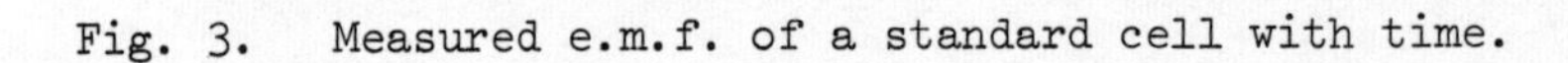
Fig. 3. Measured e.m.f. of a standard cell with time.

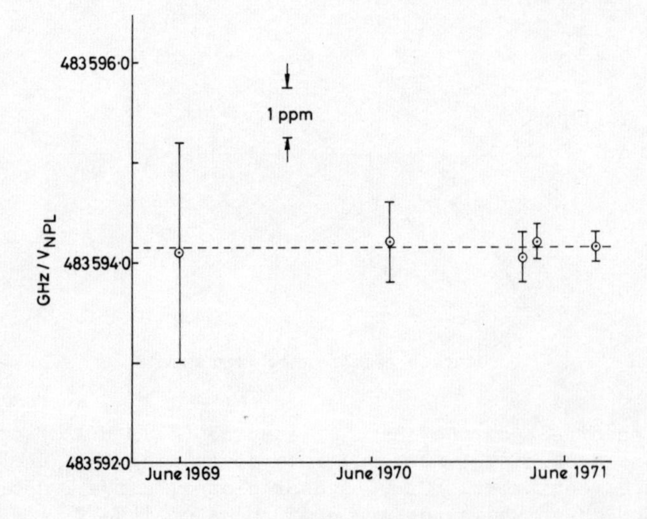

Fig. 4. Results of NPL determinations of 2e/h.

Measurements of 2*e*/*h* via the AC Josephson Effect

T. F. Finnegan, T. J. Witt,* B. F. Field, and J. Toots

Institute for Basic Standards, National Bureau of Standards, Washington, D.C. 20234

1. Introduction

The ac Josephson effect has become a well established method (1) for determining 2e/h (the ratio of twice the electron charge to Planck's constant). Determinations of 2e/h with sub-ppm (parts per million) uncertainties have been reported by several groups throughout the world. A major limitation in achieving the most precise 2e/h measurements has been the standard cells used to (a) maintain the local (voltage) standard and (b) compare the local as-maintained unit with another laboratory's unit (by the physical transfer of cells between laboratory sites). The latter source of uncertainty is particularly important if the final result of a 2e/h determination is to be critically compared with the results of other similar experiments, or combined with other accurately measured quantities to obtain values of physical constants (2). In particular, an accurate value of 2e/h can be combined with γ_p, the gyromagnetic ratio of the proton, and certain other constants to obtain a value of the fine structure constant α. In this case, both 2e/h and γ_p must be determined (either directly or indirectly) in the <u>same</u> electrical units.

In figure 1, the results of all reported sub-ppm 2e/h determinations based on measurements made in 1970 are shown. Each result has been expressed in the National Bureau of Standards (NBS) as-maintained unit of voltage using the results of the 1970 triennial international comparisons at the Bureau International des Poids et Mesures (BIPM) (3). The first sub-ppm determination of 2e/h was reported by Finnegan, Denenstein, and Langenberg (FDL) (4), at the University of Pennsylvania (UP), with a one standard deviation (1σ) uncertainty of 0.46 ppm.† Subsequently, Petley and Gallop (PG) (5) at the National Physical Laboratory (NPL); Harvey,

*Present address: Bureau International des Poids et Mesures, Sèvres, France.

† All uncertainties are expressed as one standard deviation estimates. Where applicable, unless the word "observed" is used, a judgment estimated to be equivalent to one standard deviation is meant.

Macfarlane, and Frenkel (HMF) (6) at the National Standards Laboratory (NSL); and Kose, Melchert, Fack, and Schrader (KMFS) (7) at the Physikalisch-Technische Bundesanstalt (PTB) have reported sub-ppm 2e/h determinations (in their respective as-maintained national units) with 1σ uncertainties of approximately 0.8 ppm, 0.1 ppm, and 0.4 ppm respectively (as indicated in figure 1). The final result of the series of sub-ppm 2e/h measurements made at UP by FDL is also indicated in figure 1. It has a 1σ uncertainty of 0.12 ppm.

The results are in fair to good agreement, but only if an additional uncertainty of order 0.2 ppm is added in quadrature to each result (except for the UP results which were reported in NBS units) for the uncertainty in the differences between any two national as-maintained units as determined via the 1970 BIPM intercomparisons. In particular, the difference between the NSL and final UP 2e/h results is 0.42 ± 0.16 ppm assuming that the BIPM transfers are exact; if an additional uncertainty of 0.2 ppm is included for the conversion of units, this difference becomes 0.42 ± 0.26 ppm.

The sub-ppm 2e/h measurements at UP were made over a time period covering December, 1969 to July, 1970, and the accurate comparison of the NBS voltage unit with the local UP unit was carried out over a three month period with a central date in May, 1970. (A transfer was also carried out in December, 1969; however, the uncertainty in this transfer was about three times greater than the succeeding transfers and is reflected in the significant difference in uncertainty between the preliminary result, labelled UP*, and the final result, designated UP.)

2. Recent 2e/h Measurements at the National Bureau of Standards

A series of 2e/h measurements has now been made at the site of the NBS unit of voltage with the _same_ low-voltage-comparison instruments as used in the sub-ppm measurements at UP. Since the present techniques, equipment, and procedures were very similar to those already published in detail, only the key features of the earlier experiments will be reviewed and only those changes or modifications which can affect the precision or accuracy of the new series of measurements will be considered.

A determination of 2e/h requires the accurate measurement of a frequency and a voltage. The (microwave) frequency can be readily measured and stabilized to less than 0.01 ppm, and thus it is the voltage measurement which dominates the uncertainty in a determination.

In the series of measurements reported here, a _fixed_ voltage ratio of 100:1 was used to measure a Josephson device voltage of order 10mV obtained by operating pairs of lead-lead oxide-lead tunnel junctions in series. The Josephson device voltage, which can be continuously tuned via the step number and frequency of the incident (microwave) radiation, was adjusted so that the ratio of the standard cell emf to the Josephson voltage was precisely equal to the fixed ratio produced by the voltage comparison instrument. Low voltage comparators based on two independent methods for achieving an accurate 100:1 ratio (1V:10mV) have been developed (8,9) and have been demonstrated by direct comparison experiments to agree to about 1 part in 10^8 by FDL. The first method involved a series-parallel technique (10) which depends on the fact that a chain of n nominally-equal resistors can be connected alternately in series and in parallel to obtain a ratio n^2 accurate to second order in the fractional resistor deviations about the mean. The second method involved a cascaded interchange technique, which makes use of a conventional-type voltbox (in the inverted configuration) optimized for self-calibration (11,8). The critical switching in both instruments is accomplished by commercial grade rotary switches. Furthermore, each instrument is housed in a temperature-regulated air bath and is portable. This method of construction differs drastically from the classical construction techniques of Hamon (12) and Harris, Fowler, and Olsen (13) who used mercury amalgam switches and immersed the critical circuitry in an oil bath.

For the NBS measurements, the two voltage comparators were refurbished (i.e., the mercury batteries were replaced; the current fans, voltage fans, and main resistors in the series-parallel comparator (SPC) were retrimmed, the SPC oven was rewired non-inductively, etc.) and transported to NBS. Only the SPC instrument was used to take 2e/h data; however, the two instruments were compared directly by connecting each 1V output to a separate standard cell, and then using one instrument to "measure" the 10mV output of the other. The voltage difference of the two cells measured in this manner and the value obtained by direct comparison at the 1V level agreed to 0.004 ± 0.027 ppm, well within the expected value of 0 ± 0.045 ppm.

The present series of measurements was made in a shielded room adjacent to the oil bath containing the NBS reference group of standard cells. One of the two standard cell enclosures (Serial #4130) used in the experiments at UP was transported to NBS and modified in order to improve the insulation of the standard cells and to reduce the effects of ambient temperature variations on the standard cell emfs. All 2e/h measurements were made directly with one of the three saturated cells in the enclosure. (This enclosure, with its associated switches and wiring, is termed the "local volt"; it contains three saturated and three unsaturated standard cells). Standard cell comparisons between the cells in this enclosure and those in a second shippable air enclosure were usually made immediately before, and after, each 2e/h run with a potentiometer having nV resolution. (The average standard deviation per observation was about 7nV.) The additional air enclosure was situated in the same room as the NBS reference group of cells and was compared with it on the average of about twice a week. This air enclosure was of the type regularly used in the NBS Volt Transfer Program and was measured in the same manner that the other enclosures are calibrated between trips to other laboratories.

Nine 2e/h runs were made over an approximately two month period between July 2 and August 23, 1971. The results are plotted in figure 2. The observed standard deviation of the mean (assuming equal weighting for the nine measurements) is 0.023 ppm. This statistical uncertainty arises from random variations in the thermoelectric voltages, noise in the voltage measuring circuitry, day-to-day fluctuations in the local and reference group of cells and any other randomly varying experimental parameters present in a 2e/h experiment. The observed standard deviation, s, of a single experimental run is 0.07 ppm (assuming equal weighting for the nine measurements), and is a direct measure of the precision with which the NBS reference group of standard cells can be maintained via the ac Josephson effect.

The principal sources of uncertainty are summarized in Table I. (All uncertainties are expressed as one standard deviation estimates). The following comments apply. During each run, the microwave frequency of the applied radiation was directly measured with a frequency counter whose timebase was checked at regular intervals against a reference oscillator maintained to within a few parts in 10^{10} of the U.S. Frequency Standard.

Table I

Summary of Sources of Uncertainty in 2e/h

	Uncertainty(1σ) Parts in 10^8
1. Measurement Uncertainties	
(a) Random uncertainty of the mean	2.3
(b) Frequency measurement and stability	1.0
(c) Low-voltage comparison (series-parallel comparator)	3.1
2. Local Volt Uncertainty in V_{NBS-69} (July-August, 1971)	3
RSS Total	5

The 0.031 ppm a priori estimated uncertainty in the low voltage comparison instrument was taken to be the same as that reported by FDL (8). The sources of uncertainty associated with the instrument that were expected to change or vary with time

were checked regularly. Current fan, voltage fan, and main resistor checks of the series-parallel comparator were made at frequent intervals. The leakage resistances of the various critical portions of the dc measurement circuitry were measured and found to be comparable to those observed by FDL. (Note that the uncertainty of the direct instrument comparison made at NBS is based on estimated 0.03 ppm instrument uncertainties for both the series-parallel and cascaded-interchange comparators.)

A systematic uncertainty of 0.03 ppm has been attributed to the local voltage standard due to sources such as the thermoelectric emfs in the standard cell connections, (correlated) ambient temperature changes, and possible disturbance of the standard cell directly measured during a 2e/h run. The NBS reference group of standard cells was assumed constant over the two month period of the measurements and by definition constituted the NBS as-maintained volt during this period of time. The day-to-day random uncertainty associated with standard cell comparisons between the NBS reference group of cells and groups in air enclosures is usually about 0.05 ppm.

The final root-sum-square total uncertainty in 2e/h is 0.05 ppm. But it should be emphasized here that no allowance has been made for the long term time-dependent component of uncertainty in the NBS as-maintained unit.

3. Final Result and Discussion

Our final value of 2e/h with all corrections included is

$$483.593\ 589 \pm 0.000\ 024\ \mathrm{THz/V_{NBS\text{-}69}}.$$

This is 0.27 ppm lower than the final value reported by FDL, and is 0.68 ppm lower than the reported value of HMF (using the results of the 1970 BIPM comparisons to relate the units). Both of these results, however, are based on measurements made in 1970.

The 0.27 ppm difference between the NBS and final UP result can be attributed primarily to either a shift in the NBS as-maintained unit (over a period of about 1 year), the volt transfer uncertainty, or a combination of both. (Since the same critical dc instrumentation and similar procedures were used in both of these determinations, it is very unlikely that an unsuspected systematic error, assuming one were present, would introduce a change this large). Since the assigned one standard deviation uncertainty of the May 1970 NBS-UP volt transfer was 0.12 ppm, it is conceivable that the difference arises largely from this source. However, a decrease in the NBS unit of voltage by 0.1 to 0.2 ppm cannot be unequivocally ruled out (14). Clearly, the time dependence of the various national units (based on groups of standard cells) will have to be critically considered at the current levels of accuracy of 2e/h measurements.

The 0.68 ppm discrepancy between the NBS and NSL results is of course quite disturbing. However, during the summer of 1971, under the auspices of BIPM, NBS conducted a series of direct standard cell intercomparisons between BIPM, NPL, NSL, and PTB to better establish the relationships among the various national as-maintained units. Using some of the preliminary results of these intercomparisons ($V_{NPL} - V_{NBS}$ = +1.08 ppm and $V_{NSL} - V_{NBS}$ = +0.42 ppm) (15) and the most recent (1971) values of 2e/h determined by NPL (16) and NSL (17), we find

$$2e/h\ (NPL) = 483.593\ 628\ (90)\ (0.19\ ppm)$$

and $$2e/h\ (NSL) = 483.593\ 597\ (48)\ (0.1\ ppm),$$

where the uncertainties are those assigned by the authors. The NBS, NPL, and NSL values thus appear to agree to within 0.08 ppm. Since the NSL and NPL 1971 2e/h values decreased by only 0.1 ppm and the NBS result is only 0.27 ppm less than that of UP, it is clear that the discrepancy has been removed by making new volt intercomparisons and not by the new 2e/h measurements.

4. Implied Value of the Fine Structure Constant, α.

Finally, we note that the value of the fine structure constant implied by our present result can be calculated by using Eq. (92) of Reference 2,

$$\alpha^{-1} = \left(\frac{1}{4R_\infty} \; \frac{c\Omega_{ABS}}{\Omega_{NBS}} \; \frac{\mu'_p}{\mu_B} \; \frac{2e/h}{\gamma'_p} \right)^{1/2},$$

the most recent values of the various other auxiliary constants reported by Cohen and Taylor (18), and the preliminary value of γ'_p determined by Driscoll and Olsen (19) directly in NBS electrical units. We find α^{-1} = 137.03589(14) (1.0 ppm). This is 1.40 ppm lower than the value deduced "without quantum electrodynamic theory" (WQED) by Taylor et al (2), and has an uncertainty approximately one-half that of their result. This should enable even more critical tests of QED than have been previously possible, but significant changes in the present situation are not expected (18).

Acknowledgment

We should like to thank the University of Pennsylvania for making the dc low voltage measuring instruments available to us, A. Denenstein for helpful discussions, and various members of the Absolute Electrical Measurements Section, particularly W. G. Eicke and B. N. Taylor, for assistance and discussions.

References

1. PARKER, W. H., LANGENBERG, D. N., DENENSTEIN, A., and TAYLOR, B. N., Phys. Rev. 177, 639 (1969).

2. TAYLOR, B. N., PARKER, W. H., and LANGENBERG, D. N., Rev. Mod. Phys. 41, 375 (1969).

3. TERRIEN, J., private communication. See also Metrologia 7, 78 (1971).

4. FINNEGAN, T. F., DENENSTEIN, A. and LANGENBERG, D. N., Phys. Rev. Letters 24, 738 (1970).

5. PETLEY, B. W. and GALLOP, J. C., in Proceedings of the International Conference on Precision Measurement and Fundamental Constants, edited by D. N. Langenberg and B. N. Taylor (National Bureau of Standards Special Publication 343, U.S. Government Printing Office, Washington, DC, 1971) p. 147.

6. HARVEY, I. K., MACFARLANE, J. C. and FRENKEL, R. B., Phys. Rev. Letters 25, 853 (1970).

7. KOSE, V., MELCHERT, F., FACK, H. and SCHRADER, H. J., PTB-Mitt. 81, 8 (1971).

8. FINNEGAN, T. F., DENENSTEIN, A. and LANGENBERG, D. N., Phys. Rev. B 4, 1487 (1971).

9. FINNEGAN, T. F., and DENENSTEIN, A., to be published; T. F. Finnegan, Ph. D. Thesis, University of Pennsylvania, (1971).

10. WENNER, F., J. Res. Nat. Bur. Std. 25, 229 (1940).

11. DENENSTEIN, A., Ph. D. Thesis, University of Pennsylvania, (1969).

12. HAMON, B. V., J. Sci. Instr. 31, 450 (1954).

13. HARRIS, F. K., FOWLER, H. A. and OLSEN, P. T., Metrologia 6, 134 (1970).

14. TAYLOR, B. N., private communication.

15. EICKE, W. G., private communication.

16. PETLEY, B. W., private communication and Proceedings of this Conference.

17. HARVEY, I. K., private communication and Proceedings of this Conference.

18. COHEN, E. R., and TAYLOR, B. N., private communication and Proceedings of this Conference.

19. DRISCOLL, R. L. and OLSEN, P. T., private communication and Proceedings of this Conference.

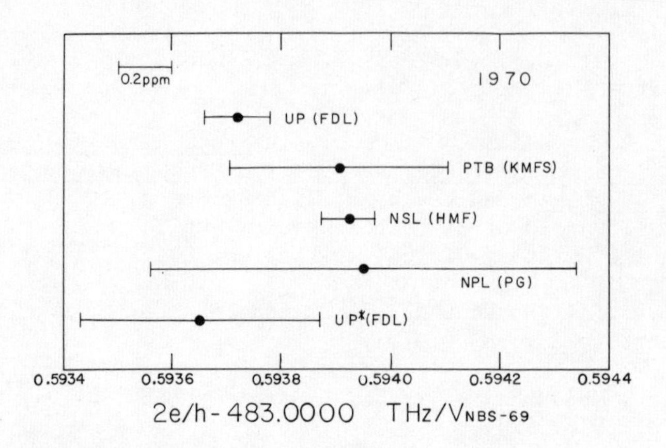

Figure 1. Comparison of the reported sub-ppm 2e/h determinations based on measurements made in 1970. All values have been expressed in NBS electrical units with the results of the 1970 BIPM comparisons. The error bars are those quoted by the authors for their values of 2e/h expressed in their respective national as-maintained units.

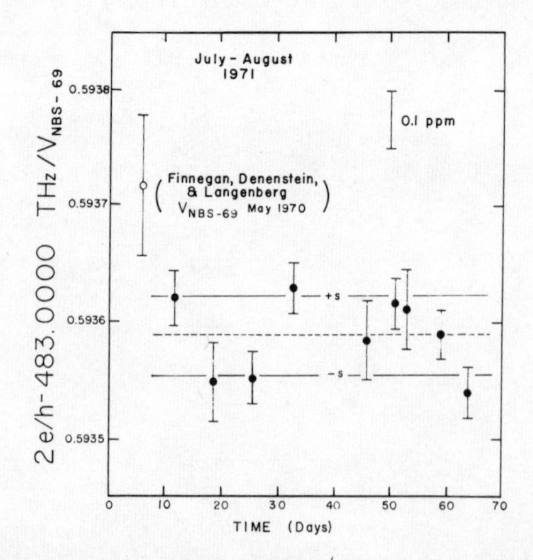

Figure 2. Experimental values of 2e/h determined at NBS during July and August, 1971, as a function of time. The one standard deviation error bars on the experimental points include the observed statistical uncertainty of the mean for the run (between 0.02 and 0.05 ppm) and an estimate of the random uncertainty in relating the measured standard cell emf during the run to the mean of the NBS reference group of cells. The final value is indicated by the dashed line; the observed (unweighted) standard deviation is indicated by the solid lines. The final value quoted by FDL (8) in NBS units is also shown for comparison.

Part 10 Rydberg Constant

Results and Possibilities in the Determination of the Rydberg Constant

L. Csillag

Central Research Institute for Physics, Budapest, Hungary

The importance of the spectra of the hydrogenic atom in the development of physical research is well known. The Sommerfeld–Dirac formula, together with the small quantum-electrodynamical corrections, describes both the positions and the fine structure of the levels exactly. The Rydberg constant, connected with it, has been determined very accurately, and we regard it as one of the fundamental constants of physics.

The value of the Rydberg is known with an accuracy of 0.1 ppm. The actual reliability of this value, however, is limited by some problems, connected with the experimental conditions.

The present value of the Rydberg is based mainly on pre-war measurements, the most accurate one being that of Drinkwater, Richardson and Williams (DRW) [1] on H_α and D_α. A reevaluation of the Rydberg, in the light of post-war results on the Lamb shift, has been performed by Cohen [2], by Martin [3] and by Taylor, Parker and Langenberg [4].

Of the most recent experiments I should like to mention the work of Masui [5] on the H_α, the measurement of Stoner [6] on H_β and H_γ using atomic beam technique, and the detailed studies of Roesler and his coworkers [7] on the 4686 Å line of He^+. An analysis of the results and possibilities in determining the Rydberg has been given by Series (8) in Gaithersburg last year.

Some years ago we also carried out accurate wavelength measurements on six lines of the Balmer series of deuterium (from D_β to D_η) using the classic Fabry–Perot techniques with photographic recording [9]. In the following I want to summarize and discuss the main results of this measurement and after that I should like to suggest some new ideas for further experiments.

The determination of the Rydberg constant from D_β–D_η

I think our experiment differs from others in two respects: first our measurement was carried out on six lines instead of one or two, and secondly high frequency ring discharge light source was used instead of a DC or AC discharge. There are some advantages and disadvantages connected with these characteristic points.

a) In our case, it was advantageous, that the overlapping of the components caused less uncertainties than it had in previous experiments.

To explain this, let us consider first the structure of the D_α line (Fig. 1). In the figure, the heavy bars represent the components, their lengths indicating the relative intensities. In a discharge light source, the lines are broadened by the Doppler-effect and overlap each other partially

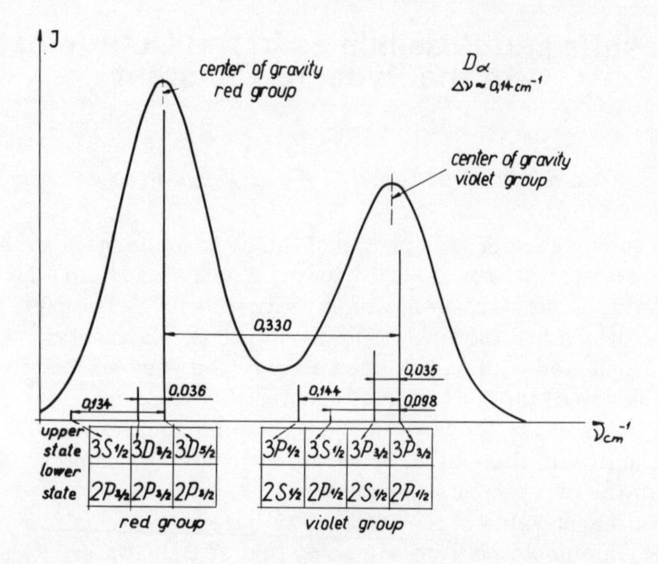

Fig. 1. The fine structure of the line D_α

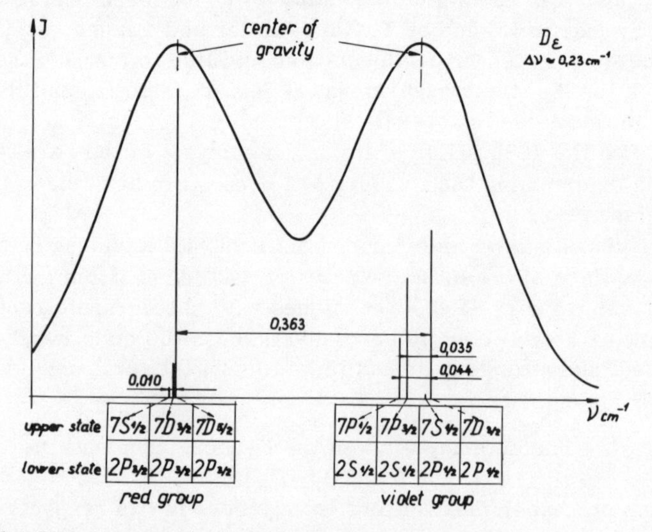

Fig. 2. The fine structure of the line D_ε

or fully, depending on their distance and the temperature. In the figure we have given the line contour as it appears at about 200°K. There are two groups of lines (a "red" group and a "violet" group), which give two maxima. By optical methods, only the positions of these maxima can be measured accurately. From these, the Rydberg can be evaluated by separating one component from the others. In a good approximation, for heavily overlapping components the maxima are given by their centre of gravity. On the basis of this assumption, and knowing the distances and relative intensities of the components, the separation can be easily effected. But this procedure is less accurate when the distances of the components are comparable with their line widths, as in the case with the D_α. In addition to this, the accuracy is further limited by the fact that the relative intensities of the components will differ from those expected theoretically. The greater the distance of the components from each other, the greater is the uncertainty in the procedure mentioned.

The situation is much more advantageous in the case of higher series members, since in that case the components of a group come closer to each other. As an example the line D_ε is shown (Fig. 2). In the red group, the effect of the two weak components on the $2P_{3/2}-7D_{5/2}$ "main" component is negligible. In the violet group, the effect of the weaker components on the $2P_{1/2}-7D_{3/2}$ "main" component must be taken into account, though all distances are very small. In any case, the position of the red group is less sensitive to possible intensity anomalies, than the blue one.

b) Another advantage of the six lines was that we had 12 independent data for the determination of the Rydberg. In this way, not only the statistical error of the measurement could be diminished, but also the effect of certain systematic error sources on the final result has been greatly reduced.

c) Let us consider now the advantages of high frequency excitation. In Fig. 3 the sketch of the light source is shown. The pyrex discharge tube of 20 cm length and 6 cm diameter is immersed in liquid nitrogen and surrounded by an external coil of ten turns excited by a 30 MHz oscillator. Pure deuterium gas of 0.025 torr pressure is circulated in the system. A small amount of water vapour, freezing on the wall of the tube, prevents the molecular recombination.

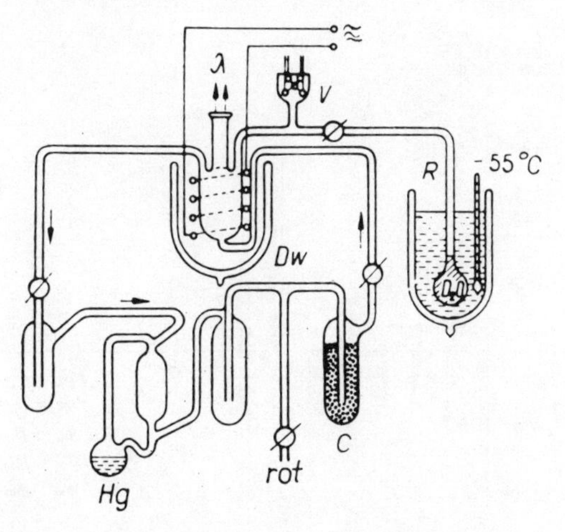

Fig. 3. Schematic diagram of the light source. Hg – mercury diffusion pump, C – active charcoal, R – D_2O recipient, V – thermocouple, Dw – Dewar flask, λ – direction of the light beam, rot – to rotation pump.

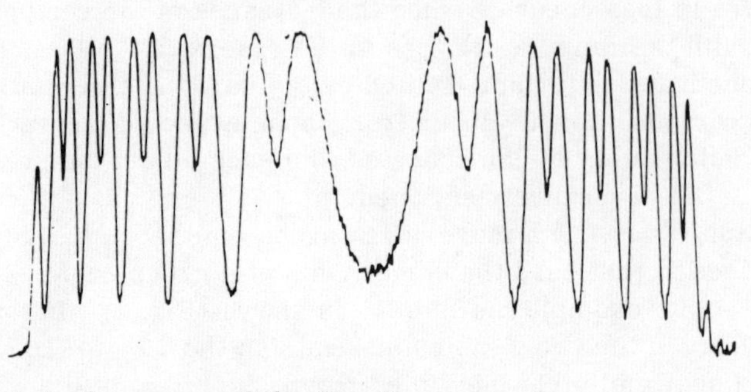

Fig. 4. FP interferogram of the line D_ε

This light source has the following features:

1) Though the excitation is weak, the produced spectrum is strong and free from molecular lines;

2) In the middle of the tube there is practically no external electric field;

3) There is no pressure effect;

4) One disadvantage is that the cooling is not very effective. The gas temperature in the middle of the tube is about 210°K. In spite of this, the two maxima for all lines can be resolved easily. As an example, Fig. 4 shows an interferogram of the D_ε line. The intensity of the saddle is 41% of the intensity of the maxima. For the purpose of comparison I would like to add that in our case the width of the D_β line was about equal to that of H_α of DRW.

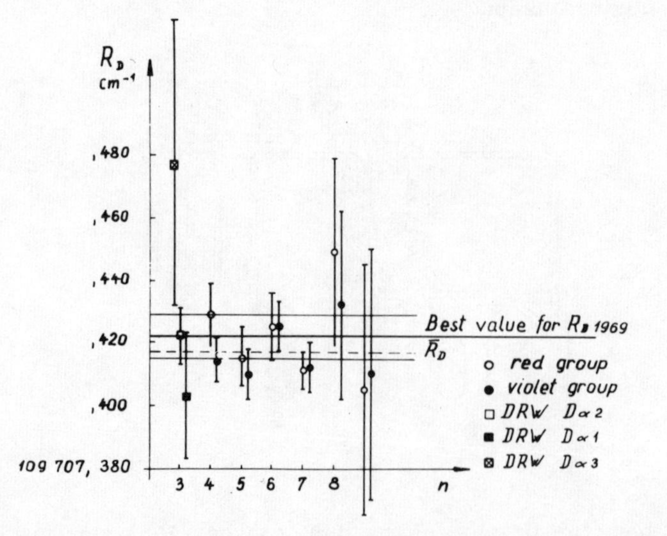

Fig. 5. Graphical representation of the Rydberg constants derived from the different Balmer lines. The straight lines indicate the measuring error.

Results

The numerical results of the measurement are summarized in Table I. This contains the measured vacuum wave numbers, the isolated wave numbers of the $2P_{3/2}-nD_{5/2}$ ("red") and $2P_{1/2}-nD_{3/2}$ ("violet") components and the values of the Rydberg constant, deduced from them.

The weighted average of R_D is:

$$\overline{R_D} = 109\,707{,}417 \pm 0{,}003 \text{ cm}^{-1}.$$

In Fig. 5 the deviations of the individual R_D-values from the average are illustrated. Generally they are less than the measuring errors. The results of DRW on D_α and the "best value for 1969" of Taylor, Parker and Langenberg [4] are also indicated. The agreement seems to be satisfactory.

TABLE I.
Measured wavenumbers and Rydberg constants for $D_\beta - D_\eta$

n	Line	Red group vac ν cm^{-1}	$2P_{3/2}-nD_{5/2}$ cm^{-1}	R_D cm^{-1}
4	D_β	20570,2177 ± 0,0017	20570,2206	109707,429 ± 0,009
5	D_γ	23038,6380 ± 0,0021	23038,6396	415 ± 0,010
6	D_δ	24379,5121 ± 0,0024	24379,5131	425 ± 0,011
7	D_ε	25188,0126 ± 0,0013	25188,0131	411 ± 0,006
8	D_ζ	25712,7710 ± 0,0066	25712,7714	449 ± 0,030
9	D_η	26072,5274 ± 0,0100	26072,5277	405 ± 0,040

n	Line	Violet group vac ν cm^{-1}	$2P_{1/2}-nD_{3/2}$	R_D cm^{-1}
4	D_β	20570,5452 ± 0,0013	20570,5684	109707,414 ± 0,007
5	D_γ	23038,9758 ± 0,0016	23038,9966	410 ± 0,008
6	D_δ	24379,8540 ± 0,0018	24379,8745	425 ± 0,008
7	D_ε	25188,3568 ± 0,0019	25188,3764	412 ± 0,008
8	D_ζ	25713,1118 ± 0,0066	25713,1314	432 ± 0,030
9	D_η	26072,8739 ± 0,0100	26072,8935	410 ± 0,040

Discussion

Let us now discuss the different sources of error. The statistical error of the measurement is only 0,03 ppm. But the actual error may be increased by possible systematic errors.

In our case, the main source of possible error is the uncertainty of the Hg^{198} wavelength standards. We used the wavelength values quoted in the literature [10, 11, 12] and since there is an uncertainty of about ± 0,0001 Å, it causes an error of ± 0,003 cm^{-1} in the R_D.

Another error source was measuring in non-standard air. Since the errors at the different lines are different in their sign, this error is not greater than 0,002 cm^{-1}.

The next source of error originated from using only one interferometer spacing. Since the phase change of the Fabry–Perot interferometer is also line dependent, the errors are averaged again, and the final error isn't greater than 0,002 cm^{-1}.

We have already discussed the problem of the evaluation of the Rydberg from the overlapping components. The error of the evaluation procedure can be estimated from the difference between the averages of R_D–s deduced from the "red" and the "violet" groups respectively. They are:

$$\overline{R_D}\,(\text{red}) = 109\ 707{,}418\ \text{cm}^{-1}$$
$$\overline{R_D}\,(\text{viol}) = 109\ 707{,}416\ \text{cm}^{-1}$$

The difference is only 0,002 cm^{-1}. Since the position of both maxima may change in the same direction too, the probable error due to overlapping is a little greater and amounts to 0,003 cm^{-1}.

The good agreement between the "red" and "violet" averages has indicated that in our experiment no significant deviations occurred in the intensity ratios of the components. This fact was proved more directly by measuring the relative intensities of the "red" and "violet" line groups. As is shown in Table II the agreement between the measured and calculated values is satisfactory. The error of the photographic measurements is about 5%.

TABLE II.
Measured and calculated intensity ratios for the red and violet groups

Line	D_β	D_γ	D_δ	D_ε	D_ζ	D_η
meas $\frac{I\ \text{red}}{I\ \text{viol}}$	1.04	1.07	1.02	1.01	1.01	1.02
calc $\frac{I\ \text{red}}{I\ \text{viol}}$	1.10	1.05	1.00	0.99	0.98	0.97

The last possible error source to be discussed is the Stark broadening. In our light source, the main cause of such perturbations could have been the electric field of the protons. But even if this field was present, its effect was not significant. This was proved by measuring the widths of the lines. The higher hydrogen states are much more sensitive to the Stark broadening, than the lower ones [13] and therefore the widths of the lines from D_β to D_ε ought to have grown faster than those given by the Doppler effect. But we measured the widths of these lines and did not find any extra broadening. Therefore no error limit can be given for this effect.

For the statistical sum of all errors: i.e. the uncertainty of the wavelength standard, the work in non-standard air, the phase error of the interferometer, the errors due to intensity anomalies and overlapping of the components and finally the statistical error of measurement, we get

$\pm 0{,}006\ cm^{-1}$. This is the real limit of the accuracy of our result. We can write:

$$\overline{R_D} = 109\ 707{,}417 \pm 0{,}006\ cm^{-1}.$$

We want to deal now with the problem of how to measure the Rydberg more accurately? The possibilities have already been thoroughly discussed by Series [8]. Now I will only mention two ways:

1) to repeat this experiment under more favorable conditions and

2) to carry out a new experiment using metastable H atoms and tunable dye laser excitation.

Ad. 1. On the basis of the above discussion it would be worth making a new accurate measurement on the lines $D_\alpha - D_\varepsilon$, using a similar light source, but with the following improvements:

a) using the Kr^{86} primary standard or another standard with great wavelength accuracy;

b) measuring in vacuum;

c) correcting the phase error of the interferometer;

d) using photoelectric recording;

e) making a computerised calculation of the deconvolution of the components and of the determination of the Rydberg.

I think with these modifications the real accuracy of the Rydberg could be increased to about $0{,}003\ cm^{-1}$.

Ad. 2. Another possibility for accurate measurements seem to be the use of the absorption of hydrogen atoms in the metastable $2S_{1/2}$ state. Here, the light source could be a tunable dye laser with narrow spectral linewidth.

This method would be very favorable, since in this case the Stark-effect and other perturbations, connected with the electric discharges, could be avoided and the line structure would become more simple as well. (Two components: $2S_{1/2}-nP_{3/2}$ and $2S_{1/2}-nP_{1/2}$ lying close to each other, with an intensity ratio of 2 : 1.)

The idea of the measurement is simple: in some way, metastable H atoms are produced and the shape of the absorption curve is determined.

It would be very convenient, if for such an experiment, the metastable atoms could be produced in an atomic beam. Much work has already been done in this field [6]. But the question is, whether the number of metastable atoms in the beam would be enough for observation?

Another possibility is again the high frequency discharge, but measuring in an afterglow period. After cutting off the exciting field, there is probably a time period, when the perturbing electric fields have already ceased to act, and the other states have also decayed, but the metastable $2S_{1/2}$ state has still a great population.

For absorption measurements, the parallelity of the laser beam is a great advantage, as absorption can be considerably increased by passing the beam several times through the tube. In this way an absorption length of 2 m can be easily achieved. For obtaining an intensity variation of 1%, a metastable density of $10^8/cm^3$ is required, which appears to be possible.

In the case of an atomic beam, the useful volume is less, but this is compensated by the sharpening of the absorption line, and therefore the conditions remain as before.

At present tunable dye lasers, covering the whole visible spectral region, are available already [14]. On the other hand using a suitable optical system, linewidths of 0,005 cm^{-1} or less can be achieved [15]. It appears that with further optical filtering it would not be difficult to reduce the linewidth to 0,0001 cm^{-1}.

It has been noted, however, that before carrying out such measurements the following problems have to be dealt with:

1) Investigation of the possibility of the production of a high density of metastable hydrogen atoms.

2) Construction of dye lasers with controlled wavelength and extreme narrow bandwidth, designed specially for such measurements.

3) Development of a high accuracy wavelength measuring technique on absorption lines.

Let us finish our discussion with referring to a measurement of 60 years ago. In 1914 Curtis [16] measured the wavelength of the first six Balmer lines of hydrogen very accurately. From his data and on the basis of the present theory the Rydberg constant can be calculated:

$$R_H \text{ (Curtis)} = 109\,677{,}59 \pm 0{,}01 \text{ cm}^{-1}.$$

Comparing this value with the present "best value" of Taylor, Parker and Langenberg [4]:

$$R_H \text{ (TPL)} = 109\,677{,}578 \pm 0{,}005 \text{ cm}^{-1}.$$

The deviation is hardly less than 0,01 cm^{-1}.

This excellent agreement over a distance of 60 years is a good demonstration of the permanence of physical measurements. The good work of our predecessors should stimulate us to make further efforts to achieve more accurate results in this field.

References

1. DRINKWATER, J. W., RICHARDSON, O. and WILLIAMS, W. E., Proc. Roy. Soc., A174, 164 (1940).

2. COHEN, E. R., Phys. Rev., 88, 353 (1952).

3. MARTIN, W. C., Phys. Rev., 116, 654 (1959).

4. TAYLOR, B. N., PARKER, W. H., and LANGENBERG, D. N.: The Fundamental Constants and Quantum Electrodynamics, Academic Press (New York–London), 58-67 (1969).

5. MASUI, T., Paper presented at the Internat. Conf. on Prec. Meas. and Fundam. Const., Gaithersburg (1970).

6. STONER, J. O., Jr., Thesis, Princeton University (1963), private communication.

7. BERRY, H. G. and ROESLER, F. L., Phys. Rev., A1, 1504 (1970).

8. SERIES, G. W., Paper presented at the Internat. Conf. on Prec. Meas. and Fundam. Const., Gaithersburg (1970).

9. CSILLAG, L., Acta Physica Acad. Sci. Hung., 24, 1 (1968).

10. BLANK, I. M., Journal of the Opt. Soc. Am., 40, 345 (1950).

11. MEGGERS, W. F. and KESSLER, K. G., Journal of the Opt. Soc. Am. 40, 737 (1950).

12. KAUFMAN, V., Journal of the Opt. Soc. Am., 52, 866 (1962).

13. BETHE, H. A. and SALPETER, E. E.: Quantum Mechanics of One- and Two-Electron Systems. Encyclopedia of Physics. ed. by Flugge, Vol. XXV, 326. Springer, Berlin (1957).

14. KAGAN, M. R., FARMER, G. I., and HUTH, B. G., Laser Focus, 26 (1968).

15. WALTHER, H. and HALL, J. L., Appl. Phys. Letters, 17, 239 (1970).

16. CURTIS, W. E., Proc. Roy. Soc., A90, 605 (1914).

A New Measurement of the Rydberg

B. P. Kibble and W. R. C. Rowley

Division of Quantum Metrology,
National Physical Laboratory, Teddington, England

and

G. W. Series

Physics Department, University of Reading, England

1. Introduction

The Rydberg constant which determines the wavelengths of spectral lines of one-electron atoms may be expressed in terms of the fundamental physical constants as

$$R_\infty = \frac{m_e e^4}{4\pi\hbar^3 c} \cdot \left(\frac{\mu_o c^2}{4\pi}\right)^2$$

and because it can be measured with relatively high accuracy it plays a key role in assessing the most probable measured values of these constants. The presently accepted value of R_∞ has a quoted error of 1 in 10^7 and the ever increasing accuracy of other measurements in such an assessment now means that this error is now only just sufficiently small, whereas previously an accuracy of 1 in 10^6 would have been sufficient.

Notwithstanding the remarkable agreement between the results of the four determinations on which the present value is based (1-4) the possible systematic errors listed below have caused the accuracy claimed to be questioned (5).

a) Unresolved blends of fine structure components were measured. Mixing and displacement of states by electric fields present in the light source (Stark effects) could have caused undetected shifts and intensity anomalies in these components.

b) Wavelength standards other than the krypton-86 606 nm primary standard were used.

c) The signal/noise and linearity of the intensity scale in the interferograms were poor because of the photographic techniques used.

d) The possible inequality of light flux distribution through the interferometer between light of the different wavelengths to be compared is not discussed.

e) In some cases there is no mention of the differing phase change at the reflecting surfaces of the interferometer for the two wavelengths to be compared.

We have therefore undertaken a redetermination in which particular attention has been paid to these sources of error.

We excite the spectrum of an atomic hydrogen isotope with a radio frequency discharge in a sealed bulb which is cooled in liquid helium and we measure the wavelength of the light arising from the n=3 to n=2 Balmer-α transition in terms of the krypton-86 standard. The fine structure of these states is illustrated at the top of Figure 1 and the components (1)-(5) from the observed transitions at the expected wavelength intervals are drawn at the bottom. The vertical height of the lines represents the expected relative intensity. In the centre of the figure is one order, obtained point by point, of a deuterium Balmer-α interferogram in which it will be seen that these observable components are broadened by the Doppler effect amongst the thermally moving atoms, and by the instrumental width of the interferometer. The peak plotted with crosses is that of the krypton-86 606 nm line, which is sufficiently narrow for the observed peak to give a good qualitative idea of the instrumental width. The metrological problem is to determine the position of any one fine structure component with respect to the krypton-86 peak, and clearly the inability, even at a low discharge temperature of a few tens of kelvins, to resolve these components fully will render some deconvolution process necessary. Nevertheless the resolution obtained is a considerable improvement on that of the previous determinations.

2. The apparatus

The light source is illustrated in Figure 2. A pyrex bulb blown on to the end of a short pyrex rod is filled with a mixture of deuterium or tritium and a little helium. In the case of tritium the latter is provided by the natural radioactive decay of the tritium to ^{3}He in the course of a few weeks. As the bulb cools down in the liquid helium the deuterium is condensed and a radio frequency discharge starts between the external capacitative electrodes in the helium in the bulb. The heat generated serves to increase the vapour pressure of the deuterium and the proportion excited can be adjusted simply by altering the radio frequency power level. In order to prevent excessive condensation of deuterium on the colder parts of the bulb remote from the discharge region these parts are lagged with solid nylon caps which have a poor thermal conductivity at liquid helium temperatures. Light from the bulb is conveyed to the outside of the standard helium storage dewar by a fused silica light pipe.

The wavelength comparator illustrated in Figure 3 centres round a pressure scanned Fabry-Perot etalon, with the centre of the fringe system imaged on a small circular hole. Light from a standard krypton-86 lamp and from the deuterium lamp is chopped at two different frequencies and enters separate collimators. Light of the appropriate spectral line from each source is selected by prisms and interference filters and is then combined to pass through the etalon, imaging lens and hole onto the photomultiplier. Two lock-in amplifiers locked to the two chopping frequencies separate the photocurrents corresponding to the intensity of each source. These are stored and punched on paper tape at intervals governed by the advancement of the precision screw driving the piston which alters the pressure of dry air in the etalon box thus scanning the fringe system.

We employ a method of substitution in order to perform the wavelength comparison. First the upper channel is calibrated with reference to the lower by sending standard light from the front of the krypton-86 lamp through the lower via the mirror M, and light from the back through the upper. Since the two collimators do not illuminate the etalon equally (the upper is in fact half the aperture of the lower) a small systematic difference is found. M is then removed and light of the wavelength to be measured enters the lower collimator, the source having been arranged to occupy the same position as the virtual image of the krypton-86 lamp in M. In this way equal illumination of the etalon and hole by the sources put in turn in front of the lower collimator is assured. The light passing through the upper collimator merely serves as a monitoring wavelength with the convenient feature that it differs in wavelength by only 2 in 10^8 from the standard light from the front of the krypton lamp.

3. The results

The typical interferogram shown in Figure 1 has a sufficiently good signal/noise ratio for us to develop a technique to determine the constituents of the compound peaks (1)+(4) and (2a)+(2b) by a process of deconvolution of the instrument function, which is known from the krypton-86 peak, followed by curve-fitting to obtain the positions and intensities of these components. Meanwhile to judge the quality of the results as they were obtained, the positions of the peaks (1) + (4) and (2a) + (2b) were measured by finding the centroid of an upper fraction of the peaks. This was then corrected, assuming the theoretical intensity ratio and displacement of the constituent components, to give the positions and hence the wavelengths of components (1) and (2b). We then calculated what departure from the value of R_∞ assumed in the tabulation of Garcia and Mack (6) would give the observed wavelengths of (1) and (2b) for each separate peak on every interferogram. These departures are plotted in Figure 4. A correction has been made for the dispersion of air between the 606 nm krypton-86 wavelength and the 656 nm Balmer-α wavelength assuming Edlen's formula (7).

The values of R_∞ implied by the arithmetic means of these results are summarized below.

$$R_\infty = 109\,737.310 + \text{cm}^{-1}$$

	(1)	(2b)
Deuterium	−0.008	−0.008
Tritium	+0.010	+0.021

The effect of possible Stark shifts has not yet been fully investigated, but an experimental attempt to assess their effect was made by altering the power level in the lamp as much as possible, thus changing the ion density and with it the local electric field seen by the emitting atoms. A change in the relative intensity of the blends (1) + (4) to (2a) + (2b) tending to the theoretical value with reduction of the power level in the lamp was observed. There do not seem to be any changes in the peak position correlated with the changes in relative intensity.

The general interferometric techniques and the effect of phase change in the etalon films were tested by measuring the wavelength of the 644 nm cadmium red line from a lamp using exactly the same technique as for the Balmer-α wavelength measurement. The results agreed to within 5 in 10^8 with that obtained by a more accurate determination using a much longer etalon with aluminium films.

4. Discussion of results

The excellent agreement between the values obtained from the two main peaks in the deuterium interferogram implies that the fine structure of deuterium is properly represented and is in good agreement with the results obtained by much more accurate radio frequency and level-crossing techniques (8). Preliminary measurements of the position of component 3a shows that this also is in its expected position. Thus there is some evidence at this early stage in the analysis that Stark effects are not too important.

The agreement between the results obtained for R_∞ from the two main peaks in the tritium interferograms is less good, but a possible explanation lies in the contamination of the deuterium and tritium lamps with hydrogen. Measurements made with a high resolution grating spectrometer show that whereas the hydrogen Balmer-α intensity was less than 3% of the deuterium Balmer-α intensity in the deuterium lamps, and moreover the choice of etalon spacer had been made to cause the fine structure patterns of the two isotopes to coincide so that the only effect of the hydrogen contamination would be to broaden the peaks slightly, the same condition did not hold for tritium. The hydrogen Balmer-α intensity of up to 15% in the tritium lamps could well have displaced the peaks slightly, and this will be corrected in the course of profile analysis of the interferograms but it is worth remarking that Kireev (9) has measured the hydrogen-tritium isotope shift and he also found an anomaly of the same sign and magnitude as that implied by our preliminary results.

For the present, from a preliminary measurement of the interferograms which may be corrected by techniques now under development, we conclude that our results with deuterium do not conflict with the value of the Rydberg quoted by Taylor *et al.* (10) in their review of values of the fundamental constants. The results with tritium are discrepant, but this may well be due to contamination of the sample with hydrogen.

5. Acknowledgements

We thank Mrs. S. Swan and Mr. B. Hinde of the NPL for processing the data presented here.

References

1. HOUSTON, W.V., Phys. Rev., 30, 608-13 (1927).

2. CHU, D.-Y., Phys. Rev., 55, 175-80 (1939).

3. DRINKWATER, J.W., RICHARDSON, O. and WILLIAMS, W.E., Proc. Roy. Soc., A174, 164-88 (1940).

4. CSILLAG, L., Acta Phys. Acad. Sci. Hungar., 24, 1-18 (1968).

5. SERIES, G.W., Proc. Int. Conf. Prec. Meas. and Fund. Constants, Gaithersburg 1970. (To be published.)

6. GARCIA, J.D., and MACK, J.E., J. Opt. Soc. Amer., 55, 654-85 (1965).

7. EDLEN, B., J. Opt. Soc. Amer., 43, 339 (1953).

8. TAYLOR, B.N., PARKER, W.H. and LANGENBERG, D.N., Rev. Mod. Phys., 41, 375-496 (1969).

9. KIREEV, P.S., Dokl. Acad. Nauk. USSR, 112, 41 (1957).

10. TAYLOR, B.N., PARKER, W.H. and LANGENBERG, D.N., loc. cit.

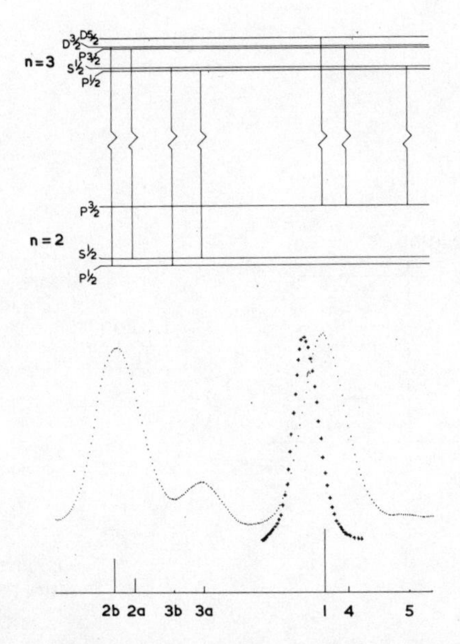

Fig. 1. The term diagram for the transitions between n=3 and n=2 in atomic hydrogen, with the expected fine structure components and a typical observed interferogram. The peak plotted with crosses is that of the ^{86}Kr standard line.

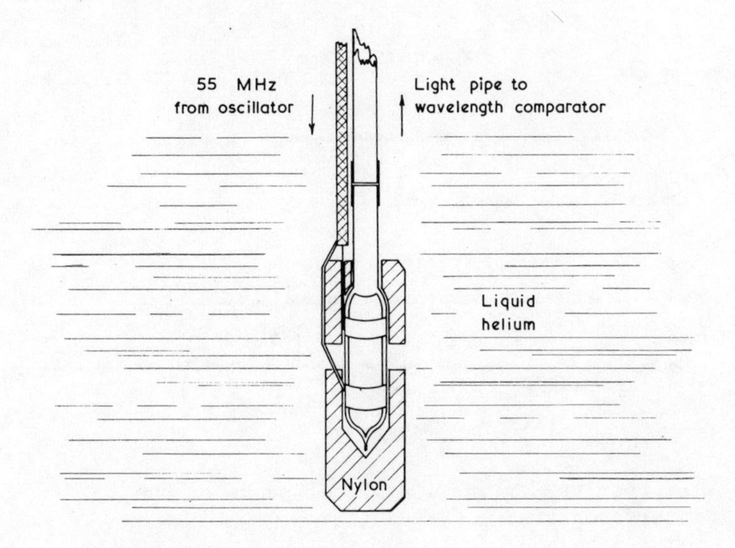

Fig. 2. The light source.

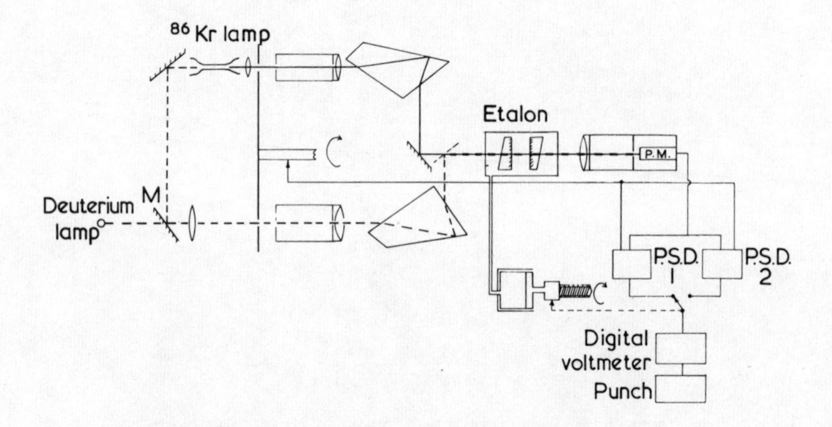

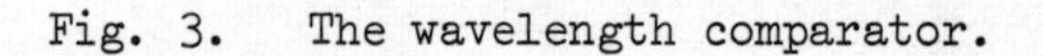
Fig. 3. The wavelength comparator.

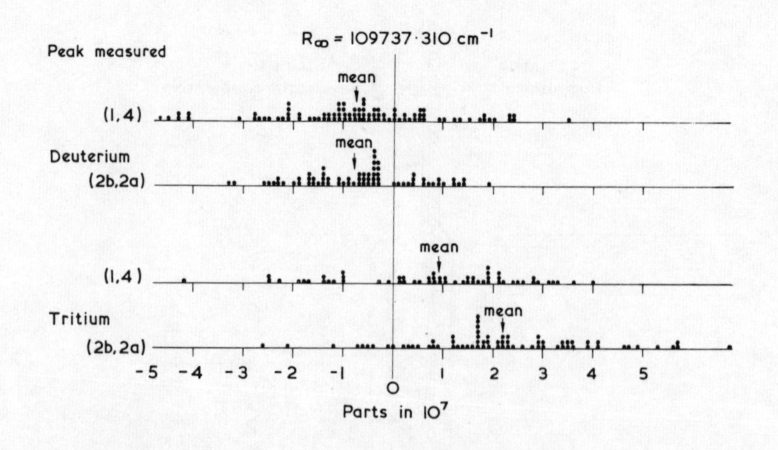

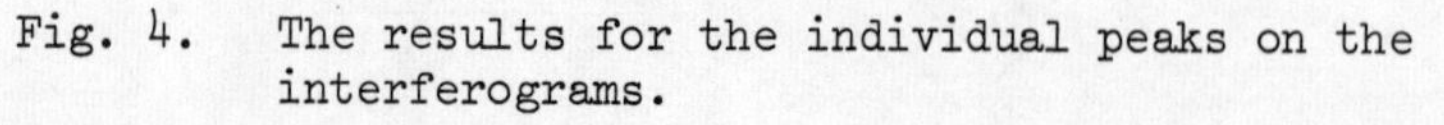
Fig. 4. The results for the individual peaks on the interferograms.

Determination of the Rydberg Constant from the HE11 $n = 3\text{-}4$ (4686Å) Line Complex

E. G. Kessler, Jr.

Institute for Basic Standards, National Bureau of Standards, Washington, D.C. 20234 U.S.A.

1. Introduction

The review of the fundamental physical constants by Taylor, Parker, and Langenberg (1) emphasized the need for a more accurate value for the Rydberg constant. At the International Conference on Precision Measurement and Fundamental Constants, Series (2) recommended that the most advanced spectroscopic techniques be used to measure the wavelengths of spectroscopically resolved fine structure components of hydrogen (and its isotopes) and ionized helium in order to redetermine the Rydberg. With the encouragement of the above individuals and Drs. William C. Martin and Frederick L. Roesler, a redetermination of the Rydberg is being pursued at NBS based on absolute wavelengths of the best resolved fine structure components of the n = 3-4 (4686Å) transition in He II. A preliminary discussion of this experiment was presented at the International Conference on Precision Measurement and Fundamental Constants (3). This paper is a progress report on that experiment.

2. The He II n = 3-4 (4686Å) Line Complex

The level structure of the He II n = 3-4 transition is shown in Fig. 1. The 13 fine structure components are spaced over a 2.5 cm^{-1} range and are numbered in order from high to low wavenumber for ease of discussion. This transition has been chosen because it is the simplest He II transition accessible to Fabry-Perot spectroscopy. High resolution studies of this line by Roesler and Mack (4), Larson and Stanley (5), and Berry and Roesler (6) have provided an understanding of this transition which is essential for the determination of absolute wavelengths. Specifically, an explanation of (or at least a method of monitoring) intensity anomalies, stark shifts, and differential Doppler shifts has resulted from these investigations.

A representative spectrum of 4686Å line excited in a liquid-nitrogen-cooled hollow cathode is shown in Fig. 2. The resolution is limited by the widths of the He II fine structure components which are anomalously large compared to widths of He I lines produced in the same discharge. Anomalously wide He II lines have been observed in an electron excited atomic beam as well as in hollow cathode discharges and have been attributed to the transfer of linear momentum from the exciting electron to the excited He^+ state during the excitation process (5,6). Berry and Roesler have improved the resolution by using liquid-helium cooling, which reduces the line widths to approximately 0.8 times the line widths obtained when liquid nitrogen was employed as a coolant. However, because of the spacing and intensity distribution of the fine structure components, wavelength measurements on components 5+6 and 9 should be more precise than measurements on the other components. A reduction in the widths of components 5+6 and 9 to 0.8 of the value shown in Fig. 2 would improve the precision of the measurements by only a small amount while significantly increasing the difficulty of the experiment. Thus liquid-nitrogen-cooling has been used throughout the experiment.

In the previous discussion of this experiment, wavelength measurements on components other than 5+6 and 9 in the n = 3-4 transition and on fine structure components in other He II transitions (n = 4-6, 6560Å and n = 4-5, 10124Å) were proposed. At the present time wavelength measurements on components other than components 5+6 and 9 in the 4686Å transition are no longer being considered.

3. Instrumentation

The instrumentation for this experiment is a pressure-scanned double-Fabry-Perot spectrometer shown schematically in Fig. 3. The ionized helium spectrum is excited in a double-anode hollow cathode. This design is necessary to correct for a differential Doppler shift of the fine structure components due to the drift of the positively charged helium ions toward the cathode (7). The drift shift has been measured as a function of cathode diameter and pressure, and it decreases with increasing cathode diameter and increasing discharge pressure. To reduce the drift shift, cathode diameters greater than 15 mm and discharge pressures greater than 0.3 torr have been used for all data. However, wavelengths are measured with the ions drifting toward and away from the spectrometer so that systematic errors resulting from any remaining drift shift may be eliminated.

The optical train for the He II radiation consists of a grating premonochromator operated at low resolution and two Fabry-Perot etalons, E_1 and E_2, in separate pressure chambers. The Fabry-Perot fringes are focused upon a circular aperture and are detected with a photomultiplier. Two etalons are necessary to obtain an instrumental width of 0.030 cm^{-1} while suppressing overlapping orders. Facilities to maintain a constant pressure difference between the two etalons are provided. More detailed discussions of Fabry-Perot spectrometers are provided in the literature (8,9,10).

The standard source is a water cooled ^{198}Hg electrodeless lamp, S. Light from the standard source can be passed through the same optical train as the He II radiation or can be passed by means of a partially reflecting mirror M through etalon E_1 and the aperture and detected by the photomultiplier. The He II and the standard source light beams are provided with choppers at two different frequencies so that the He II spectrum and the standard source can be scanned simultaneously. The photomultiplier output is sent to two lock-in detectors which detect and demodulate the signals from the two sources.

An auxiliary calibration system consists of a ^{198}Hg electrodeless lamp, L, a long spaced etalon, E_c, which is scanned simultaneously with etalon E_1, and a photomultiplier. The calibration system provides a wavenumber scale for double etalon scans of the He II 4686Å line and allows correction for nonlinearity of the scan rate.

Signals from the He II spectrum, the standard source, and the calibration lamp are recorded on a strip chart recorder and digitally punched on paper tape. Component positions, intensities, and widths are measured by reconstructing the recorded spectrum with the aid of a computer. The reconstructed spectrum is generated by substituting a numerical profile with adjustable position, intensity, and width for each component and is least-squares fitted to the recorded spectrum.

4. Experimental Procedures

The procedures for recording and analyzing data are somewhat different from those described in the earlier report (3). Neither the data recording nor the data analysis has yet been completed.

The 4686Å line is first passed through the double etalon system and recorded. Both etalons are scanned and a constant pressure difference is maintained between the etalons. A representative spectrum obtained in this manner is shown in Fig. 2. Relative positions, line widths, and relative intensities obtained from spectra such as this aid in the decomposition of overlapping single etalon scans from which absolute wavelengths are determined.

Absolute wavelengths are determined by the method of exact fractions (11) which requires recording at least two Fabry-Perot orders. A double etalon system in which both etalons are scanned cannot be used because repeated orders are suppressed. By removing etalon E_2 from the optical train, repeated but overlapped orders of the 4686Å line are obtained as is shown in Fig. 4. The decomposition of spectra such as this to obtain accurate positions of components 5+6 and 9 is a formidable task. However, the decomposition will be made easier by the use of information obtained from double etalon scans. The intensities of components 1, 2, 3, 4, 7+8, 10, 11, 12 and 13 (hereafter called minor components) relative to component 5+6 will be fixed at values observed in the double etalon scans. The widths of the minor components will be set equal to the double-etalon-scan widths adjusted for removal of etalon E_2. The minor component positions will be fixed by the etalon spacing and the measured relative positions. All parameters of components 5+6 and 9 will be permitted to vary to obtain the best least squares fit.

Repeated orders for a particular component of the 4686Å line can also be obtained with etalon E_2 in the optical train. The pressure in etalon E_2 must be adjusted for maximum transmission of the component under consideration and must then be held constant while etalon E_1 is scanned. Etalon E_2 serves as a narrow pass filter, of approximately 0.13 cm^{-1} half width, to isolate a particular component of the He II structure. A representative spectrum of component 5+6 and component 9 obtained in this way is shown in Fig. 5. Component 5+6 and component 9 are each only slightly evident in the scan of the other component. All other components are suppressed to the extent that their presence is not visually detected. The positions of components 5+6 and 9 will again be determined by computer fitting scans such as these. Two fitting procedures will be used. The first assumes that the recorded spectrum results only from component 5+6 or component 9. The second assumes that the recorded spectrum is a combination of the primarily transmitted component and the suppressed components which will be included with parameters fixed as explained above except that the intensities will be multiplied by the transmission of etalon E_2.

Two experimental procedures are used to obtain He II wavelength measurements.

Procedure A: Initially the He II fine structure components were compared to the standard lines by alternately passing the He II spectrum and the standard lines through the grating premonochromator and the etalons. Precautions were taken to avoid systematic errors associated with temperature changes of the scanning gas and etalons, mechanical instability of the etalons, and measurement of exact fractions for the standard line and the He II component at different indices of refraction (different gas pressures). A He II wavelength measurement took approximately two and one-half hours and consisted of a scan of two orders of the standard line, a scan of two orders of the He II structure, and then a second scan of two orders of the standard line. The measured order numbers of the two scans of the standard line typically differed by 1 to 2 parts in 10^7.

Procedure B: By chopping the light from the standard lamp and the hollow cathode at different frequencies, the standard and unknown wavelengths can be simultaneously scanned, thus avoiding or reducing errors related to temperature changes, mechanical instability, and measurement of exact fractions for the standard and unknown at the same index of refraction. The light beams from the standard lamp and the hollow cathode are chopped at 500 Hz and 13 Hz respectively and are combined before passing through etalon E_1 (Fig. 1). A He II wavelength measurement takes approximately 50 minutes and consists of two orders of the He II structure and two orders of the standard line recorded simultaneously. The signal/noise ratio is decreased when the signals are chopped and recorded simultaneously because 50% of the He II photons are excluded from the detector and the signal from the standard lamp adds noise to the recorded He II spectrum. In fact, the standard line must be passed through a neutral density filter which reduces its intensity to a level comparable to the He II spectrum in order to obtain a usable He II signal/noise ratio. However, this technique significantly reduces the signal/noise ratio of the standard line. In Figs. 4 and 5 the He II and standard signals were chopped. In Fig. 2 the He II signal was not chopped.

5. Preliminary Results

A large number of scans were recorded using procedure A. Thirty scans in which etalon E_2 was in the optical train were completely analyzed - twelve scans using a 4.5 mm spacer and eighteen scans using a 6 mm spacer. Eight measurements of the wavelength of component 5+6 were made. The value of the Rydberg constant obtained from these measurements is $R_\infty = 109{,}737.314 \pm 0.055\ cm^{-1}$ where the error limit represents one standard deviation.

In a completely separate experiment in this laboratory results of comparable accuracy have also been obtained from wavelength measurements of the He II n = 2-3 (1640Å) transition. This line along with germanium, silicon, and thorium standards has been photographed on a 10.7 m and a 7 m vacuum spectrograph. Two of the seven fine structure components are relatively well isolated and from them a value of $R_\infty = 109{,}737.314 \pm 0.043\ cm^{-1}$ was determined.

A number of simultaneous scans have already been obtained using procedure B with one spacer. Shifts in the measured order numbers of the standard line between successive scans are accompanied by corresponding shifts in the measured order numbers of the He II components accurate to 1 part in 10^7. These data have not been completely analyzed at present.

In conclusion, the results obtained thus far represent an accuracy of 5 parts in 10^7 for procedure A. A completely different experiment involving the 1640Å line yields results of comparable accuracy. Data analyzed to this date indicate that procedure B is capable of achieving an accuracy of 1 part in 10^7. The accuracy is at present limited by the stability of the instrument. If all stability problems were solved, the inherent line width would limit the accuracy to approximately 5 parts in 10^8.

References

1. TAYLOR, B. N., PARKER, W. H., and LANGENBERG, D. N., Rev. Mod. Phys. 41, 399 (1969).

2. SERIES, G. W., Proceedings of the International Conference on Precision Measurement and Fundamental Constants, NBS Special Publication 343 (U.S. Government Printing Office, 1971). (to be published).

3. KESSLER, JR., E. G., and ROESLER, F. L., Proceedings of the International Conference on Precision Measurement and Fundamental Constants, NBS Special Publication 343 (U.S. Government Printing Office, 1971). (to be published).

4. ROESLER, F. L., and MACK, J. E., Phys. Rev. 135, 58 (1964).

5. LARSON, H. P., and STANLEY, R. W., J. Opt. Soc. Am. 57, 1439 (1967).

6. BERRY, H. G. and ROESLER, F. L., Phys. Rev. 1A, 1504 (1970).

7. ROESLER, F. L. and DeNOYER, L., Phys. Rev. Letters 12, 396 (1964).

8. CHABBAL, R., Rev. Optique 37, 49 (1958).

9. CHABBAL, R. and Jacquinot, P., Rev. Optique 40, 157 (1961).

10. MACK, J. E., McNUTT, D. P., ROESLER, F. L., and CHABBAL, R., Appl. Opt. 2, 873 (1963).

11. MEISSNER, K. W., J. Opt. Soc. Am. 31, 405 (1941).

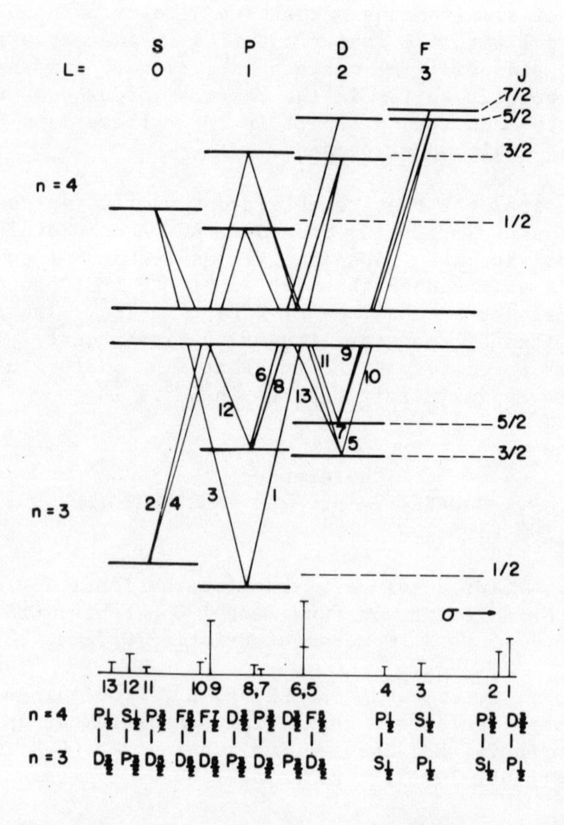

Fig. 1. Energy level diagram of the He II n = 3-4 (4686Å) transition. Theoretical relative positions and statistical intensities of the fine structure components are shown below the level diagram.

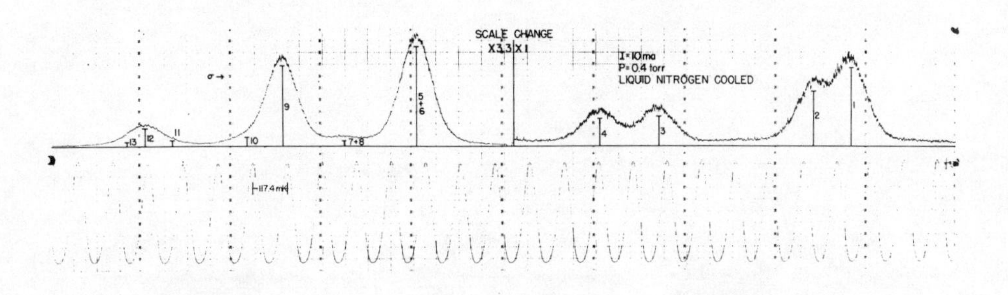

Fig. 2. The He II 4686Å spectrum excited in a liquid-nitrogen-cooled aluminum hollow cathode. The current was 10 mA and the pressure 0.4 torr. The peak to peak separation of the calibration sine curve is 0.1174 cm^{-1}.

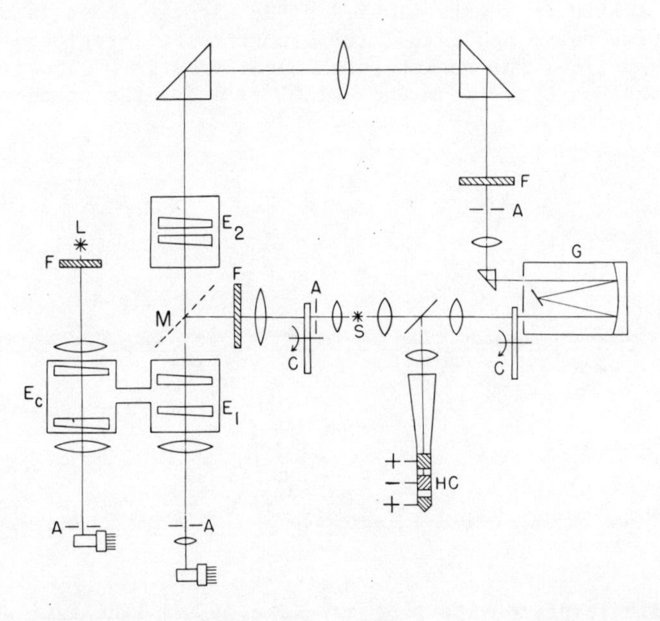

Fig. 3. Sources and spectrometer. E_1, E_2, and E_c are the resolving etalon, suppression etalon, and calibration etalon, respectively. The other parts are: F, filters; C, choppers; A, circular apertures; G, grating premonochromator; M, partially reflecting mirror; HC, hollow cathode; S, standard source; and L, calibration lamp.

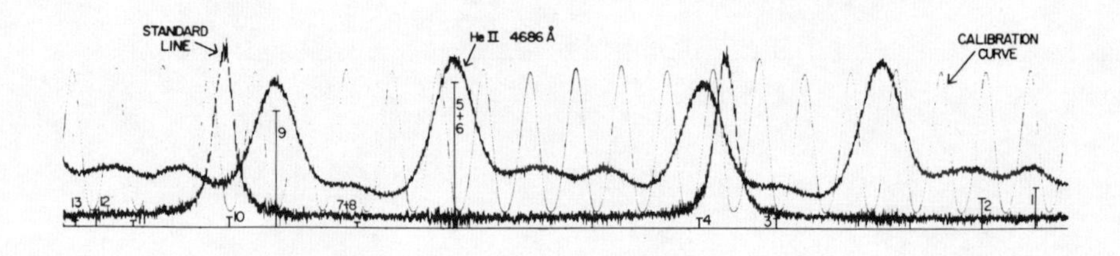

Fig. 4. Single etalon scan of the He II 4686Å line excited in a liquid-nitrogen-cooled aluminum hollow cathode. The current was 10 mA and the pressure 0.4 torr. The 13 fine structure components drawn on the figure are all in the same order. The standard wavelength is the 5461Å line excited in a water-cooled ^{198}Hg electrodeless lamp. The sine curve provides interferometric calibration of the spectrum; the peak to peak separation is 0.1174 cm^{-1} for the He II spectrum and 0.1101 cm^{-1} for the standard line.

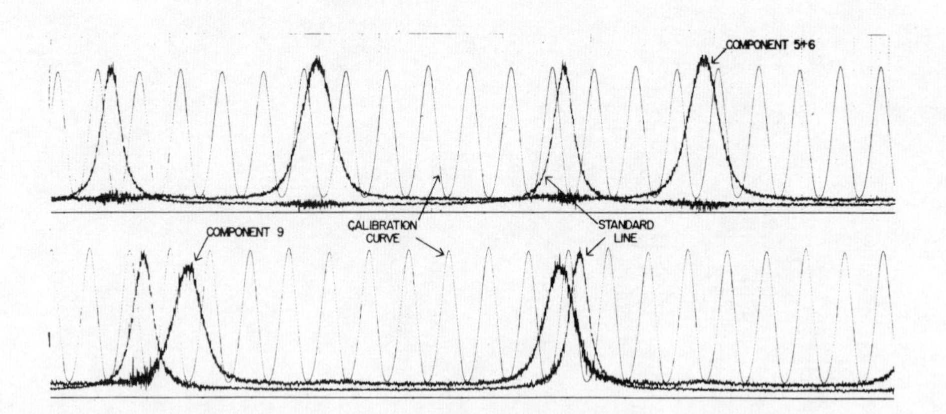

Fig. 5. Single etalon scans of the He II 4686Å line excited in a liquid-nitrogen-cooled hollow cathode. The pressure in etalon E_2 was adjusted to isolate component 5+6 and component 9 in the upper and lower scans respectively. The discharge current was 10 mA and the pressure was 0.4 torr. The standard wavelength was the 5461Å line excited in a water-cooled ^{198}Hg electrodeless discharge lamp. The interferometric calibration is the same as in Fig. 4.

Excitation-Dependent Characteristics of H and HeII Lines used in Determining the Rydberg Constant

F. L. Roesler

Dept. of Physics, Univ. of Wisconsin, Madison, Wisc. U.S.A.

1. Introduction

Series (1) has given an excellent general review of the situation concerning the Rydberg constant, and there is no need to elaborate here. Basically the problem is that most of the data on which the presently recommended value of the Rydberg constant is based was obtained more than thirty years ago, and that as a result of more recent advances in instrumental technique and in the understanding of various complications arising from the methods used in exciting hydrogen-like spectral lines there is the risk of uncertain systematic errors in the old data and the possibility of improved precision of future measurements. It is satisfying that at this conference there are two papers (2,3) from different laboratories reporting progress on new evaluations of the Rydberg constant based on measurements of deuterium and ionized helium. Since many of the technical complications in measuring hydrogen-like lines will be discussed in those papers, this paper will concentrate on the effects different excitation techniques can have on the appearance of these lines. The effects are of concern in measuring the Rydberg constant because the complicated fine structures of hydrogen-like lines are never fully resolved, and knowledge of the relative intensities and widths of the components is necessary for most satisfactory correction for blending. Moreover, it would be obviously advantageous if a method for producing very narrow hydrogen-like lines could be discovered.

Figure 1 shows the structure of the n=3-4 transition of He^+. This structure has been extensively studied (4-7), and will be the principal concern of this paper. Detailed work on other ionized helium lines (8,9) tends to confirm the conclusions drawn from the n=3-4 transition. There are thirteen allowed components making up its fine structure, and for convenience these have been numbered in order from high to low wavenumber. Figure 2 shows this structure excited in a liquid-helium-cooled D.C. hollow-cathode discharge, and observed with a scanning double-Fabry-Perot spectrometer. (See Ref. 2, Fig. 3 for an example of the experimental arrangement.) It is equal in resolution to the best that has been achieved for this transition, and shows clearly eight components. It also shows that none of the components is sufficiently free from blending that the influence of its neighbors can be fully neglected in determining its position. We shall now consider how this structure changes with changes in the excitation conditions.

2. Hollow-Cathode Excitation

The hollow-cathode excitation of ionized-helium lines has been studied in considerable detail (4-6,8,9). The most obvious feature, shown in Fig. 3, is the dependence of the relative intensities on the filling pressure of the discharge tube. At high pressures statistical relative intensities are approached, and as the pressure is reduced all components increase relative to the reference component 9, arising from the $4^2F_{9/2}$ level. This anamolous intensity behavior can be explained by the different cross sections for excitation to the 4s,p,d,f levels and the subsequent alteration of their populations towards a statistical distribution by collisions with neutral gas atoms, the latter effect being greater at greater pressures.

For current densities below about 15ma/cm^2 the relative intensities seem to have no dependence on current. For much higher current densities lines arising from upper P levels increase slightly, consistent with the relatively larger cross sections for excitation to excited P states for electron excitation of ground state ions (6).

Another anomaly, which becomes obvious only for the liquid-helium-cooled discharge, is the excessive width of the ionized-helium lines. From the results obtained with neutral helium excited in a liquid-helium-cooled hollow cathode (10), widths were expected approximately 1/4 as great as actually observed for ionized helium. Comparison of the widths of ^{3}He and ^{4}He isotopes (6) showed that the widths were proportional most nearly to the mass of the atoms, indicating that the excessive width was a result of momentum transfer from the exciting electrons. A study by Larson and Stanley (7) on electron excitation of a helium atomic beam showed similar results. Our conclusion is that the predominant excitation mode for the production of ionized-helium lines in a hollow cathode at low current densities is simultaneous ionization and excitation of neutral helium by electron impact, and that the excessive width is an inescapable result of this excitation process. Further evidence in support of this will be given later in the discussion of experiments on the excitation of helium gas by controlled electron impact.

A third and more subtle effect is the differential shift of the components (5,6) shown in Fig. 4. The excited ions accelerate in any residual electric field which exists within the hollow-cathode plasma. The longer-lived states acquire a higher velocity and hence lines originating from these states show a greater doppler shift. Changing the pressure in the hollow cathode changes the shift when the time between collisions becomes comparable to the mean life of a particular state. Pressure changes can also cause changes in the plasma field, and hence in the shifts of the components. The hollow cathode diameter also has an effect due to the geometry dependence of the internal fields: larger diameter cathodes show smaller shifts. (In Ref. 5 the cathode bore was erroneously reported as 8 cm rather than 8 mm.) The ionized-helium data obtained by Houston (11) and by Chu (12) used in the Rydberg determination probably are little affected by drift shift because of the large diameters of their hollow cathodes; however, they used currents much higher than those used in the drift shift experiments. It has been carefully determined that when a symmetrical double-anode tube is used, the average of measurements red-shifted and blue-shifted agrees with the theoretical fine-structure separation within the uncertainty in determining line centers (6,9).

It is noteworthy that no evidence for Stark or pressure shifts has been detected.

3. Excitation by Controlled Electron Impact

We have begun the study of the excitatation of ionized-helium states by controlled electron impact using a gun giving electrons with energies from threshold to several hundred eV (13). This work supports our conclusions that the most important excitation mode in the hollow cathode is simultaneous ionization and excitation from the ground state of neutral helium. The electrons excite helium gas in a liquid-nitrogen-cooled chamber, and the emitted light is observed as before with a double-Fabry-Perot spectrometer. Because of the low light levels obtained in this process, photon counting is used to achieve maximum sensitivity. The results at low pressure (.02 Torr) are shown in the bottom line of Fig. 5. The striking feature is that components 3 and 12 arising from 4s states are now among the strongest components. The lines also have the excessive width which appears to be characteristic of the electron excitation. This spectrum is essentially identical with the one obtained by Larson and Stanley in their atomic beam. When the pressure in the excitation chamber of the gun is increased, components 3 and 12 decrease rapidly in intensity, and the appearance of the spectrum becomes very similar to that of the low-pressure hollow cathode. This confirms the conclusion that collisions are responsible for the alteration of the relative intensities in the hollow cathode; however we have not yet completed the detailed study of the mechanism of the alteration. We have not yet studied the effect on the widths and relative intensities of changing electron energies because of the very weak signals; however we expect to be able to do this in the very near future when an interferometer roughly ten times more sensitive will be applied to this problem.

4. Excitation in the Afterglow of a Pulsed Hollow Cathode

Motivated by the assumption that if one could detect ionized-helium light several microseconds into the afterglow of a pulsed hollow cathode, the ions producing the light would cool sufficiently by collision with neutral helium atoms to produce narrow spectral lines, we were encouraged in our pulsed cathode excitation experiments (14) to look for a narrowing of the lines.

The total afterglow light from the 4686 Å complex rose to peak at a time delay depending on the filling pressure and magnitude of the current pulse and decayed with a decay constant of one to two microseconds. The results are consistent with two electron collisional-radiative recombination (15) of the helium double ion: $He^{++} + 2e \rightarrow He^{+}(n,\ell) + e$. One of the electrons is captured preferentially into long-lived high-n levels in Saha equilibrium at a rate determined largely by the electron temperature. Some of the electrons then cascade into lower levels which rapidly decay, sometimes passing through the n=4 levels. This process produces a unique set of relative intensities nearly independent of pressure and current pulse magnitude. The top line of Figure 5 shows strong lines arising from upper D and F states, and nearly undetectable components from upper S states. The components, however, do not appear narrowed, presumably as a result of the strong interaction between the hot electrons and the He^{++} ions. While this excitation technique offers some advantage by reducing the effects of minor components, the fact that it does not give narrower lines does not make it worth considering for Rydberg measurements. A detailed study of the line positions was not carried out. The time-delayed spectra were obtained using the scanning double-Fabry-Perot spectrometer, but with the photomultiplier activated by a pulse 1μs wide at a variable time delay after termination of the hollow-cathode current pulse.

A different intensity anomaly showed up in the pulsed spectra. Whereas in all previous work the relative intensity relationships within a multiplet were strictly obeyed, in the pulsed case the transition $4^2P_{3/2} \rightarrow 3^2S_{1/2}$ was more intense than the transition $4^2P_{1/2} - 3^2S_{1/2}$ by more than the theoretically predicted factor of two. We attribute this to a sufficient concentration of ground state He ions to effectively reduce the branching ratio to the ground state because of reabsorption. At intermediate concentrations of ground state ions, the population in the $4^2P_{3/2}$ level can become greater than the statistical ratio relative to $4^2P_{1/2}$.

It is possible that a similar effect is involved in the frequently observed intensity anomaly in the hydrogen Balmer line. With a high concentration of ground state hydrogen atoms, the transitions $3^2P_{1/2} \rightarrow 1^2S_{1/2}$ and $3^2P_{3/2} \rightarrow 1^2S_{1/2}$ are strongly reabsorbed in the discharge, resulting in an enhancement of the $3^2P_{1/2} \rightarrow 2^2S_{1/2}$ and $3^2P_{3/2} \rightarrow 1^2S_{1/2}$ transitions. However, it is surprising that a shift associated with this anomaly has been noticed only by Masui (16).

5. Photoionization-Excitation

The photoionization-excitation process, $He + h\nu \rightarrow He^{+}(n=4,\ell) + e$, can occur for photons with wavelengths less than about 164 Å. Using the University of Wisconsin electron storage ring which produces synchrotron radiation having a high photon flux at this wavelength, a study of this process has been begun (17). One of the expectations is that the smaller momentum transfer from the photon will allow the observation of very narrow lines. The prospect is clouded somewhat because of the momentum carried off by the ejected electron. However, by using prefilters of thin silicon or beryllium foils, photons with energies far above threshold can be cut out, keeping the energy and momentum of the ejected electron and excited ion small. The helium absorption cell is liquid-helium cooled, and line widths are expected to be between 0.03 cm^{-1} and 0.04 cm^{-1}, or as small as 1/3 the width observed in Fig. 2.

In this experiment, the synchrotron radiation passes directly through a prefilter and the thin plastic window of the gas cell and is absorbed by the helium. The fluorescence from the helium is sent into a 150 mm aperature double-Fabry-Perot spectrometer which uses an interference filter rather than a grating premonochromator to select the n=3-4 transition at 4686 Å. In spite of the very low light level, the large aperture interferometer makes this study feasible.

We have completed only the first phases of this experiment, in which we have measured the cross section for this process and studied the relative intensity vs. pressure variation at low spectroscopic resolution. The latter data are shown in Fig. 6. As expected, the results at low pressure are characteristic of the direct excitation cross sections, and at higher pressures the relative intensities are altered toward the statistical distribution by collisions. We expect to make experiments at higher resolution in the very near future.

6. Conclusion

Since the experimental measurements on which the Rydberg constant is based must always be corrected for some degree of blending of components, it is important to understand in detail the characteristics of the line structure being studied. These characteristics are often strongly dependent upon the method used in exciting the line. These characteristics have been extensively studied in the case of ionized helium lines, but the well-known intensity anomaly in hydrogen does not appear to have been fully understood. Resolution of the hydrogen intensity anomaly would increase the confidence in the Rydberg value derived from hydrogen measurements.

Experiments to produce significantly narrower H and He II lines have so far been unsuccessful. Even if the photoionization-excitation experiments produce lines as narrow as expected, the low signal-to-noise ratio and difficulty of the experiment will likely make this method impractical for a Rydberg determination. Development of a hydrogen atomic beam is in progress at the University of Arizona (18), and while it is not certain that significantly narrower lines will be produced with sufficient intensity to be practical for a Rydberg measurement of much higher precision than is presently expected, the method of excitation may avoid the uncertainties related to the hydrogen intensity anomaly in an evaluation of the Rydberg. Consequently, unless new and unconventional methods for measuring the Rydberg constant are developed, the value obtained from the experiments presently in progress, especially if careful attention is paid to the effects of the excitation mechanisms, should be as good as one can reasonably hope to obtain.

References

1. SERIES, G. W., Proceedings of the International Conference on Precision Measurements and the Fundamental Constants. NBS special publication No. 343, Ed. D. N. Langenberg and B. N. Taylor, U. S. Govt. Printing Office (1971).

2. KESSLER, E. G., Proceedings of this Conference.

3. KIBBLE, B. P., Proceedings of this Conference.

4. ROESLER, F. L. and MACK, J. E., Phys. Rev. 135, A58 (1964).

5. ROESLER, F. L. and DeNOYER, L., Phys. Rev. Letters 12, B508 (1964).

6. BERRY, H. G. and ROESLER, F. L., Phys. Rev. A1, 1504 (1970).

7. LARSON, H. P. and STANLEY, R. W., J. Opt. Soc. Am. 57, 1439 (1967).

8. BERRY, H. G., J. Opt. Soc. Am. 61, 123 (1971).

9. KESSLER, E. G. and ROESLER, F. L., Opt. Soc. Am., submitted.

10. BROCHARD, J., CHABBAL, R., CHANTREL, H., and JACQUINOT, P., J. Phys. Rad. 18, 596 (1957).

11. HOUSTON, V. W., Phys. Rev. 30, 608 (1927).

12. CHU, D. Y., Phys. Rev. 55, 175 (1939).

13. EVANS, W. D., University of Wisconsin Ph.D. thesis in progress.

14. KUHLOW, W. W., University of Wisconsin Ph.D. thesis, 1970.

15. COLLINS, C. B., and HURT, W. B., Phys. Rev. 179, 179 (1969).

16. MASUI, T., Proceedings of the International Conference on Precision Measurements and the Fundamental Constants. NBS special publication No. 343, Ed. D. N. Langenberg and B. N. Taylor, U.S. Govt. Printing Office (1971).

17. TRACY, D. H., University of Wisconsin Ph.D. thesis in progress.

18. STONER, J. O., Jr., private communication.

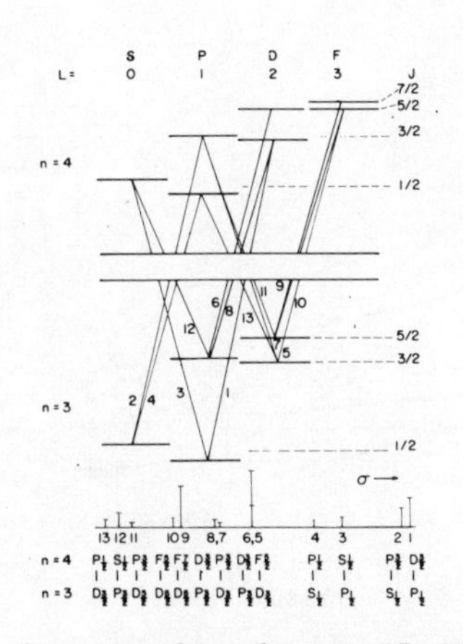

Fig. 1. Energy-level diagram for the He II 4686 Å transition. Approximate relative positions and intensities of the 13 components are shown below the level diagram. The allowed transitions are numbered 1-13 and identified at the bottom of the figure. (6)

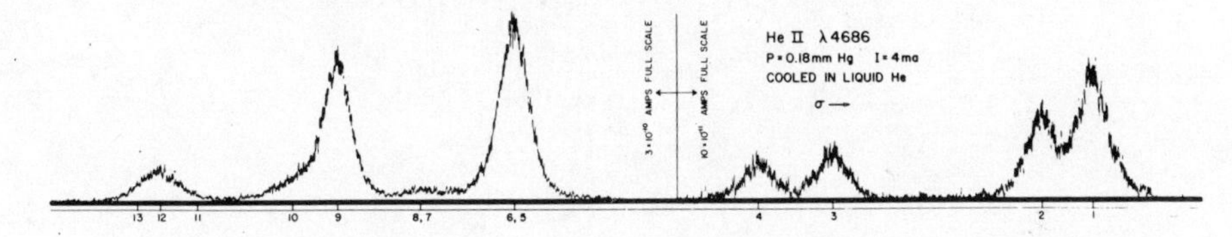

Fig. 2. The He II 4686 Å line complex observed in a liquid-helium-cooled hollow cathode. The gain has been increased a factor of three for components 1-4. (4)

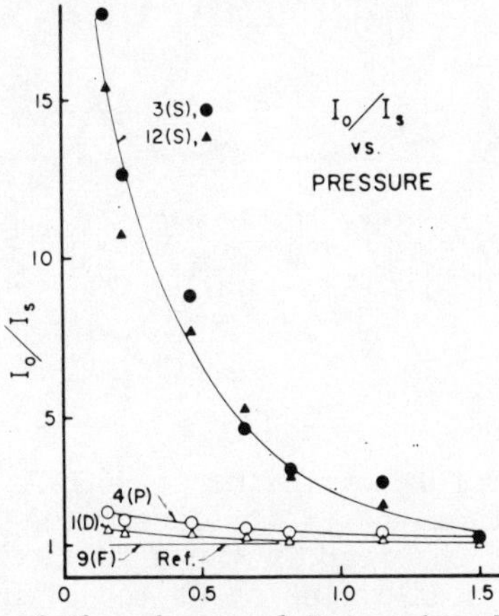

Fig. 3. The ratio of the observed Intensity I_o to the statistical intensity I_S as a function of pressure for some of the lines of the He II 4686 Å line complex observed from a liquid-nitrogen-cooled hollow cathode. The letter in parenthesis is the ℓ-value of the upper state. (4)

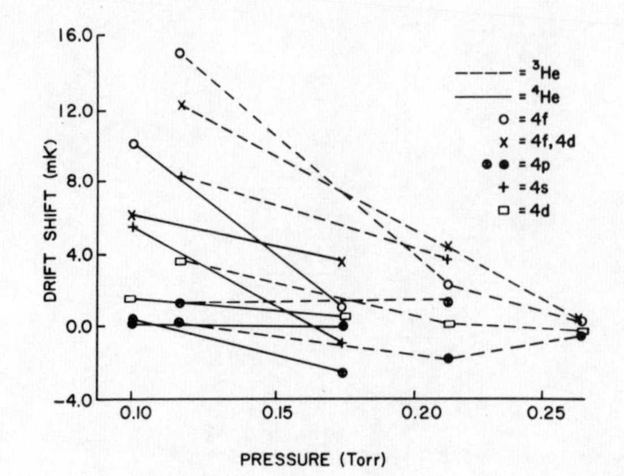

Fig. 4. Drift shift observed for components of the He II 4686 Å line complex as a function of pressure for a liquid-helium-cooled aluminum hollow cathode 1.2 mm in diameter. The n,ℓ values for the upper state of each transition are shown. (6)

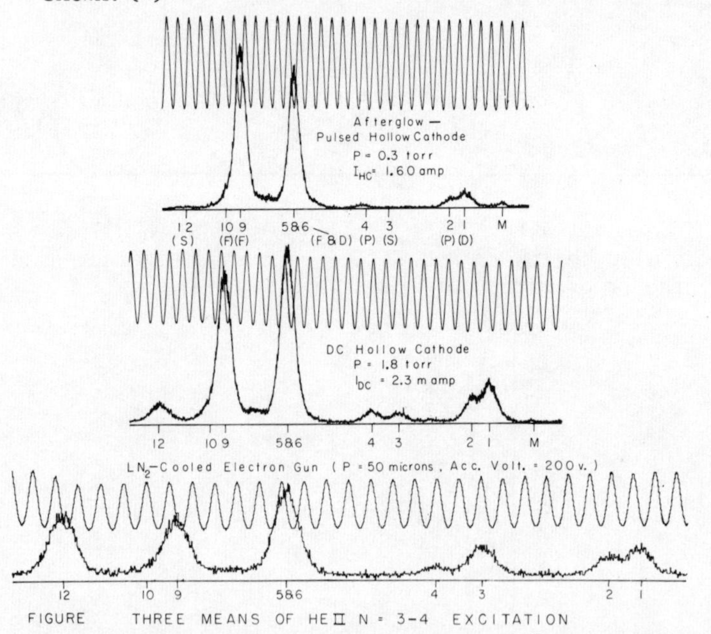

Fig. 5. Observed structure of the He II 4686 Å line complex for different methods of excitations. The top line shows the structure observed in the afterglow of a pulsed hollow cathode, the middle line shows the structure observed in a liquid-nitrogen-cooled DC hollow cathode (compare with the structure shown in Fig. 2 obtained at lower temperature and pressure), and the bottom line shows the structure observed from liquid-nitrogen-cooled helium gas bombarded with 200 eV electrons.

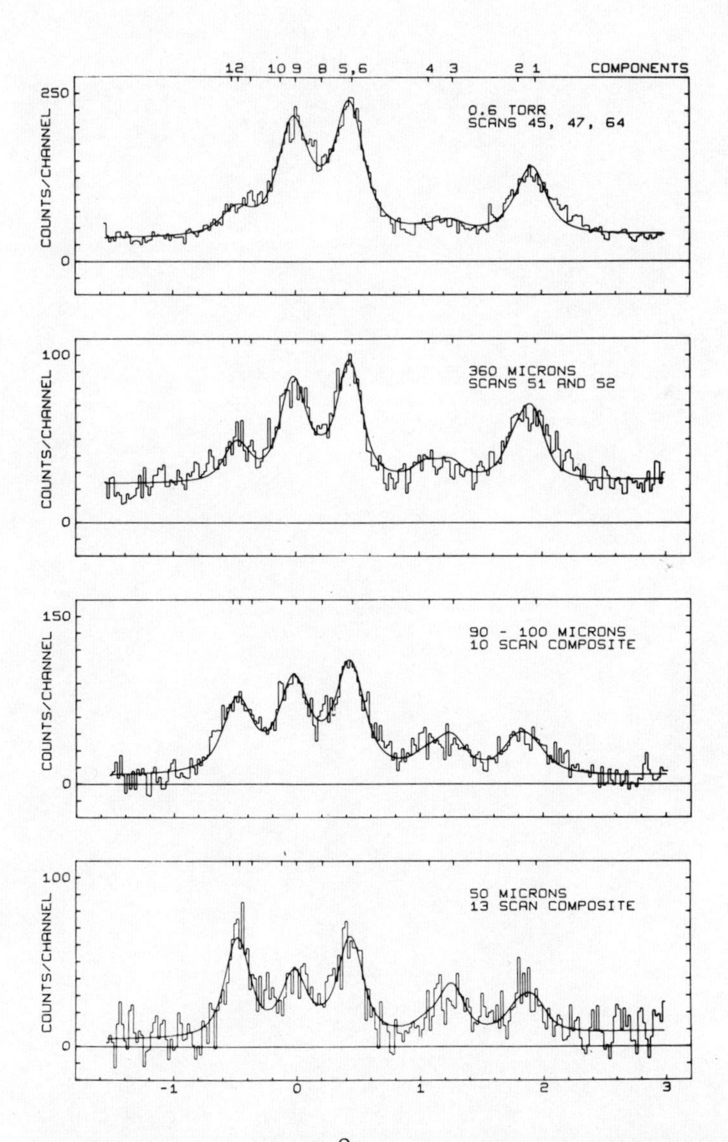

Fig. 6. Structure of the 4686 Å line of He II observed by photo-ionization-excitation of helium gas at various pressures. The composite shows the strong pressure dependence of components 3 and 12 which arise from the 4S level. The spectroscopic resolution was adjusted to be much lower than used for the other structures shown in this paper in order to achieve maximum efficiency for measuring the pressure dependence of the components. The solid line is the computed best fit to the data points.

Part 11 Magnetic Moments

Recent NPL Work on μ_P/μ_N

B. W. Petley and K. Morris

Division of Quantum Metrology,
National Physical Laboratory, Teddington, England

Abstract

Recently we have been able to scan the cyclotron resonance curve in our quadrupole omegatron by varying the applied radio frequency and keeping the magnetic field constant. This has enabled a number of small systematic effects to be investigated with a greater precision than hitherto. Our confidence in the omegatron method has been increased considerably and it appears that the cyclotron frequency can be defined meaningfully to better than one hertz when the magnetic flux density is 1.4 T. The D_2-^{4}He mass difference has also been measured.

1. Introduction

The post-war methods of measuring the magnetic moment of the proton in terms of the nuclear magneton have all made use of the method proposed by Alvarez and Bloch (1) by measuring μ_p/μ_N as the ratio of the proton spin precession frequency f_s to its cyclotron frequency f_c in the same magnetic field (2). The proton spin precession frequency may be measured quite simply by one of the standard techniques employed in nuclear magnetic resonance (n.m.r.), a field of physics which has been well studied both experimentally and theoretically. The cyclotron frequency of the proton is the more difficult to measure. A long lifetime is required to define the cyclotron frequency precisely and the frequency is also perturbed by electrostatic fields or field gradients, depending on the measurement method. These shifts must be eliminated in some way, often by making measurements on several ions of heavier masses in order to deduce the unperturbed proton cyclotron frequency.

In this paper we consider the problem of measuring f_c in our quadrupole omegatron.

2. The proton cyclotron frequency

In developing our quadrupole omegatron we have tried to construct a device in which the equations of motion of the ions may be solved theoretically and which satisfies as closely as possible the original theory developed by Sommer, Thomas and Hipple (3), while including the effects of the electrostatic trapping field that is used to increase the ion containment time to about one millisecond.

The simple theory predicts that the resonance curve observed when the ions are accelerated to hit an ion collector should be rectangular in shape and symmetric about the cyclotron frequency f_c ($Be/2\pi M$). The frequency width Δf_c is given by $E_o/\pi \; B \; R_k$, where E_o is the amplitude of the applied radio frequency electric field, B is the magnetic flux density and R_k the radius of curvature of the ion orbit at collection. Although R_k is essentially the distance of the collector from the electron beam the expression should strictly be $E_o/\pi B(R_k-r)$, where $-r_o < r < r_o$, r_o being the radius of the electron beam which generates the ions ($2r_o \sim 0.1 R_k$). Thus the finite size of the electron beam causes the resonance to be of a trapezoidal shape. Both we (4) and Fystrom (5) have recently been operating with resonance curves of this shape.

Since an electrostatic field is used to trap the ions in the magnetic field direction there are two further effects to be considered. First the ion orbit guiding centre drifts in a direction which is perpendicular to the magnetic field so limiting the ion lifetime (the ions can only be collected until they have drifted to the point where they are closer to the containing electrodes than the ion collector). The second effect of the trapping field is that the cyclotron frequency is shifted from f_c. In our omegatron this shift at 1.4 T amounts to 558.6 Hz per volt of applied trapping potential. Although this shift could in principle be eliminated by varying the applied trapping voltage, there are residual electrostatic fields present which contribute a further shift of about 30 Hz, or 6 ppm of the mass 4 u ion cyclotron frequency. These shifts should be independent of ion mass and are eliminated in our experiment by bringing ions of different mass successively to resonance (in random order). We have been operating with H_2^+, HD^+, D_2^+ and He^+ ions.

It has been usual to scan the cyclotron resonance curve by changing the magnetic field. However the precision of this method is limited to about one ppm. Moreover most of the quantities in which one is interested are most easily measured if the frequency is varied at constant magnetic field. We have recently been operating with a frequency synthesizer the use of which has considerably increased our ability to see small effects. The important requirement of using the synthesizer is that the r.f. amplitude remains constant as the frequency is changed. We have found that a change of 50 kHz is required before the r.f. amplitude, measured at the base of the omegatron, changes by as much as 0.1%. The distortion of our cyclotron resonance curves (< 2 kHz wide) is therefore negligibly small ($\sim$0.1 Hz on f_c).

3. Effect of changing the r.f. amplitude

The theory indicates that the frequency width and shape of the resonance curve should not depend on ion mass. Our measurements have shown that the shapes agree to within 2 Hz in a resonance 2 kHz wide and that the widths are the same within the precision with which the r.f. levels can be set for different mass ions (0.1 dB).

When the r.f. amplitude was changed from 0.5 to 0.9 V (r.m.s.) so increasing the width from 1 kHz to 1.8 kHz the centre frequency (f_c) changed by between 6 and 14 Hz, the amount depending on the trapping voltage. However the frequency shifts were the same for all four mass ions used, within 1.5 Hz. Thus the shifts with amplitude are mass independent to around 0.2 ppm of the mass 4 u ion cyclotron frequency. The shifts are almost certainly due to orbit drift causing the ions to average the electrostatic fields differently - the ion lifetimes are being nearly halved with the increased r.f. amplitude.

4. Trapping voltage

The theory suggests that the frequency shift with trapping voltage should be mass independent. This shift has been measured and found to be the same, 558.6 Hz per volt trapping potential, within the standard deviation of about 3.5 Hz/V. The standard deviation was made larger through the inclusion of measurements taken on different days. The correlation with the fitted straight lines was good (r = 0.9998 for the 50 observations per fitted line), so that there are unlikely to be major higher order terms. However even if there are they are likely to be the same for ions of different mass.

The above runs were taken at 1 μA electron beam current, 1.4 T and a pressure of 4 $\mu N\,m^{-2}$ (0.03 μtorr). The residual frequency shift deduced by combining measurements on different mass ions and assuming their nuclidic mass values amounted to 31.1 ± 1.8 Hz, or 6 ppm of the mass 4 u ion cyclotron frequency. If the standard deviation of 1.8 Hz is taken as an indication of the precision with which this shift may be eliminated, it indicates a precision of about 0.3 ppm for μ_p/μ_N work. However the runs were not all taken on the same day and experimentally we merely require that the 31.1 Hz shift is stable during a 15 minute period.

5. The filament current effect

In the omegatron the ions are created by a beam of ~70 eV electrons which are emitted from a heated filament. The filament heater current produces a magnetic field which shifts the cyclotron frequency. This frequency shift had only been seen previously as a difference in the values of μ_p/μ_N obtained when the heater current direction was reversed. We have now measured the change in the ion cyclotron frequency with reversal of the heater current and find that the shifts are compatible with the relation:

$$[f_c(-)-f_c(+)]_{Av} = (3.34 \pm 0.24) + (57.99 \pm 0.68)/M_u$$

where M_u is the ion nuclidic mass (^{12}C = 12u), the correlation coefficient being 0.9998. This dependence on the reciprocal of the ion mass is consistent with a magnetic field effect and is indicative of a ~3 ppm difference in the magnetic field seen by the ions for the two filament current directions.

The intercept in the fitted line indicates a shift in the average space charge for the two filament current directions. Although the intercept is probably a result of the lack of exact day-to-day reproducibility in the operating conditions (pressure etc.), it could also indicate a movement of the filament. If the filament did move the reversal procedure would not eliminate the effect. Fortunately however we have been able to locate a small n.m.r. probe sufficiently close to the omegatron to detect this small magnetic field. This probe indicates that the magnetic field certainly reverses to within 0.2 ppm with filament current direction. At higher electron currents than we use we have seen non-reversal effects of about 0.5 ppm. There is therefore a slight movement of the filament under the influence of the magnetic field, but it is negligible at our normal operating electron current.

6. Substitution correction

The omegatron method, in common with all of the other μ_p/μ_N measurements, requires the interchange of the apparatus with an n.m.r. probe. Although our omegatron uses considerably less material than is used in the mass synchrometer of Mamyrin and Frantsuzov (6) there is the possibility that in moving the omegatron the magnetic field is changed. We have made provision for the measurement of this effect by inserting an n.m.r. probe inside the omegatron when the experiment is completed. We have however been able to place an upper limit on the correction by inserting an n.m.r. probe between the omegatron and the pole-face of our electromagnet. The upper limit on the effect is about 1.5 ppm, but we continue to assign a possible error of 2 ppm to this effect.

7. Skewness

In our determination reported at the previous conference (7) the values of f_c deduced from estimates of the resonance centre at the resonance base and top differed by some 86 Hz. In spite of this we were able to obtain μ_p/μ_N values which were consistent to 2 ppm. In the present work the top and base frequencies differ by some 5 Hz, the actual amount depending on the ion mass. This skewness is partly the result of the different relativistic mass increases for ions contributing to the top and base, their final energies differing by about 10%. The remainder of the skewness is attributable to the variation of the magnetic and electrostatic fields over the ion orbit. This is unlikely to lead to an uncertainty of more than about 0.2 ppm in the value of μ_p/μ_N if measurements are made at the resonance half-height.

8. The value of μ'_p/μ_N

The results of our present series of measurements are that the value of μ'_p/μ_N is likely to lie within 2 ppm of 2.792 770. Most of our corrections are small in magnitude, the reference n.m.r. probe is close to the omegatron and a spherical pure-water sample is used for the final f_s value. As a result of using frequency scanning the standard deviation of a single measurement of μ'_p/μ_N is now 0.5 ppm instead of the 5 ppm that we had last year.

A further correction on which we would like more information is the partial pressure correction arising from the fact that the partial pressures of the ions are not the same. At the pressures at which we are operating the total pressure shift in the ion cyclotron frequency is only about 3.5 Hz. Differences in the partial pressures will therefore produce shifts that are smaller than this value. Attempts to measure the partial pressure shift have not so far been very conclusive, but its magnitude appears to be about 45 Hz per pA difference in the collected ion current. This value leads to differences of 0.5 Hz or less for the partial pressure differences that we can experience. Thus present indications are that this will not affect μ_p'/μ_N by more than about 0.3 ppm.

9. The D_2-^4He mass difference

It was found that if the 1965 mass values were used for D_2^+ and He^+ ions the values of μ_p/μ_N did not agree. We have therefore measured the D_2-^4He mass difference and obtain 25 600.12 ± 0.3 μu. The 1965 tables (8) give 25 601.3 ± 0.3 μu. The measurement of the doublet reported by L.G. Smith at this conference is 25 600.328 ± 0.008 μu (see his paper for references and discussion of other work). Our omegatron determination is therefore in excellent agreement with the value reported by Smith. (Note. The 1971 recommended mass values reported by Wapstra and Gove at this conference also indicate that the 1965 values were slightly high).

10. Dimensionless parameters

As further tests of the precision of the omegatron we have evaluated two dimensionless parameters whose values may be compared with those obtained from nuclidic mass tables. These parameters are not completely independent statistically, but are P(1) and P(2), where $P(1) = [f_c(HD)-f_c(D_2)]/[f_c(H_2)-f_c(D_2)]$, and $P(2) = [f_c(HD)-f_c(He)]/[f_c(H_2)-f_c(D_2)]$. The relativistic and partial pressure corrections were applied before taking the ratios. The values that we obtain are within 0.3 ppm of the values indicated by L.G. Smith's recent work. These parameters are a sensitive test of the omegatron, as frequency differences are present in both the numerator and denominator. Thus for μ_p/μ_N work the precision is better than the 0.3 ppm indicated above.

11. Conclusions

The general conclusion to be reached from our recent work is that the omegatron method, when used with frequency scanning, is capable of a greater precision than had been thought previously. None of the effects that we have been able to investigate so far suggests errors in excess of a part in a million. It may well be possible to approach an overall experimental error of about one ppm in the next few months, particularly if we are able to improve on the present 0.3 ppm stability of our magnetic field.

References

1. ALVAREZ, L.W. and BLOCH, F., Phys. Rev., 57, 111 (1940).

2. PETLEY, B.W., Proc. Int. Conf. Prec. Meas. and Fund. Constants, Ed. D.N. Langenberg and B.N. Taylor, Nat. Bur. Stds. Special Publication 343 (1971), p.160.

3. SOMMER, H., THOMAS, H.A. and HIPPLE, J.A., Phys. Rev., 82, 697 (1951).

4. PETLEY, B.W. and MORRIS, K., Proc. ICPMFC 1970 as in ref. (2), p.173.

5. FYSTROM, D., Phys. Rev. Letters, 25, 1469 (1970) and as in ref. (2), p. 169.

6. MAMYRIN, B.A. and FRANTSUZOV, A.A., Zh. Eksp. Teor. F.2, 48, 416 (1965) [Sov. Phys.-JETP, 21, 274 (1965)] and as in ref. (7), p.427.

7. PETLEY, B.W. and MORRIS, K., Proc. Third Int. Conf. Atomic Masses, R.C. Barber, Ed. (Univ. of Manitoba Press, Winnipeg), 461 (1968).

8. MATTAUCH, J.H.E., THIELE, W. and WAPSTRA, A.H., Nucl. Phys., 67, 1 (1965).

New Determination of the Magnetic Moment of Proton in the Units of Nuclear Magneton

B. A. Mamyrin, N. N. Arujev and S. A. Alekseenko

Academy of Sciences of USSR,
A. F. Ioffe Physico-Technical Institute, Leningrad

The magnetic moment of proton in nuclear magnetons is equal to the ratio of the nuclear magnetic resonance frequency to the cyclotron resonance frequency; it is therefore reasonable to assume that the precision of the determination of this ratio can be made very high.

The transference of this adjustment constant into the group of precisely determined constants would be of great importance, since in this case the constant in question were the first in the group of precisely determined and possessing information on proton mass:

$$\mu_p/\mu_n = \frac{C\gamma_p M_p}{e}$$

The "dramatic" situation with "low" and "high" μ_p/μ_n values and its relation to the Faraday was clearly described in the book of Taylor, Parker, Langenberg (1) and we shall not repeat it here. Still let us note that the recent measurements of Petley and Morris, Fystrom and Gubler, Richard and Roush have increased the acuteness of the problem (2) (3) (4).

With this in mind we considered that a new μ_p/μ_n measurement with the highest possible precision would be of importance.

All known methods of μ_p/μ_n determination include two sources of error - the ion trajectory change caused by magnetic field inhomogeneities and changes caused by stray electric fields in the ion drift space.

One can correct for the magnetic field inhomogeneity if the magnitude and direction of the field at every point of the ion trajectory are sufficiently well known. This was achieved in our method by the use of the magnetic resonance mass spectrometer (5) (6) (7).

The correction for stray electric fields is useful only if the distribution of the field is the same when the cyclotron frequency of ions of different e/m are being measured. Unfortunately this condition was not strictly satisfied in all our measurements; this was the case also in our earlier work.

However, the use of a secondary electron multiplier enabled us to use very small ion currents. This removed the main cause of the variability of the electric fields, which was due to the charging of metal surfaces by the bombardment of the ions, and the correction became negligible compared with the uncertainties of our 1965 determination.

The essence of new technique consists of the fact that the ion source does not require readjustment for ions of different e/m and the whole pattern of the ion beams leaving the ion source remains invariable. It was achieved by special ion source construction, where two sorts of ions with different e/m leave the source with different energies. As a result both ion beams pass through the slit of the modulator. The procedure for observing different ions consists of varying the frequency and amplitude of the r.f. modulator voltage only. To plot the extrapolation graphs we used the following pairs of ions:

$$He_4^+ - Ne_{20}^{++}; \quad He_4^+ - Ne_{20}^+; \quad He_4^+ - Ar_{40}^+$$

Using this technique we reduced drastically uncertainties in μ_p/μ_n due to extrapolation. A special screen was inserted in the chamber to disturb the stray electric fields, and this caused a large change of slope in the extrapolation graph. However this did not affect the final μ_p/μ_n value more than 0.2 ppm.

To decrease the uncertainties connected with other sources of error the following has been done:

1. The time stability of magnetic field was improved by using superstabilization of the magnetic flux in the gap of the magnet. During the time allotted to a single frequency ratio measurement ω_n/ω_c the relative change in the magnetic field did not exceed 0.2 ppm. The possibility of the small amplitude 3 Hz sinusoidal modulation of the magnetic field was provided to generate cyclotron resonance and NMR signals.

2. The homogeneity of the magnetic field was increased drastically. The inhomogeneity of the magnetic field on the circular orbit of the ions with 224 mm diameter did not exceed $\pm$ 8 ppm of that in the centre of the chamber where the NMR probe is located.

3. A new system for the precise measurement of magnetic field inhomogeneity at 24 points of the orbit for ΔH_z, ΔH_ρ and ΔH_ψ components was developed. The total time allotted for the measurement of the magnetic field inhomogeneity map was decreased. That allowed us to repeat the procedure before and after each run of measurements.

The above improvements enabled us to calculate the correction for the difference between the magnetic field H_{eff} at the orbit and the field in the vicinity of the water NMR sample with an error of less than 0.2 ppm.

4. To decrease the uncertainty of the determination of H_{eff} the cross section of the ion beam was minimized (1.5 mm^2 instead of 4 mm^2). For the same reason the volume of NMR sample was diminished from 500 mm^3 to 60 mm^3 (the inner diameter of the spherical water sample container was $\simeq$ 2.4 mm).

As shown by magnetic field gradient measurements, the uncertainty of the magnetic field in the limits of the ion beam cross section was less than 0.1 ppm.

5. The error in the correction for the magnetic field change due to replacement of the chamber in the gap of the magnet was reduced to 0.24 ppm.

6. A method of obtaining direct readings of the NMR and cyclotron resonance frequency ratio was developed. When combined with the peak superposition technique on the screen of the oscilloscope and quartz stabilization, this method enabled us to reduce the error of a single ω_n/ω_c reading to less than 0.2 ppm.

7. By using improved filtering, the effect of harmonics of the r.f. voltage was eliminated.

8. A system of automatic amplitude stabilization of the r.f. voltage was developed. As a result the stability of the cyclotron resonance signal was improved and the peak superposition procedure was made easier.

9. The influence of the cathode filament current was removed totally by the use of compensating leads. The filament current correction was less than 0.2 ppm.

All the measures mentioned above improved the resolving power of the mass spectrometer to $\simeq 50.10^3$.

The assymetry of the cyclotron resonance signal was less than 1%.

The rms uncertainty of μ_p/μ_n (including extrapolation procedure) in each set of measurements was about 0.3 ppm, and 0.65 ppm for all the sets. The average RMS correction from 11 sets of measurements was found to be 0.20 ppm.

The data of single run are shown in Table I.

In calculating the result for one pair of ions the average mean of 10 ω_n/ω_c ratio frequency readings were used. After calculation the $[\mu_p/\mu_n]^*$ value extrapolated for each pair of ions the following corrections were added:

1) The average correction for inhomogeneous magnetic field Δ_1, which was about $\pm$ 1 ppm.

TABLE I

Pairs of ions	$(\mu_p/\mu_n)^*$	Δ_1	Δ_2	$(\mu_p/\mu_n)^* + \Delta_1 + \Delta_2$
$He_4^+ - Ne_{20}^+$	2,7928070	+ 0,0000013	- 0,0000021	2,7928067
$He_4^+ - Ne_{20}^{++}$	2,7928080	+ 0,0000013	- 0,0000027	2,7928073
$He_4^+ - Ar_{40}^+$	2,7928076	+ 0,0000013	- 0,0000019	2,7928075
$He_4^+ - Ne_{20}^+$	2,7928079	+ 0,0000013	- 0,0000021	2,7928076

$$(\mu_p/\mu_n)^{**} = 2.7928073$$

TABLE II

Number of run	Data	$(\mu_p/\mu_n)^{**}$	$(\mu_p/\mu_n)^{**} + \Delta_3 + \Delta_4$
1	25.2.71	2,7928087	2,7927776
2	9.3.71	2,7928043	2,7927732
3	11.3.71	2,7928040	2,7927729
4	15.3.71	2,7928061	2,7927750
5	5.5.71	2,7928073	2,7927762
6	11.5.71	2,7928034	2,7927723
7	13.5.71	2,7928040	2,7927729
8	21.5.71	2,7928035	2,7927724
9	24.5.71	2,7928047	2,7927736
10	13.7.71	2,7928057	2,7927746
11	14.7.71	2,7928046	2,7927735

2) The relativistic correction Δ_2. For He_4^+ - Ne_{20}^+ pair Δ_2 = - 0.55 ppm.

The results for all runs are shown in Table II.

Two corrections were added to the average value of each run. They are the following:

1) The influence of the chamber on the magnetic field, which was + 2.85 ppm.

2) Modulator correction Δ_4 caused by the ion velocity change in the ion flight through the modulator. The distance between two grids of the modulator was 1.35 mm. Δ_4 in our instrument was about - 14 ppm.

Our final result is the average of all runs of measurements, without correction for the diamagnetism of water, μ_p/μ_n = 2.7927745 (12) (0.43ppm).

The RMS error indicated consists of

1) statistical uncertainty of the mean;

2) the error in the modulator correction calculation;

3) the error of measurement of the magnetic field inhomogeneity correction.

To study probable sources of error the following control experiments have been performed.

1. Varying the gaseous component partial pressures it was proved that the change of polarization potentials in the ion drift space between the modulator and the drift slit (the drift slit cuts the ion bunches out of the ion beam) did not influence the result of the measurements.

2. Varying the amplitude of the modulator voltage it was shown that even variations much greater than normal did not change the result more than 0.2 ppm.

3. The independence of the results of the chamber surface conditions was checked (all screens made of copper were coated by aquadag, etched and their configuration was varied).

4. The independence of measurement results of the amplitude of magnetic field modulation was checked.

5. The speed of chamber moving into the magnetic gap was ten times decreased (as compared with normal speed). It was found that the magnetic field profile was not affected by this procedure.

6. It was found that the massive ring-shaped chamber wall does not change the magnetic field distribution.

7. The measurements were performed with the chamber displaced by an amount which exceeded the possible errors in setting the chamber. It was found that the displacement did not affect the result of the measurements.

8. It was tested that electric fields penetrating in the drift space from the ion source region did not affect the result of measurements.

9. It was found that there is no influence of different materials used (aquadag, the metal of the screens and the NMR probe) on the result of measurements.

10. No influence of the stray magnetic fields in the laboratory on the result of measurements was found.

In future we intend to carry out several runs of measurements and additional test experiments to investigate the errors of measurements more thoroughly.

This work as we believe revealed that our technique is free of any fundamental and technical difficulties limiting further improvements in the accuracy of the measurement of μ_p/μ_n .

References

1. TAYLOR, B.N., PARKER, W.H. and LANGENBERG, D.N. "The fundamental constants and quantum electrodynamics". Academic Press, New York - London, 1969.

2. PETLEY, B.W. and MORRIS, K. Rep. Int. Conf. on precision measurement and fundamental constants, NBS, Gaithersburg, Maryland, U.S.A., August 1970.

3. FEYSTROM, D.O., Phys. Rev. Letters, 25, 1469, 1970.

4. GUBLER, H., REICHART, W., ROUSH, M., STAUB, H.H. and ZAMBONI, F. Rep. Int. Conf. on precision measurement and fundamental constants, NBS, Gaithersburg, Maryland, U.S.A. August 1970.

5. MAMYRIN, B.A., and FRANTSUZOV, A.A. Zh. Eksp. i Teor. Fiz., 48, 416, 1965.

6. MAMYRIN, B.A. and FRANTSUZOV, A.A. Proceedings of the third international conference on atomic masses. University of Manitoba Press, Winnipeg, Canada, 427, 1968.

7. MAMYRIN, B.A. and FRANTSUZOV, A.A. Soviet Physics JETP, 21, 274, 1965.

Gyromagnetic Ratio of the Proton

Progress with the NPL weak-field apparatus

P. Vigoureux

National Physical Laboratory, Teddington, England

1. General Description

The weak-field NPL apparatus for determining the gyromagnetic ratio of the proton, and for checking the ampere as maintained by the Laboratory standard resistors and Weston cells, is working satisfactorily. The principle of the method is shown in Fig 1. The field-forming coils are vertical, and, when carrying 1.018A, produce at the centre a flux density of about 1.2mT, corresponding to a precessional frequency of about 50 kHz, uniform to 0.5 in 10^6 over the volume of a sphere 40 mm in diameter. The current is provided by a 110V battery of large accumulators, and is controlled by a specially made controller of the galvanometer and photo-cell type, so that only very slight hand adjustment, if any, is required to keep the current constant at the value determined by the standard resistor and Weston cell.

The water, source of protons, is contained in a spherical bulb 40 mm in diameter surrounded by a detector coil and a polarising coil. The latter produces a flux density of some 0.1T for a few seconds. When the polarising current is switched off, the polarisation assumes the direction of the standard field, but is subsequently rotated 90° by an alternating field produced by the discharge of a capacitor in the detector circuit tuned to 50 kHz. Precession then takes place and induces into the detector coil an emf decaying with a time constant of 1 or 2 seconds. The period of this signal is measured by an electronic counter which registers the time occupied by 20 000 periods, approximately 0.4s.

Fig 2 is a photograph of the assembly, showing the two Helmholtz pairs which neutralise the Earth's magnetic field, and, inside, the field-forming coils, of copper wire wound in helical grooves in cylinders of fused silica. The outer cylinder has at each end a compensating winding of 8 turns to render **uniform** the magnetic field of the main coil of finite length. The outer cylinder also carries a Helmholtz pair forming part of a servo system to cancel the variations of the ambient magnetic field.

2. Ambient Magnetic Flux Density

(1) Steady part of ambient flux density. The currents, of approximately 100 mA, in each large Helmholtz pair, Fig 2, are provided by a separate 110V battery. They are kept constant at

predetermined values by simple current controllers of the galvanometer and photo-cell type. From time to time a flux-gate magnetometer is substituted for the coils at the centre to adjust the controller resistances if need be.

Although this adjustment is correct to approximately 1 in 1000, an accuracy of only 1 in 100 would be adequate, for on the one hand the final adjustment for the vertical component is effected by ensuring that the precessional frequency is not appreciably changed on reversal of the standard field, and, on the other hand, a horizontal residual as large as 1% of the ambient field, would introduce into the measured value of the gyromagnetic ratio an error less than 1 in 10^7.

(2) Variation of ambient flux density. The ambient field suffers variation with time, partly from natural causes, but, in the Teddington area, more so from man-made disturbances, especially d.c. electric trains. The variation of the horizontal component is comparatively small, the r.m.s. value even in the daytime being less than 1nT, or 1 in 20 000; according to the remark at the end of the last paragraph, this variation need not be compensated.

The variation of the vertical component is large, because the return rail of the electric traction system not being well insulated, allows about one third of the return current to flow below ground, forming a large loop, which produces a considerable vertical field a distance away. The result is that during rush hours the r.m.s. variation can reach 20nT. This variation is equivalent to 1 in 50 000 of the standard field, and must be compensated.

Matters are not however as bad as this figure might suggest, because the measurement of the period of precession is affected by only a part of the energy spectrum of the ambient field. This spectrum is shown in Fig 3. The mean square vertical flux density per hertz at 1 kHz is approximately 10^{-25} T^2/Hz, it is 10^{-21} at 100 Hz, reaches a maximum of just above 10^{-20} at 50 Hz, then decreases slowly to increase again probably, for periods of the order of 100s. The low-frequency part of the spectrum, say periods of 1000s or over, are compensated if, as is always done, the average of the periods of precession with the standard field directed upwards and downwards is taken. As on the other hand the time for a single measurement of period occupies 0.4s, or 20 000 periods of the precession signal, that part of the spectrum with periods shorter than, say, 0.01s, are also compensated. It is thus only the region between 0.001 Hz and 100 Hz that impairs the accuracy by producing erratic readings of period. The corresponding mean square flux density is less than the total, though still too large to go uncompensated.

Compensation has been achieved as follows: the output of a flux-gate magnetometer 6m from the centre of the apparatus, is applied to a moving-coil recorder in the ordinary way, and, in parallel with the recorder, to a d.c. amplifier which feeds a coil surrounding the magnetometer, and also a Helmholtz pair wound on the same cylinder which carries the auxiliary field-forming coil. The flux density at the flux-gate magnetometer is maintained constant by this servo control, as can be seen from Fig 4a, which shows the recorder output with and without control. The ratio of the currents required in the servo coil and in the Helmholtz pair can be calculated from their dimensions, and has also been determined by experiment. Compensation at the centre of the apparatus is not as good as at the magnetometer, presumably because the disturbances do not all come from a distance large compared with 6m, but, as Fig 4b shows, it is adequate even in the daytime.

3. Gradients of Magnetic Flux Density

Gradients of magnetic flux density at the centre of the apparatus can be due to faulty relative position of the field-forming coils, or to iron or electric currents in the vicinity of the apparatus. Their effect can be compensated by three orthogonal Helmholtz pairs with equal currents in opposition in the two coils of a pair, and can be distinguished because in the first case the gradient must be reversed when the standard field is reversed, whereas the gradients produced by electric currents or by fairly distant masses of iron are independent of the direction of the standard field. We have provided for the vertical and NS directions already, and are making gradient coils for the EW direction.

A compensating gradient is not however a remedy for faulty relative position of the coils: we use it only as an indication of the amount of adjustment required to centre them. The reason is that since the compensating gradient has to be reversed when the standard field is reversed, any small residual flux density due to asymmetry of the Helmholtz gradient pair would not be cancelled by reversal of the standard field, and therefore would introduce an error.

4. Results

As the machine built to measure the linear dimensions of the coils is still undergoing tests, we have only measured precession frequency on rare occasions, to verify that the apparatus remained in working order, and we have not taken all the care that would be taken, before a proper series of measurements, to keep the vicinity of the apparatus free from iron.

The results, based on nominal dimensions of the field-forming coils, and on an assumed coefficient of linear expansion for silica of 0.4×10^{-6}/K, have however been promising. The five results obtained lie within 1.6 in 10^6, and four of them within 0.6 in 10^6.

5. Conclusion

The results of precession frequency obtained hitherto, and

the work already done on the measuring machine, suggest that, apart from uncertainty in the knowledge of the ampere, the gyromagnetic ratio of the proton can be determined with a probable uncertainty (50% confidence) of not more than 1 in 10^6.

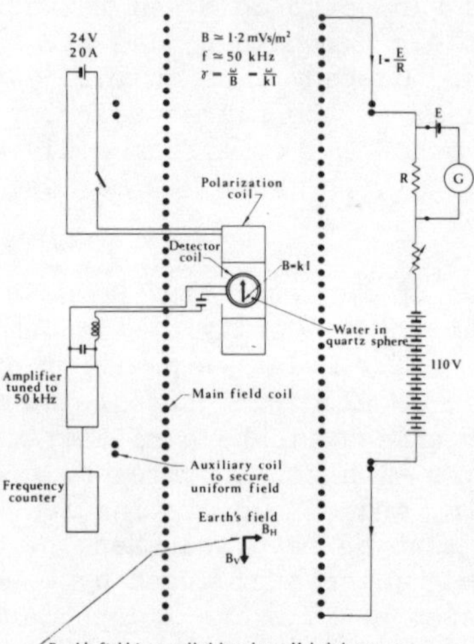

Fig. 1. Diagram of the weak-field NPL apparatus for measurement of the gyromagnetic ratio of the proton.

Fig. 2. Weak-field apparatus for measurement of the gyromagnetic ratio of the proton (Crown copyright reserved).

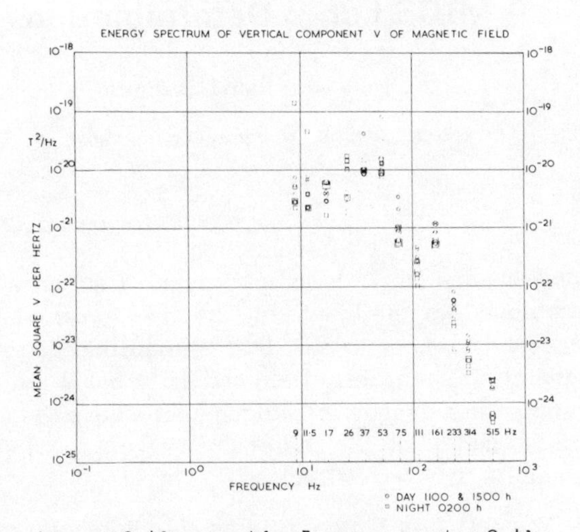

Fig. 3. Spectrum of the vertical component of the ambient magnetic flux density.

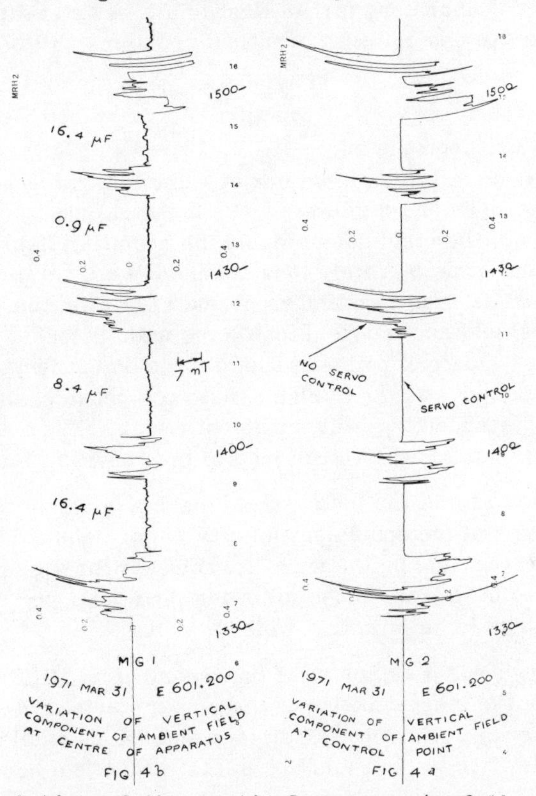

Fig. 4 Variation of the vertical component of the ambient flux density, with and without servo control, (a) at servo location 6m from centre of apparatus, (b) at centre of apparatus.

A MICS For γ'_p Determination

Ko Hara and Hisao Nakamura

Electrotechnical Laboratory, Tokyo, Japan

1. Introduction

A single layer solenoid with an appropriate field uniformity correction has been employed for the low field (about 4π gauss) determination of γ'_p. This method is in general better than a high field determination in computing accurately the magnetic field strength, but it suffers from the difficulty of finding a non-magnetic environment where the earth's field is sufficiently uniform and quiet. A rather poor signal to noise ratio of the precessional signal is another difficulty.

An MICS (Magnetically Isolated Calculable Solenoid) may eliminate these difficulties to a great extent and is now being constructed at the Electrotechnical Laboratory for its first trial. A brief explanation of its principle and a discussion of its practical problems will be given in this paper.

2. MICS

A schematic depiction of an MICS is given in Fig. 1. The main solenoid is confined in an inner shield box of rather thick high permeability material. The upper and the lower slabs make an infinite number of successive images of the main solenoid and the resulting field inside it is equivalent to that of an infinitely long solenoid (4π G/A), provided that the relative permeability (μ) is infinite. When μ is finite, the flux leaks outside the inner shield and the field inside the main solenoid reduces by a small fraction. A pair of compensation solenoids connected in series and with the same polarity as the main solenoid compensates almost exactly this reduction, provided that μ is sufficiently high. The outer shield is a rather thin frame and is the return path of the resulting flux.

Four side walls of the inner shield make a plane array of an infinite number of images of the equivalent infinitely long solenoid. The contribution to the centre field by the walls is, however, negligibly small when $\mu \gg 1$ and hence is ignored in the following, and the upper and the lower slabs are assumed to be infinitely wide.

The following is a summary of the theory of an MICS previously published (1), when the magnetization of the slab material is linear. In the geometry in Fig. 2 in which linear dimensions are normalized to the radius, the field $H_z(r, z)$ as a function of the cylindrical coordinates r and z is given as (2)

$$H_+(r, z) = H_1(r, z) + H_2(r, z) \qquad (1)$$

$$(I_i = I_0 = 1)$$

$$H_-(r,z) = H_+(r,z) - 2H_0(r,z) \tag{2}$$

$$(I_i = -I_0 = 1)$$

Here, $H_i(r,z)$ $(i = 0,1,2)$ is expressed as

$$H_i(r,z) = (2\pi/g)\sum (Z_n^m)_i z^m r^n \tag{3}$$

and $(Z_0^0)_i = 2q_1(H_i)$, $(Z_2^0) = (3/2)q_5(H_i)$, $(Z_0^2) = -3q_5(H_i)$

$$(Z_4^0)_i = -(15/8)[q_7(H_i) - (7/4)q_9(H_i)],\ (Z_2^2)_i = -8(Z_4^0)_i,\ (Z_0^4)_i = (8/3)(Z_4^0)_i$$

and g is the common pitch of the main and compensation solenoids. Quantities $q_n(H_i)$ as functions of A, C and μ are examplified in Fig. 3 and Table I. These computations are correct to the first order of $1/\mu$. It is noteworthy that the parameters of the outer shield do not appear in the expressions for the first order approximation. H_+ is seen to be a very good approximation of the field of an infinitely long solenoid. $H_+ - H_- = 2H_0$ can be determined experimentally and is compared with Fig. 3(c) to estimate the value of μ.

3. MICS as a magnetic circuit

When the hysteresis of the material of the inner and the outer shields is taken into account, a full analytical solution of an MICS is not feasible. A magnetic circuit analysis gives some physical insight, where the flux (ϕ) and the magneto-motive force (k) relation of the circuit of the material is assumed to be quadratic i.e.,

$$(r\phi)^2 - 2\alpha(r\phi)k + k^2 = k_m^2(1-\alpha^2) \tag{4}$$

When an alternating mmf $k = k_m \sin \omega t$ is applied, ϕ is solved as $r\phi = k_m \cdot \sin(\omega t + \eta)$, where $\cos \eta = \alpha$.

The hysteresis of the inner and outer shields are considered separately for the sake of simplicity. First, the magnetic circuit of an MICS, when the hysteresis of the inner shield is assumed, is shown in Fig. 4(a). R_1 is the reluctance of the main solenoid and R_2 and R_3 are those of the upper and the lower compensation solenoids. The mmfs across the main and the compensation solenoids are denoted as k_1, k_2 and k_3. It is assumed $k_1/R_1 = k_2/R_2 = k_3/R_3 = \phi$. This assumption is natural because these solenoids have a common diameter and pitch and are connected in series. The flux ϕ_x through the main solenoid is solved, when ac mmf $k_1 = R_1\phi = R_1\Phi \cdot \sin \omega t$ is applied, as

TABLE I

$q_n(H_i)$ $(A = 8,\ C = 2/3,\ \mu = 10^4)$

	q_1	q_5	q_7
H_1	$1\text{-}2.15 \times 10^{-6}$	4.7×10^{-8}	7×10^{-10}
H_2	1.94×10^{-6}	-5.0×10^{-8}	-8×10^{-10}
H_0	7.6×10^{-5}	-9.0×10^{-8}	-3×10^{-10}

$$\phi_X = \phi_2 - \phi_1 = \Phi_1 \sin(\omega t + \theta_1) \tag{5}$$

where
$$\Phi_1 = \left[1 - \frac{r_1}{R_1}\frac{r_2}{R_2 + R_3}\right]\Phi$$

and
$$\tan\theta_1 = \frac{r_1}{R_1}\frac{r_2}{R_2 + R_3}\eta_1$$

$$\cos\eta_1 = \alpha_1$$

This solution is correct to the second order when r_1, $r_2 \ll R_1$, R_2, R_3 are assumed. The phase angle θ_1 is interpreted as a measure of the hysteresis of ϕ_X against k_1. It is seen that ϕ_X is a very good approximation of $\phi = k_1/R_1$, because in general (r_1/R_1), $(r_2/(R_2 + R_3))$ are the quantities of $(1/\mu)$. The hysteresis θ_1 is reduced by a factor of $(1/\mu)^2$ compared with the original hysteresis η_1.

A similar result is obtained when the hysteresis of the outer shield is taken into account. The magnetic circuit in this case is given in Fig. 4(b) and the ϕ_X is

$$\phi_X = \phi_2 - \phi_1 = \Phi_2 \sin(\omega t + \theta_2) \tag{6}$$

$$\Phi_2 = \left[1 - \frac{r_1}{R_1}\frac{r_2}{R_2 + R_3}\right]\Phi$$

$$\tan\theta_2 = \frac{r_1}{R_1}\frac{r_2}{R_2 + R_3}\eta_2$$

$$\cos\eta_2 = \alpha_2$$

It is easily seen that by employing the outer shield and the compensation solenoid pair, the effective μ of the inner shield is multiplied by a factor $(R_2 + R_3)/r_2$ and the effective hysteresis is reduced by a factor $r_2/(R_2 + R_3)$, when an equivalent singly shielded MICS is assumed (Fig. 5). It is shown later that μ and η of a high permeability material are the order of 10^4 and 10^{-2} respectively and thus the hysteresis of a doubly shielded MICS can be ignored completely, because $\sin\theta_1$ or $\sin\theta_2$ is the ratio of the half width of the hysteresis curve along the mmf axis to the maximum amplitude of the applied ac mmf. As noted previously, this ratio under the application of an ac mmf can be interpreted as the reproducibility of the flux through the main solenoid under successive dc mmf applications. Thus a doubly shielded MICS does not only enhance the effective permeability but also reduces the hysteresis of the shielding material.

4. Residual magnetomotive force

As is well known a magnetic material has a residual mmf depending on its magnetization history. In the equivalent circuit of Fig. 4, the circuit of the material is open. In the real MICS, however, the inner and outer shields are magnetically closed circuits. This situation is shown in the equivalent circuit of Fig. 6 where the outer shield is ignored. If the material of the inner shield is uniform, i.e. $k_1/r_1 = k_2/r_2$ where k_1 and k_2 are the residual mmfs, it can easily be seen no flux due to k_1 and k_2 would flow through R. Thus, no residual field is contributed in the main solenoid due to the residual mmf, as long as the material is uniform. This is equivalent to saying a uniformly magnetized body generates a magnetic field outside it equivalent to that due to a uniform current sheet surrounding the body. In the MICS the magnetic circuit is closed and therefore no leakage field results. As an example when the material is not uniform, k_2 is assumed to be zero in Fig. 6. The fraction of the flux due to k_1 to the total flux through R is roughly $k_1/(2k)$ and k_1 is roughly $(\eta/\mu)k$. Thus, the residual field is about 1 ppm of the total field. It should be added, however, that the magnitude of the residual mmf depends on the magnetization history in a very complex manner and hence demagnetization would play an important role.

5. Magnetic property of the shielding material

The inner shield box is now being completed and its magnetic property is presented in the following. The front and back walls are removed and the resultant frame is shown in Fig. 7. The magnetization curve is recorded by a low speed B-H curve tracer and μ and $\sin\eta$ are plotted as shown in Fig. 8, 9, 10, and 11. The magnetic property depends largely on the applied field strength h_{max} and the tracing period T. The material (Supermalloy) has rather low resistivity and is rather thick and hence the eddy loss is significant even at the longest tracing period. At the same time, the peak flux density tested is between 10 and 100 gauss (these are the flux densities at which the outer shield is excited in use), and hence drift of the integrator of the tracer is significant. The flux density of the inner shield in use at H_+ state is practically zero. On account of this situation, dc and low field properties can be estimated only by extrapolation. Dashed parts in Fig. 8 to 11 are extrapolated graphically.

Fig. 8 is the measured μ vs. h_{max} curve and Fig. 9 is the μ at $h_{max} = 0$ vs. T curve where μ at $h_{max} = 0$ is extrapolated from Fig. 8. Similarly, Fig. 11 is obtained by extrapolation from Fig. 10. From these, it may be seen μ and $\sin\eta$ at dc and low applied field are about 8,000 and less than 10^{-2} respectively.

It is important to note that η/μ^2 thus obtained is as low as 10^{-10}, because this is a measure of the reproducibility of the MICS field. The value of η/μ, which is a measure of the residual field due to residual mmf, is about 2×10^{-6}.

The measured permeability shows strong nonlinearity. This is important in connexion with the permeability estimation by means of the polarity reversal of the compensation solenoid pair discussed in section 2. In H_+ state, the flux density at the inner shield is almost zero, whereas in H_- state it is finite. This situation is schematically shown in Fig. 12 where μ_+ and μ_- are the permeabilities of the inner shield material in H_+ and H_- states respectively. From Fig. 8 and 9, $(\mu_- - \mu_+)$ is estimated to be about 10^3. Hence the uncertainty of μ_+ estimated by the polarity reversal is at most 10^3. From eq. (1)

$$H_+(\mu_+) = H_+(\mu_0) + [(dH_1/d\mu) + (dH_2/d\mu)]_{\mu_0}(\mu_+ - \mu_0) \tag{7}$$

And $(dH_1/d\mu)$ and $(dH_2/d\mu)$ are graphically estimated to be $0.134\,\mu^{-2.08}$ and $-0.02\,\mu^{-2}$ respectively when A = 8, C = 2/3 and $\mu = 10^4$. Here μ_0 is the estimated permeability. Thus the uncertainty in H_+ due to nonlinearity is at most 1 ppm.

6. Conclusion

The effect of the hysteresis and nonlinearity of the shielding material on an MICS is discussed on the basis of the magnetic circuit approximation. It is found both of them have little effect on the H_+ field as long as the shielding material is uniform. The magnetic circuit approximation in this case, we believe, gives us a rather good picture on the reproducibility of the field, because in an MICS leakage flux is small and hence reluctances are well defined. The residual field due to the residual mmf may be the worst effect. If the residual field is uniform enough to generate a precessional signal, polarity reversal of the main solenoid together with the compensation pair may eliminate the effect. Discussions on this matter and others will be appreciated.

The authors wish to express their deep gratitude to Mr. T. Kobayashi for his useful discussions.

References

1. Hara, K. and Nakamura, H., Proc. Internat. Conf. on PMFC. held at NBS (Gaithersburg) in Aug. 1970.
2. Hara, K., Denkishikenjo Kenkyuhokoku, No. 560 (1957) (in Japanese) Researches of the Electrotechnical Laboratory, No. 560 (1957).

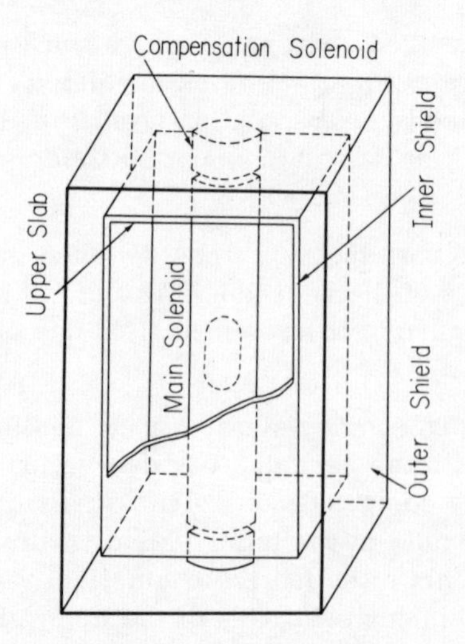

Fig. 1. Schematic diagram of an MICS.

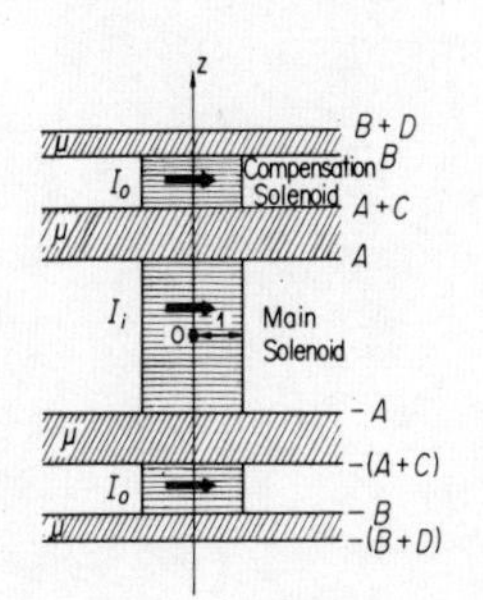

Fig. 2. Dimensions of an MICS.

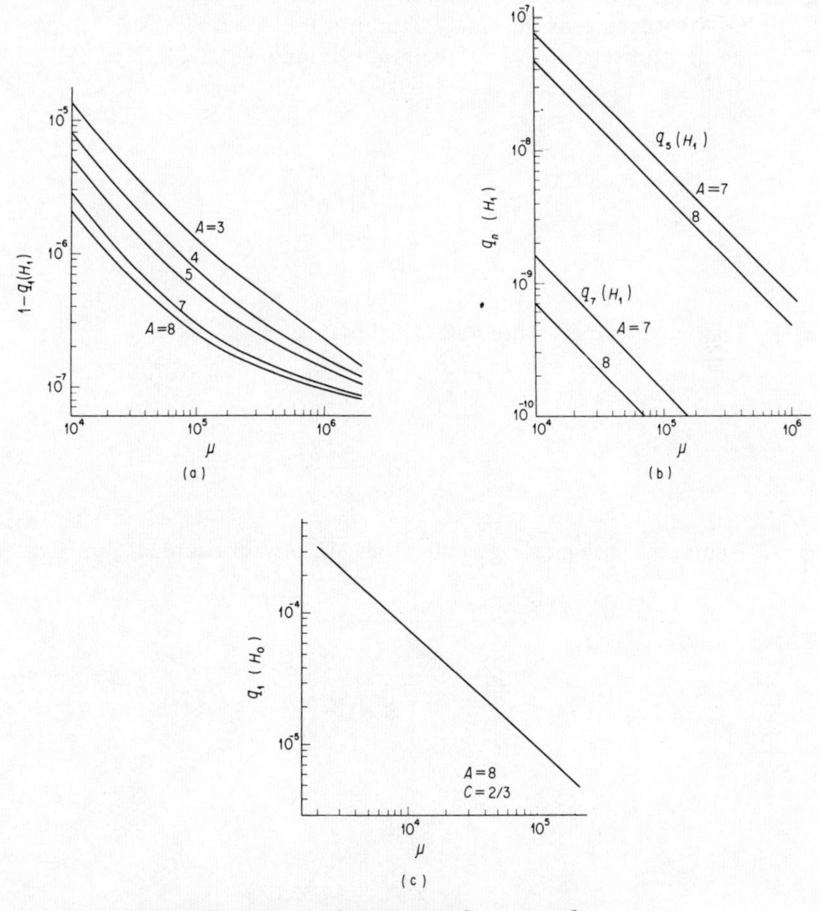

Fig. 3. $q_n(H_i)$ coefficients as functions of A, C and μ.
(a) $(1-q_1(H_1))$ as functions of A and μ.
(b) $q_n(H_1)$ as functions of A and μ.
(c) $q_1(H_0)$ as a function of $\mu(A = 8,\ C = 2/3)$.

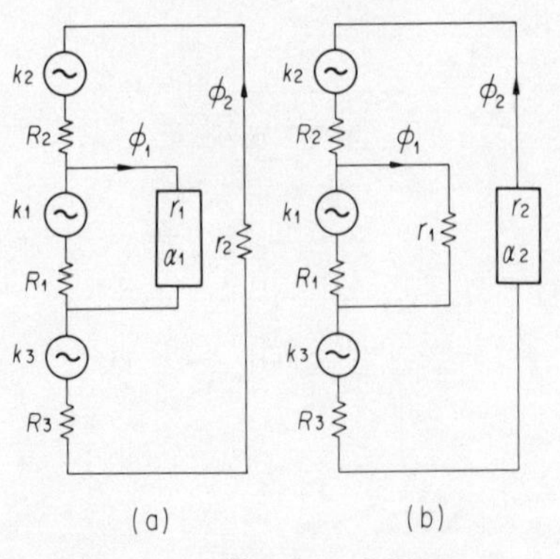

Fig. 4. Equivalent magnetic circuit of an MICS.
(a) Hysteresis of the inner shield is included.
(b) Hysteresis of the outer shield is included.

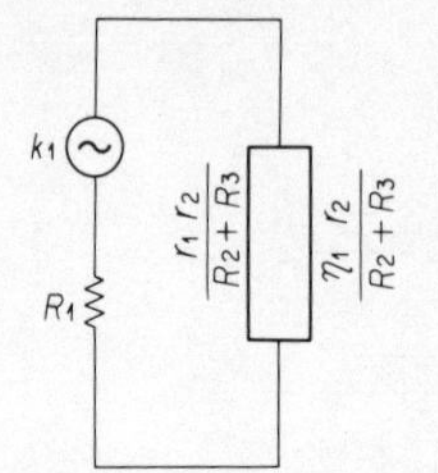

Fig. 5. Equivalent singly shielded MICS to Fig. 4(a).

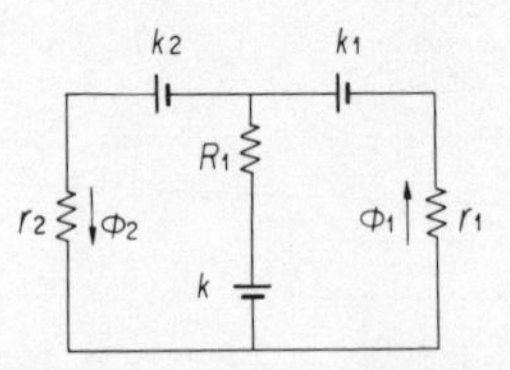

Fig. 6. Equivalent magnetic circuit of an MICS with residual mmfs as indicated.

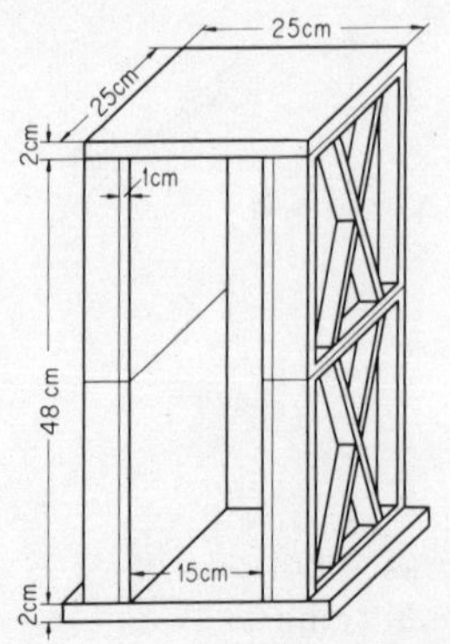

Fig. 7. Geometry of the inner shield (the front and back walls are removed for measuring the magnetic behaviour).

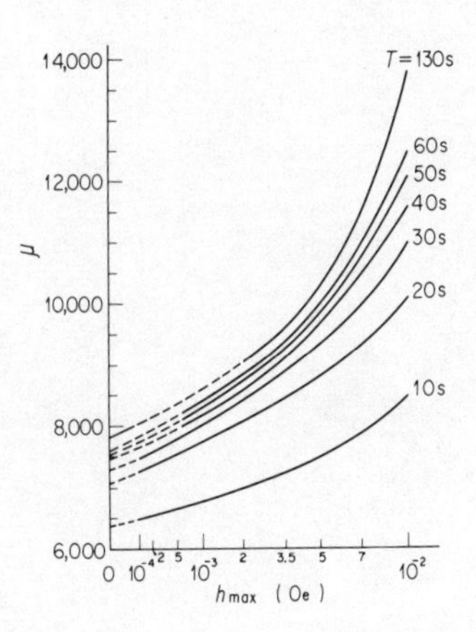

Fig. 8. μ vs. h_{max} with varying tracing period.

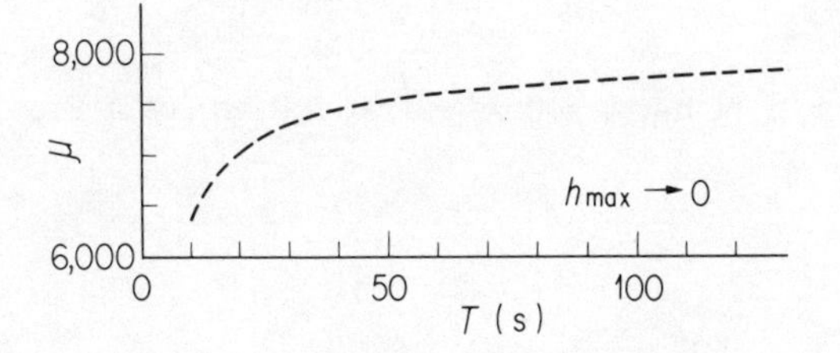

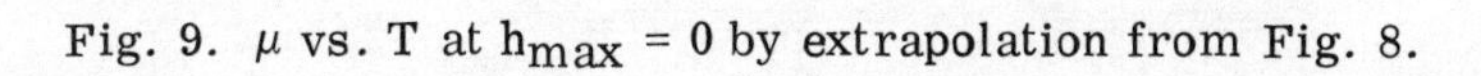
Fig. 9. μ vs. T at h_{max} = 0 by extrapolation from Fig. 8.

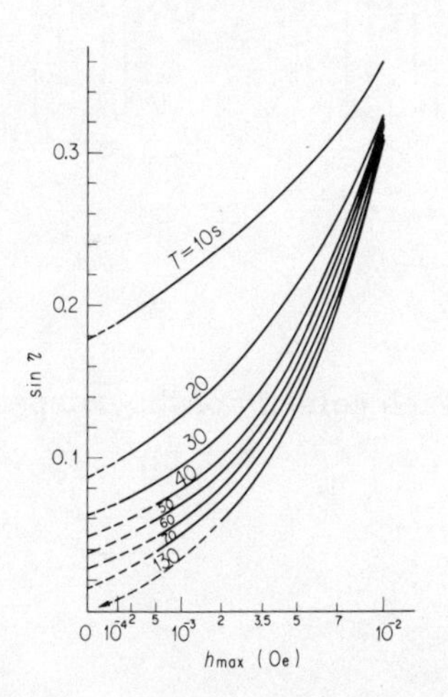

Fig. 10. $\sin\eta$ vs. h_{max} with varying tracing period.

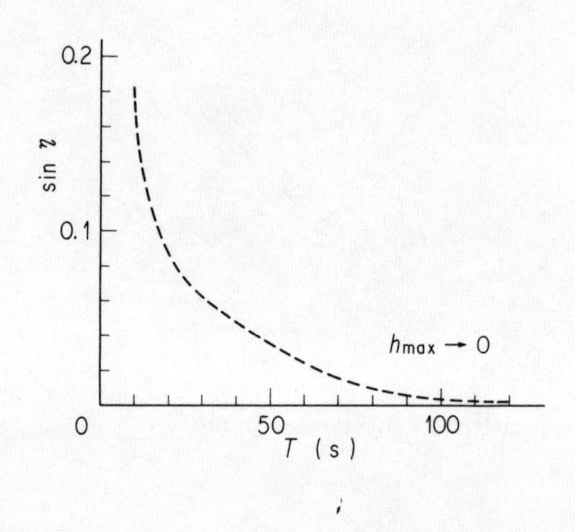

Fig. 11. $\sin\eta$ vs. T at $h_{max} = 0$ by extrapolation from Fig. 10.

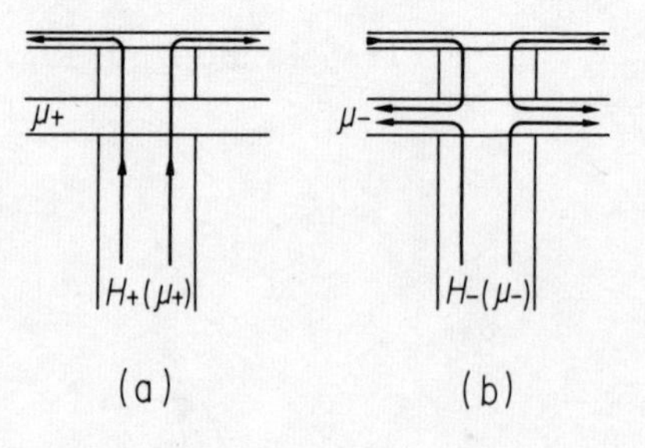

Fig. 12. Flux and μ for each polarity of the compensation solenoid pair.
(a) H_+ state.
(b) H_- state.

Determination of γ'_p at the National Bureau of Standards*

P. T. Olsen and R. L. Driscoll**

Institute for Basic Standards, National Bureau of Standards, Washington, D.C. 20234

1. Introduction

The gyromagnetic ratio of the proton, γ'_p is one of our most important fundamental physical constants. (The prime indicates that the protons are in a spherical sample of pure H_2O.) Not only is it used for calibration purposes in nuclear magnetic resonance experiments, but it plays a crucial role in determining the fine structure constant from the measurement of 2e/h via the ac Josephson effect (1). Additionally the continued measurement of γ'_p can be used to monitor as-maintained units of current (2).

The precision of γ'_p measurements at the National Bureau of Standards (NBS) has improved significantly since the early measurements of the 1950's, from several parts per million (ppm) to the present 0.1 to 0.2 ppm. Current efforts are now being directed towards improving the present 3 to 4 ppm accuracy of the experiment by an order of magnitude. It is the purpose of this paper to briefly report the progress being made in this direction. First; a general discussion of the experiment is given, including (a) use of the method of nuclear induction to determine the precession frequency; (b) a description of a new and improved series of pitch measurements using a laser interferometer; (c) a discussion of the effect of the change in original winding tension on the effective diameter of the windings; and (d) an analysis of the effect of the finite susceptibility of the solenoid and water sample support structure on the calculated magnetic field. All of the known corrections to the field are then summarized, and finally, a brief analysis of the uncertainties in the experiment is given along with a value for γ'_p.

2. Description of the Experiment

A. Method of Nuclear Induction

The method of nuclear induction was first suggested by Bloch (3)

** Retired. Present address: 10139 Cedar Lane, Kensington, Maryland 20795.

and has been discussed in the literature (4). The method is well known and widely used. For low field measurements, however, the residual field gradients must be small; a compensated solenoid (5) provides a calculable magnetic field of 1.2×10^6 nT with sufficient uniformity. The controlled current passing through the solenoid is monitored by comparing the voltage drop across a standard one ohm resistor with the emf of a saturated standard cell.

Figure 1 is a block representation of the electronic equipment used in determining the precession frequency (2). Considering the required driving field of 1.0 nT of the transmitter coil, the detecting coils or receiver coils must be carefully adjusted orthogonal with respect to the transmitter coils in order that the 4×10^{-4} nT field arising from the precessing protons can be detected. Even after careful adjustment, a quadrature leakage signal remains. This quadrature leakage signal is nulled by means of an injection signal introduced on the B channel of a differential amplifier used as a summing amplifier. The amplitude and phase of the injection signal are adjusted to their proper values with the aid of the lock-in amplifier and with the magnetic field turned off resonance field by 500 ppm. With the magnetic field restored, the protons are induced to resonate at some center frequency near 52.5 kHz. A commercial frequency synthesizer supplies drive frequency and reference frequency. With the phase of the reference properly adjusted on the two-phase lock-in amplifier, the synthesizer is frequency adjusted until the indicated dispersion is zero. Signal to noise ratio of the proton signal before the lock-in is employed is 3:1 but the resolution of the lock-in amplifier allows the center of the line to be located to within 0.1 ppm (fig. 2).

Figure 3 is a plot of the monitoring of the standard cells in enclosure 1461 (2). Here the assigned value of the resistance standard has been used throughout, the coil constant is adjusted at each point according to the observed temperature, and the proton precession frequency is measured. With these numbers and a constant value of γ_p', the effective emf of the cells is calculated. A least-squares linear fit to the data yields a standard deviation for a single observation of about 0.2 ppm.

B. Pitch Measurement

The precision solenoid, constructed in 1951, has a single layer of 1000 turns of oxygen free copper wire. The wire is laid in a lapped groove on a cylinder of fused silica. The average spacing of the turns has been measured a number of times by comparing sections of about 1/2 meter with a calibrated fused quartz end standard. This method has given useful information on the length of 1000 turns accurate to about 2 to 3 ppm. However, when attempting to find local variations from the ideal pitch by measuring small intervals, the method proved to be too inaccurate. Corrections to the field strength for pitch variation were, therefore, not made because of the observed large scatter in the data.

Recently, a commercial laser interferometer was used to determine the mean pitch as well as the local deviations from this mean more accurately than previously. The solenoid was mounted horizontally on a marble bed and the laser interferometer mounted on the same bed as near as possible to the end of the solenoid. A contacting probe with corner cube affixed was stepped along the solenoid locating wire positions with respect to a reference turn.

The contacting probe with its three feet was positioned on top of the coil. Two wedge feet in front nestled down between the same two turns of wire. The back foot (a shim) rested on top of the turns of wire some eight centimeters behind the front wedges. A cross-wise level on top of the probe assured reproducible settings on top of the solenoid while a length wise level measured the backward or forward tilt of the probe. The tilt level was calibrated by shimming the back foot and it was found that a forward tilt of the probe registering as 0.1 division on the level reduced the corner reflector-interferometer distance by 0.33 μm.

With the probe locating the space between the first and second turns of wire, the corrections for speed of light (humidity, air temperature, and atmospheric pressure) were dialed into the interferometer and the readout was reset to zero. A pair of hard-wood rails raised by a screw lifted the probe off the wire. By hand pressure the probe was slowly moved along the rails to the 900th turn and the rails were lowered, the probe nestling into the space between the turns 900 and 901. (The back foot made full length measurements impossible.)

To take into account a possible drift resulting from thermal expansions taking place during the course of measuring, a period of 90 minutes, the first and 900th positions were located alternately several times in order to provide a highly reliable beginning reference point. Linearity of drift was checked by locating four other reference points (300, 500, 600 and 800 turns). These intermediate points were checked during the course of the measurements. Every 20 turns were located and their distance from the first turn recorded. Angular positions measured around the coil were 0°, 30°, 60°, 180°, 210°, and 240°. (At present, the lead in wire support structure severely limits the possible angular positions which may be measured.) After these measurements were made (2 or 3 sets per angular position) the coil was turned end for end and the measurements repeated at the same six angular positions. A total of 20 measurement sets were taken on the solenoid.

After the level corrections were made and the reference point drift corrected for, the data were analyzed by computer. A least-squares straight line was fitted to a plot of turn-position vs. turn-number. The slope of this line is the reciprocal of the mean pitch and the residuals are the deviations of the individual turns from their "ideal" positions. A plot typical of these deviations for one angle is shown in figure 4. The pitch variation correction is calculated from these deviations via the equation given in Table I.

C. Solenoid Diameter

The diameter of the solenoid has not been measured rigorously in recent years. A plot of coil resistance vs. time indicates that the resistance is decreasing in an exponential fashion, suggesting a reduction in the original winding tension. Such a reduction would allow the coil form to swell, resulting in a larger diameter. A diameter spot check measurement was made at points previously measured. A comparison of measurements indicate that the solenoid diameter has increased by 6.4 ppm, resulting in a decrease of field strength by 0.5 ppm. Corrections to the field strength resulting from diameter variations are assumed to be a function of the rigid form, and not primarily a function of winding tension. Thus, the correction obtained previously for diameter variation was used. A comprehensive diameter measurement is planned for the near future.

D. Susceptibility

Our previous γ_p' determinations took susceptibility effects into account by prescribing a larger uncertainty to the experiment. In order to reduce this uncertainty, the susceptibilities must be determined and their effect on the strength of the magnetic field accounted for. The effect of the susceptibility of the solenoid fused silica form and of the compensating coil pyrex form can be readily calculated. The pole strength at the ends of the form are computed and their effect at the center determined by calculation. The effect of the aluminum tube which supports the water sample and pick-up coil was determined empirically. A sample of this tube was placed into the opposite end of the solenoid and positioned as a mirror image of the support tube. The precision of the γ_p' experiment was used to determine the effect, by noting the frequency difference when the added tube was present and when it was removed. The effect of the pick-up coil and wooden box surrounding the pick-up coil and water sample is yet to be determined. For the present, we take it into account by including an additional uncertainty of 0.4 ppm. (First estimates indicate that the correction is less than this.)

3. Summary of Solenoid Data

The following tables summarize the solenoid dimensions and their use in field computation (Table I), and the susceptibility and compensating coil contributions to the solenoid field strength (Table II).

4. Uncertainties

Table III summarizes the uncertainties considered. All known sources of uncertainty are included and the values given are meant to correspond to one standard deviation (6).

Table I
Solenoid Magnetic Field

$$H = \frac{\mu_o NIU}{L}\left[1 + \frac{a}{2U}\int_{-\frac{L}{2}}^{+\frac{L}{2}}\left(\frac{(2x^2 - a^2)U_a(x)}{(x^2 + a^2)^{5/2}} - \frac{3axU_x(x)}{(x^2 + a^2)^{5/2}}\right) dx\right]$$

L = Length at 25.0°C		1000.08758mm
$2a = 2\bar{a} + 2\Delta a$ at 25.0°C		279.6260mm
Overall diameter 25.0°C	280.3257mm	
Wire diameter	.6991mm	
Mean coil diameter	$2\bar{a}$= 279.6266mm	
Strain in wire	10.08 x 10^{-4}	
Current distribution effect	Δa = 1-2.4 x 10^{-6}	

$$U = 1/[1+(2a/L)^2]^{1/2} = 0.96306360$$

N = 1000 turns

$U_x(x)$, $U_a(x)$ are the observed displacement of any part of a winding in an axial or radial sense respectively, from the position it would have if the winding were uniform or "ideal".

Table II
Total Magnetic Field

Total Magnet Field 25.0°C	
Solenoid	1.21011543 x 10^6 nT/A
Correction to solenoid field	
diameter variations	+0.62
pitch variations	+1.00 ppm
susceptibility of silica form	+0.074 ppm
Compensating coil Hc	1.400739 x 10^3 nT/A
Corrections to total field	
return lead wire	+0.44 ppm
susceptibility of comp. coil form	+0.42 ppm
susceptibility of al. support tube	+0.19 ppm
Coil constant	Kc(25) = 1.21151948 x 10^6 nT/A
Temperature coefficient	Kc(t) = Kc(25)[1+0.48x10^{-6}(25.0°C-t)]

Table III
Summary of Uncertainties

Item	Uncertainty (ppm)
Pitch and pitch variation	0.72
Diameter	0.32
Diameter variation	0.20
Susceptibilities	0.45
Precession frequency	0.15
Standard cells	0.15
Standard Resistor	0.12
Return lead alignment	0.10
Temperature coefficient of coil constant Kc	0.02
Compensating coil	0.007
Earth's field	0.002
RSS Total:	0.97

Final assigned uncertainty: 2.0 ppm (see text).

The largest source of uncertainty arises from the coil spacing measurements which give the pitch, and pitch variation correction. The itemized factors which contribute to the pitch and pitch variation uncertainty are given in Table IV. The preliminary and limited nature of the present data, which leads to the expansion of the uncertainty assigned the final γ_p' result to 2 ppm, is presented in the discussion associated with the table.

Discussion (Table IV)

The stability of the laser was checked several times by comparing its wavelength to a highly stabilized laser. The interferometer laser never deviated by more than 0.03ppm. The interferometer as a unit was checked against a length standard of known dimensions and found to be accurate to at least 0.1 ppm plus 1 least count of 0.1μm. The speed of light corrections were limited by the ability to measure the temperature, humidity, and atmospheric pressure. The temperature of the air was the largest contributor to the length uncertainty because of the presence of the observer near the light path.

The environment in which the coil was measured was controlled to better than 0.5°C. The solenoid temperature as monitored never appeared to change by more than 0.2°C from day to day and never more than 0.1°C during the course of the measurement. The pitch of the solenoid, as compared to previous measurements of pitch using end standards, was within the spread (4ppm) of those determinations. Pitch variation corrections as determined before varied from +10.0ppm to -10.0ppm, so comparison is out of the

Table IV
Detailed pitch and pitch variation uncertainties

Item		Uncertainty (ppm)		Uncertainty (ppm)
Pitch of Fused Silica Solenoid and Pitch Variation Correction				
Measurement uncertainty				
Reproducibility of contacting probe		0.15		
Reading of level		0.15		
Observed s.d. of six angular positions		0.64		
	Total	0.67		0.67
Laser interferometer				
Laser stability		0.03		
Interferometer accuracy		0.10		
Laser light path (speed of light)				
Humidity		0.05		
Atmospheric pressure		0.07		
Temperature		0.20		
	Total	0.24		0.24
Solenoid temperature				0.10
			Total	0.72

Discussion (Table IV, cont'd.)

question.

As noted previously, the pitch and associated pitch variation was measured at six different angles (0°, 30°, 60°, 180°, 210°, 240°). Either three or four sets of measurements were taken at each angle, making a total of 20 sets. The observed standard deviation of the computed magnetic field obtained from the data at a particular angle (via the equation of Table I) was 0.81 ppm at 210° and 0.50 ppm or less at the other angles. The observed standard deviation of the magnetic field computed from the data obtained at all six angular positions (i.e., six values) is 0.64 ppm. If the diametrically opposite positions are first averaged (i.e., 0° with 180°, 30° with 210°, and 60° with 240°), the observed standard deviation for the field from the three values is 0.32 ppm. This suggests a slight curvature in the coil, a hypothesis which is strengthened by the observation that the observed standard deviation of the pitch when diametrically averaged (i.e., three values) is only 0.16 ppm, while the standard deviation of the average pitch for all six angular positions is 0.92 ppm. The implication is that the coil is a section of a toroid of 10^5 meter radius.

If the present limited data are assumed representative of the entire solenoid, then the statistical uncertainty of the calculated magnetic field may reasonably be taken as equal to the above 0.64 ppm standard deviation. This is what has been done in Table IV. However, the incomplete sampling of the angular positions of the solenoid (measurements were not carried out on 2/3 of the solenoid), and the fact that only every twentieth turn was measured, leaves us a bit uneasy This uneasiness increases when the possible toroidal nature of the solenoid is considered along with the scatter in the pitch data when it is not diametrically averaged. Thus, to take into account the preliminary and limited nature of the present data, we increase the total assigned uncertainty by a factor of 2 to 2 ppm (Table III). While some may consider this an overly conservative approach, there are, in our minds at least, too many unanswered questions to believe that the measurement is really good to 1 ppm (The limited nature of our present data on the diameter of the solenoid and the susceptibility corrections, to be discussed in the next two sections, also contributes to this belief.)

Table V
Detailed diameter and diameter variation uncertainties

Item	Uncertainty (ppm)	Uncertainty (ppm)
Diameter of Fused Silica Solenoid		
Measurement uncertainty		
Observed s.d.	1.1	
End standard	0.8	
End standard temperature	0.05	
Solenoid temperature	0.1	
Contact force correction	1.8	
Total	2.3	2.3
Relaxation of winding		
50% of diameter change via spot check		3.2
Wire diameter	0.5	
Current distribution	0.3	
Total	0.6	0.6
Total		4.0
Diameter Uncertainty Contribution to Magnetic Field (8%)		0.32
Uncertainty in Correction for Diameter Variation		0.20

Discussion (Table V)

Because the solenoid is a long solenoid, the effect of diameter measurement uncertainties is reduced to 8%. The measurement uncertainties are those established when the diameter was measured. The spot check measurement is too incomplete to be assumed reliable and a large uncertainty is given for this. The wire diameter would change if the resistance decrease mentioned earlier is a function of a reduction in wire stress. The computation for current distribution due to the reduction in the tension of the wire is small, 0.3 ppm, but the total correction of 2.4 ppm is not expected to be better than 10%.

Table VI
Detailed susceptibility correction uncertainties

Item	Uncertainty (ppm)
Fused silica form	0.1
Pyrex form of field uniformity compensating solenoid	0.1
Aluminum support	0.02
Undetermined (estimated)	0.4
Total	0.45

Discussion (Table VI)

Susceptibility uncertainties due to the forms and aluminum support are small and could perhaps be further reduced with additional evaluation. The undetermined uncertainty, an estimate, is based on other bodies in and around the solenoid and must be further evaluated. Calculation of the effect of the susceptibilities of some geometries are extremely difficult and true representation of geometry for empirical (experimental) evaluation is sometimes equally difficult.

Table VII
Detailed precession frequency uncertainties

Item	Uncertainty (ppm)
Crystal oscillator	0.0005
Frequency synthesizer	< 0.001
Reading frequency synthesizer	0.02
Adjusting phase of reference	0.1
Adjusting frequency to center of proton line	0.1
Cancellation of leakage signal	0.05
Total	0.15

Discussion (Table VII)

The crystal oscillator is compared to WWV by phase comparison. Periodic adjustments are made when necessary, but are seldom needed more often than once a year. The synthesizer locked to the crystal oscillator is good to 0.1% of the continuously adjustable dial. Reproducibility in reading this dial from observer to observer is 0.2 division, which corresponds to 0.002Hz in 50 kHz ($4/10^8$). Difficulties in adjusting the phase of the reference caused large scatter in the data during the early stages of our use of the nuclear induction technique. This has now been corrected. With all leakage signals nulled, and the solenoid field held constant, the frequency synthesizer is slowly changed to sweep the line. The phase of the reference is adjusted so that for any absolute value of the dispersion (+y and -y) the amplitude of the signal is the same. Once adjusted, the phase of the reference remains the same. If the line is symmetrical, this adjustment could be considered to be statistical and average out over many readings. However, if the line is not symmetrical because of circuit tuning or some other cause, this could lead to a systematic uncertainty. Thus, the last three items in Table VII are included for the present in order to take into account possible incomplete statistical averaging.

Table VIII
Detailed electrical standards uncertainties

Item	Uncertainty (ppm)	Uncertainty (ppm)
Standard cells (transfer from NBS reference group to non-mag. lab.)		0.15
Standard resistor		
Calibration	0.08	
Temperature	0.04	
Barometric pressure correction	0.02	
Load coefficient	0.08	
Total	0.12	0.12

Discussion (Table VIII)

The uncertainty quoted for the standard cells, which is quite conservative, has been verified by many transfers between the NBS reference group of standard cells and the non-magnetic laboratory. (Indeed, the two transfers on which the value of γ'_p to be reported in this paper is based agreed to within a few parts in 10^8.) The standard one ohm resistor has been calibrated many times and the values obtained have been constant to about 0.1 ppm for more than eight years. Calibrations have been made at the used power load. The temperature of the oil bath is controlled at the temperature where the temperature coefficient of the resistor is approximately zero; a temperature change of 0.5°C about this point results in a resistance change of $1/10^7$.

Table IX
Details of other factors considered in the uncertainty

Item	Uncertainty (ppm)	Uncertainty (ppm)
Field Uniformity Compensating Solenoid		
Length standard (diameter)	1.8	
Length standard (pitch)	1.4	
Observed s.d. (diameter)	2.7	
Observed s.d. (pitch)	2.1	
Temperature	0.4	
Ratio of resistance standards for dividing current between main and compensating solenoids	4.0	
Total	5.8	
Compensating solenoid uncertainty contribute to total field ($1.15\text{x}10^{-3}$)		0.007
Alignment of Return Lead		0.1
Temperature Coefficient of Solenoid Constant		0.02
Helmholtz Coils for Nulling Earth's Field		
Vertical cancellation	0.001	
Horizontal cancellation	0.001	
Total	0.002	0.002

Discussion (Table IX)

The compensation coil uncertainty is reduced by 1.15×10^{-3}, i.e., the ratio of its field to the precision solenoid's field. The uncertainty in the alignment of the return lead was very evident recently when a spot check found it to be off by 0.6mm. The effect was the sine of the angle of 0.6mm/1000mm times the return leads field strength at the center of the solenoid. This contribution reverses when the solenoid is reversed and is ever present as a systematic effect unless properly aligned.

The nulling of the earth's field is a second order effect for the off axis component. Unless the helmholtz coil current is changed by about ±20%, no evidence of its presence can be detected to within $1/10^7$.

5. Final Result

The primary improvement in the NBS γ_p' experiment over the past year is the increased accuracy with which the pitch and pitch variation correction has been determined. The result of the preliminary work reported here is

$$\gamma_p'/2\pi = 42.5760995 \times 10^6 \text{ Hz}/T_{NBS},$$

or

$$\gamma'_p = 2.67513523 \times 10^8 \text{ rad s}^{-1}/T_{NBS},$$

with an assigned uncertainty of 2.0 ppm. This should be compared with the value 2.6751301, uncertainty of 3.7 ppm, implied by the measurements carried out over the period 1958 to 1968 (1) when the latter are converted to post 1 January 1969 NBS electrical units (1).

These values have been attained from the equation

$$\gamma'_p = \frac{2\pi\nu'}{A_{NBS}\ Kc[1 + 0.48 \times 10^{-6}(25.00°C-t)]},$$

where the term in brackets takes into account the temperature dependence of the solenoid constant (Table I) and A_{NBS} is the NBS as-maintained ampere (i.e., the ratio of the NBS as-maintained units of voltage and resistance) realized during the period August 11, 1971 to August 27, 1971.

6. Acknowledgments

We wish to thank Dr. H. Ku and Mr. B. Field for assistance with the statistical analysis of the pitch measurement data, Dr. F. K. Harris for the loan of the interferometer, Mr. A. T. Funkhouser for investigating the stability of the laser, and Drs. B. N. Taylor and E. R. Williams for helpful discussions and assistance.

NOTE ADDED IN PROOF. A preliminary measurement of the bending of the precision solenoid due to its own weight indicates that the value of γ'_p given here should be decreased by 0.3 ppm. (A simple calculation of the effect agrees with the measured result to within the 0.1 ppm uncertainty of the latter.)

References

1. TAYLOR, B. N., PARKER, W. H., and LANGENBERG, D. N., Rev. Mod. Phys. 41, 375 (1969).

2. DRISCOLL, R. L., and OLSEN, P. T., Proc. Int. Conf. on Prec. Meas. and Fund. Constants, edited by D. N. Langenberg and B. N. Taylor (NBS Spec. Pub. 343, USGPO, Washington, D. C., 1971), p. 117.

3. BLOCH, F., Phys. Rev. 70, 460 (1946).

4. ABRAGAM, A., Principles of Nuclear Magnetism (Clarendon Press, Oxford, England, 1961).

5. SNOW, C., and DRISCOLL, R. L., J. Res. Nat. Bur. Stds. 69c, 49 (1965).

6. Where applicable, unless the word "observed" is used, a judgment estimated to be equivalent to one standard deviation is meant.

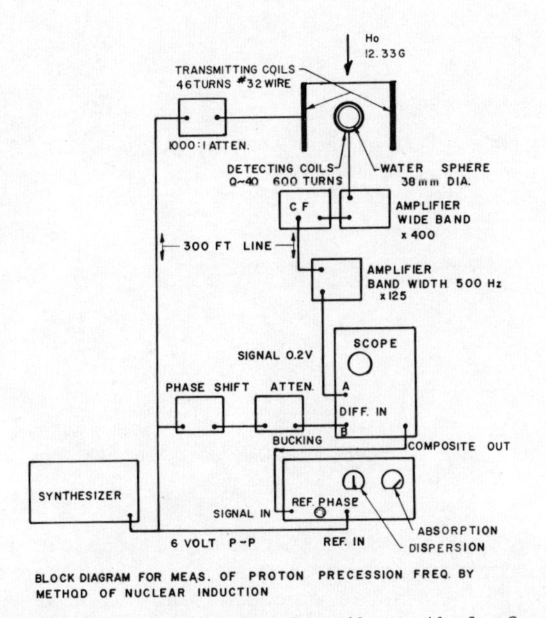

Fig. 1. Schematic diagram showing how the method of nuclear induction is used to determine the proton precession frequency.

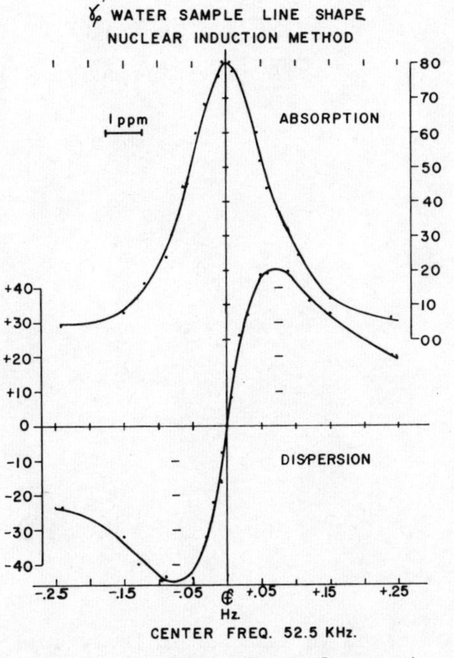

Fig. 2. Observed resonance line shape for protons in an H_2O sample as obtained via the nuclear induction technique.

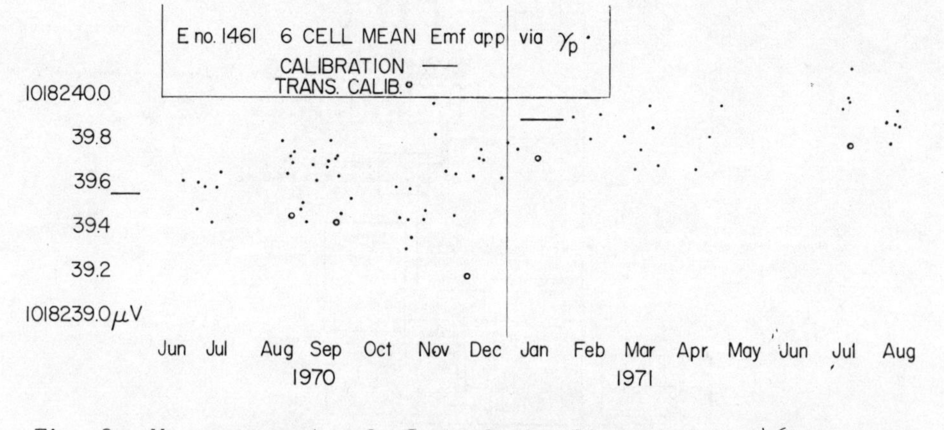

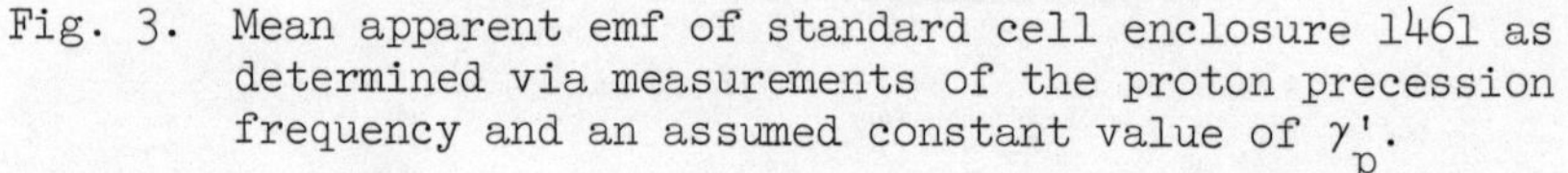
Fig. 3. Mean apparent emf of standard cell enclosure 1461 as determined via measurements of the proton precession frequency and an assumed constant value of γ_p'.

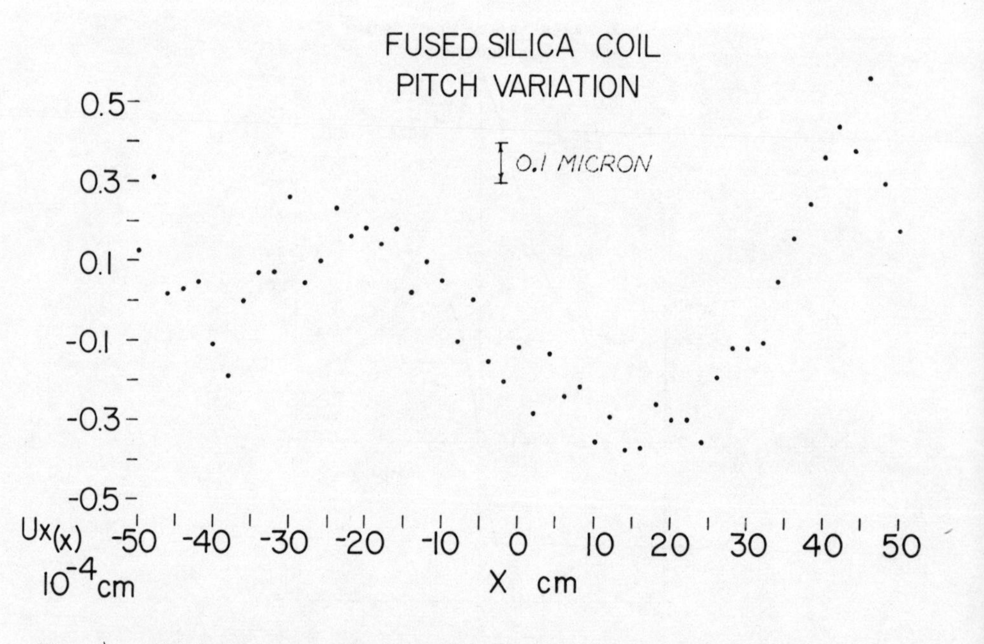

Fig. 4. Typical observed variations of individual turns from their expected or ideal positions.

The g-Factor Anomalies of the Electron and Positron

John Wesley and Arthur Rich

Physics Department, University of Michigan

1. Introduction

Since the last meeting of this conference four years ago there has been a great deal of progress on both the theoretical and experimental fronts of Quantum Electrodynamics (QED), particularly as related to the lepton g-factor anomalies (1). In this paper we will summarize the most recent work on the electron-positron anomalies and in addition we will try to assess the potential for future progress in the field.

2. Definitions and History

For an elementary quantum particle with mass M, charge Q, and intrinsic angular momentum (spin) $\vec{S}$, symmetry requires that any associated magnetic moment $\vec{\mu}$ be parallel or anti-parallel to $\vec{S}$. Dimensional analysis shows that $\vec{\mu}$ will be proportional to $(Q/Mc)\,\vec{S}$. Expressed as an equality, this relation becomes

$$\vec{\mu} = \frac{g}{2}\frac{Q}{Mc}\vec{S} \tag{1}$$

where the constant of proportionality g (the g-factor) is dimensionless. Dividing g by 2 is conventional. The magnitude of g is characteristic of the internal structure or interactions of the particle. The degree of agreement between theoretical predictions and experimental measurements of g therefore provides a means for testing the validity of the theory.

In order to place the current work in perspective, a brief historical survey of high points in the development of the subject will prove useful. In 1925 Uhlenbeck and Goudsmit* postulated that the electron had a spin of $\hbar/2$ and a g-factor of 2 in order to explain optical doublets and the anomalous Zeeman effect in alkalai spectra. Within two years Dirac was able to show that these values followed from his relativistic formulation of quantum mechanics. This value of g stood until 1947 when Kusch and Foley reported that the quantity now defined as the g-factor anomaly

*Only page references which are judged of current interest will be included.

($\underline{a} \equiv (g-2)/2$) had the non-zero value 0.001 19(5)*.

By December of 1947, Schwinger had developed a non-divergent formulation of QED that permitted him to show that the lowest order radiative correction to the electron anomaly was equal to $\alpha/2\pi = 0.001\ 16$. Hereafter the numerical value of the fine structure constant $\alpha = e^2/\hbar c$ will be taken as $\alpha^{-1} = 137.0608$ (2). Later work by Dyson showed that the anomaly (as well as other quantities obtained from QED calculations) could be expressed as a power series in α viz:

$$a = A(\alpha/\pi)+B(\alpha/\pi)^2+C(\alpha/\pi)^3+\cdots \qquad (2)$$

Note that although the coefficients (A,B; 2nd order, 4th order...) must be finite, the series may only converge asymptotically.

Calculation of the coefficients in Eq. 2 and their comparison with experiment are of vital importance as tests of QED. Unfortunately, the computations have proven rather intractable; B was first computed in 1950 (3) and recalculated in 1957 (4,5) as -0.32848, while the first published result for C, C = 1.49, was just completed this year (6).

Note that we have not distinguished electron from positron thus far. This is because, if TCP is conserved, particle and antiparticle g-factors (as well as masses and life times) are identical. Thus a comparison of measured $e^- - e^+$ g-factors provides a check of TCP invariance.

On the experimental side the years after 1947 saw continuous improvements in precision reached by work along three tracks. One track, an extension of the Kusch-Foley method, involved measurements of g itself by atomic beam techniques; the second involved a comparison of the magnetic moment of the proton in Bohr magnetons (μ_e/μ_B) with the electron moment also in Bohr magnetons (μ_e/μ_B). Both methods can be pushed to accuracies approaching a part per million (ppm), but this is only 1000 ppm of $\underline{a}$, still insufficient to test QED to fourth order.

The third track, which has so far proven most successful, was initiated by Crane at The University of Michigan. The Michigan technique was originally a measurement of the g-factor of the <u>free</u> electron but has evolved to a direct determination of $\underline{a}$ for free electrons and positrons. In our latest work, $\underline{a}$ has been measured

*The quantity enclosed in parentheses represents the uncertainty in the final digits of a numerical value. Thus, 0.001 19(5) is equivalent to $0.001\ 19 \pm 0.000\ 05$. All errors are one standard deviation unless otherwise noted.

to 3 ppm (6th order coefficient checked) with only minimal systematic corrections necessary.

Several other well developed resonance type experiments to measure $\underline{a}$ or $\underline{g}$ are also under way at this time, but all have the common feature of working with free electrons or positrons, (see section 4).

3. Theory

The purpose of this section is to review only the latest developments in the calculation of the 6th order coefficient. References 1 and 7 contain comprehensive discussions of the theory, including a description of dispersion theory estimates of C and speculation on the effects a breakdown of QED might have on $\underline{a}$.

The computational problems involved in the 6th order calculation are formidable. There are 72 distinct diagrams of the types shown in Fig. 1. The integral expressions obtained from these diagrams are 7-dimensional, and are of such complexity that the algebraic manipulations and integrations required to complete their evaluation often require the use of a computer. The necessary Feynman algebra can be handled with algebraic manipulation programs, such as the ones described in reference 8. The resultant integrals can then be evaluated by a variety of numerical techniques. All sixth order diagrams have been calculated. The total sixth-order contribution is

$$C^{(6)} = [(1.23 \pm 0.2)+(0.0554)-(0.154 \pm 0.009)+(0.36 \pm 0.04)](\alpha/\pi)^3$$

$$= (1.49 \pm 0.2)(\alpha/\pi)^3.$$

The first, and by far the most difficult term to calculate, is the sum of the vertex contributions (6); the second (9) and third (10) terms are vacuum polarization contributions, and the fourth term is the photon-photon scattering contribution (11). Several groups are now working on the vertex calculation, and it is hoped that independent checks will be forthcoming shortly. In particular, we would like to mention an alternative approach to the problem now being pursued by a group at Michigan (12) who are using Schwinger's mass operator method (13). The advantages of the mass operator formalism are that the resultant integral expressions are fewer in number and somewhat less complex, and mass and charge renormalization can be included more readily.

The final theoretical result for $\underline{a}_e$ complete through 6th order may now be written as

$$\underline{a}_e(T) = 0.5(\alpha/\pi) - 0.32848(\alpha/\pi)^2 + (1.49 \pm 0.2\ (\alpha/\pi)^3 \qquad (3)$$

The quantity $(\alpha/\pi)^3$ is about 10 ppm of $\underline{a}$, so the 6th order electromagnetic contribution to $\underline{a}_e$ is about 15 ppm. As far as non-electromagnetic contributions to $\underline{a}_e$ are concerned we note that hadronic vacuum polarization diagrams contribute about 60 ppm to a_μ (7). Their effect on $\underline{a}_e$ should scale roughly as $(m_e/m_\mu)^2$, so that strong interactions may be neglected.

4. Experiments

1. General Comments

As mentioned in section 2, all current g-2 experiments which have the possibility of ppm accuracy are performed with unbound particles. Two distinct experimental techniques may be identified. The first (precession technique) involves a direct observation of the spin precession of polarized leptons in a static magnetic field (Michigan e^-, e^+ and CERN μ^-, μ^+ experiments). The second (resonance technique) is characterized by the presence of an oscillating electromagnetic field that induces transitions between the energy levels of an electron in static electric and magnetic fields (Universities of Washington, Mainz and Stanford). Both types will be discussed in the following sections.

2. The Precession Method--General Description

The essentials of this method may be described as follows. If a particle with velocity $\vec{v}$ and rest mass m moves perpendicular to a uniform magnetic field $\vec{B}$, its orbital motion will be a uniform rotation of $\vec{v}$ at the cyclotron frequency (radians/sec) $\omega_c = \omega_o/\gamma$, where $\omega_o = eB/mc$, $\gamma = (1-\beta^2)^{-\frac{1}{2}}$. Its spin motion in the laboratory frame will be a uniform precession of $\vec{S}$ at frequency $\omega_S = (g/2)\gamma\omega_c+(1-\gamma)\omega_c$, where the $(1-\gamma)\omega_c$ term is the Thomas precession due to the acceleration associated with uniform circular motion. The relative precession of $\vec{S}$ with respect to $\vec{v}$ occurs at the difference frequency defined as $\omega_D = \omega_S-\omega_c$. Substituting the above expressions for ω_S and ω_c into ω_D we obtain the well known result

$$\omega_D = a\omega_o \tag{4}$$

The fortunate circumstance that ω_D is rigorously independent of γ in the idealized situation just described makes it possible to measure $\underline{a}$ with only minimal corrections for particle energy.

In the more general case of a particle moving with arbitrary β in inhomogeneous electric $\vec{E}(\vec{r})$ and magnetic fields, we may define the angular velocities $\vec{\Omega}_\beta$ and $\vec{\Omega}_S$ through the equations $d\hat{\beta}/dt = \vec{\Omega}_\beta x\hat{\beta}$ $(\hat{\beta} = \vec{\beta}/\beta)$ and $d\vec{S}/dt = \vec{\Omega}_S x\vec{S}$. Then the difference frequency vector $\vec{\Omega}_D$, analogous to ω_D, is

$$\vec{\Omega}_D = \vec{\Omega}_S-\vec{\Omega}_\beta = -\frac{e}{mc}\{a\vec{B}-a\frac{\gamma}{\gamma+1}\vec{\beta}(\vec{\beta}\cdot\vec{B})+[(\beta\gamma)^{-2}-a]\vec{\beta}x\vec{E}\} \tag{5}$$

The equation of motion of $\vec{S}$ in the electron rest frame is simply $d\vec{S}/dt = \vec{\Omega}_D x\vec{S}$. The instantaneous motion of $\vec{S}$ with respect to $\vec{v}$ will be a precession of $\vec{S}$ about $\vec{\Omega}_D$ with angular velocity $\omega_D=|\vec{\Omega}_D|$. It is equation (5) with its axial velocity and electric field correction terms that will be used in discussion of the actual precession and resonance experiments.

The method employed to measure a consists of an indirect determination of ω_o and a direct determination of ω_D. The quantity ω_o may be expressed as:

$$\omega_o(e^-) = \omega_P(H_2O)(\mu_p'/\mu_B)^{-1}.$$

Here $\omega_P(H_2O)$ is the NMR frequency of protons in a spherical water sample and μ_P'/μ_B is the magnetic moment of the proton in Bohr magnetons as measured in a similar sample (uncorrected for diamagnetism). This ratio as directly measured is known to better than 0.5 ppm (14) and at the fields used in our work $\omega_P(H_2O)$ may be measured to about 0.2 ppm so that determination of $\omega_o(e^-)$ to better than 1 ppm is fairly straightforward (15).

The direct measurement of ω_D constitutes the most difficult aspect of the work. To see how this is accomplished in a general manner, note that the total angle of precession of $\vec{S}$ with respect to $\vec{v}$ from time t=0 to t=T will be $\theta_D = \omega_D T$, so $\hat{S}\cdot\hat{v}$ (a quantity proportional to the lepton helicity) will vary as $\cos \omega_D T$. A device with an output proportional to $\hat{S}\cdot\hat{n}$ (a polarimeter, where $\hat{n}$ defines a fixed direction in space) can be used to measure ω_D by means of the scheme shown in Fig. 2. Assume that we prepare a polarized* lepton beam and inject it into a cyclotron orbit perpendicular to a region of uniform magnetic field $\vec{B}$. At time t=T, we measure $\hat{S}\cdot\hat{v}$ with the polarimeter. If this experiment is repeated for various values of T, the polarimeter output will vary as $\cos \omega_D T$, so that if N complete oscillations of the output are observed between $T=T_1$ and $T=T_2$, then $\omega_D = 2\pi N/(T_2-T_1)$.

The accuracy to which ω_D can be determined will increase in direct proportion to N, all other factors being equal,** so that the most obvious way to improve the precision of the measurement is to increase the time the particle spends in the magnetic field. If only a uniform magnetic field is used, any component of $\vec{v}$ parallel to $\vec{B}$ will cause the orbit to drift parallel to the field direction so that we must employ axial focusing (a magnetic mirror trap is used).

*We define polarization ($\vec{P}$) in the usual manner; i.e., $\vec{P}=\langle\hat{S}\rangle$ where the brackets indicate an ensemble average. All references to $\vec{S}$ or $\hat{S}$ are to be understood in terms of $\vec{P}$.

**If we can store particles for a time equal to or greater than the equipment repetition time then we lose intensity with increasing storage time. Statistical analysis shows that the accuracy is then proportional to $\sqrt{N}$.

3. The Michigan Electron Experiment

The method employed for the measurement of ω_D is shown in Fig. 3. The entire experiment takes place in a magnetic mirror trap with extremely small mirror ratio ($[B_{Max}-B_{Min}]/B_{Max} \cong 100$ ppm). Mott scattering is used to polarize the electrons and to analyze their final spin orientation.

The apparatus operates in a cyclic manner. One complete cycle (the unit experiment) consists of the following sequence of events: A pulse of 100 KeV electrons is scattered from the polarizing foil. Those electrons which scatter nearly perpendicular to the magnetic field (field strengths employed are 800-1200 gauss) are transversely polarized ($|\vec{P}| = 0.2$). These electrons spiral into the trapping region, which is enclosed by a pair of cylindrical electrodes. As the electrons drift across the gap between the cylinders, a momentary voltage applied to the injection cylinder causes them to lose sufficient axial velocity, so that they become permanently trapped in the magnetic field. The electrons oscillate back and forth in the trap for a predetermined time T, until a second momentary voltage pulse applied to the ejection cylinder gives them sufficient axial velocity to reach the analyzing foil. Here they undergo a second Mott scattering. Those which scatter parallel to the magnetic field are counted. The probability of scattering parallel to $\hat{z}$ is proportional to $\hat{S}\cdot(\hat{v} \times \hat{z})$. Because of the g-2 precession, the number of electrons scattered into the detector as a function of T (assuming that the number of electrons ejected from the trap is independent of T) will be:

$$R(T) = R_o\{1+\delta\cos(\omega_D T+\phi)\}. \qquad (6)$$

Here δ is the Mott asymmetry factor (typically 0.01-0.05) and ϕ is a phase constant. We determine ω_D by sampling R(T) as a function of T at two widely separated values of trapping time. By fitting the data of R(T) vs. T, the position of two maxima of R(T) can be established to be at $T=T_1$ and $T=T_2$ (Fig. 4). By taking a nominal amount of additional data between the two selected maxima, N can be established without an explicit counting of all of the intermediate cycles.

For electrons in a magnetic trap in the presence of a radial electric field E_r (in our work due in part to contact potentials and space charge) and with $\gamma^2 \ll a^{-1}$, the observed difference frequency $\langle[\omega_D]\rangle$ will be given to a sufficient approximation by

$$\frac{\langle[\omega_D]\rangle}{\langle[\omega_o]\rangle} = a - a\,\frac{\langle[v_z^2]\rangle}{2c^2} - \frac{\langle[E_r]\rangle}{\beta\gamma^2 B} \qquad (7)$$

The notation [] indicates that the enclosed quantity is to be averaged over one oscillation period in the trap, while the notation $\langle\ \rangle$ indicates a further average over the ensemble of electrons with various amplitudes of oscillation. The axial motion itself be obtained from measurements of the axial field dependence, coupled with use of the adiabatic invariance of the orbital magnetic moment. Once the time-average quantities are calculated as a function of the axial oscillation amplitude, the ensemble averages can be evaluated from direct measurements of the amplitude distribution of the trapped electrons.

We must now consider evaluation of the $\langle[E_r]\rangle$ term in Eq. 7. The size of the electric field effects makes them a major source of difficulty both in our work and in the resonance experiments. For example the value of E_r expected in our apparatus is about 10-100 mV/cm corresponding to an $[E_r]$ of about 1-10 mV/cm and finally a shift in $\langle[\omega_D]\rangle$ of 3-10 ppm. This electric field is too small to be measured directly, especially under operating conditions. Consequently, one must resort to an extrapolation procedure to determine its effect. Equation 7 can be rearranged in the form

$$a' = \frac{\langle[\omega_D]\rangle}{\langle[\omega_o]\rangle}\left\{1 + \frac{\langle[v_z^2]\rangle}{2c^2}\right\}. \tag{8}$$

Here all quantities on the right-hand side are experimentally measurable or calculable, and a' is the apparent value of the anomaly i.e., $a' = a-\langle[E_r]\rangle X$ with $X = (\beta\gamma^2 B)^{-1}$. If $\langle[E_r]\rangle$ is independent of X, $\underline{a}$ can be obtained by measuring a'(X) at several values of X and extrapolating the results to X=0. (Fig. 5).

We see from the figure that the extrapolation in our work causes a systematic shift of less than 2 standard deviations. Final error in the experiment is 3 ppm (15), in agreement with theory but disagreeing with earlier work by Wilkinson and Crane (16) as revised by Rich (17). Results of all recent experiments are shown in Fig. 6.

4. The Michigan Positron Experiment

The method employed in the positron work is similar to that used for the electron experiments except for the means used to polarize the positrons and to analyze their helicity. Positrons are obtained from a Co^{58} source and, because of parity nonconservation in beta decay, are longitudinally polarized ($|\vec{P}|=v/c\cong 0.7$). Economic considerations limit the maximum source strength to about one Curie of Co^{58} (15% positron decay ratio). For a source of this strength, the average number of positrons trapped per machine cycle is only about 10^{-3}. If Mott scattering were used as a polarimeter, a single data run would take over 10 years.

Fortunately, a more efficient method is available in the form of a positron polarimeter first suggested by V. L. Telegdi (18).

The device is based on the formation and subsequent decay of positronium in a strong magnetic field. Its overall efficiency is about 10%, versus 10^{-2}% for Mott scattering. It is this increased efficiency which makes the positron experiment practical.

In spite of the gain in counting rate due to the new polarimeter, the principal difficulties in the positron work still arise from low beam intensity. In order to increase the data collection rate, the relative depth of the magnetic trap (10^4 ppm) was 100 times greater than that used in the electron experiments. The low counting rate also made it impractical to measure a' at several values of X therefore no electric field extrapolation was possible. However, on the basis of the electric fields encountered in Michigan electron experiments of similar geometry, the electric field correction in the positron experiments is estimated to be small compared to final statistical uncertainty in ω_D. Statistical uncertainty contributed approximately 90% of the final error of 950 ppm (19) in a_{e+} (0.95 ppm in g_{e+}). This experiment, when taken together with the latest muon results (see the article by F. J. M. Farley in these proceedings) constitutes a check of TCP invariance for leptons at the 1 ppm level.

5. The Resonance Method

Here we distinguish between the work done at Washington (20, 21) and Mainz (22,23) with a uniform magnetic field and electric quadrupole trap and that being carried out by Fairbank and collaborators at Stanford in a shaped magnetic field. Our treatment will emphasize the principles of the experiments as opposed to details of technique since the level of accuracy reached is not yet competitive with the precession experiments.

5A. The Washington-Mainz Experiments

The W-M field configuration (a Penning trap) consists of a uniform axial magnetic field $\vec{B} = B_o\hat{z}$, and a superimposed electric quadrupole field generated by a pair of hyperbolic electrodes that surround the storage region (Fig. 7). With a voltage V_o applied between the electrodes, the electrostatic potential in the trapping region is $V(r,z) = (V_o/r_o^2)(r^2/2 - z^2)$ where $r^2 = x^2+y^2$. The vector potential is $\vec{A}(x,y) = (B_o/2)(-y\hat{x}+x\hat{y})$. The eigenvalues of the nonrelativistic Hamiltonian,

$$H = (\vec{p}+e\vec{A}/c)^2/2m_e - eV + (1+a)(e/2m_ec)\vec{\sigma}\cdot\vec{B}, \tag{9}$$

have been calculated exactly (24). They are

$$E(n_B,n_E,n_{EB},s_z)/\hbar = (n_B+\tfrac{1}{2})\omega_B + (n_E+\tfrac{1}{2})\omega_E - (n_{EB}+\tfrac{1}{2})\omega_{EB} - (1+a)\omega_o s_z, \tag{10}$$

where n_B, n_E, and n_{EB} are integer quantum numbers, $s = \pm\frac{1}{2}$,

$\omega_E = (2eV_o/mr_o^2)^{\frac{1}{2}}$, $\omega_B = \omega_o - \omega_{EB}$, $\omega_{EB} = \omega_o/2 - (\omega_o^2/4 - \omega_E^2/2)^{\frac{1}{2}}$, and $\omega_o = eB/m_e c$.

The motion in the trap consists of three decoupled motions; (1) cyclotron motion at the perturbed cyclotron frequency $\omega_B(2\pi \times 12\ \text{GHz})$, (2) a slow drift of the cyclotron orbit center about z at the magnetron frequency $\omega_{EB}(2\pi \times 200\ \text{MHz})$, and (3) an axial oscillation at frequency $\omega_E(2\pi \times 40\ \text{MHz})$. The numerical values given are approximate frequencies for $B_o = 4$ kG, $V_o = 10$ V, and $r_o = 0.8$ cm. For the experiments discussed, $\omega_E \ll \omega_B$, and therefore $\omega_{EB} = \omega_E^2/\omega_o$.

Transitions between eigenstates of H can be induced by appropriate types of oscillating electromagnetic fields. Three types of transitions of interest and the required fields are:

(1) Cyclotron orbit transitions: $\Delta n_B = \pm 1$, $\omega = \omega_B$.
This is an electric dipole transition, induced by a field of the type $\vec{E} \propto E_1 \hat{x} \cos \omega_B t$.

(2) Spin flip transitions: $\Delta s_z = \pm 1$, $\omega = \omega_S = (1+a)\omega_o$.
This is a magnetic dipole transition, induced by a uniform magnetic field of the type $\vec{B} \propto B_1 \hat{x} \cos \omega_S t$.

(3) Direct g-2 transitions: $\Delta n_B = \pm 1$, $\Delta s_z = \pm 1$, $\omega = \omega_S - \omega_B = a\omega_o + \omega_{EB} = \omega_D'$.
This is a combination of (1) and (2), induced by a field of the type $\vec{B} \propto B_2(x\,\hat{x} + y\,\hat{y} - 2z\,\hat{z}) \cos \omega_D' t$.

Note that in case (3), the oscillating field must have a gradient perpendicular to $\hat{z}$ in order to induce the desired transition. The reason for this is that in the electron rest frame a field with a linear x dependence will appear to oscillate at the cyclotron frequency ω_B while a field which is already oscillating at ω_D' will have components oscillating at $\omega_B + \omega_D'$ and $\omega_B - \omega_D'$. Since $\omega_B + \omega_D' = \omega_S$, a spin flip can be induced.

Before going on to discuss the measurement of a we note that spin-orbit and relativistic terms are not included in Eq. 9. These terms have been considered in detail (23). Their net effect is to cause a fractional shift in the various transition frequencies of order $T/m_e c^2$, where T, the energy of the stored electrons, is 2eV or a 4 ppm shift in the Mainz experiment and 0.01 eV or an 0.01 ppm shift in the Washington experiment.

Finally we note that the orbital and spin motion in the Penning trap can be derived by straight-forward application of the macroscopic equations of motion used in the discussion of the precession technique. Since the principal quantum number n_B is of order $10^3 - 10^4$, there is no explicit need for a quantum mechanical solution. The quantum mechanical solution does, however, have the advantages of being both simple and exact.

In order to measure a both the Mainz and Washington groups measure ω_D', although each uses a different technique to detect the spin flip. In the Mainz work ω_D' is measured as a function of V_o, and the result is extrapolated to $V_o=0$ (a 5000 ppm extrapolation) with the intercept being interpreted as ω_D (Fig. 8). A direct measurement of ω_S now permits one to obtain a as $\omega_D/(\omega_S-\omega_D)$. The final result of the experiment was accurate to 260 ppm with the major source of error being the linewidth of the g-2 transition.

In the Washington experiment the frequencies ω_B, ω_{EB} and ω_E are measured directly. One may thus obtain ω_o as $\omega_o = \omega_B+\omega_{EB}$ and finally determine a from the relation

$$\underline{a} = (\omega_S-\omega_o)/\omega_o = (\omega_D' - \omega_{EB})/\omega_o \ .$$

Since typical frequencies in the experiment are (in MHz); $\omega_o/2\pi = 2 \times 10^4$, $\omega_E/2\pi = 60$, $\omega_D'/2\pi = 25$, and $\omega_{EB} = 0.08$, we see that the electrostatic correction term ω_{EB} is of order 2500 ppm of ω_D'. The major uncertainty in the experiment was in measurement of ω_{EB} since accidental destruction of the ion trap prevented a detailed investigation of systematic effects associated with the cyclotron and g-2 resonances. Consequently, the uncertainty in ω_{EB} was conservatively estimated to be 2 kHz, resulting in a final uncertainty in a of 80 ppm. The actual widths of the resonances (Fig. 9), and the accuracy to which ω_{EB} could be determined in a single data run was considerably less.

5B. The Stanford Experiment

In the limit of zero electric potential, the energy eigenvalues of H (Eq. 9) corresponding to the state (n_B, s_z) become

$$E(n_B, s_z) = \hbar\omega_o\{(n_B+\tfrac{1}{2})+(1+a)s_z\}+ \frac{p_z^2}{2m_e} \ . \tag{11}$$

In the above, $n_B = 0,1,2,\ldots$, and $s_z = \pm\frac{1}{2}$. These states are the Landau levels of a spin $\frac{1}{2}$ particle in a magnetic field. The states corresponding to $n_B = 0$ and $n_B = 1$ can in principle be used to measure a to high accuracy.

The level structure of the lowest Landau levels, together with the transitions of interest, are shown in Fig. 10. The states are labeled with the notation (n_B, s_z). The energy of the ground state $(0,-\frac{1}{2})$ is approximately -10^{-8} eV per kG of magnetic field, while the energy of the first pair of higher states is approximately $+10^{-5}$ eV per kG of field. The transition $(1,-\frac{1}{2})\leftrightarrow(0,-\frac{1}{2})$ is a cyclotron transition at frequency ω_o, while the transition $(0,\frac{1}{2})\leftrightarrow(0,-\frac{1}{2})$ is a spin transition at frequency $\omega_S = (1+a)\omega_o$. Using $\hbar\omega_o = 2\mu_B B_o$, Eq. 11 can be rewritten as

$$E(n_B, s_z) = \{(n_B + \frac{1}{2}) + (1 + a)s_z\}\, 2\mu_B\, B_o + \frac{p_z^2}{2m_e} \ .$$

Here $\mu(n_B, s_z)$ can be identified as the total magnetic moment of the state (n_B, s_z), i.e., the sum of the orbital moment $2(n_B+\frac{1}{2})\mu B$ and the intrinsic electron moment $(1+a)\mu_B$. Note that the moment of the ground state $(0,-\frac{1}{2})$ is negative and small $(-a\mu_B)$ as opposed to the larger positive moments (of order μ_B for the higher states). The experiment described below is based on this feature (25).

The entire apparatus (Fig. 11) operates at liquid helium temperatures, the magnetic fields being generated by superconducting solenoids. A 0.1- msec pulse of electrons is produced by a tunnel cathode located in a region of inhomogeneous magnetic field ($B \cong 6$ kG). The axial gradient causes the electrons to experience a force $F_z = \mu\, \partial B_z/\partial z$. Consequently, the ground state electrons ($\sim$1 electron/pulse, axial energy $\sim 10^{-8}$ eV) are decelerated slightly as they drift away from the cathode while those electrons in higher states experience a strong acceleration. The electrons now drift into a region of homogeneous magnetic field (4 kG) approximately 1 meter long. Electrons in higher states drift through this region in less than 0.3 msec, while the ground electrons require up to 30 msec. Thus a time of flight technique may be used to identify them.

The drifting electrons pass through the ends of a microwave cavity located in the homogeneous region. A second region of higher magnetic field separates the output end of the cavity from an electron multiplier detector. For electrons in the ground state, the high field region acts as a slight axial potential trough, but for the higher state electrons, it acts as a potential barrier. Therefore, electrons that undergo a transition from the ground state to a higher state will be unable to reach the detector. The technique actually employed to detect resonance is identical to the Ramsey method used in molecular beam work, i.e., microwave fields in either end of the cavity are π out of phase. The observed resonance line will therefore consist of a broad cavity resonance with a much narrower maximum in the center. The linewidth of the central spin and cyclotron resonances is expected to be approximately $1/\pi N$, where N is the total number of cyclotron or spin rotations in the cavity. For a drift time of 10^{-2} sec and a field of 4 kG, this gives a relative linewidth of about 3×10^{-9}, corresponding to an accuracy of 3 ppm in a. Knight (25) has proposed, however, that the resonance line may be split by about $1{:}10^3$, although achieving this accuracy would place extreme stability requirements on the experimental apparatus. Work at Stanford is currently concerned with detecting the induced transitions and with a technique for thermalizing positrons from a beta source so that an improved measurement of a_{e^+} can also be made (26).

5. Critical Analysis and Prospects for Future Progress

A. Precession Experiments

Error in the Michigan g-2 work was due, in about equal parts, to uncertainty in the time average field experienced by the ensemble of electrons, statistical error in determining ω_D, and error caused by the electric field extrapolation (15). We have considered, in a rather preliminary way, the possibility of an extension of the technique to fields of order 10 kG and a 10 ppm trap. We conclude that with an almost fully polarized electron beam such an experiment might reach an accuracy of 0.5 ppm.

The primary limitation on the positron work has always been intensity. It is probably possible to improve the analyzer efficiency to some degree but order of magnitude increases in the trapped beam intensity would constitute a real breakthrough in this area.

B. The Resonance Experiments

The W-M effort is limited primarily by the electric field extrapolation, resonance line widths, and perhaps relativistic effects (Sec. 4.5A). Difference frequency resonances have exhibited widths of from 10 to 100 ppm. Since these experiments have been preliminary demonstrations of new techniques, rather than concerted precision measurements, detailed lineshape theories have not been developed or applied, and little line splitting has been attempted. However, unless the linewidth can be reduced to a few ppm, future experiments will require formulation and verification of a satisfactory lineshape theory.

Turning attention to the electric field perturbations, we note once again the large size of the effect, 5000-15000 ppm (Mainz) and 2800 ppm (Washington). There is no objection to such an extrapolation, provided the assumptions underlying the extrapolation are correct. For example, in the presence of a non-quadratic potential due to either space charge or non-uniform surface potentials on the trapping electrodes, the expressions for the various frequencies of motion (Eq. 10) must be modified, and the extrapolation of ω_D' to V_o=0 does not necessarily yield $a\omega_o$. Also, ω_{EB}' is no longer independent of R_{EB}. Therefore, if the applied rf fields are not uniform, additional complications will be introduced into the lineshape theory. Walls (27) estimates that the fractional shift in ω_{EB} due to the space charge field from 10^4 stored electrons could be as large as 10^{-2}, but is more probably of order 10^{-3}-10^{-4}. The effect is certainly not significant at the 100 ppm level, but may be important in more accurate measurements where careful studies of ω_D and ω_{EB}' as a function of the number of trapped electrons will be necessary (Sec. 4.5A).

Since the Stanford experiment is still in a preliminary stage systematic effects associated with the technique, if any, remain to be discovered. Here we merely note that the effects of dc electric fields must be shown to be negligible. Preliminary tests show that the electric fields on the axis of the OFHC copper drift tube used to enclose the drift region have no significant effect on the axial motion of 10^{-8} eV electrons. It remains to be demonstrated, however, that the spin and cyclotron frequencies are independent of the conditions inside the drift tube to the projected accuracy of the experiment.

C. Conclusions

As of the date of completion of this article (August 1971) agreement between the various low and high energy predictions of QED theory and their respective experiments is excellent. In particular, the g-factor anomalies of the electron and muon constitute a verification of QED at the 3 ppm (6th order) and 300 ppm (4th order) levels respectively.

The only cloud on the horizon is the revised Wilkinson-Crane experiment which is three standard deviations below the recent theoretical and experimental results. In this connection, we would like to emphasize again, however, that the 6th order confrontation of theory and experiment for the electron rests entirely on the work of one theoretical and one experimental group. We feel it to be of utmost importance that the work reported to date be verified. It is particularly gratifying to note that on the experimental side, the resonance techniques being employed are completely different from the precession method in use at Michigan.

Acknowledgements

We thank S. Brodsky, R. Carroll and E. Yao for discussions concerning section 3. The Michigan g-factor experiments have been generously supported by the United States Atomic Energy Commission.

References

1. For a review of the current status of QED, see S. J. Brodsky and S. D. Drell, Ann. Rev. Nucl. Sci. 20, 147 (1970).

2. B. N. Taylor, W. H. Parker, and D. N. Langenberg, Rev. Mod. Phys. 41, 375 (1969).

3. R. Karplus and N. Kroll, Phys. Rev. 77, 536 (1950).

4. C. M. Sommerfield, Phys. Rev. 107, 328 (1957); Ann. Phys. 5, 26 (1957).

5. A. Petermann, Helv. Phys. Acta. 30, 407 (1957).

6. M. J. Levine and J. Wright, Phys. Rev. Letters 26, 1351 (1971).

7. J. Bailey and E. Picasso, Prog. Nucl. Phys. Vol. 12 D. M. Brink and J. H. Mulvey (ed.), (Pergamon Press, London), (1970).

8. M. J. Levine, J. Comput. Phys. 1, 454 (1967); A. C. Hearn, Stanford University Report ITT 247 (1969).

9. J. A. Mignaco and E. Remidi, Nuovo Cimento 60A, 1519 (1969).

10. S. J. Brodsky and T. Kinoshita, Phys. Rev. D3, 356 (1971).

11. J. Aldins, T. Kinoshita, S. J. Brodsky, and A. Dufner, Phys. Rev. D1, 2378 (1970).

12. R. Carroll and E. Yao (private communication).

13. J. Schwinger, Proc. Natl. Acad. Sciences 37, 452 (1951).

14. E. Klein, Z. Physik 208, 28 (1968).

15. J. C. Wesley and A. Rich, Phys. Rev. Letters 24, 1320 (1970).

16. D. T. Wilkinson and H. R. Crane, Phys. Rev. 130, 852 (1963).

17. A. Rich, Phys. Rev. Letters 20, 967 (1968).

18. L. Grodzens, Prog. Nucl. Phys., Vol. 7, O. R. Frisch (ed.) (Pergamon Press, London), 219 (1959). The first demonstration of the effect was by L. Dick, L. Feuvrais, and V. L. Telegdi, "Aux En Provence Intern. Conf. on Elem. Particles", Vol. VI, 295 (1961).

19. J. Gilleland and A. Rich, Phys. Rev. Letters 23, 1130 (1969).

20. H. G. Dehmelt and F. L. Walls, Phys. Rev. Letters 21, 127 (1968).

21. F. L. Walls, Ph.D. Thesis (Univ. of Washington, 1970) (unpublished).

22. G. Graff, F. G. Major, R. W. H. Roeder, and G. Werth, Phys. Rev. Letters 21, 340 (1968).

23. G. Graff, E. Klempt, and G. Werth, Z. Physik 222, 201 (1969).

24. A. A. Sokolov and I. G. Pavlenko, Optics and Spectroscopy XXII, 1 (1967).

25. L. V. Knight, Ph.D. Thesis (Stanford Univ. 1965) (unpublished).

26. B. Kincaid (private communication).

27. F. L. Walls (private communication).

(a) VERTEX GRAPHS: $\delta a_e = (1.23 \pm 0.2)(\frac{\alpha}{\pi})^3$

+46 Others

(b) VACUUM POLARIZATION: $\delta a_e = (-0.099 \pm 0.009)(\frac{\alpha}{\pi})^3$

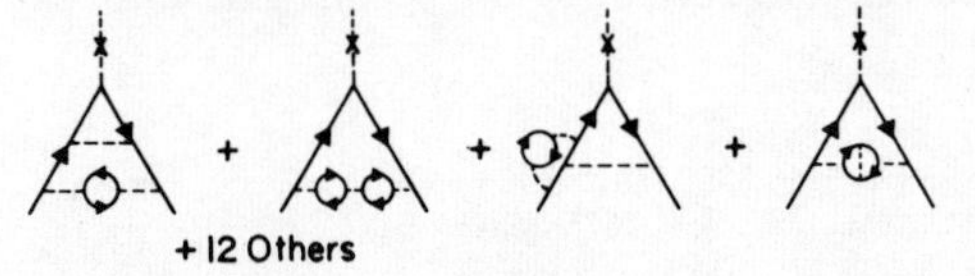

+12 Others

(c) PHOTON-PHOTON SCATTERING: $\delta a_e = (0.36 \pm 0.04)(\frac{\alpha}{\pi})^3$

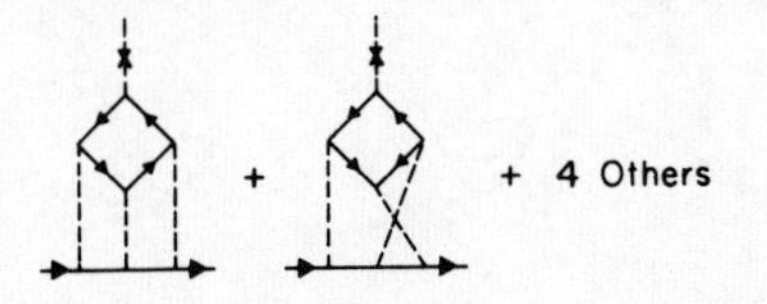

Fig. 1. Sixth Order Graphs and Their Contributions to the Electron Anomaly.

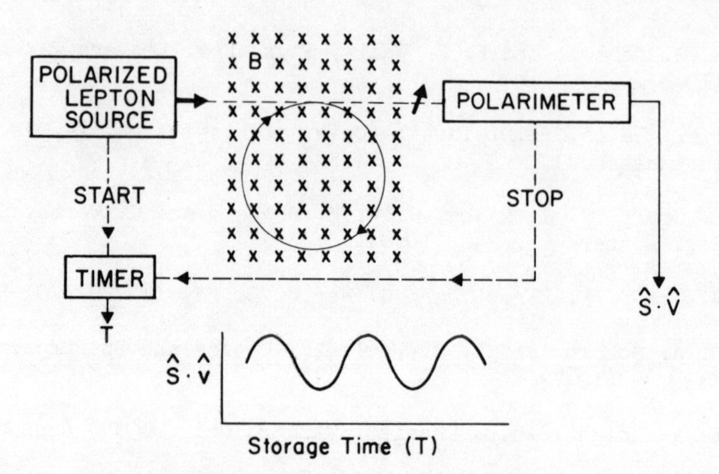

Fig. 2. Schematic Outline of the Precession Experiments.

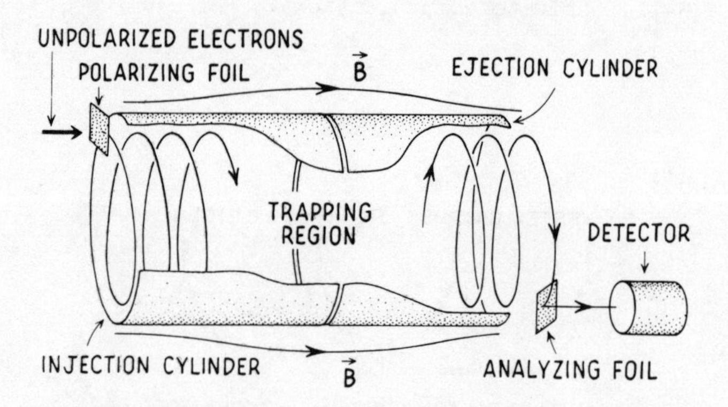

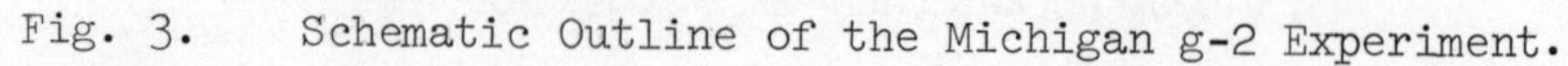
Fig. 3. Schematic Outline of the Michigan g-2 Experiment.

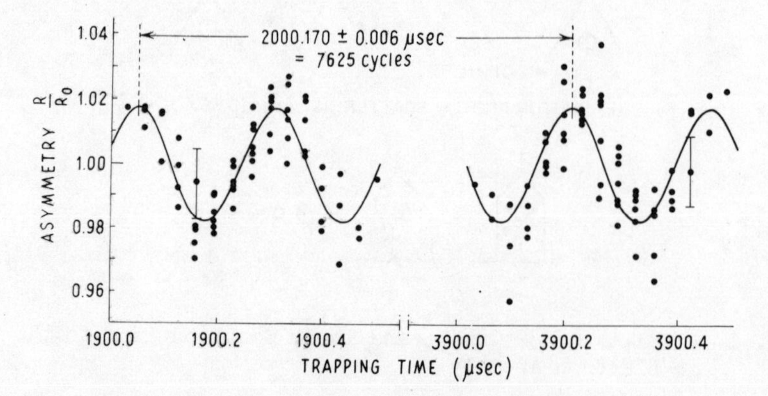

Fig. 4. Data From the Michigan g-2 Experiment.

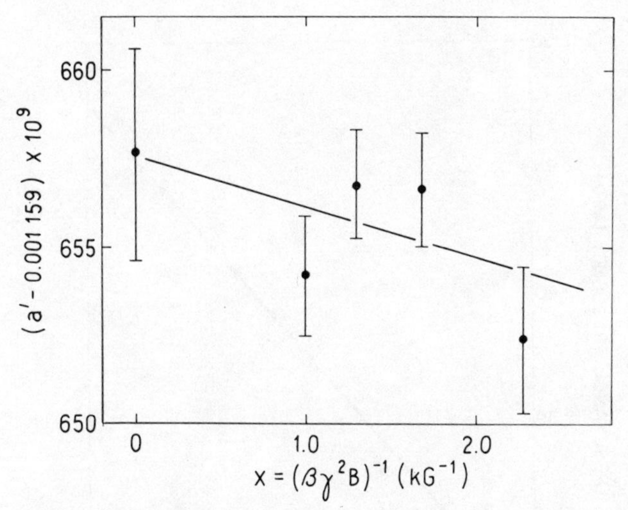

Fig. 5. Electric Field Extrapolation in the Michigan Experiment.

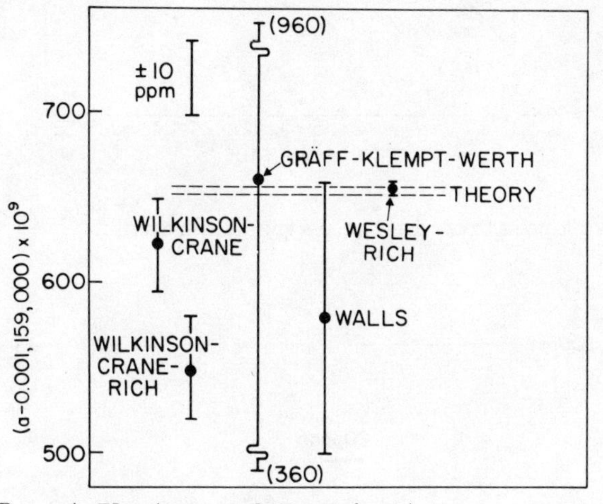

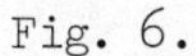
Fig. 6. Recent Electron g-2 Experiments.

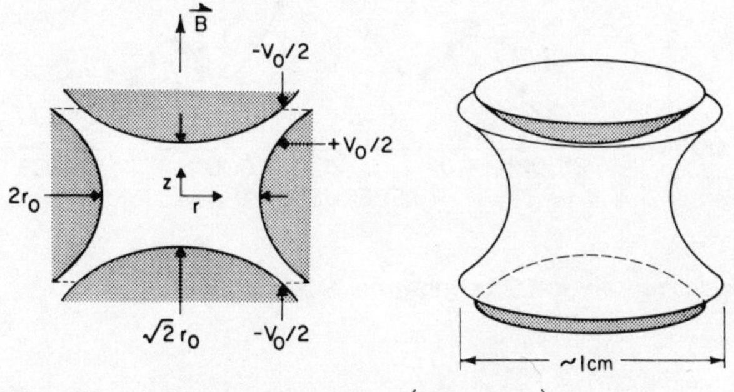

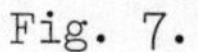
Fig. 7. The Electric Quadrupole (Penning) Trap.

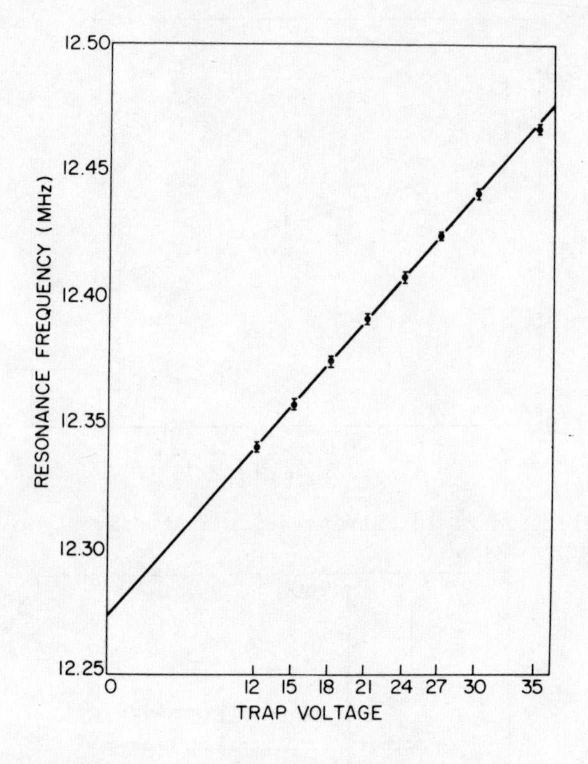

Fig. 8. Extrapolation in the Mainz Experiment.

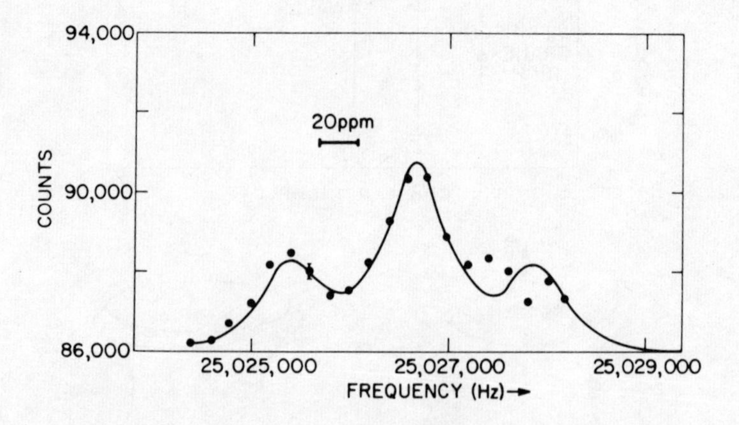

Fig. 9. Washington g-2 Resonance Signal.

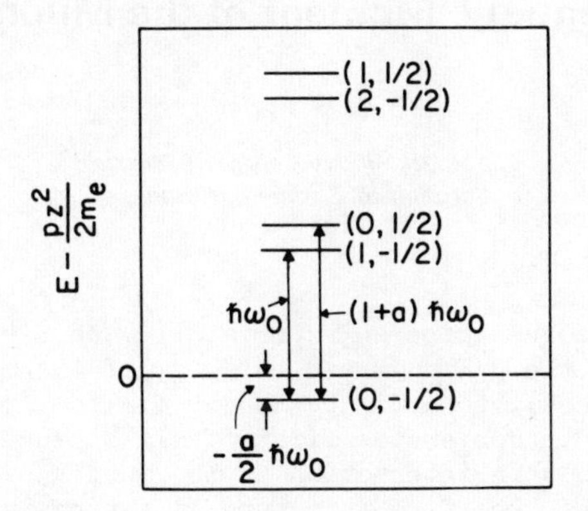

Fig. 10. The Lowest Landau Levels for an Electron in a Magnetic Field.

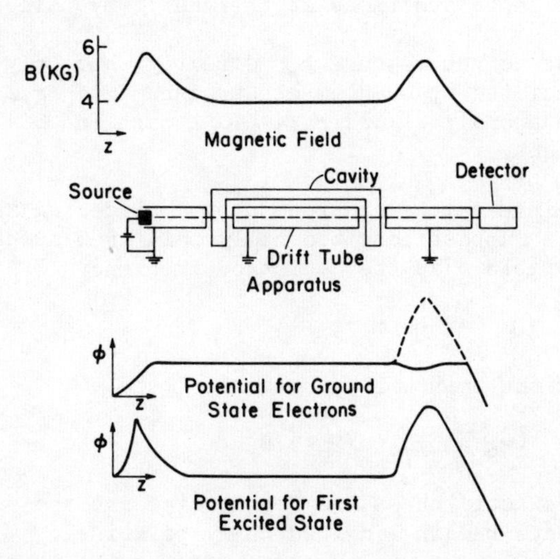

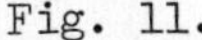
Fig. 11. Schematic Outline of the Stanford Experiments.

Precise Measurement of the Anomalous Magnetic Moment of the Muon

F. J. M. Farley

The Royal Military College of Science,
Shrivenham, Swindon, Wiltshire,
England

I am going to describe an experiment(1) carried out at CERN which determines the gyromagnetic ratio g_μ of the muon, (and hence its magnetic moment), to ± 0.3 parts per million (ppm). This number is relevant to your conference first of all because you will be considering the spin-spin interaction of the muon and electron in muonium. Here the hyperfine structure splitting is proportional to $g_\mu \alpha^2$ so an accurate knowledge of g_μ is essential for a determination of α by this method. The experiment also establishes that the muon is a point-like particle, obeying the standard quantum-electrodynamics, and that it has no extra detectable interaction or structure which might perturb the energy levels in muonium. I will return to these questions at the end of my talk.

In fact we do not measure g_μ directly, but only the small correction predicted by quantum-electrodynamics: $g_\mu = 2(1+a_\mu)$ where a_μ, the anomaly of order $\alpha/2\pi$ (1 part in 800), is determined by the experiment.

The principle may be understood by considering the non-relativistic case. Suppose longitudinally polarised muons are turning in a magnetic field B at the cyclotron frequency

$$\omega_c = eB/m_o c, \tag{1}$$

Simultaneously the spin will precess at

$$\omega_s = g_\mu (e/2m_o c)B \tag{2}$$

If g = 2 exactly these two frequencies are the same and so the muons will always remain longitudinally polarised. As however g_μ is greater than 2, the spin precesses faster than the momentum vector at relative angular velocity

$$\omega_a = a_\mu (e/m_o c)B \tag{3}$$

It can be shown that this formula is also valid for muons which have relativistic velocities. So by recording the polarisation direction as a function of time a_μ can be determined.

The magnetic field is measured in terms of the corresponding proton precession frequency ω_p. The value of $(e/m_o c)$ for the muon is obtained from the ratio of muon to proton precession frequency in the same field, $\lambda \equiv \omega_s/\omega_p$. This is known(2) for muons at rest to an accuracy of a few ppm. We then have

$$(1 + a_\mu)\ (e/m_o c)B = \lambda(1 + \varepsilon)\ .\ \omega_p \tag{4}$$

where the correction $(1 + \varepsilon)$ is due to the fact that our muons are in vacuum while the protons are in water.

Eqns (3) and (4) give

$$\omega_a/\omega_p = a_\mu(1 + a_\mu)^{-1}\lambda(1 + \varepsilon) \tag{5}$$

which is the ratio determined by our measurements.

To obtain an accurate value of ω_a we need to record data for a large number of precession cycles; but on the other hand the time available is limited by the muon lifetime at rest which is τ = 2.2 μsec. To offset this limitation we work in a high magnetic field (17 kG) to give large ω_a, and use muons of high laboratory momentum (1.3 GeV/c) so that the lifetime is lengthened by relativistic time dilation to 27 μsec.

Fig. 1 shows the weak focusing ring magnet, 5 metres in diameter, continuous in azimuth with n = 0.13. The injection of muons is accomplished as follows. A beam of 10 GeV protons is ejected from the CERN Proton Synchrotron (PS) and focussed onto a target inside the storage ring. Here ~70% of the protons interact, creating π^- of ~1.3 GeV/c which start to turn in the ring. The pion lifetime is such that in one turn 20% of the π^- decay. The exactly forward decay creates μ^- of 0.1% higher momentum, and these muons, together with undecayed π^- and stable particles from the target, will eventually hit the target assembly and be lost. However, decay at small forward angles gives rise to μ^- of slightly lower momentum, and some of these fall into orbits which miss the target assembly and remain permanently stored in the ring. About 200 muons are stored per cycle of the PS. They are forward polarised because they come from the forward decay of the pions in flight.

The detection of the polarisation as a function of time depends on the anisotropy of the electrons emitted when the muon decays. When the muon decays in flight the decay electron has less momentum and tends to come out of the magnet on the inside. It then hits a counter consisting of layers of lead and scintillator, producing a shower and giving a pulse of light which is proportional to its energy. By setting an appropriate detector threshold we arrange to count only electrons of energy greater than 700 MeV. This implies that the decay electron must have been emitted more or less forward in the muon rest frame. Therefore, as the muon spin rotates according to eqn. (3) the counting rate shows an exponential decay at the dilated lifetime, plus a sinusoidal modulation at frequency ω_a,

$$N(t) = N_o e^{-t/\tau} \left\{ 1 - A \cos(\omega_a t + \phi) \right\} \quad (6)$$

Typical results are shown in fig. 2.

The overall result obtained by making a least-squares fit of eqn. (6) to the data is

$$a_\mu^{exp} = (116\ 616 \pm 31) \times 10^{-8} \quad (7)$$

This must be compared to the theoretical value(3)

$$a_\mu^{theory} = \alpha/2\pi + 0.7658\,\alpha^2/\pi^2 + (21.8 \pm 1.1)\,\alpha^3/\pi^3$$

$$+ \text{ strong interaction correction}$$

$$= (116\ 588 \pm 2) \times 10^{-8} \quad (8)$$

based on $\alpha^{-1} = 137.03608 \pm 1.9$ ppm. The modification due to virtual ρ, ω and ϕ - meson contributions to the photon propagator, (strong interaction correction + 6.5×10^{-8}), is included.

This gives

$$a_\mu^{exp} - a_\mu^{theory} = (28 \pm 31) \times 10^{-8} \quad (9)$$

Thus the anomaly, which is a pure quantum effect, is confirmed to (250 ± 270) ppm; and the g-factor itself is determined to ± 0.31ppm.

If the muon has a structure, giving a form factor

$$F(q^2) = 1 - q^2/\Lambda_\mu^2 \quad (10)$$

the calculation leading to eqn. (8) would be modified giving a correction

$$\Delta a_\mu / a_\mu = - 4m_\mu^2 / 3\Lambda_\mu^2 \quad (11)$$

From the absence of this effect we deduce $\Lambda_\mu > 7$ GeV to 95% confidence.

This is currently the best test of the validity of Q.E.D. for the muon, and a good confirmation of the validity of the theory at high q^2.

Furthermore, any unknown field coupled to the muon would make its own contribution to a_μ. Again from the absence of such effects we can put limits on the existence of unknown couplings; for example for a neutral vector boson of mass M and coupling constant $f^2 \equiv g^2/4\pi$ we have

$$f^2/M^2 < (6 \times 10^{-4})/M^2_{nucleon} \quad (12)$$

To see what this limit implies for the $\mu^+ - e^-$ system (muonium) we can use the electron (g-2) experiment to put a corresponding limit on the unknown vector boson coupling to the electron. We then find that the maximum fractional perturbation of the Rydberg constant R for muonium is

$$\Delta R/R \sim (3\pi/\alpha)(\Delta a_\mu . \Delta a_e)^{\frac{1}{2}} (q^2/m_\mu m_e) < 10^{-10} \tag{13}$$

where q^2 corresponds to the Bohr momentum.

Thus the (g-2) measurements for electron and muon imply that muonium is an ideal atom of two point-like particles, with no appreciable interaction other than quantum-electrodynamics: this makes it an excellent system for the measurement of fundamental constants.

Another by-product of this experiment is the verification of time dilation in a circular orbit to 1% for $\gamma = 12$, thus giving a direct experimental check on the relativistic "twin paradox".

I would like to thank Drs. E. Picasso and B. E. Lautrup for many very relevant remarks and discussions.

REFERENCES

1. J. BAILEY, W. BARTL, G. von BOCHMANN, R. C. A. BROWN, F. J. M. FARLEY, H. JÖSTLEIN, E. PICASSO and R. W. WILLIAMS, Phys. Letters 28 B, 287 (1968).

   J. BAILEY, W. BARTL, G. von BOCHMANN, R. C. A. BROWN, F. J. M. FARLEY, M. GIESCH, H. JÖSTLEIN, S. van der MEER, E. PICASSO and R. W. WILLIAMS, Nuovo Cimento (in the press).

2. J. F. HAGUE, J. E. ROTHBERG, A. SCHENCK, D. L. WILLIAMS, R. W. WILLIAMS, K. K. YOUNG and K. M. CROWE, Phys. Rev. Letters 25, 268 (1970).

   D. P. HUTCHINSON, F. L. LARSEN, N. C. SCHOEN, D. I. SOBER and A. S. KANOFSKY, Phys. Rev. Letters 24, 1254 (1970).

3. For a recent review see B. E. LAUTRUP, A. PETERMAN and E. de RAFAEL, "Recent developments in the comparison between theory and experiments in quantum electrodynamics", CERN TH.1388, August 1971, Physics Reports (in the press).

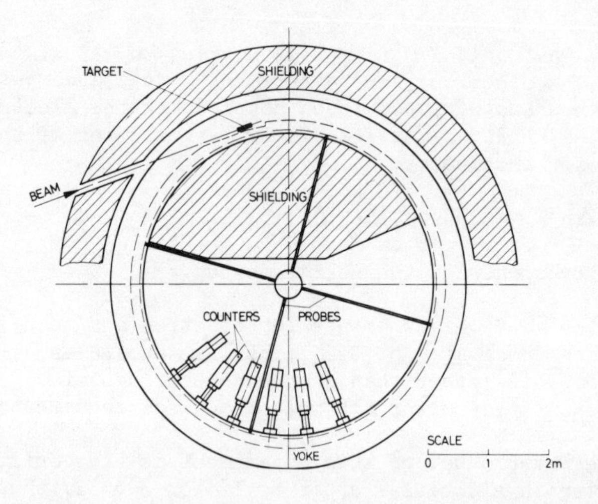

Fig. 1. CERN muon storage ring, 5 metres diameter. NMR magnetometer probes in 4 positions are used to monitor the magnetic field which is stabilized with respect to a fifth probe, (not shown), to a few ppm.

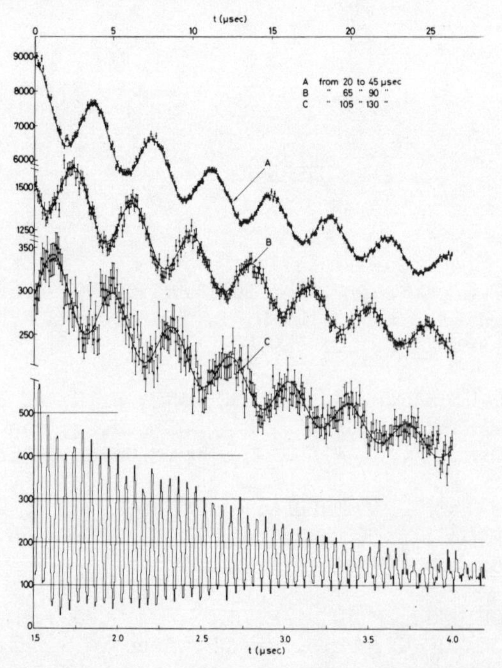

Fig. 2. Counting rate of decay electrons from stored muons versus storage time t: A, B and C are sections of the decay curve showing the (g-2) modulation. The lower curve shows modulation at the rotation frequency and enables the mean radius of the muons to be determined.

Part 12 Miscellaneous Constants

Lattice Spacing Measurements and Avogadro's Number

M. Hart

Department of Physics, University of Bristol, Bristol, England

I. G. Morgan

National Physical Laboratory, Teddington, Middlesex, England

1. Introduction

The present unsatisfactory situation concerning the value of Avogadro's number has been briefly discussed by the authors in an earlier paper (1) in which was described a proposed method of direct measurement based on a new technique of X-ray interferometry (2,3) which provides a means for absolute determination of the lattice spacing of crystals.

In outline, the method is to measure the absolute lattice parameter of a silicon sample by means of X-ray and optical interferometry. With knowledge of the perfection of the silicon this can be related to the volume of the unit cell without loss of accuracy. The density of silicon crystals will be determined by direct measurements of volume and mass of large pieces of silicon crystal. The lattice parameters of the density samples and the X-ray interferometer can be mapped and compared with great accuracy: double crystal topographs, obtained with high order Bragg reflections and short X-ray wavelengths can be used to examine large areas of crystal with high strain sensitivity and spatial resolution. Multiple Bragg reflection diffractometry, which has a strain sensitivity of 10^{-9} can be used to transfer lattice parameter measurements from one sample to another. Combining these results with a knowledge of the abundances and atomic masses of the three stable isotopes of silicon leads to a determination of Avogadro's number.

Apparatus for making the lattice spacing measurements has recently been completed and these measurements are about to be made. The present paper discusses some aspects of the density determination.

For a definitive measurement of density, the sample requirements are few in number; the object must be physically and chemically homogeneous or, if inhomogeneous, then it is essential that quantitative mapping of the inhomogeneity is undertaken. At the 1 ppm level these requirements have so far been met only by liquid samples (4,5). In the case of solids, adequately sensitive mapping procedures have been developed in the last few years for single crystal samples and these form the basis of our measurement of the density of silicon. At present we merely outline the general principles of an experiment which is in progress.

In essence we are measuring the ratio of the mass to the volume of a block of silicon. Clearly the measurement of linear dimensions is the more difficult and for this reason we must use the largest crystals available.

Float-zoned silicon, such as we have used in our X-ray interferometers, can be obtained in diameters up to 30 or 40 mm. It is free of dislocations and is extremely pure with homogeneously dispersed residual impurities (6). Samples made by different manufacturers have the same lattice spacing to within a few parts in 10^7 (7). Design calculations (discussed later) indicated that larger samples would be advantageous. We have therefore concentrated our attention on Czochralski-grown silicon which is known to contain oxygen as an impurity. Dislocation free material is available commercially up to 70 or 80 mm diameter and has been made in much larger diameters in research laboratories. In typical crystals the concentration of oxygen causes lattice parameter variations of a few ppm. We have recently completed a double crystal topographic camera capable of surveying crystals up to 140 mm square in a single exposure.

2. Double crystal diffraction topographs

Figure 1 is a double crystal topograph (8,9) of a silicon crystal grown by the International Business Machines Corporation. The crystal was grown in the [001] direction (vertical) and the upper edge is a section through the decanted interface. Large thermal gradients set up during decanting cause bands of dislocations to propogate back into the otherwise dislocation free crystal, there setting up long range strains. An intensity change of 1% corresponds to a local change of Bragg angle equal to 1.3×10^{-7} radians in this picture. The extinction depth is approximately 4 μm so that this topograph contains information only from the layer within perhaps 10 μm of the crystal surface. Clearly the lower part of this crystal is eminently suitable for use as a density sample. We have obtained several crystals of similar quality from other manufacturers.

After this preliminary survey, blocks were cut from the dislocation free region and then all surfaces were examined by the same technique followed by transmission diffraction topography using the Borrmann effect. The oxygen concentration can also be independently mapped using the absorption band at 9 μm wavelength (10).

Having obtained suitable material, cut and lapped into blocks whose dimensions are measured by optical interferometry, we have developed lapping procedures which do not cause harmful dilations of the worked surface layer. Figure 2 is a 'zebra pattern' (11) of a typical lapped wafer. The central stripe was chemically polished to a depth of 10 μm. This removes the lapping damage and provides a perfect crystal reference area. The horizontal bands obtained with this double crystal technique are essentially contour lines of equal Bragg angle. The spacing between contours is

6.0 sec. arc in this case and the absence of a step in the contours at the edge of the reference region indicates that the perfect crystal Bragg angle and the lapped crystal Bragg angle are equal to better than 1 ppm. In this topograph the extinction depth is 3.8 μm.

3. Scheme for density measurement

To establish a value for the density of silicon we propose to make absolute measurements of mass and linear dimensions of rectangular samples cut from these single crystals. Samples having masses of the order of 100 g and linear dimensions of the order of a few centimetres are being prepared by hand lapping.

The linear dimensions must be measured with an accuracy of a few parts in 10^7 using optical interferometry. It is unrealistic to assume that systematic errors, e.g. wringing film thicknesses and surface reflection phase losses can be held within these limits for measurements on a single sample. We aim to circumvent such difficulties by measuring the differences of mass and dimensions between the members of pairs of samples having different volumes but equal surface areas. As shown in figure 3 equivalence of surface area is the condition that common systematic errors of measurement of linear dimensions cancel out. Figure 3 also indicates the typical forms of the samples. Typical nominal dimensions for a pair are 35 mm x 35 mm x 50 mm with a mass of 140 g and 30 mm x 13 mm x 100 mm with a mass of 90 g. We can determine the mass difference of about 50 g with an accuracy of 5 parts in 10^7 or better using a precision balance. For the length measurements we shall employ the standard technique of residual fractions in an interferometer using four visible wavelengths of cadmium and mercury to make absolute measurements at a selection of points near to the corners of the specimens. Here we can resolve 3 parts in 10^7 on all the dimensions other than the 13 mm dimension for which we should anticipate 6 parts in 10^7.

The samples clearly cannot be finished with a precision of geometrical form comparable with the accuracy of measurement required and it is therefore necessary to perform a detailed mapping of the variations in linear dimensions over the surfaces. For this we shall employ two independent techniques, optical interferometry and pneumatic gauging.

A most significant source of error in the volume determination is breakdown of the edges and corners of the specimens. Lapping of the specimens now being carried out indicates that edge breakdown can be kept within an envelope corresponding to a 10 μm chamfer but such a chamfer, if ignored, could alone contribute an error approaching 1 part in 10^6. It will thus be necessary to make a photogrammetric study of the edges of the finished specimens to assess, and correct for, the lost material. (It may be remarked that the total edge lengths of the members of the sample pair are similar. This has the effect of reducing the error to the extent that edge breakdown may follow a similar pattern in both members but this cannot be assumed so without photogrammetric study). Were we to work with cylindrical specimens the effect of edge breakdown would be trivial and this has been considered, but it is not possible to obtain either the necessary accuracy of diameter measurement, or the necessary accuracy of lattice parameter mapping with cylindrical samples.

To obtain a precision of geometrical form compatible with the requirements for dimensional measurement, e.g. flatness and parallelism of the surfaces within 0.1 μm overall, involves appreciable mechanical working of the surfaces by lapping but figure 2 showed this to be unimportant. We have yet to study in detail chemical pollution of the surfaces but we anticipate no difficulty here. Likely surface impurity, e.g. SiO_2, is of similar density to silicon and the differencing of samples of the same surface area is favourable to the reduction of error from this source.

In our silicon lattice spacing and density programme we aim at an accuracy of 1 in 10^6 but in obtaining Avogadro's number from these results alone we face a limitation, in the present state of the art, in the uncertainty to be associated with the atomic mass of the silicon. To determine the atomic mass to an accuracy of 1 in 10^6 we should need to know the relative abundances of the three isotopes Si^{28}, Si^{29} and Si^{30} to 0.001%. We are currently studying the possibilities for measurements of relative abundance in our samples but it seems at present doubtful whether better than 0.01% can be obtained, this corresponding to about 6 in 10^6 for the weighted mean atomic mass.

However, we can make error free (to a few parts in 10^9) transfers of lattice parameter to other materials (12) which may be mono-isotopic elements or compounds of single isotopes and the densities of these can be referred to the absolute value obtained for silicon. For these subsidiary measurements only small samples are necessary.

1. HART, M., MILNE, A.D., MORGAN, I. and CURTIS, I., International Conference on Precision Measurement and Fundamental Constants, National Bureau of Standards, U.S.A. 1970.

2. BONSE, U. and HART, M., Appl. Phys. Lett., 6, 155 (1965).

3. HART, M., J. Phys. D, 1, 1405 (1968).

4. COOK, A.H., Phil. Trans. Roy. Soc. (Lond.), A254, 125 (1961).

5. COOK, A.H. and STONE, N.W.B., Phil. Trans. Roy. Soc. (Lond.), A250, 279 (1957).

6. HART, M., Science Progress Oxf., 56, 429 (1968).

7. BAKER, J.A., TUCKER, T.M., MOYER, N.E. and BUSCHERT, R.C., J. Appl. Phys., 39, 4365 (1968).

8. BONSE, U., Z. Physik, 153, 278 (1958).

9. BONSE, U., Direct observations of imperfections in crystals, John Wiley & Sons, New York, 431.

10. KAISER, W., KECK, P. and LANGE, C., Phys. Rev. 101, 1264 (1956).

11. RENNINGER, M., Zeit. f. ang. Phys., 19, 20 (1965).

12. HART, M., Proc. Roy. Soc. A309, 281 (1969).

Fig. 1. (0$\bar{1}$0) surface cut 20 mm from the axis of a large silicon crystal grown by I.B.M. using the Czochralski method. 440 Bragg reflection of Cu Kα radiation. Field 70 mm x 100 mm.

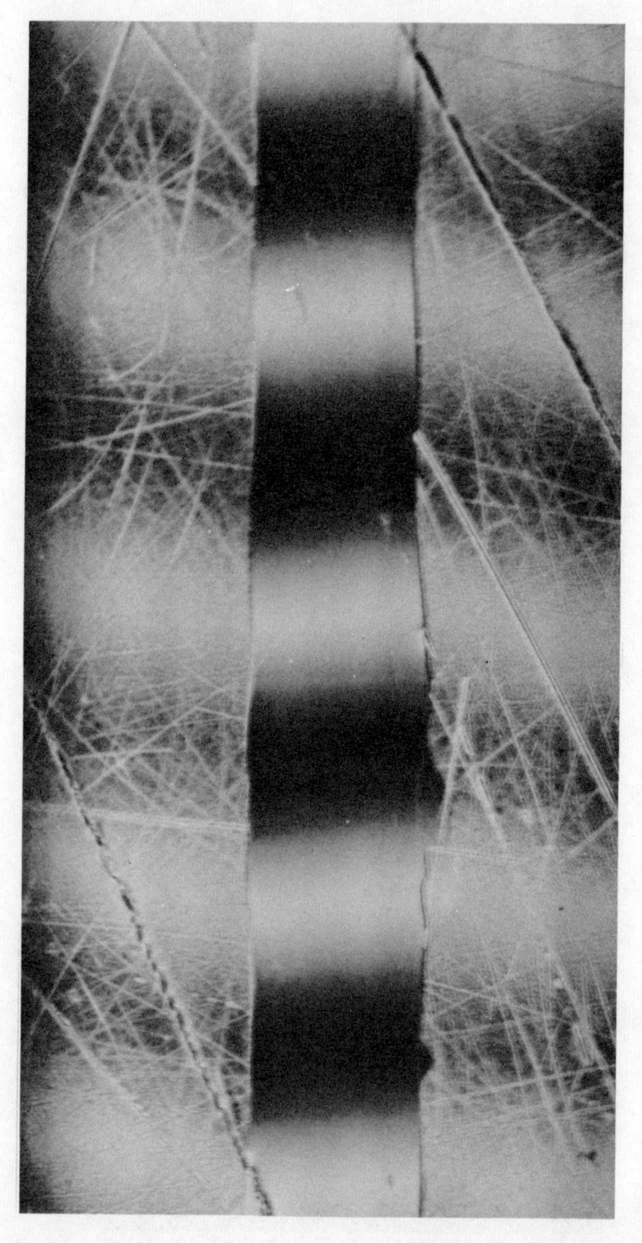

Fig. 2. Bragg angle contour map ('zebra' pattern) of a lapped silicon wafer. The central zone is chemically polished to be free of damage. Cu Kα radiation, 333 Bragg reflection. Field 8 mm x 15 mm.

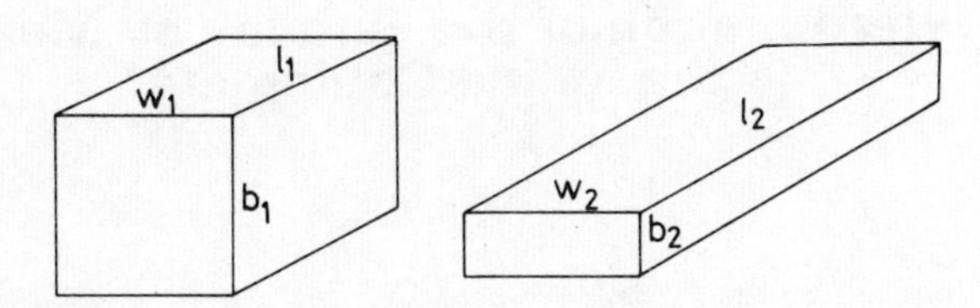

VOLUME = V_1 VOLUME = V_2

SURFACE AREA = A_1 SURFACE AREA = A_2

FOR MEASURED DIMENSIONS w, l & b SUBJECT TO A COMMON SYSTEMATIC ERROR δ,

$$V_1 = w_1 l_1 b_1 \overset{+}{(-)} [b_1 l_1 + l_1 w_1 + w_1 b_1] \times \delta \quad \text{(TO FIRST ORDER)}$$

$$= w_1 l_1 b_1 \overset{+}{(-)} \tfrac{1}{2} A_1 \times \delta$$

$$V_2 = w_2 l_2 b_2 \overset{+}{(-)} \tfrac{1}{2} A_2 \times \delta$$

FOR $A_1 = A_2$ THEN $V_1 - V_2 = w_1 l_1 b_1 - w_2 l_2 b_2$

FIGURE 3

Fig. 3. Reduction of effect of systematic errors in linear measurement by taking differences between two samples of the same surface area.

Determination of the Faraday by Means of the Iodine Coulometer

V. E. Bower

Institute for Basic Standards, National Bureau of Standards, Washington, D.C. 20234

1. Introduction

There have been many efforts (1, 2, 3, 4, 5) to measure the Faraday by direct electrochemical methods. Of these only the experiments employing the oxidation of iodide (3), the electrodeposition of silver (4), and the electrodissolution of silver (5) have ever been viewed seriously as determinations of a fundamental physical constant.

By reason of the uncertainty of the contributions of included matter in the silver deposit, the silver deposition coulometer is objectionable (6). The work of Bates and Vinal on the iodine coulometer was performed almost 60 years ago. There remains, then, as a direct, modern electrochemical determination of the Faraday, only the experiment using the dissolution of silver in perchloric acid which was performed by Craig and his colleagues 10-15 years ago at the National Bureau of Standards (5).

Now, one may calculate an indirect value of the Faraday by means of the relation:

$$F = \gamma_p' M_p/K^2(\mu_p'/\mu_n), \tag{1}$$

where γ_p' is the gyromagnetic ratio of the proton, μ_p'/μ_n is the protonic nuclear magnetic moment in nuclear magnetons, M_p is the atomic mass of the proton, and K is the ratio of the NBS ampere to the absolute ampere. F and γ_p' are measured in terms of the NBS as-maintained ampere. (If γ_p' is determined via the high field method, K^2 is omitted.)

At another point in this conference, Cohen and Taylor (7) remark on the adjustment of the fundamental constants and give a comparison of recent directly determined μ_p'/μ_n values. Against these is compared the indirect value of μ_p'/μ_n calculated through equation (1) from the electrochemically determined value of Craig. One may simply state that the results of the two methods, physical and electrochemical, are discrepant.

This situation suggests, a) a new determination by means of a metal dissolution coulometer and, b) a determination by means of a very different chemical system. At the National Bureau of Standards we have been following the second course and are engaged in the determination of the Faraday by means of the iodine coulometer.

At this conference we report the methods employed up to the present and we reveal the results of the study up to this time. Our values are few and are reported merely to indicate the direction events seem to be taking. They are not to be interpreted as final. Furthermore, they represent only the anode reaction.

It is considered a virtue of the iodine coulometer that it is in principle reversible in the sense that both anode and cathode reactions are governed by the relation

$$3I^- \rightleftharpoons I_3^- + 2e \, , \tag{2}$$

the reaction going to the right at the anode and the reverse at the cathode. Provided the difficult determination at the cathode can be made, the reaction provides an internal check on the experiment.

An additional virtue of the iodine coulometer is that iodine may be purified with very little trouble by a sublimation process which removes heavy metals,halogens, cyanide etc. (8), to the point where the contaminants remaining are at the part per million level. Furthermore, iodine is a mononuclidic element whose atomic mass is known to parts in 10^8 (9), so that in this experiment no ambiguity arises from atomic weight errors or uncertainties in isotope abundance ratios.

Moreover, the iodine-iodide electrolysis may be well separated from the electrolysis of the solvent. This may be demonstrated by voltametric measurements of the type advocated by Lingane (10,11). The decomposition of the solvent should contaminate the principal reaction by no more than one part in 10^6 or 10^7 over a broad range of concentrations and current densities.

A defect detracting from this array of virtue is the fact that iodine has a substantial vapor pressure which makes its quantitative manipulation tedious and difficult. A further disadvantage of the iodine coulometer is that the iodine generated in the coulometer must be measured with a reagent which has been standardized with pure iodine. This whole process requires two or more complicated chemical analyses with intrinsic errors at each step. This may be contrasted with the silver-silver perchlorate coulometer which derives its elegance from the fact that the mass transfer in the measurement is determined, in principle, by two weighings.

2. Methods

The apparatus and the methods used are described in detail elsewhere (12) and will be only briefly reviewed here.

The method for measuring the iodine generated in the coulometer is as follows. A solution of arsenious acid is standardized with a freshly sublimed sample of iodine. The arsenious acid is then used to titrate the iodine generated in the coulometer. The value for the titer of the arsenious acid solution at the time it is used in the coulometric measurement is determined from interpolation by linear functions fitted piecewise to the titer vs. time data by the method of least squares. The standard deviation of a single titration about a least squares line is about 22ppm or 2.1C mole^{-1} in the Faraday.

The principal features of the titration may be briefly stated at this point. Freshly sublimed iodine is weighed in a tared dish. The dish has a flat-ground lip upon which rests a coverglass secured by a platinum wire clip. The loss of iodine in the weighing process is negligible. Enough stock arsenious acid to react with all but 0.2% of the iodine is weighed out directly from a storage flask into a tared flask resting on a balance pan. The dish containing the iodine is carefully placed in the flask containing the arsenious acid solution, the clip is released, and the iodine and arsenious acid react. The small excess of iodine is titrated to the endpoint with a solution of arsenious acid about 1/200 the strength of the regular stock solution. The endpoint is determined amperometrically by a method similar to that of Ramsay.(13).

For the titration of the iodine generated in the coulometer the standardized stock solution is weighed out to within 0.2% of the amount required. The iodine solution is drained and washed down through tubes in the bottom of the coulometer into the flask containing arsenious acid below. The titration is completed amperometrically as above.

The number of moles of iodine generated may then be calculated.

In addition to the amount of material generated in the coulometer, we must know accurately the time interval of the experiment. We must also know accurately the current passed through the coulometer during that time.

The time is measured by counting every ten pulses on the 1 kHz standard frequency mains of the National Bureau of Standards. In an experiment generating 1 g of iodine or more, the time error is estimated to be 1ppm.

The current is regulated by a commercial current stabilizer working on a voltage feedback principle. With proper choice of feedback resistors and associated apparatus this instrument may be made to keep the current constant to parts in 10^7 provided the feedback resistance is adjusted periodically. The actual value of the current is obtained by balancing the IR drop across a resistor against a thermostated standard Weston cell. The current thus measured is thought to be in error by no more than one ppm.

The current-time product divided by the number of moles of iodine generated yields the Faraday.

3. Results and Discussion

Table I lists the results of the experiments of Craig, the

derived value calculated as in the introduction above using Cohen and Taylor's new adjustment of the physical constants, and our own results up to the present.

TABLE I

Values of the Faraday, F

	F, C mole^{-1}	σ_m	σ_i	σ
Craig (5,7)	96,486.51	0.5	1.6	1.06
This work	96,483.05	0.9	1.8	1.56
Adjusted value (7)	96,484.23			0.65

σ_m = standard deviation of the mean

σ_i = standard deviation of a single observation

σ = estimate of total error

The value given for Craig's silver coulometer determination of the Faraday is based upon nine experiments, and our own tentative result upon merely four. The standard deviations of a single observation are comparable and our own standard deviation (1.8 C mole $^{-1}$) is obviously largely attributable to the error in the titer (2.1 C mole $^{-1}$). Our values seem low and tend, up to this point, to support the derived value. The random errors are large enough that further similar measurements may with no great improbability yield a final value that exceeds the derived value by as much as it now falls short. We should be surprised, however, to obtain a result so high as to be concordant with that of the silver coulometer.

4. References

1. HERROUN, E.F., Phil.Mag (5) 40, 91 (1895)

2. KREIDER, D.A., Am.J.Sci(4), 20, 1 (1905)

3. BATES, S.J. and VINAL, G.W., J.Am.Chem.Soc. 36, 916 (1914)

4. International Conference on Electrical Units and Standards, 1908: Minutes and Verbatim report of the Meeting of the Delegates. His Majesty's Stationer's Office, London, page 73(1909)

5. CRAIG, D.N. et al., J.Research NBS 64A, 381 (1960)

6. HAMER, W.J., ibid. 72A, 435 (1968)

7. COHEN, E.R., and TAYLOR, B.N., private communication, and these proceedings.

8. MACINNES, D.A. and PRAY, A.R., No 1 del Supplemento 6, Serie X 232 (1957)

9. MATTAUCH, J.H.E., THIELE, W., and WAPSTRA, A.H., Nucl.Phys. 67, 1 (1965)

10. LINGANE, J.J., and ANSON, F.C., Anal.Chim.Acta 16, 165 (1957)

11. LINGANE, J.J. and KENNEDY, J.H., ibid. 15, 465 (1956)

12. BOWER, V.E., in Proceedings of the International Conference on Precision Measurement and Fundamental Constants, edited by D.N. Langenberg and B.N. Taylor, (NBS Special Publication No. 343, (U.S. GPO Washington, D.C., 1971), p.147.

13. RAMSAY, W. et. al., Anal.Chem. 22, 232 (1950)

The Gravitational Constant G

R. A. Lowry, W. R. Towler, H. M. Parker, A. R. Kuhlthau and J. W. Beams

University of Virginia, Charlottesville, Virginia 22903

I. Introduction

The gravitational interaction between any two particles of matter is stated in Newton's law of gravitation

$$F = G \frac{m_1 m_2}{d^2}$$

where F is the force of attraction between any two particles of matter in the universe having masses of m_1 and m_2, d is the distance between the particles, and G is the constant of proportionality. Gravitational interaction possesses two rather unique properties: great universality and extreme weakness. Both of these features contribute to the difficulty of measuring the gravitational constant, G.

The universality of G is evident from the fact that every particle of mass is coupled directly to every other particle of mass in the universe by a force of attraction, and from the fact that careful experiments have failed to show any variation of G with such things as the nature and magnitude of the attracting masses, their states of chemical combination, their temperatures, or the amount of matter placed between the attracting masses (1,2,3). The extreme weakness of the gravitational interaction can be illustrated by the fact that the gravitational attraction between two spheres each with a mass of 10 kg and their centers separated by 0.15 m is approximately 3×10^{-11} newtons or about 3×10^{-13} times the force of gravity on each sphere. Thus the experimenter is somewhat restricted in available techniques, and it is not possible to isolate or shield the test masses from the gravitational fields and gradients produced by all of the other masses in the universe.

Considering the great precision of the astronomical measurements of the paths of celestial bodies one might think that somehow G could be found from these data. However, it turns out that the product GM, where M is the mass of the body, is obtained from such astronomical observations instead of G or M. Consequently it is necessary to have an independent determination of G if the mass M and the mean density of the earth or other celestial bodies are to be found. Experimental methods of measuring G have fallen into three general classes: 1) comparison of the earth's pull on a body with that of the attraction of a large natural mass such as a mountain or other topographical features; 2) comparison of the earth's pull on a body with that of a known mass as in the common balance experiments; 3) measurement of the force between known masses in the laboratory. Up to the present time only 2) and 3) of the above classification have given reliable values of G.

Table I gives a partial list of some values reported.

It is difficult to estimate the precision of the values obtained before the work of Poynting (1891) and of Boys (1895). However, an examination of the results of the last half dozen workers listed in Table I shows that the variation of the values obtained for G is substantially larger than what might be expected from the precision of the data indicated in the papers. Recent analyses (4,5) of the data of Heyl and his associates which is usually considered to be the most accurate of that listed in Table I give $(6.670 \pm 0.015) \times 10^{-11} Nm^2kg^{-2}$ for G.

In 1965 the authors proposed[6] a new experimental method for the determination of G, which showed promise of much improved precision. Briefly the principle of the method is illustrated in Fig.1.

Two large spherical masses (tungsten spheres) are mounted on a rotary table which can be driven about its axis of rotation by a specially designed electric motor. Also mounted from the same rotary table is an airtight chamber in which a small horizontal cylinder made of copper is suspended by means of a quartz torsion fiber fastened to the top of the cylinder and accurately hanging in the axis of rotation. This small horizontal cylinder is commonly

TABLE I

Name	Year	Method	G($10^{-11} Nm^2kg^{-2}$)
Cavendish	1798	Torsion-balance(deflection)	6.754
Reich	1838	Torsion-balance(deflection)	6.61
Baily	1842	Torsion-balance(deflection)	6.475
Von Jolly	1881	Common balance	6.465
Wilsing	1889	Metronome balance	6.596
Poynting	1891	Common balance	6.698
Boys	1895	Torsion-balance(deflection)	6.6576
Braun	1896	Torsion-balance(oscillation)	6.6579
Eötvös	1896	Torsion-balance(oscillation)	6.65
Richarz	1898	Common balance	6.685
Burgess	1901	Torsion-balance(deflection)	6.64
Heyl	1930	Torsion-balance(oscillation)	6.670 ± 0.005
Zahradnicek	1932	Torsion-balance(resonance)	6.659 ± 0.02
Heyl and Chrzanowski	1942	Torsion-balance(oscillation)	6.673 ± 0.003

called the small mass system as contrasted to the large spheres which are referred to as the large mass system.

The gravitational interaction between the two mass systems tends to align the longitudinal axis of the small mass cylinder with the centers of the two large spheres, changing the angle, θ. However, the angle θ never changes appreciably because a beam of light from a source mounted on the rotary table is reflected from a mirror mounted on the axis of the quartz fiber and falls on a photocell, also mounted on the rotary table. The photocell senses the minute changes in θ, and produces an "error" signal which is used to drive the motor which rotates the table, thus maintaining θ constant. With the angular separation, θ, remaining constant, the small mass system experiences a constant torque, which in turn causes a constant angular acceleration of the rotary table. This acceleration can be determined very accurately by measuring the period of the rotating table, and can be shown to be a direct measure of G.

This method possesses three novel features which contribute to the potential for improved accuracy. First, the interaction force of the two masses is manifested in an acceleration (change of frequency) rather than in a deflection. The effect of the interaction is cumulative and can be integrated over a long period of time, thus improving precision (the noise level is also automatically reduced). Second, the two mass systems rotate about an axis many times during a measurement, and hence the effects of gravitational fields or field gradients caused by extraneous masses are effectively cancelled except for higher order effects. Third, the coordinates or positions of the interacting masses with respect to a rotating coordinate system do not change during a measurement and hence their distances can be accurately determined.

2. The Determination of G

A theoretical analysis(6) of the gravitational torque system gives:

$$G = \frac{\alpha \pm \dot{\omega}_{w/o}}{A(1 + B + C \ldots)} \tag{1}$$

where

$$A = \frac{3M}{R^3} \frac{\left[\frac{1}{3} - \frac{1}{4}\left(\frac{2a}{L}\right)^2\right]\sin 2\theta}{\left[\frac{1}{3} + \frac{1}{4}\left(\frac{2a}{L}\right)^2 + \left(\frac{2}{L}\right)^2 \frac{I_s}{m}\right]}, \tag{2}$$

$$B = \frac{5}{6}\left(\frac{L}{2R}\right)^2 \frac{\left[\frac{1}{5} - \frac{1}{2}\left(\frac{2a}{L}\right)^2 + \frac{1}{8}\left(\frac{2a}{L}\right)^4\right]}{\left[\frac{1}{3} - \frac{1}{4}\left(\frac{2a}{L}\right)^2\right]}(7\cos^2\theta - 3), \tag{3}$$

$$C = \frac{1}{48}\left(\frac{L}{2R}\right)^4 \frac{\left[\frac{1}{7} - \frac{3}{4}\left(\frac{2a}{L}\right)^2 + \frac{5}{8}\left(\frac{2a}{L}\right)^4 - \frac{5}{64}\left(\frac{2a}{L}\right)^6\right]}{\left[\frac{1}{3} - \frac{1}{4}\left(\frac{2a}{L}\right)^2\right]} \tag{4}$$

$$\times[1386\cos^4\theta - 1260\cos^2\theta + 210],$$

and

G = The Newton Gravitational Constant,

α = Measured angular acceleration with spheres on the table,

$\dot{\omega}_{w/o}$ = Angular acceleration with spheres removed,

M = Mass of one sphere,

R = Distance from axis of rotation to center of mass of the spheres,

m = Mass of small mass system cylinder,

a = Radius of small mass system cylinder,

L = Length of small mass system cylinder,

I_s = Moment of inertia of small mass system stem,

θ = Angle between the longitudinal axis of the small mass system cylinder and the imaginary line of centers of the spheres.

The three terms in the denominator of Eq. 1 correspond to the first three non-zero terms resulting from an expansion of the gravitational potential of the small mass system cylinder in terms of Legendre polynomials. Additional terms can be calculated if necessary, but they do not contribute to the present level of precision. It is important to note the R^3 dependence in Eq. 1.

The experimental task is to determine $\alpha, \dot{\omega}_{w/o}, R, M, a, L, I_s, m$, and θ as accurately as possible. Five of these (a, L, I_s, M, and m) can be considered as constants of the apparatus and can be measured directly, independently of the experimental observations. The other four ($\alpha, \dot{\omega}_{w/o}, R$, and θ) must be determined as part of the observation. A typical set of values, including estimated errors, is given in Table II.

TABLE II

M	$= 10.49012 \pm .00007$ kg
R	$= 7.6178 \pm .0005$ cm
a	$= .1985 \pm .0003$ cm
L	$= 3.8105 \pm .0004$ cm
m	$= 4.051205 \pm .000005$ gm
I_s	$= .0445103$ gm cm^2
θ	$= .7835 \pm .0002$ rad
α	$= (4.3299 \pm .0022)\times10^{-6}$ rad/s^2
$\dot{\omega}_{w/o}$	$= (-0.3431 \pm .0022)\times10^{-6}$ rad/s^2

The procedure in carrying out this experiment consists of first assembling the small mass system in the gas tight chamber and properly positioning it. The quartz fiber is adjusted to coincide with the axis of rotation and the tracking optical lever is set so that the effective twist in the fiber is as close to zero as possible. The chamber is evacuated, filled with helium and sealed. The tungsten spheres are placed in their mountings and are positioned equidistance from the axis of rotation in the same horizontal plane as the axis of the cylinder of the small mass system and with the centers of mass of the two spheres on a line through and perpendicular to the axis of rotation. The entire apparatus, located in a small room, is then allowed to come to temperature equilibrium with the servo system operating and the rotary table stationary. This usually requires one day or more. The temperature of the room is controlled to a point just above the apparatus in a manner which produces a slight positive vertical gradient.

After the apparatus has reached temperature equilibrium the absolute value of R is determined by comparison with a precision gauge block using a pair of white light interferometers in conjunction with an electronic indicator. In this manner R is determined to within $\pm 5 \times 10^{-5}$ cm. The apparatus is then allowed to stand until late at night when there are few people in the building and the community is relatively quiet. The observations are then begun by releasing the rotary table which is supported on a gas bearing and locking the optical sensor and servo system into a high gain mode of operation for tracking the small mass system. The rotary table and mass systems are then allowed to accelerate for 10 to 20 revolutions with each revolution being accurately timed to determine α. Next with the servo system returned to a low gain mode the spheres are carefully removed, usually without changing the speed of the rotary table appreciably. The servo is then returned to the high gain mode and the table is tracked for another 10 to 20 revolutions to determine $\dot{\omega}_{w/o}$. The procedure of placing the spheres in the mounts and removing them is repeated several times without stopping the rotary table. Finally the rotary table is stopped and R is again measured. These measurements have shown that during the process of removing and replacing the spheres on the table R does not change more than 5×10^{-4} cm. Greater precision requires stopping the observations and remeasuring R each time the spheres are replaced on the table.

The acceleration measurements shown in Fig. 2 are typical of very recent observations. The same data are displayed in Fig. 3 with the acceleration scale magnified by a factor of 50. The first few revolutions after removing or replacing the spheres are neglected because of large vibrations of the small mass system. Even with the greatest of care, placing the spheres on the table results in some slight mechanical shock and several revolutions are required before the vibrations damp out. As would be expected, the small mass system is very sensitive to vibration.

Seven sets of observations similar to Fig. 3 have been obtained since July 22, 1971 and in all cases α and $\dot{\omega}_{w/o}$ show a dependence on angular velocity. Possible sources of this dependence being presently investigated include magnetic fields fixed in the laboratory, variations in the temperature of the apparatus, and the circulation and acceleration of the gas in the chamber containing the small mass system. This velocity dependence was not noted in our earlier measurements but may have been masked by the noise in the acceleration measurements at that time. In any event by proper treatment of the data it is now possible to determine a value for $\dot{\omega}$ with a statistical uncertainty for a single set of observations of 1 part in 3000.

Recent observations are summarized in Table III. We have not had time to perform a careful analysis of these data or to completely check for systematic errors. The values of G shown in Table III were computed from Eq. 1 to normalize the data for comparisons and are not presented as absolute determinations of G for obvious reasons. None of the observations was made with magnetic shielding around the small mass chamber. Each run represents some change in operating conditions, either planned or accidental. For example run 7/22/71 represents a larger value of R. The table was rotating in the opposite direction during run 8/17/71. Higher temperatures and poor temperature stability existed during run 8/19/71 because of a failure of the building air conditioning system. There was a very large twist in the fiber during run 8/21/71, hence a large $\dot{\omega}_{w/o}$. A large number of observations of this type will need to be made before the best operating conditions can be determined and, hopefully, all non-gravitational interactions with the small mass system identified and properly accounted for.

3. Summary

The most attractive feature of the present apparatus is its high sensitivity for measuring very small forces (torques). It is expected that the sensitivity can be improved further by replacing the quartz fiber with a magnetic suspension. The major known

TABLE III

(M, a, L, m, θ, and I_s same as given in Table II)

Date	$\dot{\omega}$ 10^{-6} rad/s^2	R cm		G 10^{-11} m^{2}kg^{-2}
7/22/71	4.4388	7.768 ± .003		6.6942
8/13/71	4.7022	7.61798	± .00005	6.6857
8/16/71	4.6731	7.61819		6.6448
8/17/71	4.6738	7.61820		6.6459
8/19/71	4.6304	7.61813		6.5840
8/20/71	4.6941	7.61815		6.6746
8/21/71	4.7071	7.61827		6.6934

difficulties associated with this method of determining G are:

1. The effect of the gas in the small mass system chamber. Eventually the experiment will have to be done in a vacuum, as originally proposed.

2. The effect of magnetic fields. It was intended that the small mass system be made from high purity oxygen free (diamagnetic) copper but unfortunately the present cylinder was found to be paramagnetic. A new small mass system is being constructed from pure oxygen free copper and magnetic shielding of the small mass system is planned. Other materials are being considered for the small mass system.

3. Temperature stability. This has been greatly improved but may again become a problem when greater precision is attempted.

4. Vibrations. Because of the extreme sensitivity of the small mass system to vibrations the present experiments were performed with the small mass chamber filled with gas. Other methods of damping the small mass system are being considered. The coupling of torque to the small mass system through vibration is unknown. Fortunately the errors in the determination of acceleration resulting from vibration induced variations in timing can be reduced to an arbitrarily small value by averaging over an adequate number of revolutions of the rotary table.

The value we previously reported (6), $G = (6.674 \pm 0.012) \times 10^{-11} Nm^2/kg^2$, will remain our best until enough observations can be accumulated to permit the gravitational forces to be reliably isolated.

4. References

1. VON EÖTVÖS, R., PEKAR, D., and FEKETE, E., Ann. Physik 68, 11 (1922).

2. CHAMPION, F.C. and DAVY, N., Properties of Matter (Phil. Library, Inc., New York, 1959). 3rd edition.

3. DICKE, R.H., The Theoretical Significance of Experimental Relativity (Gordon and Breach, New York, 1964), pp. 3-13.

4. Recommended Prefixes; Defined Values and Conversion Factors; General Physical Constants, US Natl. Bur. Std. Misc. Pub. US Government Printing Office, Washington, D.C. (1963).

5. STEPHENSON, L.M., Proc. Phys. Soc. (London) 90,601 (1967).

6. BEAMS, J.W., KUHLTHAU, A.R., LOWRY, R.A., PARKER, H.M., Bull. Am. Phys. Soc. 10, 249 (1965); ROSE, R.D., PARKER, H.M., LOWRY, R.A., KUHLTHAU, A.R., BEAMS, J.W., Phys, Rev. Lett. 23, 655 (1969); TOWLER, W.R., PARKER, H.M., KUHLTHAU, A.R., BEAMS, J. W., Proceedings of Int. Conf. Precision Meas. and Fund. Const., 3-7 Aug. 1970.

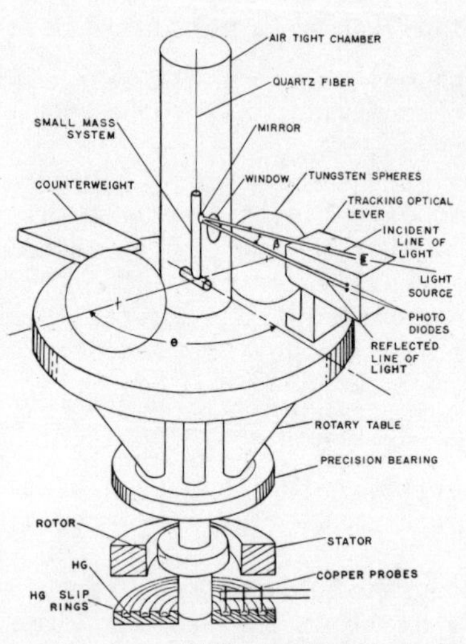

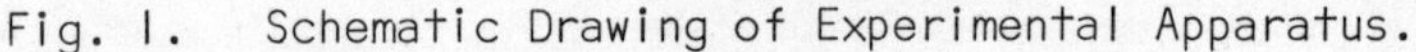
Fig. 1. Schematic Drawing of Experimental Apparatus.

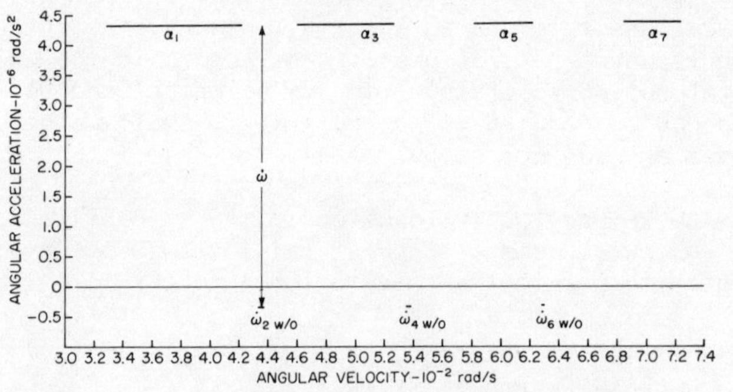

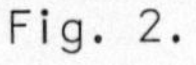
Fig. 2. Measurements of Angular Accelerations α, $\dot{\omega}_{W/O}$ - August 16, 1971.

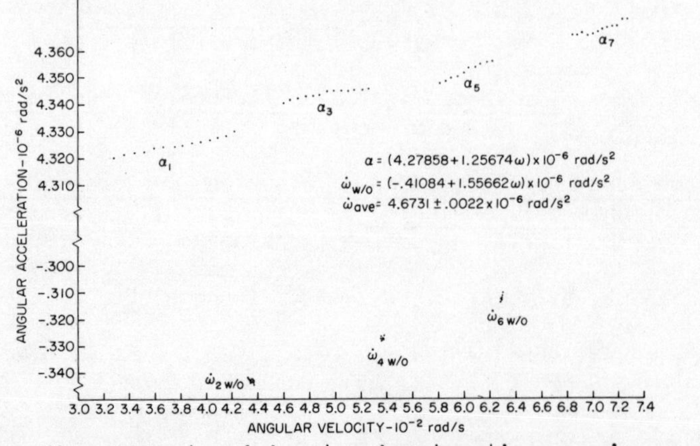

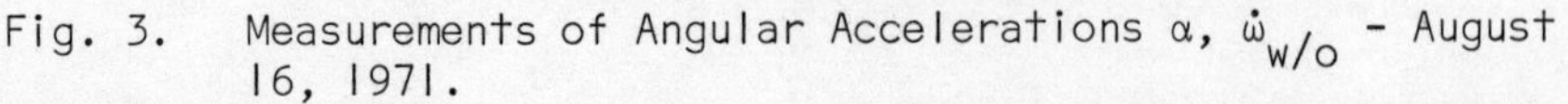
Fig. 3. Measurements of Angular Accelerations α, $\dot{\omega}_{W/O}$ - August 16, 1971.

A Proposal for a New Determination of the Gas Constant

T. J. Quinn

Division of Quantum Metrology,
National Physical Laboratory, Teddington, England

1. The gas constant

The gas constant, R_o, appears as one of the fundamental constants in physics. Its presently accepted value is based on measurements leading to the normal molecular volume of an ideal gas, V_o, which, at a pressure of one standard atmosphere, P_o, at the temperature of the freezing point of water, T_o, is related to R_o by the equation of state of an ideal gas

$$P_o V_o = R_o T_o . \tag{1}$$

Many determinations of V_o were made by a number of workers over the period 1924 to 1941, notably G.P. Baxter and H.W. Starkweather (1924-1926) (1), E. Moles and collaborators (1934-37) (2) and T. Batuecas, F.L. Casado and G.C. Malde (1939-41, published in 1952) (3). All these results were obtained from measurements of the density of oxygen. At that time oxygen was the only gas which could be used since it was unique in having its molecular weight defined. A detailed review of all the experimental results has been given by Batuecas (4) and will not be repeated here as there seems little doubt that the present value of V_o, and hence R_o, is the best one that can be obtained from the existing experimental data.

However it is worth pointing out that our present value of R_o derives essentially from only one experiment repeated with minor modifications by a number of workers. The possibility of there being an undetected systematic error characteristic of oxygen density measurements cannot be dismissed. The difficulties inherent in gas density measurement are well illustrated in the various papers referred to above. It is of interest to note that even in the most recent work, that of Casado and Batuecas upon which our value of R_o is largely based, the scatter of the experimental results is large. Fig.1 shows a histogram of the sixty-nine separate determinations of V_o made by Casado and Batuecas using four glass bulbs at pressures from one third to one atmosphere.

A knowledge of the gas constant (or Boltzmann constant) is required in any determination of thermodynamic temperature, made for example either by means of gas thermometry, in which R_o enters directly or by means of thermal radiation measurements in which Boltzmann's constant is required in either Planck's or Stefan's equation. A further method of determining T recently developed at both NBS and NPL is based on measurements of the velocity of sound V in a gas for which

$$V^2 = \frac{\gamma_o R_o T}{M} \quad \text{pressure} \rightarrow o \tag{2}$$

in which γ_o is the ratio of the specific heats at zero pressure and M is the molecular weight.

Any of these methods could in principle be reversed and used to determine R_o (or k) if the temperature could be specified independently. It is only possible to do this at, or close to, the defining fixed point of temperature namely the triple point of water 273.16 K. This for the time being rules out most thermal radiation methods as the temperature is too low.

As a result of the experience gained at NPL and NBS on acoustic thermometry together with the unsatisfactory situation with regard to the value of R_o being based solely on oxygen density measurements, an attempt is being made at NPL to determine R_o from velocity of sound measurements at a temperature of 273.15 K.

2. Velocity of sound measurements

The method to be used is that of acoustic interferometry in which the wavelength of sound propagated in a cylindrical tube by an electromagnetically driven diaphragm is obtained from measurements of the distance between successive positions of a reflector for which the gas column comes into resonance. The instrument and measurement techniques to be used are similar to those which have been developed at NPL for the measurement of low temperatures. These have recently been fully described (5, 6) together with a detailed discussion of the theory of acoustic interferometry.

The acoustic interferometer which is under construction at NPL for the determination of the gas constant is illustrated in Fig.2. The main differences between it and the low temperature instrument are:

(a) the whole instrument has been designed to be outgassed at a temperature of 150 °C,

(b) there is no active temperature control since it is simply to be immersed in an ice bath at 273.15 K,

(c) the diameter of the cylindrical cavity is 3 cm rather than 2 cm and is proportionally longer,

(d) the whole of the mechanism for moving the reflector is enclosed in the gas chamber so that there is no net change in volume on moving the reflector.

There are a number of possible sources of error which have to be considered in the measurement of the velocity of sound in a gas using an acoustic interferometer. For an interferometer having a cylindrical cavity there is a frequency, which depends on the gas, below which only the plane wave or zeroth mode of propagation is possible. Above this frequency non plane-wave modes are possible each one having its own phase velocity, the number of such modes increasing with frequency. It has been found that the problem of separating the phase velocity of a particular mode from the measured group velocity is very difficult. For this reason it has been found advantageous to operate at frequencies below the cut-off of the first higher mode. This cut-off frequency is related solely to the diameter of the cylindrical cavity, d, and for the first higher mode is equal to 1.73d. With a cylinder of diameter 3 cm this means that the cut-off frequency for all gases at about 273 K is equal to or less than about 20 kHz.

The result of operating the interferometer at relatively low frequencies to avoid the intractable problem of the higher mode is, unfortunately, to introduce other difficulties arising from boundary-layer effects at the walls of the cylinder. These boundary-layer effects in cylinders have been studied over many years and there is now general agreement both from theory and experiment that the magnitude of the effect on both the velocity and absorption coefficient is a function of the square root of the frequency. Recent experiments using the NPL low temperature interferometer have supported this. There is doubt however as to the form of the constant of proportionality which includes such factors as the thermal conductivity, viscosity and density of the gas. Some of this doubt must stem from errors in the numerical value of some of these parameters resulting in disagreement between calculated and measured boundary layer effects. Fortunately it is possible to avoid using the constant of proportionality in making boundary-layer corrections to the measured velocity. This is so because the effect on the velocity is inversely proportional to frequency and the effect on the absorption coefficient is proportional to frequency, both effects having the same constant of proportionality. Thus from experimental measurements of absorption coefficient the velocity correction can be calculated following a method first used by Smith and Harlow (7). The method is described in detail in reference (5). There are further boundary-layer effects which result from losses during reflection at both the transducer and reflector, but these can be dealt with without difficulty.

Equation 2 refers, of course, to an ideal gas; for a real gas the relation can be written down as a virial expansion of terms in pressure having equation 2 as the first (and pressure independent) term. For practical purposes equation 2 can be considered as the velocity of sound in a real gas at vanishingly low pressure and is obtained from an extrapolation to zero pressure of a series of measurements made at decreasing pressures.

The choice of gas for the experiment is not now limited to oxygen since oxygen no longer occupies the special place which it once did in the system of atomic weights. The factor which limits the choice is now the uncertainty in the isotopic content of a particular gas. There are a number of reasons why a heavy gas is preferable; the boundary-layer effects even when working at constant wavelength rather than constant frequency decrease with increasing molecular weight of the gas. Furthermore the effect on the mean molecular weight of the gas of small additions of the lighter gases resulting from degassing of the material of the interferometer is clearly less for a heavy gas. The acoustic coupling between transducer and gas, which ideally should be strong to give high sensitivity, depends on the product of gas density and velocity of sound. This product increases with molecular weight. For this reason we are examining the possibility of using some of the heavier gases such as argon, krypton or even xenon for the gas constant determination.

3. The uncertainty to be expected in the result

At this stage it is not possible to predict the uncertainty due to random errors in measurement which will be present in the final result since these will depend on noise sources and as yet unknown instabilities in the system. The systematic errors if the heavier gases such as argon, krypton or xenon are used are likely to be those arising from uncertainties in the isotopic content of the samples; while if helium is used the relatively poor acoustic coupling between the gas and the transducer may cause difficulties. It is expected however that the random errors in this new determination of the gas constant will be somewhat less than those of the previous determinations based upon oxygen density measurements.

References

1. BAXTER, G.P. and STARKWEATHER, H.W., Proc. Nat. Acad. Sci., 10, 479 (1924); ibid., 12, 699 (1926).

2. MOLES, E., TORAL, T. and ESCRIBANO, A., Trans. Faraday Soc., 35, 1439 (1939).

3. CASADO, F.L. and BATUECAS, T., An. Fis. Quim. (Madrid), 48B, 5 (1952).

4. BATUECAS, T., Proc. 2nd Int. Conf. on Nuclidic Masses (Vienna 1963), ed. W.H. Johnson (Springer-Verlag, Vienna-New York, 1964), p.139.

5. COLCLOUGH, A.R., Proc. of 5th Symposium on 'Temperature: its Measurement and Control in Science and Industry' held in Washington in July 1971 (to be published).

6. COLCLOUGH, A.R., Acustica, 23, 93 (1970).

7. SMITH, D.H. and HARLOW, R.G., Brit. J. Appl. Phys., 14, 102 (1963).

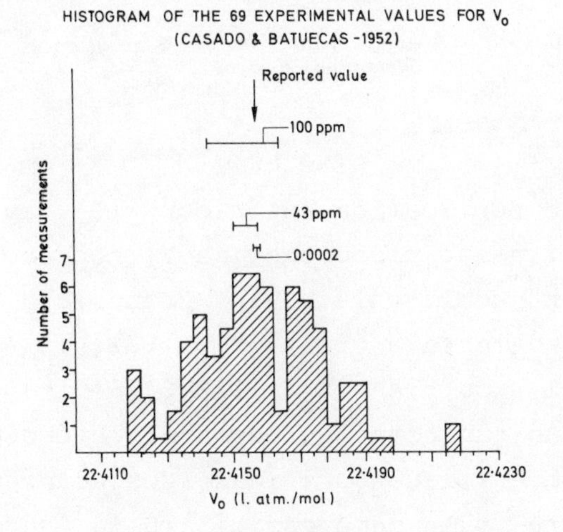

Fig.1. Histogram of the 69 experimental values of V_0 obtained by Casado and Batuecas 1952 (ref.3).

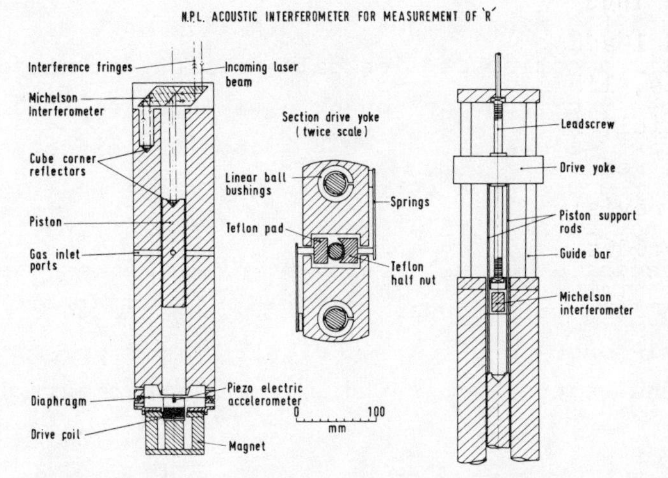

Fig.2. The acoustic interferometer to be used at NPL for the determination of the gas constant. The whole assembly shown here to be in a gas-tight box, immersed in ice at 0 °C, and filled, in the first instance, with helium and later with the heavier monatomic gases.

Volume normal moléculaire, V_o, des gaz à l'état idéal

T. Batuecas

Cons. Super. Investig. Cientif. Université. Santiago de Compostela. Espagne

INTRODUCTION

Dans une publication antérieure (1), ayant pour but la connaissance rigoureuse de $\underline{V}_o$, on parvenait à la conclusion que la valeur plus probable de cette <u>constante fondamentale</u> était: $V_o \mp (22{,}413_0 \pm 0{,}0004)$ lit.-atm./mol., ou bien: $V_o = (22413{,}6 \pm 0{,}4)$ cm^3.atm./mol. Rappelons que cette valeur avait été déduite, moyennant une critique judicieuse des données modernes sur la <u>densité</u> et la <u>compressibilité</u> (à 0ºC et sous de faibles pressions) du gaz O_2.

La nouvelle révision numérique -plus soignée et étendue- concerne les gaz O_2 et N_2, en accord avec ce que, dès 1963, nous avions suggeré (1), savoir: Qu'après l'adoption de l'échelle unifiée des masses atomiques, l'oxigène ne jouant plus le rôle d'élément referentiel, l'évaluation de $\underline{V}_o$ pouvait s'éffectuer en ayant recours à un autre gaz, notamment le N_2.

La révision a été accomplie en faisant appel à deux procédés de calcul, rélevant de la <u>méthode des densités limites gazeuses</u> (2). L'un des procédés éxige la connaissance de $\underline{L}_p$, à 0ºC et sous trois pressions -aun moins- entre 1 et 0 atm., ce que, moyennant la formule:

$$L_o \pm p.L_p.A_o^1 - L_p = 0 \qquad (1),$$

permet de calculer, par moindres carrés, et la <u>densité limite</u> (L_o) et <u>l'écart à la loi de Boyle</u> (A_o^1) du gaz en étude. Or, $\underline{L}_o$ étant connue, en adoptant pour <u>masse</u>

moléculaire la valeur dans l'échelle unifiée, soit de M_{O_2}, soit de M_{N_2}, la formule

$$M = V_o . L_o \qquad (2),$$

fait alors possible d'évaluer V_o.

L'autre procédé de calcul, dont nous avons fait usage, a pour base la formule:

$$M = V_o . L_1 \, (1 - A_o^1) \qquad (3),$$

où L_1 désigne la masse du litre à 0ºC et sous 1 atm. standard. Pourtant, les valeurs alors utilisées pour A_o^1 ont été celles déduites par C.S. Cragoe (3), pour l'un ou l'autre des gaz O_2 et N_2, en évaluant les pentes (slopes) des isothermes pv-p, a 0ºC dans l'intervalle 1-100 atm. Rappelons que les pentes (slopes) des isothermes sont pratiquement égales, mais de signe opposé aux coéfficients A_o^1.

I

Éxaminons maintenant, quelque peu en détail, l'application de ce qui précéde, tout d'abord au gaz O_2, puis à l'N_2.

Les recherches modernes sur la densité de l'oxygène, à 0ºC et dans l'intervalle 1-0 atm., sont chronologiquement énumerées: Celles de G.P. Baxter et H.W. Starkweather (4), celles de E. Moles et collab. (5), enfin celles de T.Batuecas, F.L.Casado et G.García Malde (6).

Les recherches de Baxter et Starkweather, de par les soins extrêmes apportés à la préparation et purification de l'O_2, ainsi qu'aux mesures, doivent être qualifiées de magistrales. Pourtant, un éxamen attentif montre que toutes les mesures des chercheurs américains n'ont pas une égale précision. C'est le cas notamment des 26 valeurs de L_1 à 1 atm., obtenues avec de l'O_2 preparé par électrolyse et purifié seulement par voie chimique, valeurs qui sont légèrement plus fortes que les 45 autres obtenues en opérant sur de l'O_2 préparé par trois procédés (électrolyse, MnO_4K,ClO_3K) et bien purifié, d'abord chimiquement, puis condensé et fractionné. Or, les écarts individuels, par rapport à la moyenne générale, des 26 valeurs, étant quatre fois plus forts environ que ceux des autres 45 (ou plutôt 42, comme nous les verrons plus loin), ceci justifie que nous ayons pris ces dernière comme définitives, laissant de côté les 26 premières, que nous jugeons préliminaires.

Certes, Baxter et Starkweather n'avaient pas exclue aucune de leurs mesures (sauf une) à 1 atm. Pourtant, il semble plus correct et donc préférable d'appliquer un critère limitatif, où seules doivent être prises en considération les valeurs de L_p qui soient comparables à tous les égards.

Notre révision numérique a comporté: La récalculation des valeurs de L_p (compte tenu de l'adsorption) pour chacune des pressions, 1, 3/4, 1/2 et 1/4 atm., en vue de les rapporter à la gravité correcte du Harvard's Coolidge Laboratory (g=980,385, au lieu de g=980,399 gal.) et à g_o= 980,665 gal. Puis, le calcul des erreurs et, en premier lieu, de l'erreur probable, qu'il faut connaître pour appliquer le critère de Chauvenet, afin de savoir quelles valeurs doit-on admettre et, par suite, quelle est la moyenne plus probable. D'après ce critère, il résulte que, des 45 valeurs à 1 atm. prises en considération pour L_p, 3 doivent être exclues, tandis que toutes les valeurs à 3/4, 1/2 et 1/4 atm. restent valables.

La Table I montre les moyennes de L_p concernant les quatre pressions, p_{og}, étudiées par Baxter et Starkweather. À noter que ces valeurs de la pression ne sont pas identiques à celles des chercheurs américains,

TABLE I

(Mesures de Baxter et Starkweather sur la densité du gaz O_2)

p_{og}(atm.)	L_p(réc.)gr./lit.	
0,99996	$(1,42903 \pm 0,00001_2)$	(Moyenne de 42 valeurs)
0,74995	$(1,42871 \pm 0,000\,1_9)$	(" de 20 ")
0,49997	$(1,42837 \pm 0,00002_6)$	(" de 12 ")
$0,24998_4$	$(1,42804 \pm 0,000\,3_6)$	(" de 10 ")

les petites différences ayant pour origine les changements de g qui agissent, en premier lieu, sur p_{og} et ensuite sur L_p, celle-ci étant définie par rapport à p_{og}.

Avec les valeurs de cette Table et la formule (1), on déduit par moindres carrés: $A_o^1 = 0,00092_7$ et L_o = = $(142770 \pm 0,00005_0)$ gr./lit. (L'erreur précédente a été calculée par la formule de propagation des erreurs). Or, L_o étant connue, ainsi que: $M_{O_2} = (31,9988 \pm 0,0002)$, la formule (2), après substitution, fournit pour le volume normal moléculaire: $V_o = (22,412_7 \pm 0,0009_3)$ lit.-atm./mol.

Un nouveau résultat pour cette constante fondamentale est obtenu moyennant la formule (3), après substitution de la valeur: L_1(réc.) = $(1,42903 \pm 0,00001_2)$ -tirée des mesures à 1 atm. de Baxter et Starkweather- et de celle: $A_o^1 = (0,00095_1 \pm 0,00009)$ déduite par Cragoe (3), à partir des isothermes aux pressions élevées. C'est ainsi que l'on trouve: $V_o = (22,413_3 \pm 0,0005_o)$ lit.-atm./mol^{-1}.

Voyons maintenant les recherches de E. Moles et collab., où des grands éfforts furent accomplis pour préparer de l'O_2 très pur et aussi pour atteindre une grande rpécision dans leurs mesures. L'oxygène obtenu par deux procédés (MnO_4K, chlorates) et ayant subi une purification chimique très poussée, ne fut pas condensé

TABLE II

(Mesures de Moles et collab. sur la densité du gaz O_2)

p_{og}	L_p(réc.)gr.-lit.
0,99994	1,42903
0,66663	1,42858
0,49997	1,42838
0,33331	1,42813

ni distillé.

Notre révision numérique s'est bornée à récalculer les valeurs moyennes de L_p, consignées par E. Moles(5), afin de les rapporter à la gravité correcte du Laboratoire à Madrid (g = 979,939, au lieu de 979,953 gal.) et à g_o. Dans le Table II sont indiquées les moyennes récalculées de L_p, concernant les quatres pressions, p_{og}, étudiées, sans indication des arreurs, qui ne figurent pas dans la publication originale (5).

Par un calcul analogue à celui antérieurement esquissé, les précédentes valeurs fournissent, par moindres carrés: $A_o^1 = 0{,}00093_4$ et $L_o = 1{,}42770$ gr./lit., puis à l'aide de celle-ci et la formule (2): $V_o =$ $= 22{,}412_9$ lit.-atm./mol.

D'autre part, les mesures à 1 atm. de Moles et collab. conduissant à la moyenne: L_1(réc.) = 1,42903 gr./lit., la formule (3), après substitution de cette valeur et celle de Cragoe: $A_o^1 = 0{,}00095_1$, fournit pour le volume normal moléculaire: $V_o = 22{,}413_3$ lit.-atm.-mol.$^{-1}$; résultat identique à celui tiré des mesures de B. et Starkw.

Nos recherches, en collaboration avec Casado et García Malde, sur la densité et la compressibilité du gaz O_2, ayant été commentées ailleurs (6), nous ne dirons ici que le strictement indispensable. L'oxygène, préparé par deux méthodes (MnO_4K, ClO_3K), subit une purification chimique très soignée, mais ne fut pas condensé ni distillé. L'ensemble des valeurs de L_p à

1, 2/3 et 1/3 atm. ont été récalculées, afin de rapporter à la gravité correcte: g = 980,404 gal., expérimentalement déterminée dans notre Laboratoire à Compostela. L'application du critère de Chauvenet a permis de constater, qu'une des valeurs à 1 atm. devait être exclue, tandis que le reste concernant les trois pressions étudiées, doivent être acceptée. La Table III montre les valeurs récalculées de p_{og} et L_p ayant trait à nos mesures.

Avec les valeurs de la Table, on déduit par moindres carrés: $A_o^1 = 0{,}00097_2$ et $L_o = (1{,}42765 \pm 0{,}00004)$, puis, moyennant cette valeur et la formule (2): $V_o = (22{,}413_6 \pm 0{,}0009)$ lit.-atm.-mol^{-1}.

Nous avons jugé utile de grouper dans la Table IV, les résultats obtenus pour V_o, d'après nos calculs.

La moyenne pondérée de ces résultats:

$$V_o = (22{,}413_2 \pm 0{,}0007) \text{ lit.-atm.-mol}^{-1}$$

représente, à notre avis, la valeur plus probable que l'on puisse déduire pour le volume normal moléculaire, en prenant comme base l'O_2.

TABLE III

(Mesures de F.L.Casado et T. Batuecas sur la dansité du gaz O_2)

p_{og} (atm.	L_p (réc.) gr./lit.	
1,00016	$(1{,}42890 \pm 0{,}00002_8)$	(Moyenne de 19 mesures)
0,66678	$(1{,}42846 \pm 0{,}00002_0)$	(" 24 ")
0,33299	$(1{,}42807 \pm 0{,}00002_1)$	(" 25 ")

TABLE IV

V_o(lit.-at./mol^{-1})	Auteurs
$(22{,}412_7 \pm 0{,}0009_3)$	G.P. Baxter et H.W. Starkweather
$(22{,}413_3 \pm 0{,}0005_0)$	G.P.Baxter, H.W.Starkweather et C.S. Cragoe
$(22{,}412_9)$	E. Moles
$(22{,}413_3)$	E. Moles et C.S. Cragoe
$(22{,}413_6 \pm 0{,}0009_1)$	T. Batuecas et F.L. Casado

II

L'oxygène n'étant plus l'étalon des masses atomiques, l'évaluation de V_o pourra se faire, dès lors, en ayant recours à quelqu'un autre gaz, pourvu que certaines conditions soient remplies. Le choix du gaz N_2 est, à notra avis, particulièrement favorable, en raison de ses caractéristiques. Tout d'abord, avec ses deux isotopes ^{14}N et ^{15}N -dont le premier a une abondance qui depasse le 99%- l'élément a une masse atomique bien connue: N = (14,0067 ± 0,0001) et practiquement invariable (7). En second lieu, l'adsorption du gaz N_2 par les surfaces de verre, doit être négligeable à 0ºC. Enfin, contrairement à l'oxygène, qui diminue assez rapidement la tension superficielle du Hg et en déforme ses menisques, le gaz N_2 n'agit pas sur le mercure, ce qui est très intéressant pour les mesures de pression(8).

Une belle confirmation de ce qui précède, -nous allons le voir- est fournie par les mesures de Baxter et Starkweather sur les densités du N_2, à 0ºC et à 1, 2/3 et 1/3 atm. Le gaz, obtenu par deux méthodes(NO_2NH_4, oxydation du NH_3 par CuO), après une bonne purification chimique, fut condensé et distillé.

Comme pour l'oxygène, le révision numérique des mesures de Baxter et Starkweather sur le gaz N_2, a exigé la récalculation de toutes les valeurs de L_p à

1, 2/3 et 1/3 atm., afin de les rapporter à la gravité correcte: g = 980,385 gal.et à g_o. La révision a per mis de vérifier que, parmi les 18 valeurs à 2/3 atm.,7 surpassent la plus faible de celles trouvées à 1 atm. Aussi bien, malgré la précision atteinte par les chercheurs américains, la moyenne $L_{2/3}$ -marquée d'un astérisque sur la Table V- est inapplicable, ce nous semble, pour des calculs rigoureux. Par conséquent, le calcul par moindres carrés, conduissant aux valeurs: A_o^1=0,00041$_8$ et: L_o = 1,24993 gr./lit., ainsi que le résultat que l'on pourrait en déduire: V_o=22,4120 lit.-atm.-mol^{-1}, seront écartés de nos considérations.

Un résultat fort remarquable est cependant obtenu, avec la moyenne à 1 atm.:L_1(réc.)=(1,25044 ± 0,00001$_4$)et celle indiquée par Cragoe(3): A_o^1=(0,00045$_3$ ± 0,000003), car après substitution dans la formule (3), on deduit: V_o=(22,4131 ± 0,0007$_5$)lit.-atm.-mol^{-1}, valeur presque identique à celle consignée ailleurs,:V_o=(22,413$_2$ ± 0,0007), en prenant pour base l'O_2.

Vue l'extrème concordance des deux résultats,nous avons adoptée comme valeur plus probable du volume normal moléculaire des gaz à l'état idéal, la moyenne ponderée:

$$V_o = (22{,}413_2 \pm 0{,}0007)\ \text{lit.-atm.-mol}^{-1},$$

que l'on déduit à partir des mesures regardant les gaz O_2 et N_2.

En terminant, nous sommes heureux d'annoncer que les recherches en cours dans notre Laboratoire, sur la densité -à 0ºC et entre 1-1/3 atm.- concernant le gaz N_2, ont en vue d'apporter une nouvelle contribution à la connaissance rigoureuse de V_o.

TABLE V

(Mesures de Baxter et Starkweather sur la densité du gaz N_2)

p_{og}(atm.)	L_{pog}(réc.)gr./lit.	
0,99993$_6$	(1,25044 ± 0,00001$_4$)	(Moyenne de 18 valeurs)
0,6666$_2$	*(1,25031 ± 0,00001$_2$)	(" 18 ")
0,3333$_1$	(1,25009 ± 0,00002$_5$)	(" 14 ")

BIBLIOGRAPHIE

1.- Batuecas, T., Proceed. Second Intern. Confer.Nuclid. Mass. 1963. "Nuclidic Masses", 139, (1964).

2.- Batuecas, T., Rev. Real Acad. Cienc.(Madrid), 65, 451 (1971).

3.- Cragoe, C.S., J. Res. Nat. Bur. Stand., 26, 495 (1941).

4.- Baxter, G.P. et Starkweather, H.W., Proc. Nat.Acad. Sci., 10, 479 (1924); ibid., 12, 699 (1926).

5.- Moles, E., Les déterm. phys. chim. poids moléc. atom. des gaz, "Collect. Scient. Inst. Intern. Coopér. Intell." 1-75, Paris (1938).

6.- Batuecas, T. et García Malde, G., An. Soc. Esp. Fis. Quim., 46 517 (1950); Casado, F.L. et Batuecas, T., ibid. 48, 4 (1952).

7.- Greenwood, N.N., Chem. Brit., 6, 119 (1970).

8.- Batuecas, T., An. Soc. Esp. Fís. Quím., 21, 259 (1923).

9.- Baxter, G.P. et H.W. Starkweather, Proc. Nat.Acad. Sci., 12, 703 (1926).

Part 13 Evaluation

A Reevaluation of the Fundamental Physical Constants

E. Richard Cohen

North American Rockwell Science Center,
1049 Camino Dos Rios, Thousand Oaks, California 91360

and

B. N. Taylor

Institute for Basic Standards, National Bureau of Standards,
Washington, D.C. 20234

1. Introduction

This paper is a progress report on our current efforts to revise and update the comprehensive review of the fundamental physical constants by Taylor, Parker, and Langenberg (1), including their set of best or recommended values. That such an updating is necessary just two years after their review appeared is due to the extraordinary amount of new experimental and theoretical work which has since been completed. Here, we very briefly summarize the experimental and theoretical evidence, with emphasis on the new results which have become available within the last two years, and discuss various treatments of the data. However, no new set of recommended constants is given since such a set will necessarily require the inclusion of the new data which has become available at this Conference.

2. Review of Data

A. The More Precise Data

With the advent of sub-ppm measurements of 2e/h via the ac Josephson effect, the distinction between auxiliary constants and stochastic input data becomes somewhat blurred. Thus, for the moment we choose to divide the data into two categories, "The More Precise Data" and "The Less Precise Data", with the dividing line at the one ppm uncertainty level. (All uncertainties in this paper will correspond to one standard deviation unless otherwise noted.*)

(1) Comparisons of As-Maintained Electrical Units. During 1969-1970, a new triennial comparison of electrical units took place at BIPM (central date: 1 February 1970). The results are given in Table I (2). Also given are the 1 January 1969 changes which were made by the various countries in their as-maintained electrical units in order to bring them into better agreement with their absolute or SI definitions. Throughout the present work, all quantities requiring electrical units will be expressed in post 1 January 1969 units. When deemed necessary, an uncertainty of order 0.1 ppm will be assumed for the measured differences between the units as-maintained by BIPM and those maintained by the various countries.

TABLE I

Relation between the units of emf and resistance as maintained by various countries and BIPM as of 1970, and the changes made by the various countries in their as-maintained units of emf and resistance on 1 January 1969. [$X_{LAB} = X_{BIPM} + \Delta\mu X$, and $X_{69} = X_{old} + \Delta\mu X$, where X = V or Ω. See Ref. 3 for details concerning the USSR volt which was actually changed on 1 January 1970.]

		1970 BIPM Comparison		1 January 1969 changes	
Lab	Country	ΔμV	ΔμΩ	ΔμV	ΔμΩ
DAMW	E. Germany	2.49	2.10	0	0
PTB	W. Germany	-0.26	0.33	-10.4	-5.1
NBS	U.S.A.	0.17	0.03	-8.4	0
NSL	Australia	0.00	0.29	-16.2	3.8
NRC	Canada	0.10	-0.47	-8.0	2.7
LCIE	France	0.23	0.30	-6.1	12.2
IEN	Italy	0.04	----	-10.1	0
ETL	Japan	0.51	-0.19	-8.3	0
NPL	Great Britain	0.69	0.31	-13.0	3.7
IMM	U.S.S.R.	2.16	-0.01	-16.0	0
BIPM	-------	0	0	-11.0	0

(2) Velocity of Light, c. The value given in Table II is the one adopted by IUGG, URSI, and IAU. The value reported more recently by Simkin et al (12), $c = 2.9979256(11)\times10^{8}$ m/s (0.37 ppm) was measured via the same method and is essentially identical. The geodimater results (1) are not included because they too are in good agreement with the Froome measurement and because of the difficulty in reliably assessing the systematic uncertainties present in the base-line measurements.

(3) Ratio of Absolute Ohm to NBS Ohm, Ω/Ω_{NBS}. $c^2\Omega/\Omega_{NBS}$ follows from Thompson's 1969-1970 measurements of the NSL ohm via the calculable capacitor method and the data of Table I. A 0.2 ppm uncertainty (13) was assumed for the random and systematic contributions to the absolute ohm mechanical and electrical measurements, and a 0.1 ppm uncertainty for each laboratories' transfer with BIPM. (Recall $c^2\Omega/\Omega_{NBS}$ is independent of c.) We choose at present to continue Taylor et al's practice of working in NBS electrical units because of the number of fundamental constants experiments which have been carried out in terms of these units. In addition, the other basic arguments given in Ref. (1) in support of these units are still valid.

(4) Acceleration Due to Gravity, g. The g values given at the Commerce Building site (CB) and at the British Fundamental Station (BF) are the final results of the measurements of Faller and Hammond (6) and include some allowance for transfer uncertainties

(0.03 mgal per transfer; 1 mgal = 10^{-5} m/s^2). The deviation of the (old) Potsdam system gravity net from the absolute gravity net, 13.811±0.070 mgal, is based on the more accurate measurements of Sakuma at Sevres A. Sakuma (14) reports g(SA) = 980925.9490 ± 0.0054 mgal, in excellent agreement with Faller and Hammond's result, g(SA) = 980925.960±0.041 mgal. Note that g_P (Kharkov) = the value of g at Kharkov on the Potsdam system. As suggested in Ref. (1), a one ppm uncertainty in g(Kharkov) is assumed.

(5) Magnetic Moment of the Electron in Units of the Bohr Magneton, μ_e/μ_B. The quantity μ_e/μ_B follows directly from the new experimental determination by Wesley and Rich (7) of the anomalous moment of the electron. The latter is in good agreement with the present theoretical result as recently given by Levine and Wright (15).

TABLE II
Summary of the more precise data

Author, Reference Publication Date	Quantity and Units	Value	Uncertainty (ppm)
Froome (4), 1958	c; 10^8 m·s^{-1}	2.9979250(10)	0.33
Thompson(5), 1971	$c^2\Omega/\Omega_{NBS}$ 10^{16} m^2·s^{-2}	8.9875566(22)	0.24
	Ω/Ω_{NBS}	1.00000026(71)	0.71
Faller, Hammond (6), 1971	g(CB);mgal	980104.234(70)	0.071
	g(BF);mgal	981181.860(50)	0.051
	g(Kharkov);mgal	g_P(Kharkov)-13.811	1.0
Wesley, Rich(7),1971	μ_e/μ_B	1.0011596577(35)	0.0035
Kleppner(8),1971	μ_p/μ_B	0.001521032211(17)	0.011
Lambe(9);1959	μ'_p/μ_B	0.00152099322(10)	0.066
	$\sigma(H_2O)$;ppm	25.637(67)	0.067
Gove, Wapstra(10), 1971	M_p^*	1.007276485(14)	0.014
	$1 + m_e/M_p$	1.000544617(10)	0.010
	$1 + m_e/M_d$	1.0002724439(5)	0.005
	$1 + m_e/M_\alpha$	1.0001370934(3)	0.003
Taylor, Parker, Langenberg(1), 1969	R_∞; 10^7 m^{-1}	1.09737312(11)	0.10
Finnegan, Denenstein, Langenberg(11), 1971	2e/h;THz/V_{NBS}	483.593718(60)	0.12

(6) Magnetic Moment of the Proton in Units of the Bohr Magneton, μ_p/μ_B. The value of μ_p/μ_B has been derived from the measurement of $g_j(H)/g_p(H)$ by Kleppner and co-workers(8) after

T

correction for bound state effects according to the recent theory of Grotch and Hegstrom (16). The latter is substantiated by the good agreement found between the theoretical and experimental values for the hydrogen-deuterium g factor ratios (16). The value for μ'_p/μ_B (recall the prime means for protons in a spherical sample of H_2O) is derived from the measurement of $g_j(H)/g_p(H_2O)$ by Lambe (9), and the bound state corrections of Grotch and Hegstrom. (Note that μ_e/μ_B is required to obtain both μ_p/μ_B and μ'_p/μ_B from the measured g factor ratios). The diamagnetic shielding correction for protons in water, $\sigma(H_2O)$, follows from the ratio of the two magnetic moments. It has been assumed throughout that the Grotch-Hegstrom corrections to $g_j(H)$ and $g_p(H)$ have an uncertainty of $3/10^9$ due to additional but as yet uncalculated higher order terms (15).

(7) Atomic Masses and Mass Ratios. The atomic masses and mass ratios given are calculated as in Ref. (1) but with the following modifications: (a) Anticipating the result of our adjustment, we use for μ'_p/μ_n, 2.792775 with a liberally assigned uncertainty of about 20 ppm. Thus we find $M_p/m_e = (\mu'_p/\mu_n)/(\mu'_p/\mu_B) =$ 1836.152 (33). (b) The required values for the atomic masses of H, D and He were taken from the new, 1971 Mass Tables of Wapstra and Gove (10), but from those tables in the appendix which include the recent data of Smith (17). [The appendix values are M^*_H = 1.00782505(1), M^*_D = 2.01410182(2), and M^*_{He} = 4.00260294(27), which may be compared with Smith's actual measured values, M^*_H = 1.007825029(5), and M^*_D = 2.014101771(10).] While there are discrepancies between some of the older data and those of Smith, they are only of order 0.2 ppm and can be ignored for the moment.

(8) Rydberg Constant. The value for R_∞ is that adopted by Taylor et al (1). It includes the fairly recent result of Csillag (18) from the Balmer series in deuterium, R_∞ = 109737.307 ± 0.007 cm^{-1} (uncertainty expanded from 0.003 cm^{-1} for possible systematic effects), and the result of Taylor et al's detailed analysis of the older data, R_∞ = 109737.317±0.007 cm^{-1}. Although the recent result of Masui (19) from a two beam interferogram of Hα radiation, R_∞ = 109737.327±0.004 cm^{-1}, might indicate this adopted value is too low, we choose not to change it at this time because of the large uncertainties inherent in all of the Rydberg constant measurements to date (1,20), and the fact that at least two measurements (to be discussed at this Conference) are currently underway in order to significantly reduce these uncertainties.

(9) Josephson Effect Value for 2e/h. We use for the present the value of 2e/h obtained by Finnegan et al (11) because of its high accuracy and because it was measured in terms of NBS units. It is in excellent agreement with the pioneering but less accurate measurements of Parker et al (21), and in good agreement (when realistic allowances are made for interlaboratory volt transfer uncertainties) with the three other sub-ppm measurements in the literature; NPL (22): 483.59413(15) THz/V_{NPL} (0.31 ppm); NSL (23): 483.59384(5) THz/V_{NSL} (0.10 ppm); and PTB (24): 483.5937(2) THz/V_{PTB} (0.41 ppm). When converted to NBS units via the data of Table I (no uncertainty assumed) they become (in units of THz/V_{NBS}):

NPL	NSL	PTB
483.59388(15)	483.59392(5)	483.59391(20)

Only the NSL result would appear to be inconsistent with the Finnegan et al value. However, the preliminary result of a recent direct intercomparison of the NSL and NBS volts via a shippable, temperature regulated standard cell enclosure indicates that V_{NSL}/V_{NBS} is closer to 1.0000004 than to the value 0.99999983 implied by the data of Table I. While removing the discrepancy, it points up the fact that it is very difficult to compare different 2e/h measurements because of the uncertainties in the relationships between the different as-maintained volts. Similar direct comparisons are also being carried out between NBS, NPL and PTB and should provide sufficiently accurate results to enable all of the 2e/h measurements (including the new values to be reported at this Conference) to be pooled together. This will be done in the future by carrying out a least-squares adjustment on the sub-set of data consisting of the various 2e/h measurements and the measured differences in the units of voltage maintained by the laboratories involved. The output of such an adjustment will be a single value of 2e/h (expressed in one of the laboratories units) and the difference between the units of voltage in question.

B. The Less Precise Data

Table III gives the less precise data we shall consider, but which does not require quantum electrodynamic (QED) theory for its analysis. (Note that all of the data discussed so far also fits this category.) We follow Taylor et al's practice of categorizing the data in this way in order to be able to derive a WQED set of constants (without quantum electrodynamic theory) for the purpose of unequivocally testing QED. [See Ref. (1) for a detailed discussion of this approach.] Table IV will give the less precise QED data.

Table III
Summary of the less precise WQED data

Author, Reference Publication Date	Value	Uncertainty (ppm)
1. Ratio of the NBS Ampere to the Absolute Ampere, $K \equiv A_{NBS}/A$		
1.1 Driscoll, Olsen, 1968	1.0000017(97)	9.7
1.2 Driscoll, Cutkosky, 1958	1.0000008(77)	7.7
1.3 Vigoureux (25), 1965,1970	1.0000005(60)	6.0
2. Faraday Constant, F ($10^7 A_{NBS} \cdot s \cdot kmole^{-1}$)		
2.1 Craig, Hoffman, Law Hamer, 1958	9.648651(66)	6.8
3. Proton Gyromagnetic Ratio, γ_p'(low field)($10^8 Hz/T_{NBS}$)		
3.1 Driscoll, Olsen, Bender 1958-1968	2.6751301(99)	3.7
3.2 Vigoureux, 1962	2.675122(16)	5.8
4. Proton Gyromagnetic Ratio, γ_p'(high field)($10^8 Hz/T_{NBS}$)		
4.1 Yagola, Zingerman, Sepetyi, 1962-66	2.675127(20)	7.4
5. Magnetic Moment of the Proton in Units of the Nuclear Magneton, μ_p'/μ_N		
5.1 Sommer, Thomas, Hipple, 1949-51	2.792711(60)	21
5.2 Sanders et al, 1955-63	2.792701(73)	26
5.3 Boyne, Franken, 1961	2.792832(55)	20
5.4 Mamyrin, Frantsuzov, 1965	2.792794(17)	6.2
5.5 Petley, Morris (26), 1970	2.792754(23)	8.2
5.6 Fystrom (27), 1970	2.792783(16)	5.7
5.7 Gubler et al (28), 1970	2.792771(36)	13
6. Shortwavelength Limit, hc/e ($V_{NBS} \cdot kxu$)		
6.1 Spijkerman, Bearden, 1964	12373.25(41)	33
7. Avogadro Number, $N\Lambda^3$ ($10^{26} kmole^{-1}$)		

7.1 Bearden, 1965	6.05972(23)	37
7.1 Henins, Bearden, 1964	6.059768(95)	16
8. X-Unit Conversion Factor, Λ.		
8.1 Bearden, 1931	1.002030(38)	38
8.2 Henins (29), 1970	1 0020655(100)	10
9. Electron Compton Wavelength, $\lambda_C(10^{-3}$kxu)		
9.1 Knowles, 1962	24.21263(92)	38
9.2 Knowles, 1964	24.21417(36)	15
9.3 Van Assche et al (30), 1970	24.21317(77)	32

The following general comments apply to these numbers. (1) Except when otherwise indicated, the values given are taken from Ref. (1). The later should also be consulted for the appropriate original references, since here we only list the papers which do not appear in the bibliography of Ref. (1). (2) Quantities involving the electrical units have been reexpressed in terms of post 1 January 1969 units using the data of Table I. (3) Quantities requiring a value for the local gravitational acceleration [K, γ_p'(high field)] have been recalculated using the data of Table II. However, since the preliminary and final results of Faller and Hammond differ by less than 0.1 ppm, and since Taylor et al used the preliminary results, any changes are negligible. Similarly, Taylor et al used 13.810 mgal rather than 13.811 mgal to compute g(Kharkov). (4) We follow Ref. (1) and use the x-unit scale defined by $\lambda(\text{CuK}\alpha_1) \equiv$ 1.537400 kxu. The following specific comments also apply to the numbers given in Table III.

(1.3) In 1969-70, Vigoureux (25) remeasured the NPL Ampere using the NPL current balance. He found $A_{NPL}/A = 1.0000023$. This may be compared with his result reported in 1965(actually carried out in 1962-63), 1.0000004, where the latter has been obtained from the value given by Taylor et al (1), 1.0000170, and the data of Table I. The 1.9 ppm difference between the 1965 and 1970 values is not inconsistent with the approximate 0.7 ppm statistical standard deviation of each of the measurements. (The systematic uncertainties common to both are about 8 times larger). This is especially true when one considers that the dimensions of the balance coils were not remeasured during the course of the 1969-70 experiments. Rather, the dimensions obtained at the time of the earlier measurements were used. (A comparison of the calculated and measured differences of the forces exerted by the two coil systems of the balance did indicate that the coil dimensions could not have changed significantly. Minor improvements involving the beam suspension and scale pans, and a test of the symmetry of the balance beam, were also carried out.)

In view of the fact that no new dimensional measurements were made during 1969-70, we choose not to discard the older result in favor of the new one. On the other hand, the new work should not be ignored entirely since it does represent a significant amount of effort. Thus, we weight the old and new results in the ratio 2:1 to obtain the value given in Table III. [Since the uncertainty in the experiment arises primarily from systematic effects, it remains unchanged and is as given in Ref. (1).]

2.1 The new values for the atomic masses of ^{107}Ag and ^{109}Ag as given by Wapstra and Gove (10) [but with Smith's new data (31)] were used to obtain a new value for the atomic mass of silver [$M^*(Ag) = 107.86833(23)$(2.1 ppm)] and hence a revised value for the faraday. However, the change in F from this source is entirely negligible.

3.1 A new improved value of γ_p' obtained at NBS should be available at the time of this Conference.

3.2 Vigoureux is also currently undertaking a new measurement of γ_p' at NPL and will report on its progress at this Conference. We note that we do not include in Table III the results of Hara et al (1), and Studentsov et al (1) for the reasons given in Ref. (1).

4.1 In 1970, Kibble and Hunt (31) reported the result of a high field γ_p' experiment at NPL. They found $\gamma_p' = 2.675075(43) \times 10^8$ Hz/T_{NPL} (16 ppm), which becomes via the data of Table I, $\gamma_p' = 2.675076(43) \times 10^8$ Hz/T_{NBS} (16 ppm.) This result differs from the other γ_p' values in Table III by about 1.2 to 1.3 standard deviations, which is quite reasonable. However, we do not consider this new value because Kibble and Hunt (32) have emphasized that their result must be regarded as highly preliminary since it was obtained with a prototype apparatus constructed for the express purpose of gaining experience with the general technique.

5.1 This is the revised result recently given by Fystrom, Petley, and Taylor (33), which is based on their study of the original data notebooks used in the Sommer, Thomas, and Hipple experiment. As Fystrom et al suggest, we adopt the uncertainty originally given by Sommer et al interpreted as a standard deviation.

5.4 A new determination by this group is scheduled to be reported at this Conference.

5.5 The 1967 result reported by Petley and Morris (34), $\mu_p'/\mu_N = 2.792746(52)$ (19 ppm), is not included since their more accurate 1970 result (Table III) obtained with improved apparatus, replaces it. A report on this group's most recent work is scheduled for this Conference.

5.6 This work was first reported after the Taylor et al review appeared. Fystrom's experiment employed an Omegatron similar to that used by Sommer et al but with several important refinements. The measurements were carried out in a field of 1.6 T.

5.7 The 1967 Marion and Winkler (35) reaction energy result, $\mu_p'/\mu_N = 2.79260(13)$ (45 ppm), has not been included since it has been replaced by the closely related velocity gauge-magnetic analyzer result reported in 1970 by Gubler et al (Table III). At the time their value was reported (Summer 1970), Gubler et al noted that several systematic effects were under investigation. However, they did not believe their value would change by more than 15 ppm. To take this possibility into account, we add one half of 15 ppm root-sum-square to their originally quoted 11 ppm uncertainty, thereby obtaining the uncertainty given in Table III. While this procedure will suffice for the present preliminary work, additional information will be required before the Gubler et al result can be seriously considered for inclusion in any final adjustment.

8.2 This new measurement of Λ by Henins was carried out by determining the absolute wavelength of the $AlK\alpha_{1,2}$ unresolved doublet via a plane grating, and by measuring the same wavelength with respect to $CuK\alpha_1$ via a crystal spectrometer. (The original 7ppm probable error given by Henins was converted to a standard deviation by multiplying by 1.48).

9.1 We include this result for completeness only (no real use will be made of it) since Knowles has emphasized (36) that it was at best a preliminary experiment. [The calcite lattice spacing was never measured for both calcite crystals used in the experiment; Bearden's measurement of a few mm^3 of one crystal (total diffraction volume 62.5 cm^3) cannot be assumed representative of the average crystal.] Item 9.2 is the only measurement of λ_C by Knowles which may be taken at all seriously (36).

9.2 As suggested in the paper of Van Assche et al (30), the result given in Ref. (1) has been reduced by 1.3 ppm due to the difference between $\lambda(WK\alpha_1)$ generated via naturally occurring W and $\lambda(WK\alpha_1)$ when it is produced in the decay of ^{182}Ta to ^{182}W.

9.3 This new result was obtained in a manner quite similar to that for item 9.2, i.e., by comparing the annihilation radiation to the $WK\alpha_1$ line generated in the decay of ^{182}Ta to

^{182}W via a bent crystal diffraction spectrometer. (We have also reexpressed Van Assche et al's original result in terms of the scale defined by $\lambda(\mathrm{CuK}\alpha_1)$ = 1.573400 kxu.)

Table IV
Summary of the QED data

	Author, Reference Publication Date	Value	Uncertainty (ppm)
10.	Ground State Hyperfine Splittings (MHz)		
10.1	Hhfs, Hellwig et al (37), 1970	1420.405751768(2)	1.4×10^{-6}
10.2	Mhfs, De Voe et al (38), 1970 (Chicago)	4463.3022(89)	2.0
10.3	Mhfs, Crane et al (39), 1971 (Yale)	4463.311(12)	2.7
11.	Electron g Factor Anomaly, a_e		
11.1	Wesley, Rich (7), 1971	0.0011596577(35)	3.0
12.	Fine Structure (MHz)		
12.1	ΔE_H, Metcalf et al, 1968	10969.127(95)	8.7
12.2	$(\Delta E\text{-}S)_H$, Cosens, Vorburger (40), 1970	9911.173(42)	4.3
12.3	$(\Delta E\text{-}S)_H$, Shyn et al (41), 1971	9911.250(63)	6.4
12.4	$(\Delta E\text{-}S)_H$, Kaufman et al, 1969	9911.377(26)	2.6
12.5	S_H, Robiscoe, Shyn (1,42) 1970	1057.896(63)	60
12.6	S_D, Cosens, 1968	1059.282(64)	60

The first general comment concerning the numbers in Table III applies to these numbers as well. The following specific comments also apply.

10.1 This is the most recent hydrogen maser measurement of ν(Hhfs). While it is $18/10^{12}$ less than the value of Vessot et al given in Ref. (1) (some 15 times the standard deviation assigned the later value), the change is of course entirely negligible.

10.2 This is the last published result of the Chicago group and was obtained using their double resonance technique at the "magic" field 1.135 T. [A more recent and accurate value using a different method will be reported shortly (43).]

10.3 The value given here has just been reported by the Yale group and has been obtained by extrapolation to zero pressure using a quadratic term in the equation relating ν(Mhfs) to the pressure of the stopping gas. The experiments were carried out in a magnetic field of 1μT and a gas pressure of 10 atmospheres.

To obtain a value of the fine structure constant from ν(Mhfs) requires knowledge of the two quantities μ_μ/μ_p and m_μ/m_e. μ_μ/μ_p has been measured most accurately by the University of Washington-Lawrence Radiation Laboratory group [Hague et al (44)] by stopping muons in various chemical environments in a magnetic field of 1.1T. [The actual quantity determined is ω^*_μ/ω'_p, the ratio of the muon to proton precession frequency, the former in the environment in question (indicated by the asterisk) and the latter for protons in H_2O.] Varying the environment is of great importance since Ruderman (45) has proposed that the formation of the complex $(H_2O\text{--}\mu\text{--}H_2O)^+$ could reduce the muon shielding (compared with the 25.637 ppm for protons in H_2O) by 15 to 20 ppm. However, Hague et al in their determinations found no evidence of the Ruderman effect. Specifically, this was shown by measuring ω^*_μ in H_2O and NaOH. (In 0.1N NaOH, the μ^+ would be neutralized in $<10^{-10}$ s, suppressing the formation of the complex.) No significant difference was observed, nor was a significant difference observed when the muons were stopped in methylene cyanide, $CH_2(CN_2)$. The final result of all of their measurements is given as

$$\omega_\mu/\omega_p = \mu_\mu/\mu_p = 3.183347(9) \quad (2.8 \text{ ppm}). \qquad (1)$$

[In obtaining this value from ω^*_μ/ω'_p, Hague et al took into account the small chemical shifts (i.e., shielding) arising from the difference in zero point binding energies of proton and muon in the various possible molecular species (e.g., μHO, μH, etc). Hague et al estimate -1.8 ± 2.0 ppm in H_2O and NaOH, and + 0.5 ± 1.5 ppm in $CH_2(CN_2)$, all relative to protons in H_2O.]

Equation (3) is in excellent agreement with the similar result of Hutchinson et al (46), $\omega'_\mu/\omega'_p = 3.183362(30)$ (9.4 ppm), which becomes after applying the -1.8 ± 2.0 ppm correction of Hague et al,

$$\mu_\mu/\mu_p = 3.183356(31) \quad (9.6 \text{ ppm}). \qquad (2)$$

Equations (3) and (4) are also consistent with the result of De Voe et al (38) obtained from their ν(Mhfs) double resonance experiment. The later workers actually obtain the quantity $g_j(M)/g'_\mu$ where $g'_\mu = 2\mu_\mu/\mu_B$. To obtain g_s/g'_μ from which μ_μ/μ_p

may be derived, some assumption has to be made about the pressure shift of $g_j(M)$. Following a suggestion of T. Herman that $g_j(M)$ should be decreased by 11 ppm under their experimental conditions De Voe et al obtain $\mu_\mu/\mu_p = 3.183337(13)$, where no uncertainty allowance for the pressure shift is made.

We shall for the present use only the result of Hague et al because of its high accuracy compared with that of Hutchinson et al, and the yet unresolved pressure shift problem associated with the result of De Voe et al.

Following Ref. (1), we obtain m_μ/m_e from the equation

$$m_\mu/m_e = \frac{(g_\mu/g_s)\,(\mu_e/\mu_p)}{(\mu_\mu/\mu_p)} = 206.76821(58) \qquad (2.8 \text{ ppm}) \;, \tag{3}$$

where we have used the Wesley-Rich value for g_s, the CERN result (47) for g_μ, $[g_\mu/2 = 1.00116616(27)\ (0.27\ \text{ppm})]$, and a value for μ_e/μ_p derived from the Kleppner measurement of $g_j(H)/g_p(H)$ and the Grotch-Hegstrom (16) bound state corrections. $[\mu_e/\mu_p = 658.2106878(72)\ (0.011\ \text{ppm})]$.

11.2-11.6 These are the final values as given in the quoted reference and differ somewhat from the preliminary values given in the "Notes Added in Proof" Section of Ref. (1). [See this Section for items 11.5 and 11.6 as well.] We have not listed any of the results obtained in the early 1950's by Lamb and co-workers [see Ref. (1)] because of their age, the obvious incompatability of some of it with the more recent results, and the difficulty of assessing the implication of the kinematic correction of Robiscoe and Shyn (42) for Lamb's measurements of S_H and S_D.

12.1 Using the recent theoretical result for a_e given by Wright and Levine (15),

$$a_e(\text{theory}) = \frac{1}{2}\frac{\alpha}{\pi} - 0.328479\left(\frac{\alpha}{\pi}\right)^2 + (1.49 \pm .20)\left(\frac{\alpha}{\pi}\right)^3 \,, \tag{4}$$

we find that the Wesley-Rich experimental value implies

$$1/\alpha(a_e) = 137.03581(51)\ (3.7\ \text{ppm}). \tag{5}$$

However, no other group has checked Wright and Levine's calculation of the sixth order term. (Actually, only 28 of the 40 distinct graphs contributing to $a_e^{(6)}$ were calculated by Wright and Levine since the contribution of the other 12 had previously been obtained by other workers.) Consequently, while it is used here for our preliminary work, it may be prudent to wait for confirmation of their calculation before considering $1/\alpha(a_e)$ for inclusion in any final adjustment.

We do not include the old Crane value for a_e [as revised by Rich--see Ref. (1)], because it is an order of magnitude more uncertain than the Wesley-Rich result and because of the difficulty of correcting a result a posteriori.

3. Analysis of Data

A. Inconsistencies Among Data of the Same Kind

Table I. As noted in our discussion of the 2e/h data, there may be some difficulties at the 0.5 ppm level with some of the volt differences implied by the data in Table I. This problem will hopefully be resolved when the direct transfers being carried out by NBS are completed.

Table II. Most of the data in this table have already been discussed via the comments following it. In summary, there are no discrepancies between these data and other measurements of the same quantities which exceed 0.1 to 0.2 ppm. This is entirely negligible for our purposes. [See Ref. (1) for further discussion.]

Table III. It is apparent that the absolute ampere data are quite consistent among themselves, as are the two low field γ_p' measurements. This is also true of the two $N\Lambda^3$ values, and the two values of Λ. Other values for these quantities which are not listed, e. g., the new Kibble-Hunt high field γ_p' result, and the values given in Ref. (1), are also in reasonable agreement with these numbers. [See Ref. (1) for further details.]

The μ_p'/μ_N situation has improved considerably since the Taylor et al review due to the revision of the Sommer et al omegatron result and the new measurements of Petley and Morris, Fystrom, and Gubler et al. A least squares adjustment (i.e., a weighted average) of items 5.1 through 5.7 yields a Birge ratio of 0.95, which is quite acceptable. The largest normalized residual is 1.14 for item 5.1.

A questionable area is (as it was for Taylor et al) λ_C. As noted, item 9.1 should be ignored, but yet this result is in better agreement with the new value of Van Assche than it is with Knowles' own more accurate result, item 9.2. Furthermore, both items 9.1 and 9.2 are consistent with the Josephson effect value of α and the value of Λ implied by the $N\Lambda^3$ data. [See Ref. (1).] Fortunately, the λ_C problem is probably of little concern at present because of the large uncertainties in the experiments; they have little impact on determining a best value of α or Λ and can be ignored in any final adjustment. For the present, they are retained for purposes of investigation.

Table IV. The Hhfs is known to such high accuracy that any discrepancies in the last few digits from measurement to measurement are irrelevant. The two muonium hfs measurements are in excellent agreement, as are the three values of μ_μ/μ_p (See the comments following Table IV).

The three values of $(\Delta E - S)_H$, 12.2, 12.3, and 12.4, are somewhat inconsistent. While the difference between 12.2 and 12.3 is about one standard deviation of the difference, the difference between 12.4 and 12.2 is 4 standard deviations of the difference. The difference between 12.4 and 12.3 is 1.9 standard deviations. (The Birge ratio of the weighted average is 3.0, but reduces to 1.0 when 12.4 is deleted). The cause of this discrepancy is at present unknown, and may imply that the uncertainty assigned 12.4 is unrealistically low.

B. Inconsistencies Among Dissimilar Data

Table III. The values of K, γ'_p (low) and γ'_p (high) are obviously all in good agreement when one realizes that $K^2\gamma'_p$ (high) = γ'_p (low), and notes that the weighted mean of the two low field values exceeds the high field value by only 0.11 ppm.

The big problem [as in Ref. (1)] is the inconsistency between the faraday F and the μ'_p/μ_N data. [Recall $(\mu'_p/\mu_N) = M^*_p\,\gamma'_p\,(\text{low})/FK^2$; omit K^2 if γ'_p (low) is replaced by γ'_p (high).] The adjusted value for μ'_p/μ_N using items 1.1 through 4.1 is 2.792713(25) (8.9 ppm), in considerable disagreement with the weighted average of 5.1 through 5.7, 2.7927787(96) (3.4 ppm). If the adjusted value is included in the weighted average, the average becomes 2.7927702(90) (3.2 ppm), with a Birge ratio of 1.28. The normalized residual of the adjusted value in this average is 2.31 and accounts for nearly half of the χ^2. Unfortunately, the x-ray data are too uncertain to contribute to the resolution of this discrepancy, although if anything, they do tend to favor F.

Table IV. All of these data may be compared by deriving values of the fine structure constant via the appropriate theoretical equations. The results are

10.1	$1/\alpha$(Hhfs)		=	137.03591(35)	(2.6 ppm)
10.2	$1/\alpha$(Mhfs)	[Chicago]	=	137.03636(24)	(1.7 ppm)
10.3	$1/\alpha$(Mhfs)	[Yale]	=	137.03622(27)	(1.9 ppm)
11.1	$1/\alpha$(ae)		=	137.03581(51)	(3.7 ppm)
12.1	$1/\alpha(\Delta E)$		=	137.03546(59)	(4.3 ppm)
12.2	$1/\alpha(\Delta E\text{-}S)$		=	137.03574(31)	(2.3 ppm)
12.3	$1/\alpha(\Delta E\text{-}S)$		=	137.03519(46)	(3.4 ppm)

12.4	$1/\alpha(\Delta E\text{-}S)$	=	137.03427(20)	(1.5 ppm)
12.5	$1/\alpha(SH)$	=	137.0363(30)	(22 ppm)
12.6	$1/\alpha(SD)$	=	137.0350(30)	(22 ppm)
	$1/\alpha(WQED)$	=	137.03607(21)	(1.5 ppm)

For purposes of comparison, we also give the value of $1/\alpha(WQED)$ which is representative of the data of Tables II and III. This value, which arises primarily from the Josephson effect measurement of 2e/h, changes only a few tenths ppm depending on whether the x-ray data are included, how the faraday- μ_p/μ_N discrepancy is handled, etc.

In deriving the above values, we have followed Ref. (1), but the new corrections to the theoretical expression for the Hhfs and Mhfs (48,49) have been included, as well as the recent corrections to the theoretical expression for the Lamb shift (50). With the advent of the latter calculation and its high accuracy, i.e., 0.011 MHz in S_H, one standard deviation, it becomes appropriate to disregard the experimental value of S_H and derive a value of the fine structure constant directly from the theoretical expression for $(\Delta E\text{-}S)_H$. We follow this approach throughout. [Recall how the discrepancy between theory and experiment was discovered to arise from a miscalculated term (51). Reference (50) summarizes the situation up to the present time.]

It is clear from the derived α values that all of the QED data are reasonably consistent except for item 12.4, the Kaufman et al $(\Delta E\text{-}S)_H$ measurement. The weighted average of all of the values [excluding $\alpha(SH)$ and $\alpha(SD)$] is 137.03556(10) (0.74), with a Birge ratio of 2.9 (53). (The residual of item 12.4 is 6.3). Deleting 12.4 shifts the average to 137.03598(12) (0.85 ppm) and the Birge ratio to 1.08. Deleting both 12.3 and 12.4 yields 137.03603(12) (0.88 ppm) and R = 0.87. We note without comment that Wyler's theoretical value (52) for $1/\alpha$ is 137.03608245.

C. Analysis of Data via Least Squares Adjustments

We now turn our attention to data analyses via complete least squares adjustments involving all of the data. There are several ways in which this may be done. One approach is to take 2e/h as an auxiliary constant as well as $c^2\Omega\ \ /\Omega_{NBS}$ (the uncertainties in these quantities are respectively 0.12 and 0.24 ppm), and to take c as an adjustable parameter so that (a) the 2e/h data and QED data remain uncorrelated, and (b) the effect of the comparitively large uncertainty in c may be taken into account. If this is done, then the adjustable constants are $1/\alpha$, K, $N\Lambda$, c. But since the Froome value for c has become universally adopted(i.e., by URSI, IUGG, and IAU), one would hardly want to produce a set of recommended values of the constants which did not hold to this adopted value unless there was strong motivation to do so, such as a change in the value of c by more than two standard deviations.

No such variation is indicated by the existing data. To circumvent this difficulty, the adjusted solution can be constrained so that c remains at its input value, i.e., the adopted one. (The actual mechanics of how this may be accomplished will be discussed in a future publication).

An alternate approach is to take α^{-1}, e, K, N, and Λ as the adjustable constants [as in Ref. (1)], use $[c\Omega/\Omega_{NBS}]$ (2e/h) as an input datum with an uncertainty equal to the root-sum-square of the uncertainties of the individual quantities (i.e., 0.43 ppm), and ignore the effect of the uncertainty of c on the QED data since it is an order of magnitude less than these uncertainties. In any event, the two approaches yield essentially the same results and may be considered equivalent for our present purposes.

Before stating the results of adjustments involving all of the data listed, it should be pointed out that three items may be discarded at the start: item 8.1, the Bearden value of Λ, because of its age and very large uncertainty compared with the new result of Henins, item 8.2; item 9.1 because, as noted previously, it cannot be seriously considered as a determination of a fundamental constant (36); and item 9.2 because it is so extremely discrepant with all of the other data-its normalized residual is typically in excess of 3 in any reasonable adjustment (1).

With these deletions, we find that an adjustment involving all of the listed data yields χ^2 = 74.0 for 22 degrees of freedom, and a Birge ratio of 1.83. The single most discrepant item is 12.4, the Kaufman et al measurement of $(\Delta E\text{-}S)_H$. It accounts for over half of χ^2. Deleting it reduces χ^2 to 21.5 for 21 degrees of freedom with a Birge ratio of 1.01. The most discrepant remaining item is then 12.3, the Shyn et al measurement of $(\Delta E\text{-}S)_H$ (residual of 1.7). Deleting it reduces χ^2 to 18.5 for 20 degrees of freedom and a Birge ratio of 0.96. Clearly, the data are reasonably compatible with or without the Shyn et al result.

The largest remaining strain in the system (after 12.4 and 12.3 are deleted) arises, as would be expected, from the faraday and μ_p'/μ_N data. Deleting the former yields a χ^2 of 13.6 for 19 degrees of freedom and a Birge ratio of 0.85. If the faraday is included but all of the μ_p'/μ_N data deleted, then χ^2 = 6.0 for 13 degrees of freedom with a Birge ratio of 0.68. This reduction reflects the fact that there are minor inconsistencies among the μ_p'/μ_N data themselves and that the x-ray data are a bit more compatible with F than with the μ_p'/μ_N data.

We conclude this section by noting that the range of $1/\alpha$ for all of these adjustments is 137.03555(10) (0.74 ppm) to 137.03604(12) (0.89 ppm). With the deletion of items 12.3 and 12.4, the range narrows to between 137.03602(12) (0.88 ppm) and the 137.03604 value.

4. Summary and Conclusions

Although much new data has become available since the review of Taylor et al, the dominant problem present at that time is still present today, namely, the inconsistency of the Craig et al value for the faraday and the various μ_p'/μ_N measurements. One solution to the problem would be to expand the errors of the quantities involved (F and μ_p'/μ_N) so as to make them consistent. A factor of 1.3 would accomplish this. By contrast, the additional new measurements of μ_p'/μ_N and the revision of the Sommer et al result have now removed the internal inconsistencies among the μ_p'/μ_N data which were present in 1969.

The other problem which faced Taylor et al concerned the QED data. These authors believed that only the hydrogen hyperfine splitting and its theoretical equation were sufficiently reliable to warrant inclusion in a least squares adjustment to obtain a set of best or recommended constants. The situation has of course improved considerably since then with the resolution of the Lamb shift discrepancy by Applequist and Brodsky, the improved measurement of a_e by Wesley and Rich along with the calculation of the $a_e^{(6)}$ term in the theoretical expression for a_e by Wright and Levine, the improved muonium hyperfine splitting measurements and determinations of μ_μ/μ_p, and finally, the three new measurements of $(\Delta E\text{-}S)_H$. Of all of these results, only the measurement of the latter quantity by Kaufman et al would seem discrepant. Thus, serious consideration must be given to possible inclusion of the new QED data. However, expurgation of the Kaufman et al result would seem in order.

The third most important question facing Taylor et al was how to handle the x-ray data. Because of its overall inconsistency and high uncertainty, it was discarded with the exception of the two $N\Lambda^3$ values given in Table III. This choice completely eliminated the x-ray data from having an effect on the values of the constants other than Λ. Now, however, there are two new x-ray measurements, one of Λ and one of λ_C, which must be considered for possible inclusion in any final adjustment. Since they tend to support the Craig et al faraday, it would seem that some error expansion might be necessary.

Beyond what has just been listed, few other problems would seem to remain; all other difficulties are at the few tenths ppm level which is of little importance at the present level of uncertainty for the majority of the constants of interest. Some of the results scheduled to be reported at this Conference (alluded to throughout the present paper) may change this situation, or hopefully may shed some light on the aforementioned difficulties. Only time will tell.

Finally, for the interest of the reader, we list for the more important constants the anticipated changes from the final recommended values of Taylor et al based on the data presently available. Specifically we, use all of the data listed except for 12.4, 12.5, 12.6, 8.1, 9.1, and 9.2, and we expand no errors. The Birge ratio is 1.01; $\chi^2 = 21.53$ for 21 degrees of freedom.

Constant	Approximate Change, ppm
α^{-1}	-0.4
e	-5
h	-9
m_e	-13
N	-12
M_p/m_e	+22
F	-5
μ_p'/μ_N	+22

The dominant changes of course are due to the inclusion of all of the μ_p'/μ_N data, not simply those values consistent with the faraday as was done in Ref. (1). If the faraday were discarded, then some of the changes would increase significantly, e.g., the change in N would become -21 ppm and the change in F, -26 ppm.

References

* Where applicable, unless the word "observed" is used, a judgment estimated to be equivalent to one standard deviation is meant.

1. TAYLOR, B. N., PARKER, W. H., and LANGENBERG, D. N., Rev. Mod. Phys. 41, 375 (1969).

2. TERRIEN, J., private communication. See also Metrologia 7, 78 (1971).

3. ROZHDESTVENSKAYA, T. B., and SHIROKOV, K. P., Izmeritel. Tekhn. 8, 6 (1970). [Translation: Meas. Tech. 1970, 1120.]

4. FROOME, K. D., Proc. Roy. Soc. (London) A247, 109 (1958).

5. THOMPSON, A. M., private communication, 1971.

6. HAMMOND, J. A., and FALLER, J. E., J. Geophys. Res., in press.

7. WESLEY, J., and RICH, A., Phys. Rev., in press.

8. KLEPPNER, D., Proc. Int. Conf. on Prec. Meas. and Fund. Constants, edited by D. N. Langenberg and B. N. Taylor (NBS Spec. Pub. 343, USGPO, Washington, DC, 1971), p. 411.

9. LAMBE, E. B. D., Ph. D. thesis, Princeton University, 1959 (unpublished); Polarisation, Matière et Rayonnement (Societé Francoise de Physique, Paris), p. 441.

10. WAPSTRA, A. H., and GOVE, N. B., Nuclear Data Tables 9, 265 (1971).

11. FINNEGAN, T. F., DENENSTEIN, A., and LANGENBERG, D. N., Phys. Rev. B4, 1487 (1971).

12. SIMKIN, G. S., LUKIN, I. V., SIKORA, S. V., and STRELENSKII, V. E., Izmeritel. Tekhn. 8, 92 (1967). [Translation: Meas. Tech. 1967, 1018.]

13. THOMPSON, A. M., Metrologia 4, 1 (1968).

14. SAKUMA, A., Proc. Int. Conf. Prec. Meas. and Fund. Constants, edited by D. N. Langenberg and B. N. Taylor (NBS Spec. Pub. 343, USGPO, Washington, DC, 1971), p. 447.

15. LEVINE, M. J., and WRIGHT, J., Phys. Rev. Letters 26, 1351 (1971).

16. GROTCH, H., and HEGSTROM, R. A., Phys. Rev. A4, 59 (1971).

17. SMITH, L. G., Phys. Rev. C4, 22 (1971).

18. CSILLAG, L., Acta Phys. Acad. Sci. Hung. 24, 1 (1968).

19. MASUI, T., Proc. Int. Conf. Prec. Meas. and Fund. Constants, edited by D. N. Langenberg and B. N. Taylor (NBS Spec. Pub. 343, USGPO, Washington DC, 1971), p. 83.

20. SERIES, G. W., ibid., p. 73

21. PARKER, W. H., LANGENBERG, D. N., DENENSTEIN, A., and TAYLOR, B. N., Phys. Rev. 177, 639 (1969); ibid. B1, 4500 (1970).

22. PETLEY, B. W., private communication. This result is his most recent and supersedes his previously reported values.

23. HARVEY, I. K., MACFARLANE, J. C., and FRENKEL, R. B., Phys. Rev. Letters 25, 853 (1970).

24. KOSE, V., MELCHERT, F., FACK, H., and SCHRADER, H.-J., PTB-Mitt. 81, 8 (1971).

25. VIGOUREUX, P. and DUPUY, N., NPL Current Balance 1969/1970, National Physical Laboratory Report QU16, December 1970.

26. PETLEY, B. W., and MORRIS, K., Proc. Int. Conf. Prec. Meas. and Fund. Constants, edited by D. N. Langenberg and B. N. Taylor (NBS Spec. Pub. 343, USGPO, Washington DC, 1971), p 173.

27. FYSTROM, D., Phys. Rev. Letters 25, 1469 (1970). Also, Proc. Int. Conf. Prec. Meas. and Fund. Constants, edited by D. N. Langenberg and B. N. Taylor (NBS Spec. Pub. 343, USGPO, Washington DC, 1971), p. 169.

28. GUBLER, H., et al, ibid., p. 177.

29. HENINS, A., ibid., p. 255.

30. VAN ASSCHE, P. H. M., et al, ibid., p. 271.

31. GOVE, N. P., private communication.

32. KIBBLE, B. P., and HUNT, G. J., ibid., p. 131.

33. FYSTROM, D., PETLEY, B. W., and TAYLOR, B. N., ibid., p. 187.

34. PETLEY, B. W., and MORRIS, K., Proc. Third Int. Conf. Atomic Masses, R. C. Barber, Ed. (University of Manitoba Press, Winnepeg, 1968), p. 461.

35. MARION, J. B., and WINKLER, H., Phys. Rev. 156, 1062 (1967).

36. KNOWLES, J. W., private communication.

37. HELLWIG, H., et al, IEEE Trans. Instr. IM-19, 200 (1970).

38. DE VOE, R. et al, Phys. Rev. Letters 25, 1779 (1970).

39. CRANE, T. et al, Phys. Rev. Letters 27, 474 (1971).

40. COSENS, B. and VORBURGER, T., Phys. Rev. A 2, 16 (1970).

41. SHYN, T. W. et al, Phys. Rev. A 3, 116 (1971).

42. ROBISCOE, R. T., and SHYN, T. W., Phys. Rev. Letters 24, 559 (1970).

43. TELEGDI, V., private communication.

44. HAGUE, J. F. et al, Phys. Rev. Letters 25, 628 (1970).

45. RUDERMAN, M., Phys. Rev. Letters 17, 794 (1966).

46. HUTCHINSON, D. P., et al, Phys. Rev. Letters 24, 1254 (1970).

47. BAILEY, J., et al, Phys. Letters 28B, 287 (1968).

48. FULTON, T., OWEN, D. A. and REPKO, W. W., Phys. Rev. Letters 26, 1351 (1971).

49. COLE, R. S., and REPKO, W. W., Bull. Am. Phys. Soc. 16, 849 (1971).

50. ERICKSON, G. W., Phys. Rev. Letters 27, 780 (1971).

51. APPLEQUIST, T., and BRODSKY, S. J., Phys. Rev. A 2, 2293 (1970).

52. WYLER, A., Compt. Rend. 269, 743 (1969).

53. Because the uncertainty in μ_{μ}'/μ_p is 2.8 ppm, the two muonium alpha values cannot both be included since this 2.8 ppm uncertainty would be counted twice. Thus, we first combine via a weighted average the two values of ν(Mhfs) and then obtain a single value of $1/\alpha$(MHfs): 137.03631(22) (1.6 ppm). This approach will be used throughout as required.

Part 14 Conference Summary

AMCO-4 Conference Summary by Dr. Cohen

We have had a rather full week and, I hope, one which has been interesting to all present, as it has been interesting to me. I will try to review very quickly what we have proven this last week, highlighting those points that seem to be the most important things that have come from the conference. Rather than go chronologically session by session, I will first review the fundamental constants sessions more or less chronologically and then the mass data, since this seems to have more continuity than trying to follow the jumping back and forth from session to session, as the conference itself proceeded.

The conference started out with discussions of the speed of light. I think the most interesting and promising papers there were the ones by Barger and Giacomo which promise ultimately a connection between the frequency of microwaves and the wavelength of visible light, which will tie together these two constants. This opens up the question of what should be done when one can measure both the wavelength and the frequency of a single atomic transition, whether it be optical or microwave. Does this lead us to defining the velocity of light, or if you wish, the equivalent, defining the unit of length in terms of the distance that light will travel in vacuum in a time which is defined by an atomic transition? These questions are certainly important but are perhaps not yet to be discussed, at least not at this conference.

We also heard of the significant improvements in μ_p/μ_N, the magnetic moment of the proton, which I just referred to in my other talk, and we have been given promises of the direct measurement of lattice spacings of crystals by a direct comparison with the krypton wavelength – that word "direct" is apparently a misnomer, it merely implies that we don't have to go through X-ray standards and the uncertainties in the X unit – but as we have seen, the path from lattice spacings to optical wavelengths is not quite as direct (or as easy) as one might hope it to be.

We've also heard about the measurements of the fine structure constant from atomic spectroscopy, both optical and microwave. Dr. Lea gave a very interesting and a very full report of the status of these measurements, some of which, the ones on the lower levels of the atomic system, give us information on the value of α. The transitions to the higher levels then form very precise tests of the validity of quantum electrodynamics. As far as we have been able to measure and as far as we've been able to compute, I think the conclusion to be reached is that quantum electrodynamics is an adequate description of the atomic systems to the extent at least to which we are able to test that theory at the present time.

We have seen significant advances in the measurement of 2e/h to the point where 2e/h as measured in terms of maintained units of the volt is essentially an auxiliary constant and, if we are not yet in a position to define the volt in terms of a Josephson frequency, we are certainly in a position to use the Josephson frequency as a monitor on the permanence and stability of the standard voltage as maintained at national laboratories around the world. We've heard about the interesting new approach to the measurement of the gas constant and I can only hope that this work will continue, and give us numbers in the near future.

We ended the sessions on the fundamental constants with two aspects of the gravitational interaction, one the very precise measurements of the local acceleration of gravity to an accuracy which now seems to enable us to measure variation in g which hopefully are of only geophysical and not cosmological significance. At the same time we've heard of the new approach to the measurement of G, at a significantly lower level of precision than Dr. Sakuma's part in 10^8, but a vitally significant experiment, and one which hopefully can make significant improvements in the accuracy with which G is known at the present time.

On the other side of the conference, we have seen important new measurements and increases in accuracy in atomic mass determinations to the extent where atomic binding effects are now becoming important, significant, and almost determining, in some of the measurements. The history of this is that at Vienna it was first suggested that we ought to be very careful as to what it is that's being measured – are we measuring masses of nuclei or masses of neutral atoms? At Winnipeg we began to see experiments in which the distinction becomes important and how we have seriously to consider what it is that is being measured and what it is that we are implying in our mass tables. I think that it is clear from the definitions that the mass tables refer to the masses of a neutral atom in its ground state. The experimenter must now be careful in analyzing his data to be sure that he reduces his data to that form, taking into consideration the various ionization states or excited states that the atoms may be left in essentially at the end of his experiment, recognizing that there is indeed often additional ionization or coulomb energy that gets readjusted in the atom following the separation of nuclear fragments in a nuclear reaction.

In alpha energies we've now reached accuracies of a few tens of electron volts, and here also, atomic effects are becoming important in defining what we mean by a quoted decay energy. In beta decays we are approaching accuracies of the order of electron volts in decay energies and these decays now are becoming important, particularly with the very significant increases in the precision of mass determinations which are possible with Professor Smith's new instrument which is now measuring masses in some cases to tens of electron volts.

The steady improvement in gamma ray energy measurements has reached the point where there has been serious question as to what it is people are measuring because they don't have adequate standards against which to express their measurements, and hopefully the IUPAP commissions will look to this problem and we can expect eventually, and perhaps soon, getting at least recognized secondary standards so that we have consistency in that area.

In the mass formulas we have seen activity, but I have not recognized significant advances. The mass formulas are fitting data which is more precise and more voluminous, and therefore we are able to get perhaps more precise indications of where it is that the mass formulas need improvement. I think we must recognize, however, that there are two applications of mass formulas, one is as an interpolation to provide systematics to fill in gaps where measurements are unavailable, the other is to provide indications of theoretical significance in, for example, shell structure or deformed nucleii effects. I'm not sure that these two objectives are necessarily to be satisfied by the same mass formula. Ultimately, of course, if we have a complete description, these will evolve down to the same formula, but this is certainly beyond present considerations, and we will have to learn a lot more about nu lear structure and nuclear theory before, I think, we reach that point. We are seeing, in connection with the mass formula, increased experimental data on unstable nuclei in the nuclei far from the center of the stability valley. This I think is of vital importance to really getting an understanding of the structure of the nucleus since the stability line itself is only a very small cross-section or a very restricted sampling of the nuclear surface.

I think I should also say some words about the essentially extra curricular activities. I've been trying to determine what the difference is between an informal supper and a formal dinner, and it seems to be that a formal dinner the menu is printed – at an informal supper the menu is typed! With these remarks I hope I have properly summarized the conference – and I will now turn the stage over to Professor Wapstra who will have some closing remarks.

INDEX